MACHINE DESIGNERS REFERENCE

Jennifer Marrs, P.E.

Industrial Press, Inc.
New York

Library of Congress Cataloging-in-Publication Data

Marrs, Jen.
    Machine designers reference / Jen Marrs.
      p. cm.
    Includes bibliographical references and index.
    ISBN 978-0-8311-3432-7 (hard cover)
    1. Machine design–Handbooks, manuals, etc.
    2. Mechanical engineering–Handbooks, manuals, etc. I. Title.

    TJ230.M37 2011
    621.8'15–dc23

2011030875

Industrial Press, Inc.
989 Avenue of the Americas
New York, NY10018

Sponsoring Editor: John Carleo
Interior Text and Cover Design: Janet Romano
Developmental Editor: Robert Weinstein

SOLIDWORKS is a registered trademark of Dassault Systémes SolidWorks Corporation

Many pieces of art have be redrawn by Lineworks, Inc.
Composition was created by Lapiz Digital

10 9 8 7 6 5 4 3 2 1

BRIEF TABLE OF CONTENTS

About the Authors xiii

Acknowledgments xv

Introduction xvii

Chapter 1:   Design and Analysis 01

        1.1: Design of Machinery 03
        1.2: Engineering Units. 15
        1.3: Equations 18

Chapter 2:   Ergonomics and Machine Safety
             (Co-written with E. Smith Reed, P.E.) 35

        2.1: Ergonomics. 38
        2.2: Machine Safety: Design Process. 47
        2.3: Machine Safeguarding 51
        2.4: Other Safety Issues. 65
        2.5: Recommended Resources 71

Chapter 3:   Dimensions and Tolerances 93

        3.1: Limits, Fits, and Tolerance Grades 95
        3.2: Tolerances on Drawings, and GD&T 116
        3.3: Tolerance Stack-Ups (Written by Charles Gillis, P.E.). . . . 127

Chapter 4:   Precision Locating Techniques
             (Written by Charles Gillis, P.E.) 179

Chapter 5:   Pins, Keys, and Retaining Rings 215

Chapter 6: Pipe Threads, Threaded Fasteners, and Washers 279

      6.1: Pipe and Port Threads 281

      6.2: Threaded Fasteners and Washers 285

Chapter 7: Welds and Weldments 357

Chapter 8: Materials, Surfaces, and Treatments 369

      8.1: Materials 371

      8.2: Surface Finish 391

      8.3: Heat Treatment 398

      8.4: Surface Treatment 407

Chapter 9: Force Generators 413

      9.1: Springs 415

      9.2: Pneumatics 466

      9.3: Electric Motors 482

Chapter 10: Bearings 501

      10.1: Plain Bearings 503

      10.2: Rolling Element Bearings 510

      10.3: Linear Bearings 534

Chapter 11: Power Transmission Devices 543

      11.1: Shafts 547

      11.2: Shaft Couplings 563

      11.3: Gears (Written by Gregory Aviza) 574

      11.4: Gearboxes 614

      11.5: Belts and Chains 627

            11.5.1: Belts 629

            11.5.2: Chains 655

      11.6: Lead, Ball, and Roller Screws 664

Chapter 12: Machine Reliability and Performance 681

COMPLETE TABLE OF CONTENTS

1. Design and Analysis 01

    1.1. Design of Machinery 03
- Recommended Resources. 03
- Functional Design Specification 03
- Research 08
- Synthesis and Conceptual Design 09
- Detail Design and Analysis 10
- Factors of Safety. 13
- Critical Considerations: Design of Machinery 14
- Best Practices: Design of Machinery 14-15

    1.2. Engineering Units 15
- Recommended Resources. 15
- Engineering Units. 15
- Unit Conversions 17

    1.3. Equations 18
- Recommended Resources. 18
- Tables of Equations 18-33

2. Ergonomics and Machine Safety. 35

    2.1. Ergonomics 38
- Recommended Resources. 39
- Body and Workspace Dimensions 39
- Body Dimensions 39
- Workspace, Clearances, Enclosures, and Access Openings 39
- Operator Physical Capabilities and Limitations:
  Reaching / Grasping / Moving / Lifting. 41
- Critical Considerations: Ergonomics 46

    2.2. Machine Safety: Design Process .47
- Recommended Resources. 48
- Risk Assessment. 48
- Risk Reduction 50
- Critical Considerations: Machine Safety: Design Process. 51

    2.3. Machine Safeguarding 51
- Recommended Resources. 53
- Guards. 53
- Protective Devices 58
- Procedural Safeguarding: Information, Instructions and Warnings . . 62
- Warnings Included in the Manual 63
- Warning Labels Posted on the Machine 64
- Critical Considerations: Machine Safeguarding 65

2.4.   Other Safety Issues 65

- Recommended Resources. 65
- Emission of Airborne Substances or Modification of Surrounding Atmosphere. 66
- Emission of Radiation, Intense Light, Vibration, Heat 66
- Emission of Noise 66
- Hand / Arm Vibration 67
- Whole Body Vibration 67
- Machine Use in Explosive Atmospheres 68
- Moving the Machine 68
- Machine Stability 68
- Lubrication 69
- Danger Warning Alarm Signaling: Audible Signals and Visual Signals 69
- Lockout / Tagout Requirements. 69
- Emergency Stop Devices 70
- Critical Considerations: Other Safety Issues 71

2.5.   Recommended Resources 71

- Governmental Regulations, Statutes, Codes, and Publications 72
- Books 74
- Industry Safety Standards. 76
- Internet Web Sites 84
- Application of Literature to Design Topics 85

3.   Dimensions and Tolerances. 93

3.1.   Limits, Fits, and Tolerance Grades 95

- Recommended Resources. 95
- Types of Fits and Their Limits. 95
- Machining Tolerances 102
- Limits of Size Data. 102
- Critical Considerations: Limits, Fits, and Tolerance Grades 116
- Best Practices: Limits, Fits, and Tolerance Grades 116

3.2.   Tolerances on Drawings, and GD&T 116

- Recommended Resources. 116
- Implied Tolerances. 117
- Geometric Dimensioning and Tolerancing (GD&T). 117
- Critical Considerations: Tolerances on Drawings, and GD&T 126
- Best Practices: Tolerances on Drawings, and GD&T 127

3.3.   Tolerance Stack-Ups 127

- Recommended Resources. 128
- Design Practice. 128
- The Tolerance Stack-Up Chain 129
- Preliminary Tolerance Assignment 133
- Analysis and Assignment Methods 134

- Practical Applications. 146
- Summary. 174
- Critical Considerations: Tolerance Stack-Ups 176
- Best Practices: Tolerance Stack-Ups 177

4. Precision Locating Techniques 179
- Recommended Resources. 181
- Design Requirements 181
- The Two-Hole Problem 182
- Datum Reference Frame. 185
- Design Process. 187
- Precision Locating Techniques 191
- Critical Considerations: Precision Locating Techniques 212
- Best Practices: Precision Locating Techniques 212

5. Pins, Keys, and Retaining Rings. 215
- Recommended Resources. 217
- Pins and Keys in Shear. 217
- Pins. 219
- Keys 223
- Retaining Rings 225
- Critical Considerations: Pins, Keys, and Retaining Rings 226
- Best Practices: Pins, Keys, and Retaining Rings 227
- Component Data. 227

6. Pipe Threads, Threaded Fasteners and Washers. 279
  6.1. Pipe and Port Threads 281
  - Recommended Resources. 281
  - Standards 281
  - Critical Considerations: Pipe and Port Threads 282
  - Best Practices: Pipe and Port Threads 282
  - Pipe and Port Thread Dimensions 282
  6.2. Threaded Fasteners and Washers. 285
  - Recommended Resources. 285
  - Fastener Threads. 286
  - Tap Drills 288
  - Fastener Types, Materials, and Selection. 294
  - Grades and Strength of Fasteners. 295
  - Torque of Fasteners 298
  - Critical Considerations: Threaded Fasteners and Washers. 298
  - Best Practices: Threaded Fasteners and Washers. 299
  - Component Information 301
  - Component Data. 308

7.   Welds and Weldments. 357

- Recommended Resources. 359
- Weld Types. 360
- Weld Symbols 361
- Weldment Drawings. 364
- Materials and Treatments. 365
- Critical Considerations: Welds and Weldments. 367
- Best Practices: Welds and Weldments 368

8.   Materials, Surfaces, and Treatments. 369

8.1.   Materials. 371

- Recommended Resources. 371
- Metals Nomenclature 371
- Material Properties. 372
- Common Material Choices. 376
- Materials Data 382
- Critical Considerations: Materials. 390
- Best Practices: Materials. 391

8.2.   Surface Finish 391

- Recommended Resources. 391
- Surface Finish Symbols 392
- Surface Finish and Tolerance. 394
- Surface Finish Data 395
- Critical Considerations: Surface Finish 398
- Best Practices: Surface Finish. 398

8.3.   Heat Treatment. 398

- Recommended Resources. 399
- Hardness. 399
- Heat Treatment Processes 401
- Heat Treatment and Distortion. 403
- Hardness from Heat Treating. 403
- Critical Considerations: Heat Treatment. 406
- Best Practices: Heat Treatment. 407

8.4.   Surface Treatment 407

- Recommended Resources. 407
- Common Surface Treatment Types 408
- Selection of Surface Treatments 410
- Critical Considerations: Surface Treatment 412
- Best Practices: Surface Treatment 412

9.   Force Generators 413

    9.1.   Springs 415

- Recommended Resources 415
- Spring Types 415
- Spring Materials 418
- Helical Coil Spring Terminology 422
- Helical Coil Spring Fatigue 423
- Helical Coil Compression Springs 426
- Helical Compression Spring Catalog Selection Steps 429
- Helical Compression Spring Design Steps 437
- Helical Coil Extension Springs 440
- Helical Extension Spring Catalog Selection Steps 443
- Helical Extension Spring Design Steps 447
- Helical Coil Torsion Springs 450
- Helical Torsion Spring Catalog Selection Steps 451
- Helical Torsion Spring Design Steps 454
- Belleville Spring Washers 456
- Belleville Spring Washer Catalog Selection Steps 459
- Belleville Spring Washer Design Steps 463
- Critical Considerations: Springs 465
- Best Practices: Springs 465

    9.2.   Pneumatics 466

- Recommended Resources 466
- Pressure and Regulation 467
- Pneumatic Circuits 467
- Pneumatic Symbols 468
- Air Actuators 468
- Sizing Air Actuators 472
- Calculating C_v 474
- Pneumatic Valves 476
- Sizing Valves 480
- Critical Considerations: Pneumatics 480
- Best Practices: Pneumatics 481

    9.3.   Electric Motors 482

- Recommended Resources 482
- Electrical Power 482
- Motor Terminology 483
- AC Motors 484
- DC Motors 489
- Electric Motor Controls 490
- Electric Motor Frames and Enclosures 493
- Electric Motor Sizing 494
- Critical Considerations: Electric Motors 499
- Best Practices: Electric Motors 499

10. Bearings 501

   10.1. Plain Bearings 503
- Recommended Resources. 503
- Lubrication of Plain Bearings 503
- Bearing Materials for Boundary Lubrication. 506
- Boundary Lubricated Sleeve Bearings. 506
- Sleeve Bearing Selection Procedure for Boundary Lubrication 509
- Critical Considerations: Plain Bearings 510
- Best Practices: Plain Bearings. 510

   10.2. Rolling Element Bearings 510
- Recommended Resources. 511
- Lubrication, Seals, and Shields 511
- Bearing Characteristics. 512
- Bearing Types. 514
- Radial Ball and Roller Bearing Dimensions and Tolerances 520
- Bearing Arrangements 525
- Loads on Bearings 527
- Bearing Load Ratings and Life Expectancy 529
- Bearing Selection Procedure for a Rotating Shaft Application. 531
- Critical Considerations: Rolling Element Bearings 533
- Best Practices: Rolling Element Bearings 534

   10.3. Linear Bearings 534
- Recommended Resources. 534
- Lubrication 535
- Plain Linear Bearings. 535
- Rolling Element Linear Bearings. 537
- Critical Considerations: Linear Bearings 542
- Best Practices: Linear Bearings 542

11. Power Transmission Devices 543

   11.1. Shafts 547
- Recommended Resources. 547
- Methods of Attachment 547
- Materials and Treatments. 548
- Deflection, Stress, and Fatigue. 549
- Critical Speeds 556
- Shaft Design Procedure 558
- Critical Considerations: Shafts. 562
- Best Practices: Shafts. 562

   11.2. Shaft Couplings 563
- Recommended Resources. 563
- Attachment to Shafts 563
- Shaft Misalignment 564

- Coupling Types 564
- Flexible Coupling Types 567
- Coupling Selection 570
- Critical Considerations: Shaft Couplings 573
- Best Practices: Shaft Couplings 573

11.3. Gears 574
- Recommended Resources. 574
- Terms and Definitions 575
- Gear Types 582
- Gear Trains 582
- Shaft Attachment Methods. 589
- Gear Quality Ratings 589
- Backlash 591
- Materials and Treatments. 595
- Lubrication and Wear. 596
- Spur Gears 599
- Spur Gear Selection and Sizing 599
- Helical Gears 606
- Bevel Gears 607
- Worm Gears 608
- Critical Considerations: Gears 612
- Best Practices: Gears. 612

11.4. Gearboxes. 614
- Recommended Resources. 614
- Gearbox Characteristics 614
- Loads on Gearboxes 616
- Gearbox Selection 619
- Critical Considerations: Gearboxes. 626
- Best Practices: Gearboxes. 626

11.5. Belts and Chains 627
- Recommended Resources. 627
- Drive Calculations 627
  - 11.5.1   Belts 629
    - Flat Belt Drive Design and Selection Procedure 632
    - V-Belt Drive Design and Selection Procedure 636
    - Synchronous Drive Design and Selection Procedure 648
    - Critical Considerations: Belts 652
    - Best Practices: Belts. 653
  - 11.5.2   Chains. 655
    - Chain Selection and Sizing. 658
    - Critical Considerations: Chains 662
    - Best Practices: Chains 663

11.6. Lead, Ball, and Roller Screws 664
- Recommended Resources. 664

- Screw Characteristics 664
- Screw Stresses, Deflection, and Buckling 669
- Critical Speed 670
- Linear Motion Screw Types 671
- Lead Screws 673
- Ball Screws 674
- Lead or Ball Screw Selection Procedure 675
- Critical Considerations: Lead, Ball, and Roller Screws 680
- Best Practices: Lead, Ball, and Roller Screws 680

12.   Machine Reliability and Performance 681
- Recommended Resources 683
- Machine Reliability 683
- Failure Modes, Effects, and Criticality Analysis 685
- Safety Category 693
- Manufacturing Equipment Performance 695
- Condition Monitoring 698
- Critical Considerations: Machine Reliability and Performance 698
- Best Practices: Machine Reliability and Performance 699

ABOUT THE AUTHORS

Jennifer Marrs, P.E. has been working in industry as a mechanical design engineer for more than 18 years. Her focus has mainly been the design and analysis of high-speed assembly machines and related systems, but she has also worked as a product designer, manufacturing engineer, and forensic engineer. Jennifer holds a BSME from Worcester Polytechnic Institute, an MSME from Northeastern University, and volunteers with the mechanical engineering programs at both WPI and Dartmouth College. Jennifer has a successful consulting practice and is a licensed Professional Engineer in New Hampshire, Vermont, and Massachusetts. She is also a registered U.S. patent agent. Her employers and clients include Gillette, Millipore, FujiFilm Dimatix, and Green Mountain Coffee Roasters. Mrs. Marrs is currently on the Executive Committee of her local ASME subsection and holds one international patent.

Contributing Authors:

Gregory Aviza has over 18 years of industrial experience focusing on the design, building, and international commissioning of precision high-speed assembly automation. He has also worked as a product designer for the consumer goods industry and currently holds five U.S. and four international patents. He holds a BSME and MSME from Worcester Polytechnic Institute. Gregory currently works for the Gillette division of Procter & Gamble in the Front End Development Group where he is focused on developing the next generation of product and equipment designs. He can be reached through his Linked-In profile: *www.linkedin.com/in/aviza*

Charles A. Gillis, P.E. has over 15 years of machine design experience, and currently works as a mechanical engineer for Gillette, the Blades & Razors division of Procter & Gamble. During this time, Charles has designed automated machinery for manufacturing Gillette's blade and razor products. He has designed and put into service numerous pieces of manufacturing equipment consisting of complex mechanisms of precision assembled components. He has designed equipment to manufacture the Gillette Mach3®, Venus®, Sensor3®, and Fusion® product lines. Charles holds a BSME from Worcester Polytechnic Institute and an MSME from Northeastern University. He is a licensed Professional Engineer in Massachusetts. In addition, Charles is a Geometric Dimensioning & Tolerancing instructor for Worcester Polytechnic Institute's Corporate and Professional Education department.

E. Smith Reed, P.E. is a forensic engineering consultant with over 30 years experience designing, testing, and putting into service mobile power machinery and industrial equipment. He is a Board Certified Forensic Engineer and is licensed as a Professional Engineer in Minnesota, New Hampshire, Alabama, South Carolina, and Florida (Mechanical, Industrial and

Manufacturing Engineer). He has industrial/manufacturing and design engineering experience with Honeywell, Inc., Toro Co., Tennant Co., and Vermont Castings, Inc. Smith, a BSME graduate from the University of Arkansas, holds four U.S. patents. He is a member of several national engineering societies, including the Human Factors and Ergonomics Society, and serves on the board of directors of the National Academy of Forensic Engineers (formally affiliated with NSPE), also serving as its President.

ACKNOWLEDGMENTS

My heartfelt gratitude and admiration goes out to my contributing authors, the team at Industrial Press, and the many individuals who provided content or guidance.

Gregory Aviza, contributing author, wrote the section on gears. He also read most of my early drafts, provided ideas, and gave advice that significantly improved the material. He was an essential resource and sounding board throughout this project.

John Carleo, Editorial Director at Industrial Press, believed in this project when it was nothing more than an idea. Over the course of a year and a half, he helped the idea grow significantly in scope. He always rapidly got me what I needed to keep improving and expanding the content.

Charles A. Gillis, P.E., contributing author, wrote the chapter on precision locating and the section on tolerance analysis. His expert contributions to this book were essential and greatly appreciated.

Christopher J. McCauley, Senior Editor of *Machinery's Handbook,* spent countless hours providing practical guidance and technical support. He also patiently provided me with many images and other items from *Machinery's Handbook*. His advice and feedback was extremely valuable.

E. Smith Reed, P.E., contributing author, wrote most of the chapter on safety and ergonomics. His expert contribution to this book was substantial and essential to the high quality and broad content of the chapter.

Janet Romano, Production Manager/Art Director at Industrial Press, did a fantastic job designing the layout and appearance of this book. Her expertise and hard work is visible on every single page, as well as on the cover.

Robert Weinstein, freelance developmental editor for Industrial Press, copyedited the manuscript and checked the final page proofs.

I would like to give additional thanks to the following people who provided either expert advice or essential material:

Herb Arum, SDP/SI *(www.sdp-si.com)*
William Bollig, WITTENSTEIN *(www.wittenstein-us.com)*
Michael J. Brown, A123 Systems, Inc. *(www.A123Systems.com)*
Robert S. Clippard, CFPPS, Clippard Instrument Laboratory, Inc. *(www.clippard.com)*
Donald A. Cottrill
C. Wes Cross, Westinghouse Electric Company, LLC *(www.westinghousenuclear.com)*
Bruce Curry, Penna Flame Industries *(www.pennaflame.com)*

Wendy Earle, SKF USA, Inc. *(www.skf.com)*

Patrick Esposito, Misumi USA, Inc. *(www.misumiusa.com)*

Sam Feller, MIT Lincoln Laboratory *(www.ll.mit.edu)*

Glenn Frazier, Rockwell Automation, Inc. *(www.rockwellautomation.com)*

Nikki Groom, igus, Inc. *(www.igus.com)*

John Halvorsen, SMC Corporation of America *(www.smcusa.com)*

Todd Kanipe, The Precision Alliance *(www.tpa-us.com)*

Dr. Kevin Lawton, University of North Carolina, Charlotte *(www.uncc.edu)*

Bill McCombe, Curtis Universal Joint Company *(www.curtisuniversal.com)*

Eric J. Mann, Graduate Student, Dartmouth College *(www.dartmouth.edu)*

Corey Maynard, P.E., Proctor & Gamble *(www.pg.com)*

Miriam Metcalfe, WITTENSTEIN *(www.wittenstein-us.com)*

Robert L. Mott, P.E., Emeritus Professor, University of Dayton *(www.udayton.edu)*

Bill Nartowt, United County Industries Corp. *(www.countyheattreat.com)*

Dr. Catherine Newman, Industrial Product Design & Development Consultant

Robert L. Norton, P.E., Worcester Polytechnic Institute *(www.wpi.edu)*

Bill Oliver, Minuteman Controls *(www.minutemancontrols.com)*

John Slocum, SolidWorks Corp. *(www.solidworks.com)*

Jason Sicotte, Associated Spring Barnes Group Inc. *(www.asbg.com)*

Clifford M. Stover, P.E., California State Polytechnic University, Pomona *(www.csupomona.edu)*

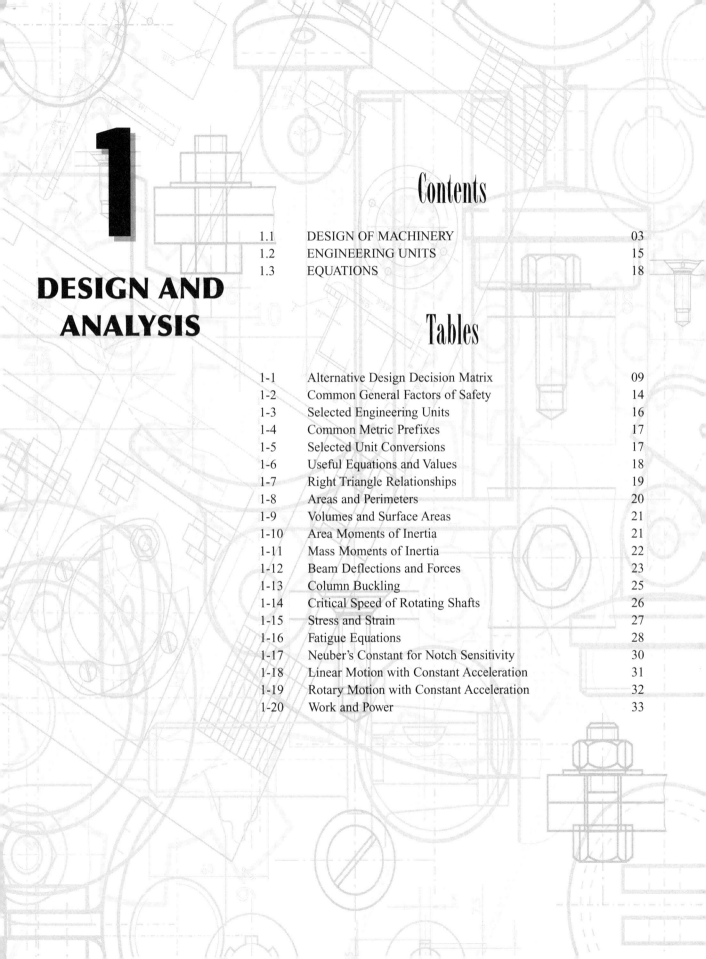

1

DESIGN AND ANALYSIS

Contents

1.1	DESIGN OF MACHINERY	03
1.2	ENGINEERING UNITS	15
1.3	EQUATIONS	18

Tables

1-1	Alternative Design Decision Matrix	09
1-2	Common General Factors of Safety	14
1-3	Selected Engineering Units	16
1-4	Common Metric Prefixes	17
1-5	Selected Unit Conversions	17
1-6	Useful Equations and Values	18
1-7	Right Triangle Relationships	19
1-8	Areas and Perimeters	20
1-9	Volumes and Surface Areas	21
1-10	Area Moments of Inertia	21
1-11	Mass Moments of Inertia	22
1-12	Beam Deflections and Forces	23
1-13	Column Buckling	25
1-14	Critical Speed of Rotating Shafts	26
1-15	Stress and Strain	27
1-16	Fatigue Equations	28
1-17	Neuber's Constant for Notch Sensitivity	30
1-18	Linear Motion with Constant Acceleration	31
1-19	Rotary Motion with Constant Acceleration	32
1-20	Work and Power	33

Section 1.1

DESIGN OF MACHINERY

As part of the design process, the machine designer must consider safety, capacity, function, material, method of manufacture, form, cost, assembly, adjustment, reliability, and many other factors. The design process is iterative, and the steps can be revisited in any order as needed to refine the design and resolve problems. This section addresses the typical stages of a machine design project, conducted in this order with iterations: design specification, research, synthesis and conceptual design, then detail design and analysis. Factors of safety should always be considered and should be appropriately applied.

RECOMMENDED RESOURCES

- Chironis, Sclater, *Mechanisms and Mechanical Devices Sourcebook,* 2nd Ed., McGraw-Hill Inc., New York, NY, 1996
- Collins, Busby, Stabb, *Mechanical Design of Machine Elements and Machines,* 2nd Ed., John Wiley & Sons, Inc., Hoboken, NJ, 2010
- R. L. Norton, *Design of Machinery,* 5th Ed., McGraw-Hill Inc., New York, NY, 2012
- R. L. Norton, *Machine Design: An Integrated Approach,* 4th Ed., Prentice Hall, Upper Saddle River, NJ, 2011
- S. Pugh, *Total Design: Integrated Methods for Successful Product Engineering,* Addison-Wesley Publishing Company Inc., 1991
- D. Shafer, *Successful Assembly Automation,* Society of Manufacturing Engineers, Dearborn, MI, 1999
- J. Skakoon, *The Elements of Mechanical Design,* ASME Press, New York, NY, 2008

FUNCTIONAL DESIGN SPECIFICATION

The first step in any new design process is to define what the device or part must do, with what frequency, under what conditions, and according to what requirements. It is helpful to prioritize these specifications to aid in evaluating trade-offs. Design specifications drive basic decisions like component or system type, structure, function, and allowable failures. When developing design specifications, it can be helpful to answer the following questions about the device or part:

General Considerations:

- Are there any domestic laws, codes, or standards governing the device?
- Must the device comply with laws and standards of any other countries?
- Is there a cost target or limit for the device?
- Is there a size or weight limit?
- Will there be shipping requirements?
- How many devices are likely to be made?

Functional Considerations:

- What does the device do?
- What is the input to the device?
- What is the output of the device?
- What loads/forces are involved?
- What movements are required and in what time period?
- What sort of contact is expected between the device and the target work piece or user?
- What sorts of intelligence or detection must the device include?
- Can the device's function be broken into discrete functional elements?
- What physical principles are involved?
- What is the life expectancy of the device?
- Should the device be adjustable and/or repairable?

Ergonomic Considerations:

- What sort of access is required to service or install the device?
- Will parts of the device be lifted by a person at any point?
- What will be the skill level of the operator?
- Is repetitive strain injury a risk for this device or tool?
- Is failure of this part or device a safety hazard?

Environmental Considerations:

- What forces will the device be exposed to?
- Will the device be exposed to extremes of temperature?

- Will the device be exposed to fluids, dirt, or other contaminants?
- Will the device be exposed to gases other than air?
- Must the device be food safe, wash-down, FDA compliant, or meet other cleanliness standards?
- What sources of power (human, electrical, fluid, etc.) are available to run the device?
- What are the maintenance capabilities where the device is installed?
- How much space can the device occupy?
- What is the carbon footprint of the device?

When writing a specification for a machine or device that is to be built by a supplier, it is doubly important to get every known or expected detail about the device in writing to guide the design and avoid misunderstandings. In the case of complex machinery, a functional specification can be 50 to 100 pages long. Every organization typically has its own preferred format and contents.

The following is a simplified example of a functional design specification for a fictional shrink wrap machine. In this example, the possibility of using a heat tunnel has already been eliminated, so it is part of the specification that a heater must be moved into position temporarily to heat the shrink wrap. If a heat tunnel was still a possibility, the functional part of the specification would be much more vague to allow for either possibility.

MACHINE: 161B Shrinkwrap Machine
INPUT: Product 161B, enclosed in shrinkwrap material (25 lb weight) presented in trays of 2
OUTPUT: Fully shrinkwrapped product 161B placed in trays of 2
CYCLE TIME: 30 sec. (2 parts per minute)
REQUIRED UPTIME: 95%
LIFESPAN: 10 years
OPERATORS: 0.33 (1 per 3 machines)
OPERATOR TYPE: trained, English or German speaking, good reading skills
MAX OVERALL SIZE: 8 feet square, up to 15 feet tall
STANDARDS COMPLIANCE: EU standards, US standards
DEVICES TO BE MADE: 6 in United States, 6 in Germany
UNITS: metric

INSTALLATION TYPE: Industrial environment, permanent, lagged to floor, no weight restrictions

FACILITIES: 480VAC, 3-Phase, Compressed Clean Dry Air 90psi

MACHINE FUNCTIONS:

1. Product is taken from tray and placed on conveyor in specific orientation.
2. Product moves into position under heater.
3. Heater lowers to product.
4. Heater dwells to shrink the wrapper.
5. Heater raises to safe position.
6. Product moves to cooling position.
7. Product dwells until cool.
8. Product moves to exit position.
9. Product is placed on finished goods tray in specific orientation.

FUNCTIONAL COMPONENTS:

- ENTRY DEVICE
    - Takes product from tray in specific orientation
    - Moves product to shrinkwrap machine entry position while maintaining orientation
    - Must be automatic (not manual feeding)
- PART CONVEYANCE
    - Moves product from entry position to heat position, cool position, and exit position while maintaining orientation of product
    - Contact with product must be limited to areas specified on print.
    - No rubbing or sliding contact with product is allowed.
- HEATER
    - Heater must be 10mm from product and maintains 300°C temperature, according to experimentation.
    - Heater must be in position 15 seconds to shrink the wrapper, according to experimentation.
    - Heater must cover 50mm-diameter flat area and 20mm down the sides of the product in order to effectively shrink the wrapper, according to experimentation.
- COOLING
    - Must be accomplished with ambient air conditions — no air jets or chillers are allowed in this application
    - Product must cool from 300°C to 100°C before being placed in tray.

- ○ Cooling to 100°C takes 25 seconds after 300°C heater removed, according to experimentation.
- EXIT DEVICE
  - ○ Takes product from machine exit position in specific orientation
  - ○ Moves product from exit position to tray while maintaining orientation
  - ○ Must be automatic (not manual removal)

INTELLIGENCE:

- Heater temperature control
- Machine not ready until heater temperature setpoint is reached
- Interlocking guard door for safe operator access
- Heater position error shuts down machine
- Conveyor jam sensing
- Operator panel must have machine performance data.

SAFETY REQUIREMENTS:

- Standards: US, EU
- Company requirements
  - ○ Fixed or interlocking guards around perimeter
  - ○ Roof on guard enclosure
  - ○ Noise level limit at operator position 80dB
  - ○ Warning label attached to thermal hazard
  - ○ Light tower 9 feet off the ground featuring green, yellow, and red at the top

SPECIAL REQUIREMENTS:

- Must allow conversion to run product 162B with 1-hour changeover time
- Operator must be able to access heater chamber from front AND back of machine.

The functional design specification is an organized list of what is known about the machine at this early stage. It should contain all the known constraints and expectations, but should be written in a way that avoids placing new constraints. For example, the product conveyance subsystem is not called an indexing conveyor here because it has not yet been decided what form the conveyance should take. An indexing conveyor is most likely, but at this point in the design process it is still open for consideration. Continuous conveyors, or pallet systems, are also possible conveyance solutions.

From this specification, the designer can begin to develop a timing diagram for the machine and determine how much time is allowed for each function. That is often the next step toward conceptual design of automatic machinery. It is important to note that the design specification details what the device must do, but not how it must be done. The question of how will be addressed in the conceptual and detail design stages. Concepts for the conveyance system and the individual devices that perform various machine operations (such as lowering and raising the heater) can then be developed and evaluated based on their ability to meet the functional design specification.

RESEARCH

The importance of research cannot be overstated. Research can be conducted in many ways, and the earlier the better. Some research methods are: reading books, searching the web, researching patents, attending seminars or tours, experimenting, and investigating the mechanical devices one is surrounded by every day.

Research of applicable theory, laws, codes, and standards is essential to any machine design project. Resources useful to this endeavor are textbooks, handbooks, trusted websites, and seminars. Research into existing devices can save the designer time and money. Commercially available solutions and items may be examined, as well as patents and mechanisms sourcebooks. Be aware that active patents must not be copied without proper licensing. Manufacturer's websites can be a wealth of information, as are the following:

Occupational Safety and Health Administration Website: **www.osha.gov**
American National Standards Website: **www.ansi.org**
United States Patent and Trademark Office: **www.uspto.gov**
Machine Design Website: **www.machinedesign.com**

Research can also take the form of experimentation, taking measurements or simply 'getting a feel for' aspects of the device or its function. The following is a short list of common tools that have proven particularly useful to the author. This is not a complete list by any means.

Rule, or Scale: Scales are useful when measuring physical objects or environmental dimensions. A scale is particularly helpful when thinking about ergonomics, access, or visualizing physical part size.

Calipers: A set of calipers is useful for measuring or visualizing thicknesses, depths, or diameters. Calipers can take measurements where the traditional ruler cannot. Calipers can be purchased from any tooling supply company.

Force Gauge: A force gauge is essential for taking measurements, "getting a feel" for forces in your design, and thinking about ergonomics. Sizing springs and other light force generators is much easier when using a force gauge to physically measure the required force/load. It is also useful when simply feeling a given force to get a sense of what you are specifying. Force gauges can be expensive, but are extremely useful during the design process.

SYNTHESIS AND CONCEPTUAL DESIGN

Synthesis of elements of prior art is the 'bread and butter' of the machine design industry. Synthesis of existing designs is usually less costly and less risky than original design. Many machine design jobs require synthesis of existing methods, with adaptation or customization. To excel at synthesis, one needs to know about a large number of mechanical devices and methods and be capable of creative visualization. The topic of design is addressed in many great books. Consult the recommended resources for more information on design and the creative process.

It is usually beneficial to conceptualize more than one solution to the design problem and then choose between alternatives based on priorities and performance. One helpful tool for choosing between designs is a decision matrix, or Pugh matrix, like that shown in Table 1-1. In this example, the shaded fields are values that are entered. A weight value, typically between 1 and 10, is assigned each design specification or characteristic of the design.

Table 1-1: Alternative Design Decision Matrix

Design Specification	Weight	Concept 1		Concept 2		Concept 3	
		Ability	Score	Ability	Score	Ability	Score
Simplicity	5	2	10	1	5	1	5
Cost	3	1	3	1	3	2	6
Ease of Use	8	2	16	1	8	1	8
Specification A	2	1	2	1	2	1	2
Specification B	10	1	10	2	20	1	10
Specification C	6	2	12	1	6	1	6
Specification D	3	1	3	1	3	1	3
Total Score:			56		47		40

Each concept is rated on its ability to meet each specification. These values can be from 1 to 10, from 1 to 3, from zero to 3, or any meaningful assortment of numbers. For simplicity's sake, Table 1-1 uses weights from 1 to 10 and ability values from 0 to 2. No zeros were assigned in this case because each concept that made it into the matrix met all criteria. Scores are then calculated for each entry by multiplying the weight by the ability value. The scores for each concept are then summed to give a total score for each concept. The highest score in this case is considered the best concept.

Concepts can be broad or specific, and many choices between concepts must be made for most designs. At the very top level, a conceptual choice for an assembly machine may be the choice between a single machine and a group of modular machines that transfer in-process components between them. Another conceptual choice may be whether to use indexing motion or continuous motion for a machine. Conceptual decisions can also be as specific as deciding between an air cylinder system and a ball screw and motor system for accomplishing some movement. Some other common examples of choices made between concepts are: cam drive vs. servo positioning drive; gears vs. timing belt; photoelectric sensor vs. proximity sensor; bulk conveyor vs. nested conveyor; compliant tooling vs. compliant product holder; quality sampling vs. 100% inspection; rolling element bearing vs. plain bearing; etc. Decisions at all levels must take into account cost, risk, safety, ergonomics, reliability, maintainability, accuracy, etc.

DETAIL DESIGN AND ANALYSIS

This stage of the design process is where machine designers generally spend the majority of their time. It is during detail design that the materials and methods of manufacture are decided, the precision locating configuration is designed, fasteners are selected, the forms of the parts take shape, surface finishes and treatments are specified, and the design is analyzed in detail regarding strength, function, safety factors, and compliance with design specifications. Consider materials and methods of manufacture early in detail design, because both influence things like joint configurations and part shapes. Sizing and analysis should always be conducted using factors of safety. Machine performance and reliability should be planned for, and analytical techniques like Failure Modes, Effects, and Criticality Analysis (FMECA) should be undertaken in the detail design phase if not earlier.

FMECA is covered in Chapter 12. The following are some best practices for detail design:

ASSEMBLIES AND SYSTEMS

- Consider accuracy, stresses, failure modes, wear, environmental exposure, operating temperature, life expectancy, spare parts, assembly procedure, alignment, and maintenance early in the design process.
- Design one or more forms of overload protection into machinery to reduce the consequences of a jam or crash.
- Use the smallest number of fasteners required for the application. The use of three screws instead of four, for example, reduces cost and assembly time. Use the same size screws wherever possible in an assembly to save cost and ease the assembly process.
- Always allow room for assembly, service, and adjustment. Tools, hands, and line of sight must be accommodated. If possible, design such that systems can be assembled and serviced from one side. Provide inspection points with easy access for all critical or at-risk components.
- Plan for insulation against galvanic corrosion when fastening dissimilar metals. Many coatings are capable of insulating materials against galvanic action. Helical coil threaded inserts with special coatings can act as galvanic insulators when a screw and part are made from dissimilar metals.
- When the orientation of a part at assembly is critical, design the parts so that they cannot be assembled incorrectly. The Japanese term for this is "Poka-yoke," which means "mistake-proofing." This is often accomplished by making the mounting or locating holes asymmetric.
- Use kinematic principles to exactly constrain parts. Never overconstrain parts.
- For easier setup, design an assembly such that each alignment direction is isolated from the others. This is an extension of the exact constraint principle in which a single-degree constraint prevents movement of a part in only one direction. Separate constraints, independently adjustable (if adjustment is needed), should be provided for each degree of freedom.
- Design setup gauges for your assemblies if alignment is critical. Make setup gauges open-sided so that parts can be pushed into the gauge along each adjustment direction. Avoid setup gauges that require that

a pin or feature drop into a hole because these tend to be much harder to work with.

- Keep the ratio of length divided by width (bearing ratio) of all sliding elements above 1.5. This will maximize precision, reduce wear, and reduce the tendency to jam.

- Design assemblies to use stock items like commercial parts, or use parts common to other assemblies. This will usually shorten design time and save fabrication cost.

PARTS

- Consider the method of manufacture, stresses, failure modes, wear, environmental exposure, and life expectancy early in the design process.

- Place material in line with the path of forces through a part or assembly. Remove material that does not see any stresses, if weight is more critical than cost.

- Design symmetrical parts when possible. Avoid asymmetric parts that look symmetric, because this can cause assembly mistakes.

- Design parts for multiple uses when possible. For example, instead of a left and right gib in an assembly, it will be more cost effective to use two identical gibs.

- Use purchased or standardized components whenever possible to save cost.

- Specify through holes instead of blind holes where possible unless the part thickness is extremely large. Through holes are often more cost effective than blind holes, especially if threaded.

- For cost and weight reduction, apply liberal tolerances and rough surface finishes where possible.

- Minimizing setups during machining will save cost. If possible, design parts to be machined from one side only.

- Design parts to match existing stock material sizes when possible to save cost. Reduce machining when possible.

- When a part is to be mass produced, consider near-net-shape casting as a cost effective alternative to machining.

- To prevent interference, chamfers should be applied to outside edges that fit snugly into machined pockets in other parts. Use small chamfers with loose tolerances to reduce cost.

These tools, in addition to those mentioned earlier, have been useful to the author during detail design:

<u>"Feeler" or Thickness Gauges:</u> A set of feeler gauges is useful for measuring and setting gaps during set up and maintenance of tools and mechanical devices. They are also useful in the office for visualizing sheet metal or other thin parts. The stiffness of stainless steel sheet metal of various thicknesses can be understood by playing with a stainless steel "feeler" gauge of the selected thickness. "Feeler" or thickness gauges can be purchased at most tooling or automotive supply companies.

<u>Surface Finish Comparator:</u> Several manufacturers offer a set of surface finish comparators with samples of common surface finishes arranged on a card. These are useful for understanding the look and feel of the different surface finishes.

<u>Screw Selector Slide Chart:</u> These slide charts are available from a variety of manufacturers. They are a convenient and fast way of looking up fastener dimensions and related information.

FACTORS OF SAFETY

Factors of safety in machinery design are used to represent the risk of failure of a component, part, or system. Factor of safety of a part, device, or system is its theoretical capacity divided by the maximum of what is expected. In machinery design, factor of safety is often defined as the maximum safe load (or stress) for a component divided by the expected maximum load (or stress) on the component. It can also be expressed as a maximum safe speed divided by the maximum expected service speed, maximum overturning moment divided by expected moment, or some other measure of failure or risk.

Sometimes factor of safety is dictated by laws or codes. When the designer is free to set a safety factor, some common values are provided in Table 1-2. It is customary to assign higher safety factors in situations where risk or uncertainty is higher. It is also customary to assign factors of safety to brittle materials that are double that for ductile materials. Higher safety factors generally result in designs that are heavier, larger, more costly, and more powerful. In cases where this must be avoided, safety factors must be kept relatively low and steps taken to reduce uncertainty to a level where the lower safety factor is acceptable. For reference, light industrial machinery is often designed with a factor of safety around 2. Critical components like bearings are often designed with a larger factor of safety, commonly between 3 and 5.

Table 1-2: Common General Factors of Safety

Safety Factor	Application
1.3 - 1.5	For use with highly reliable components or materials where loading and environmental conditions are not severe, and where weight is an important consideration.
1.5 - 2	For applications using reliable components or materials where loading and environmental conditions are not severe.
2 - 2.5	For use with ordinary components or materials where loading and environmental conditions are not severe.
2.5 - 3	For less tried and for brittle materials where loading and environmental conditions are not severe.
3 - 4	For applications in which component or material properties are not reliable and where loading and environmental conditions are not severe, or where reliable components or materials are to be used under difficult loading and environmental conditions.
4+	For applications with a high degree of uncertainty, high risks, or where unreliable components or materials are to be used where loading and environmental conditions are severe.

CRITICAL CONSIDERATIONS: Design of Machinery

- Safety is of paramount importance in all designs. Understand and obey all applicable laws, codes, and standards.
- Apply appropriate safety factors when conducting design and analysis.
- The design specification should account for all design requirements and foreseeable needs.
- Research and test. This will improve your results and help prevent mistakes.

BEST PRACTICES: Design of Machinery

- Avoid overly descriptive terms early in the design specification phase. They can limit creativity.
- State and question assumptions made during the specification and design process.

(Continued on next page)

- Build your list of known mechanical devices to maximize your synthesis capabilities. Work to improve your understanding and 'feel for' forces, distances, and machining methods.
- Allow enough time for research and conceptualization. Ensure that the work environment allows long periods of concentration. This is often a challenge in the modern office.
- Come up with multiple solutions for every problem and select the best one. Use a decision matrix to aid the selection process.
- When visualization stalls, some designers find it helpful to begin drawing or modeling the known surfaces, parts, and points in an assembly. A second round of visualization after accurately modeling what is known is often more fruitful.

Section 1.2

ENGINEERING UNITS

Proper units are critical when performing measurement and analysis. Most engineering books contain some discussion of units and unit conversions. In addition, unit conversion tools abound on the Internet. This section provides an overview of the most commonly used engineering units and conversions.

RECOMMENDED RESOURCES

- M. Lindeburg, *Mechanical Engineering Reference Manual for the PE Exam*, 11th Ed., Professional Publications, Inc., Belmont, CA, 2001
- Oberg, Jones, Horton, Ryffel, *Machinery's Handbook*, 28th Ed., Industrial Press, New York, NY, 2008
- Unit Converter Express Online: **www.unitconverters.net**
- Unit Conversion Tools Online: **www.unit-conversion.info**
- Online Conversion Website: **www.onlineconversion.com**
- Unit Conversion Website: **www.unitconversion.org**

ENGINEERING UNITS

Consistency of units is essential in all engineering calculations. Checking units will often turn up problems with a calculation. In the United States, both the Imperial (inch or foot) and SI (metric) systems are used. Some commonly used engineering units in both systems can be found in Table 1-3.

Table 1-3: Selected Engineering Units

Measurement	Imperial Units		Metric Units
	ips Basis	**fps Basis**	
Force	lbf		$N = kg \cdot m/s^2$
Length	in	ft	mm
Volume	in^3	ft^3	mm^3
Time	s		s
Density (ρ)	lbf/in^3	lbf/ft^3	kg/m^3
Mass	$bl = lbf \cdot s^2/in$	$sl = lbf \cdot s^2/ft$	kg
Weight	lbf		$N = kg \cdot m/s^2$
Pressure	$psi = lbf/in^2$	$psf = lbf/ft^2$	$Pa = N/m^2$
Torque or Moment	$in \cdot lbf$	$ft \cdot lbf$	$N \cdot m$
Stress (σ or τ)	$psi = lbf/in^2$	$psf = lbf/ft^2$	$Pa = N/m^2$
Area Moment of Inertia (I)	in^4	ft^4	m^4
Mass Moment of Inertia (I)	$in \cdot lbf \cdot s^2$	$ft \cdot lbf \cdot s^2$	$kg \cdot m^2$
Velocity	in/s	ft/s	m/s
Acceleration	in/s^2	ft/s^2	m/s^2
Angular Velocity (ω)	rad/s		rad/s
Angular Acceleration (α)	rad/s^2		rad/s^2
Energy	$in \cdot lbf$	$ft \cdot lbf$	$J = N \cdot m$
Power	$in \cdot lbf/s$	$ft \cdot lbf/s$	$W = N \cdot m/s$
Frequency	Hz = cycle/s		Hz = cycle/s

There are two common ways to represent Imperial units in the United States: the inch-pound-second system (ips) and the foot-pound-second system (fps). The ips system is more commonly used in machinery design, though certain calculations are often done in the fps system. It is important to note that the unit of pounds mass (lbm) is numerically equal to pounds force (lbf) and must be divided by the gravitational constant before it is used in equations calling for mass. Imperial units of mass are blobs (ips basis) or slugs (fps basis). The values of gravitational acceleration in the Imperial system are as follows:

ips: $g_c = 386.4$ lbm·in/lbf·s^2 (or blobs)
fps: $g_c = 32.174$ lbm·ft/lbf·s^2 (or slugs)

The metric system, or SI system, is a base-10 system of units that uses prefixes to designate orders of magnitude. The basis of length in the metric system is the meter (m). Common metric prefixes can be found in Table 1-4. Most metric machinery design is done using millimeters (mm) as the standard measure of length.

Table 1-4: Common Metric Prefixes

Prefix	Symbol	Multiplier
giga	G	10^9
mega	M	10^6
kilo	k	10^3
deka	da	10
deci	d	10^{-1}

Prefix	Symbol	Multiplier
centi	c	10^{-2}
milli	m	10^{-3}
micro	μ	10^{-6}
nano	n	10^{-9}
pico	p	10^{-12}

UNIT CONVERSIONS

Conversion between Imperial and metric units is a fact of life for many designers. Some common engineering unit conversions are listed in Table 1-5.

Table 1-5: Selected Unit Conversions

Multiply	By	To Obtain
atm	1.01325	bar
atm	14.696	psi
atm	101.3	kPa
bar	0.9869	atm
cm	0.03281	ft
cm	0.3937	in
ft	0.3048	m
ft·lbf	1.35582	J
ft·lbf	1.3558	Nm
g/cm^3	1000	kg/m^3
g/cm^3	62.428	lbm/ft^3
hp	33000	ft·lbf/min
hp	550	ft·lbf/sec
hp	0.7457	kW
in	2.54	cm
J	0.73756	ft·lbf
kg	2.20462	lbm
kPa	0.00987	atm
kW	737.6	ft·lbf/sec

Multiply	By	To Obtain
kW	1.341	hp
lbf	4.4482	N
psi	0.06805	atm
psi	6894.8	Pa
lbm	0.4536	kg
lbm/ft^3	0.01602	g/cm^3
lbm/ft^3	16.018	kg/m^3
m	3.28083	ft
m/sec	196.8	ft/min
MPa	1	N/mm^2
N	1	$kg·m/s^2$
N	0.22481	lbf
Nm	0.7376	ft·lbf
Pa	0.000145	psi
Pa	1	$kg/m·s^2$
Pa	1	N/m^2
W	0.7376	ft·lbf/sec
W	0.00134	hp
W	1	J/s

EQUATIONS

Calculation is an essential part of engineering design. It is critical that units be consistent, the correct equations used for the application, and appropriate factors of safety applied. Consult trusted references for calculation guidance. Some common machinery design calculations include geometric relationships, moments of inertia, stress, strain (deformation), and fatigue analysis.

RECOMMENDED RESOURCES

- Beer, Johnston, **Mechanics of Materials,** McGraw-Hill Inc., New York, NY, 1981
- M. Lindeburg, **Mechanical Engineering Reference Manual for the PE Exam,** 11th Ed., Professional Publications, Inc., Belmont, CA, 2001
- R. L. Norton, **Machine Design: An Integrated Approach,** 4th Ed., Prentice Hall, Upper Saddle River, NJ, 2011
- Oberg, Jones, Horton, Ryffel, **Machinery's Handbook,** 28th Ed., Industrial Press, New York, NY, 2008

The relationships in Tables 1-6 through 1-20 are particularly helpful for machinery design. Many of these equations are used in the context of later chapters, but are reproduced here as a quick reference.

Table 1-6: Useful Equations and Values

Equation of a line with slope m, and y intercept b	$y = mx + b$
Equation of a circle of radius r, centered at (h, k)	$(x - h)^2 + (y - k)^2 = r^2$
Radians to degrees conversion	$\dfrac{radians}{\pi} = \dfrac{\deg}{180}$
Standard Acceleration of Gravity	$g_c = 386.4 \dfrac{lbm \cdot in}{lbf \cdot s^2}$ $g_c = 32.174 \dfrac{lbm \cdot ft}{lbf \cdot s^2}$ $g_c = 9.807 \dfrac{m}{s^2}$
Newton's Second Law of Motion	$F = ma$
Linear Thermal Expansion	$\delta = \Delta L = \alpha(\Delta T)L_o$ α = Coefficient of linear thermal expansion L_o = Original length of part

Table 1-7: Right Triangle Relationships
Units: angles are in degrees

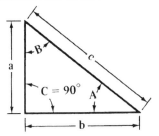

As shown in the illustration, the sides of the right-angled triangle are designated a and b and the hypotenuse, c. The angles opposite each of these sides are designated A and B, respectively.

Angle C, opposite the hypotenuse c is the right angle, and is therefore always one of the known quantities.

Sides and Angles Known	Formulas for Sides and Angles to be Found		
Side a; side b	$c = \sqrt{a^2 + b^2}$	$\tan A = \dfrac{a}{b}$	$B = 90° - A$
Side a; hypotenuse c	$b = \sqrt{c^2 - a^2}$	$\sin A = \dfrac{a}{c}$	$B = 90° - A$
Side b; hypotenuse c	$a = \sqrt{c^2 - b^2}$	$\sin B = \dfrac{b}{c}$	$A = 90° - B$
Hypotenuse c; angle B	$b = c \times \sin B$	$a = c \times \cos B$	$A = 90° - B$
Hypotenuse c; angle A	$b = c \times \cos A$	$a = c \times \sin A$	$B = 90° - A$
Side b; angle B	$c = \dfrac{b}{\sin B}$	$a = b \times \cot B$	$A = 90° - B$
Side b; angle A	$c = \dfrac{b}{\cos A}$	$a = b \times \tan A$	$B = 90° - A$
Side a; angle B	$c = \dfrac{a}{\cos B}$	$b = a \times \tan B$	$A = 90° - B$
Side a; angle A	$c = \dfrac{a}{\sin A}$	$b = a \times \cot A$	$B = 90° - A$

Table 1-8: Areas and Perimeters
Units: angles are in degrees unless otherwise noted

GENERAL TRIANGLE:		$Area = \dfrac{bh}{2} = \dfrac{ab\sin\theta}{2}$
PARALLELOGRAM:		$Area = bh = ab\sin\theta$
TRAPEZOID:		$Area = \dfrac{(b_1 + b_2)h}{2}$
CIRCLE:		$Area = \pi r^2$ $Circumference = 2\pi r$
CIRCULAR SEGMENT:		(ϕ is in radians) $Area = \dfrac{1}{2}r^2(\phi - \sin\phi)$ $c = 2r\sin(\dfrac{\phi}{2}) \quad \phi = \dfrac{s}{r}$
CIRCULAR SECTOR:		(ϕ is in radians) $Area = \dfrac{1}{2}\phi r^2 = \dfrac{1}{2}sr$ $c = 2r\sin(\dfrac{\phi}{2}) \quad \phi = \dfrac{s}{r}$

Table 1-9: Volumes and Surface Areas

SPHERE:		$Volume = \dfrac{4\pi r^3}{3}$ $Surface = 4\pi r^2$
RIGHT CIRCULAR CYLINDER:		$Volume = \pi r^2 h$ $Curved\ Surface\ Area = 2\pi rh$
RIGHT CIRCULAR CONE:		$Volume = \dfrac{\pi r^2 h}{3}$ $Curved\ Surface\ Area = \pi rs = \pi r\sqrt{r^2 + h^2}$

Table 1-10: Area Moments of Inertia

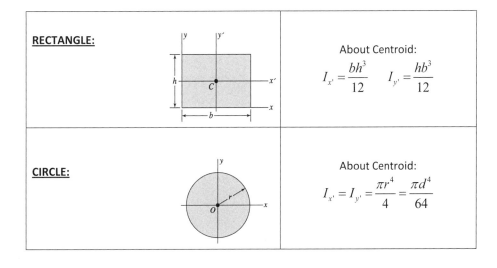

RECTANGLE:		About Centroid: $I_{x'} = \dfrac{bh^3}{12} \qquad I_{y'} = \dfrac{hb^3}{12}$
CIRCLE:		About Centroid: $I_{x'} = I_{y'} = \dfrac{\pi r^4}{4} = \dfrac{\pi d^4}{64}$

Table 1-11: Mass Moments of Inertia

RECTANGULAR PLATE:	$I_x = \dfrac{m(a^2 + b^2)}{12}$ $I_y = \dfrac{m(a^2 + c^2)}{12}$ $I_z = \dfrac{m(b^2 + c^2)}{12}$
SOLID CYLINDER:	$I_x = \dfrac{mr^2}{2}$ $I_y = I_z = \dfrac{m(3r^2 + l^2)}{12}$
HOLLOW CYLINDER:	$I_x = \dfrac{m(a^2 + b^2)}{2}$ $I_y = I_z = \dfrac{m(3a^2 + 3b^2 + l^2)}{12}$

Table 1-12: Beam Deflections and Forces
Units: always check all units for consistency

$E =$ Modulus of elasticity	$I =$ area moment of inertia
CANTILEVERED, POINT LOAD 	$y_x = \left(-\dfrac{P}{6EI}\right)(2L^3 - 3L^2 x + x^3)$ $y_{max} = -\dfrac{PL^3}{3EI}$ at x = 0 $\qquad \theta_l = \dfrac{PL^2}{2EI}$ End Slope Reactions: $R_r = P \qquad\qquad M_x = -Px \qquad M_r = -PL$
CANTILEVERED, DISTRIBUTED LOAD $W = wL$	$y_x = \left(\dfrac{w}{24EI}\right)(3L^4 - 4L^3 x + x^4)$ $y_{max} = -\dfrac{wL^4}{8EI}$ at x = 0 $\qquad \theta_l = \dfrac{wL^3}{6EI}$ Reactions: $R_r = W = wL \qquad M_x = \dfrac{-wx^2}{2} \qquad M_r = \dfrac{-wL^2}{2}$
SIMPLY SUPPORTED, POINT LOAD 	$y = \dfrac{Pa^2 b^2}{3EIL}$ at x = a $y_{max} = \left(\dfrac{0.06415Pb}{EIL}\right)(L^2 - b^2)^{\frac{3}{2}}$ at $x = \sqrt{\dfrac{a(L+b)}{3}}$ $\theta_l = -\dfrac{Pab\left(1+\dfrac{b}{L}\right)}{6EI} \qquad \theta_r = \dfrac{Pab\left(1+\dfrac{a}{L}\right)}{6EI}$ Reactions: $R_l = \dfrac{Pb}{L} \qquad R_r = \dfrac{Pa}{L}$ $M_{lx} = \dfrac{Pbx}{L}$ if x < a $\qquad M_{rx} = \dfrac{Pa(L-x)}{L}$ if x > a
SIMPLY SUPPORTED, DISTRIBUTED LOAD $W = wL$	$y_x = \left(-\dfrac{w}{24EI}\right)(L^3 x - 2Lx^3 + x^4)$ $y_{max} = \dfrac{5wL^4}{384EI}$ at x = L/2 $\theta_l = \dfrac{-wL^3}{24EI} \qquad \theta_r = \dfrac{wL^3}{24EI}$ Reactions: $R_l = R_r = \dfrac{wL}{2} \qquad M_l = M_r = \left(\dfrac{w}{2}\right)(x^2 - Lx)$

continued on next page

Table 1-12: Beam Deflections and Forces (Continued)

E = Modulus of elasticity	I = area moment of inertia

OVERHUNG POINT LOAD

$$y_{tip} = (a+b)\left(\frac{Pa^2}{3EI}\right) \text{ at P}$$

$$y_{up} = (0.06415)\left(\frac{Pab^2}{EI}\right) \text{ at x = 0.4226b}$$

Reactions:

$$R_l = \left(\frac{P}{b}\right)(b+a) \qquad R_r = \frac{-Pa}{b}$$

$$M_a = Px_a \qquad M_b = \left(\frac{Pa}{b}\right)(b-x_b)$$

FIXED SUPPORTS, POINT LOAD

Deflection at Load:

$$y = \frac{2Pa^3(L-a)^2}{3EI(2a+L)^2}$$

Reactions:

$$R_l = \frac{Pb^2}{L^3}(2a+L) \qquad R_r = \frac{Pa^2}{L^3}(3L-2a)$$

$$M_l = \frac{-Pa(L-a)^2}{L^2} \qquad M_r = \frac{-Pa^2(L-a)}{L^2}$$

FIXED SUPPORTS, DISTRIBUTED LOAD

$$y_x = \frac{-wx^2}{24EI}(L-x)^2 \qquad y_{max} = \frac{-wL^4}{384EI} \text{ at center}$$

Reactions:

$$R_l = R_r = \frac{W}{2} = \frac{wL}{2} \qquad M_l = M_r = \frac{-WL}{12} = \frac{-wL^2}{12}$$

$$W = wL$$

Table 1-13: Column Buckling
Units: always check all units for consistency

E = Modulus of elasticity	I = Area moment of inertia	S_y = Yield strength (tensile)

TEST FOR A SLENDER COLUMN	$$\frac{\pi^2 E}{\left(\dfrac{L_e}{k}\right)^2} \leq \left(\frac{1}{2}\right) S_y \qquad k = \sqrt{\frac{I}{A}}$$
EFFECTIVE COLUMN LENGTH L_e AS A FUNCTION OF UNSUPPORTED COLUMN LENGTH L a) b) c) d) e) f)	a) Both ends fixed: $L_e = 0.5L$ b) One end fixed, one end pinned: $L_e = 0.7L$ c) One end fixed, one end free to translate: $L_e = L$ d) Both ends pinned: $L_e = L$ e) One end fixed, one end free: $L_e = 2L$ f) One end pinned, one end free to translate: $L_e = 2L$
SLENDERNESS RATIO	$$SR = \frac{L_e}{k} \qquad k = \sqrt{\frac{I}{A}}$$
EULER'S FORMULA **CRITICAL BUCKLING LOAD FOR SLENDER COLUMN**	$$P_{cr} = \frac{\pi^2 EI}{L_e^2} = \frac{\pi^2 EA}{\left(\dfrac{L_e}{k}\right)^2} \qquad k = \sqrt{\frac{I}{A}}$$
CRITICAL BUCKLING LOAD FOR AN INTERMEDIATE COLUMN	$$P_{cr} = A\sigma_{cr} = A\left[S_y - \left(\frac{1}{E}\right)\left(\frac{S_y}{2\pi}\right)^2 \left(\frac{L_e}{k}\right)^2 \right]$$ $$k = \sqrt{\frac{I}{A}}$$

Table 1-14: Critical Speed of Rotating Shafts
Units: length in inches, force in pounds (lbf), speed in RPM

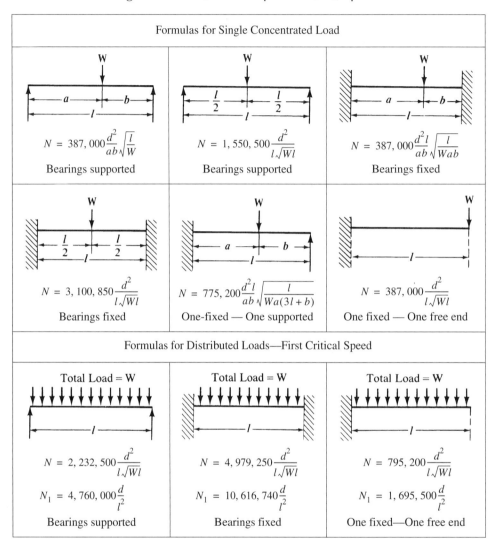

| Formulas for Single Concentrated Load |

$$N = 387,000 \frac{d^2}{ab}\sqrt{\frac{l}{W}}$$

Bearings supported

$$N = 1,550,500 \frac{d^2}{l\sqrt{Wl}}$$

Bearings supported

$$N = 387,000 \frac{d^2 l}{ab}\sqrt{\frac{l}{Wab}}$$

Bearings fixed

$$N = 3,100,850 \frac{d^2}{l\sqrt{Wl}}$$

Bearings fixed

$$N = 775,200 \frac{d^2 l}{ab}\sqrt{\frac{l}{Wa(3l+b)}}$$

One-fixed — One supported

$$N = 387,000 \frac{d^2}{l\sqrt{Wl}}$$

One fixed — One free end

| Formulas for Distributed Loads—First Critical Speed |

Total Load = W

$$N = 2,232,500 \frac{d^2}{l\sqrt{Wl}}$$
$$N_1 = 4,760,000 \frac{d}{l^2}$$

Bearings supported

Total Load = W

$$N = 4,979,250 \frac{d^2}{l\sqrt{Wl}}$$
$$N_1 = 10,616,740 \frac{d}{l^2}$$

Bearings fixed

Total Load = W

$$N = 795,200 \frac{d^2}{l\sqrt{Wl}}$$
$$N_1 = 1,695,500 \frac{d}{l^2}$$

One fixed—One free end

Table 1-15: Stress and Strain
Units: always check all units for consistency
Be sure to consult the recommended resources if combined stresses are present.

E = Modulus of elasticity	I = Area moment of inertia	A = Cross-sectional area

AXIAL TENSION OR COMPRESSION STRESS

$$\sigma = \frac{F}{A}$$

AXIAL STRAIN

$$\varepsilon = \frac{\sigma}{E} = \frac{F}{AE}$$

DEFORMATION FROM AXIAL TENSION

$$\delta = \Delta l = \frac{Fl}{AE}$$

DIRECT SHEAR STRESS

Single Shear   Double Shear

$$\tau_{AVG} = \frac{F}{A_s} \quad \text{Single Shear}$$

$$\tau_{AVG} = \frac{F}{2A_s} \quad \text{Double Shear}$$

SHEAR STRESS

(shearing force diagram)

For Beams of Rectangular Cross Section

$$\tau = \frac{3V}{2A} \qquad V = \sum_{\substack{one \\ end}} F_i$$

For Beams of Solid Circular Cross Section

$$\tau = \frac{4V}{3A}$$

BENDING STRESS

(bending moment diagram)

$$\sigma_{MAX} = \frac{Mc}{I} \qquad \sigma_x = \frac{-My}{I}$$

c = Distance from neutral axis to surface

BEARING STRESS

$$\sigma_b = \frac{F}{A_b} = \frac{F}{td}$$

ELASTIC STRESS OF TORSION

$$\tau_{MAX} = \frac{Tr}{J}$$

$$\tau = \frac{Ty}{J}$$

y = Distance from neutral axis

$$J = \frac{\pi d^4}{32}$$

J = Polar area moment of inertia

THREE DIMENSIONAL EQUIVALENT STRESS (VON MISES STRESS)
In terms of applied stresses

$$\sigma' = \sqrt{\frac{(\sigma_x - \sigma_y)^2 + (\sigma_y - \sigma_z)^2 + (\sigma_z - \sigma_x)^2 + 6(\tau_{xy}^2 + \tau_{yz}^2 + \tau_{xz}^2)}{2}}$$

Table 1-16: Fatigue Equations

MEAN AND ALTERNATING LOADS	$F_m = \dfrac{F_{max} + F_{min}}{2}$ $\qquad F_a = \dfrac{F_{max} - F_{min}}{2}$
MEAN AND ALTERNATING BENDING STRESSES	$\sigma_a = k_f \sigma_{a,nom}$ $\qquad \sigma_m = k_{fm}\sigma_{m,nom}$
MEAN AND ALTERNATING TORSIONAL SHEAR STRESSES	$\tau_a = k_f \tau_{a,nom}$ $\qquad \tau_m = k_{fm}\tau_{m,nom}$
FATIGUE STRESS CONCENTRATION FACTORS Alternating and mean. k_f and k_{fm} must be calculated separately for shear and bending. q = Notch sensitivity factor $\sqrt{a}$ = Neuber's constant (Table 1-17) r = Notch radius (inches) k = Static stress concentration factor k_f = Fatigue stress concentration factor S_y = Yield strength (tensile)	$k_f = 1 + q(k-1)$ $\qquad q = \dfrac{1}{1+\dfrac{\sqrt{a}}{\sqrt{r}}}$ $k \cong 1.5$ for large shoulder fillet with $\dfrac{r}{d} \cong 0.17$ $k \cong 2.5$ for sharp shoulder fillet with $\dfrac{r}{d} \cong 0.03$ $k \cong 2$ for milled keyseat (worst case type) $k \cong 3$ for retaining ring groove $k_{fm} = k_f$ If $k_f\left|\sigma_{max}\right| < S_y$ $k_{fm} = \dfrac{S_y - k_f \sigma_a}{\left|\sigma_m\right|}$ If $k_f\left|\sigma_{max}\right| > S_y$ $k_{fm} = 0$ If $k_f\left|\sigma_{max} - \sigma_{min}\right| > 2S_y$
VON MISES COMPONENTS OF ALTERNATING AND MEAN STRESSES. Three dimensional stress state	$\sigma'_a = \sqrt{\dfrac{\left(\sigma_{a,x}-\sigma_{a,y}\right)^2 + \left(\sigma_{a,y}-\sigma_{a,z}\right)^2 + \left(\sigma_{a,z}-\sigma_{a,x}\right)^2 + 6\left(\tau_{a,xy}^2 + \tau_{a,yz}^2 + \tau_{a,xz}^2\right)}{2}}$ $\sigma'_m = \sqrt{\dfrac{\left(\sigma_{m,x}-\sigma_{m,y}\right)^2 + \left(\sigma_{m,y}-\sigma_{m,z}\right)^2 + \left(\sigma_{m,z}-\sigma_{m,x}\right)^2 + 6\left(\tau_{m,xy}^2 + \tau_{m,yz}^2 + \tau_{m,xz}^2\right)}{2}}$

continued on next page

Table 1-16: Fatigue Equations (Continued)

ESTIMATED ENDURANCE STRENGTH With correction factors for ductile steel $S_{ut} \cong 68$ kpsi (469 MPa) for SAE 1020 CR $S_{ut} \cong 55$ kpsi (379 MPa) for SAE 1020 HR $S_{ut} \cong 91$ kpsi (627 MPa) for SAE 1045 CR $S_{ut} \cong 82$ kpsi (565 MPa) SAE 1045 HR $S_{ut} \cong 95$ kpsi (655 MPa) for SAE 4140 annealed $S_{ut} \cong 148$ kpsi (1020 MPa) for SAE 4140 normalized Heat treatment and manufacturing condition affect ultimate strength.	$S_e = C_{load} C_{size} C_{surf} C_{temp} C_{reliab} S_e'$ $S_e' \cong 0.5 S_{ut}$ when $S_{ut} < 200$ kpsi (1400 MPa) $S_e' \cong 100$ kpsi (700 MPa) when $S_{ut} \geq 200$ kpsi d = diameter for round cross section $d = \sqrt{\dfrac{0.05bh}{0.0766}}$ for rectangular cross section with width b and height h. $C_{load} = 1$ for bending, 0.7 for axial loading $C_{size} = 1$ for d ≤ 0.3 in (8 mm) $C_{size} = 0.869 d^{-0.097}$ for 0.3 in < d ≤ 10 in (d in inch) $C_{surf} = 1.34(S_{ut})^{-0.085}$ for ground surface (S_{ut} in kpsi) $C_{surf} = 2.7(S_{ut})^{-0.265}$ for machined or cold rolled surface (S_{ut} in kpsi) $C_{temp} = 1$ for T ≤ 840°F (450°C) $C_{reliab} = 0.868$ for 95% reliability $C_{reliab} = 0.620$ for 99.9999% reliability
FATIGUE FACTOR OF SAFETY (INFINITE LIFE) Determined using the modified Goodman diagram for ductile materials. It is assumed that the ratio of alternating to mean stress is constant over time. S_e = Estimated endurance limit S_{ut} = Ultimate tensile stress	$SF = \dfrac{S_e}{\sigma_a' + \left(\dfrac{S_e}{S_{ut}}\right)\sigma_m'}$

Table 1-17: Neuber's Constant for Notch Sensitivity

Material	S_{ut} (ksi)	Bending $\sqrt{a}\ \left(\sqrt{inch}\right)$	Torsion $\sqrt{a}\ \left(\sqrt{inch}\right)$
Steels	50	0.130	0.093
	55	0.118	0.089
	60	0.108	0.080
	70	0.093	0.070
	80	0.080	0.062
	90	0.070	0.055
	100	0.062	0.049
	110	0.055	0.044
	120	0.049	0.039
	130	0.044	0.033
	140	0.039	0.031
	160	0.031	0.024
	180	0.024	0.018
	200	0.018	0.013
	220	0.013	0.009
	240	0.009	0.004
Annealed Aluminums	10	0.500	
	15	0.341	
	20	0.264	
	25	0.217	
	30	0.180	
	35	0.152	
	40	0.126	
	45	0.111	

Table 1-18: *Linear Motion with Constant Acceleration*
Units: always check all units for consistency

To Find	Known	Formula	To Find	Known	Formula
Motion Uniformly Accelerated From Rest $(V_o = 0)$					
S	a, t	$S = \frac{1}{2}at^2$	t	S, V_f	$t = 2S \div V_f$
	V_f, t	$S = \frac{1}{2}V_f t$		S, a	$t = \sqrt{2S \div a}$
	V_f, a	$S = V_f^2 \div 2a$		a, V_f	$t = V_f \div a$
V_f	a, t	$V_f = at$	a	S, t	$a = 2S \div t^2$
	S, t	$V_f = 2S \div t$		S, V	$a = V_f^2 \div 2S$
	a, S	$V_f = \sqrt{2aS}$		V_f, t	$a = V_f \div t$
Motion Uniformly Accelerated From Initial Velocity V_o					
S	a, t, V_o	$S = V_o t + \frac{1}{2}at^2$	t	V_o, V_f, a	$t = (V_f - V_o) \div a$
	V_o, V_f, t	$S = (V_f + V_o)t \div 2$		V_o, V_f, S	$t = 2S \div (V_f + V_o)$
	V_o, V_f, a	$S = (V_f^2 - V_o^2) \div 2a$	a	V_o, V_f, S	$a = (V_f^2 - V_o^2) \div 2S$
	V_f, a, t	$S = V_f t - \frac{1}{2}at^2$		V_o, V_f, t	$a = (V_f - V_o) \div t$
V_f	V_o, a, t	$V_f = V_o + at$		V_o, S, t	$a = 2(S - V_o t) \div t^2$
	V_o, S, t	$V_f = (2S \div t) - V_o$		V_f, S, t	$a = 2(V_f t - S) \div t^2$
	V_o, a, S	$V_f = \sqrt{V_o^2 + 2aS}$			
	S, a, t	$V_f = (S \div t) + \frac{1}{2}at$			
V_o	V_f, a, S	$V_o = \sqrt{V_f^2 - 2aS}$			
	V_f, S, t	$V_o = (2S \div t) - V_f$			
	V_f, a, t	$V_o = V_f - at$			
	S, a, t	$V_o = (S \div t) - \frac{1}{2}at$			

Meanings of Symbols

S = distance moved in feet
V_f = final velocity, feet per second
V_o = initial velocity, feet per second
a = acceleration, feet per second per second
t = time of acceleration in seconds

Table 1-19: Rotary Motion with Constant Acceleration
Units: angles in radians

To Find	Known	Formula	To Find	Known	Formula
		Motion Uniformly Accelerated From Rest ($\omega_o = 0$)			
θ	α, t	$\theta = \frac{1}{2}\alpha t^2$	t	θ, ω_f	$t = 2\theta \div \omega_f$
	ω_f, t	$\theta = \frac{1}{2}\omega_f t$		θ, α	$t = \sqrt{2\theta \div \alpha}$
	ω_f, α	$\theta = \omega_f^2 \div 2\alpha$		α, ω_f	$t = \omega_f \div \alpha$
ω_f	α, t	$\omega_f = \alpha t$	α	θ, t	$\alpha = 2\theta \div t^2$
	θ, t	$\omega_f = 2\theta \div t$		θ, ω_f	$\alpha = \omega_f^2 \div 2\theta$
	α, θ	$\omega_f = \sqrt{2\alpha\theta}$		ω_f, t	$\alpha = \omega_f \div t$
		Motion Uniformly Accelerated From Initial Velocity ω_o			
θ	α, t, ω_o	$\theta = \omega_o t + \frac{1}{2}\alpha t^2$	α	$\omega_o, \omega_f, \theta$	$\alpha = (\omega_f^2 - \omega_o^2) \div 2\theta$
	ω_o, ω_f, t	$\theta = (\omega_f + \omega_o)t \div 2$		ω_o, ω_f, t	$\alpha = (\omega_f - \omega_o) \div t$
	$\omega_o, \omega_f, \alpha$	$\theta = (\omega_f^2 - \omega_o^2) \div 2\alpha$		$\omega_o, \theta\ t$	$\alpha = 2(\theta - \omega_o t) \div t^2$
	ω_f, α, t	$\theta = \omega_f t - \frac{1}{2}\alpha t^2$		$\omega_f, \theta\ t$	$\alpha = 2(\omega_f t - \theta) \div t^2$
ω_f	ω_o, α, t	$\omega_f = \omega_o + \alpha t$		*Meanings of Symbols*	
	ω_o, θ, t	$\omega_f = (2\theta \div t) - \omega_o$			
	ω_o, α, θ	$\omega_f = \sqrt{\omega_o^2 + 2\alpha\theta}$		θ = angle of rotation, radians	
	θ, α, t	$\omega_f = (\theta \div t) + \frac{1}{2}\alpha t$		ω_f = final angular velocity, radians per second	
ω_o	ω_f, α, θ	$\omega_o = \sqrt{\omega_f^2 - 2\alpha\theta}$		ω_o = initial angular velocity, radians per second	
	ω_f, θ, t	$\omega_o = (2\theta \div t) - \omega_f$		α = angular acceleration, radians per second, per second	
	ω_f, α, t	$\omega_o = \omega_f - \alpha t$		t = time in seconds	
	θ, α, t	$\omega_o = (\theta \div t) - \frac{1}{2}\alpha t$			
t	$\omega_o, \omega_f, \alpha$	$t = (\omega_f - \omega_o) \div \alpha$		1 degree = 0.01745 radians	
	$\omega_o, \omega_f, \theta$	$t = 2\theta \div (\omega_f + \omega_o)$			

Table 1-20: Work and Power

Units: distance in feet, force in pounds (lbf,) power in foot pounds per second

To Find	Known	Formula	To Find	Known	Formula
S	P, t, F	$S = P \times t \div F$	K	F, S	$K = F \times S$
	K, F	$S = K \div F$		P, t	$K = P \times t$
	t, F, hp	$S = 550 \times t \times hp \div F$		F, V, t	$K = F \times V \times t$
V	P, F	$V = P \div F$		t, hp	$K = 550 \times t \times hp$
	K, F, t	$V = K \div (F \times t)$	hp	F, S, t	$hp = F \times S \div (550 \times t)$
	F, hp	$V = 550 \times hp \div F$		P	$hp = P \div 550$
t	F, S, P	$t = F \times S \div P$		F, V	$hp = F \times V \div 550$
	K, F, V	$t = K \div (F \times V)$		K, t	$hp = K \div (550 \times t)$
	F, S, hp	$t = F \times S \div (550 \times hp)$			
F	P, V	$F = P \div V$			
	K, S	$F = K \div S$			
	K, V, t	$F = K \div (V \times t)$			
	V, hp	$F = 550 \times hp \div V$			
P	F, V	$P = F \times V$			
	F, S, t	$P = F \times S \div t$			
	K, t	$P = K \div t$			
	hp	$P = 550 \times hp$			

Meanings of Symbols:[a]

$S =$ distance in feet
$V =$ constant or average velocity in feet per second
$t =$ time in seconds
$F =$ constant or average force in pounds
$P =$ power in foot-pounds per second
$hp =$ horsepower
$K =$ work in foot-pounds

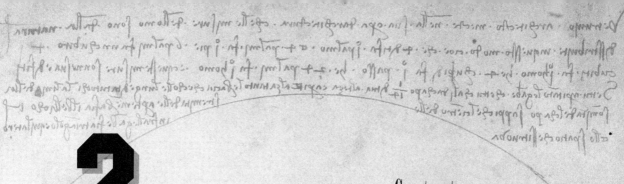

2

ERGONOMICS AND MACHINE SAFETY

**Co-Written with
E. Smith Reed, P.E.**

Contents

2.1	ERGONOMICS	38
2.2	MACHINE SAFETY: DESIGN PROCESS	47
2.3	MACHINE SAFEGUARDING	51
2.4	OTHER SAFETY ISSUES	65
2.5	RECOMMENDED RESOURCES	71

Tables

2-1	Body Dimensions, Inch	41
2-2	Body Dimensions, Metric	42
2-3	Access Openings	44
2-4	Point-of-Operation Protection: Barrier Guards	57
2-5	Point-of-Operation Protection: Operator Protective Devices	59
2-6	OSHA Noise Exposure Limits	67
2-7	Selected Codes and Standards for Machinery Safety	85

Figures

2-1	Body Dimension Illustrations	40
2-2	Clearances for Work Spaces	43
2-3	Reaching — Grasping — Moving Illustrations	46
2-4	Guard Openings vs. Distances from Point of Danger	55

Today, successful machines or products must not only perform their intended functions, and perform to customers' expectations, but they must do so without creating unnecessary hazards that could cause personal injury or property damage. Although no machine is perfectly safe (nothing is *perfectly* safe), machines are expected to be *reasonably* safe. They are expected to meet not only safety standards required by governmental codes and regulations, but also standards and criteria found in voluntary industry standards, as well as applicable international standards. In addition, they are expected to consider information and guidance found in textbooks, handbooks, and design manuals, and guidance found in various industry publications (technical journals, published seminar papers, magazine articles, etc.). Even industry customs and practices should be considered when addressing machine safety.

Machines (products in general) need to be designed to minimize, within reason, the possibility of personal injury or property damage. Machine designers must anticipate situations and events that are reasonably foreseeable relative to individuals operating the machine, individuals working on or around the machine, and individuals simply in the vicinity of the machine. They must consider the various ways the machine could cause or become involved in a mishap and the various ways a person could get hurt. Designers must consider the possibility / likelihood of a person getting *caught, nipped, pinched, drawn-in, trapped, entangled, crushed, struck, run over, cut, sheared, abraded, punctured, injected, jolted, vibrated, dragged, flung, wrenched, radiated, burned, scalded, blinded, deafened, poisoned, sickened, asphyxiated, shocked, electrocuted, overexerted* and *overextended,* as well as *losing a grasp, losing footing, losing balance, slipping, tripping, misstepping, falling on, falling into* and *falling off.* In addition, machine designers must consider the machine's potential to cause harm to animals *(unintentionally becoming caught, injured or killed)*, property *(being struck, contaminated, chemically altered, weakened, flooded, overloaded or burned)*, or the environment *(unacceptably polluting or unacceptably altering environmental or biological systems)*. The best and most effective time to assure a machine is reasonably safe is during the design process, not after.

Ergonomics is closely tied with machine safety. Ergonomics, sometimes referred to as 'human factors,' is a technical discipline concerned with (relative to design of machinery) interrelationships between humans and equipment. Ergonomics is applied for the purpose of integrating human capabilities and limitations with equipment characteristics, not only to reduce operational and safety issues and problems, but also to improve human-work effectiveness. The types of problems that can be mitigated by the application of human factors include: the failure to perform a required task, performing a task that should not have been performed, the failure to recognize a hazardous condition, a wrong decision in response to a problem, inadequate or inappropriate response to a contingency, or poor timing to a response, to name a few. For a machine to be reasonably safe for its operators or bystanders, ergonomics (human factors) must be part of the design process.

The subjects of machine safety and ergonomics have been written about in a number of well-researched books. In addition, industry safety standards have been created through collaborations of experts in their fields. It should be noted that the books and standards cited at the end of this chapter are not the only literature to which one should refer. They represent, however, a useful cross-section to which a machine designer can refer to understand today's expectations. The reader must understand that the material in this chapter is necessarily brief and general in nature. Hopefully, this introduction will encourage every designer to read further and seek a better understanding of the subjects. Section 2.5 of this chapter provides a list of recommended resources.

ERGONOMICS

Section 2.1

Ergonomics is a discipline that studies humans, their characteristics, capabilities, and limitations, in relationship with their surroundings. Relative to machine design, it is the science of designing a workspace environment and machine interface to fit the user. Like safety, ergonomics is a broad topic that is the subject of many excellent texts. Poor ergonomics can result in either acute injury or long-term repetitive use injury. It is essential to consider not only the physical geometry and arrangement of workspaces, but also the forces involved and the frequency in which they are encountered during a given workday. Actions like bending, reaching, grasping, lifting, turning, positioning, applying force, and releasing all must be conducted with proper ergonomics to achieve efficiency and assure a safe workplace.

Ergonomics issues one must consider when designing equipment and work space are:

- Operator body and workspace dimensions
- Operator performance capabilities and limitations

This section is meant to be an introductory reference only. The reference books and industry standards cited at the end of this chapter provide detailed information for the reader on these subjects.

 RECOMMENDED RESOURCES

- A list of books and industry standards are provided in Section 2.5 of this chapter.

BODY AND WORKSPACES DIMENSIONS

The figures and tables in this section represent only one set of data; many sets are available that generally represent information that has been accumulated by various groups and individuals over the years. Care must be taken to choose and refer to the appropriate information source for any given design task at hand.

BODY DIMENSIONS

Anthropometry is the study of body dimensions for the purpose of understanding human variations. Figures 2-1, 2-2, and related information are examples of anthropometric information available from various sources. The information in Tables 2-1 and 2-2 provide basic body dimensions representative of the adult civilian population in the United States. This type of information is particularly useful when designing workstations and controls and when determining various task assignments.

WORKSPACE, CLEARANCES, ENCLOSURES AND ACCESS OPENINGS

Machine workspaces for routine operation as well as maintenance and repair must be designed to accommodate operators as well as maintenance personnel. Interior access openings must be sized appropriately for the various purposes for accessing the area, whether for normal machine operation or for maintenance or repair. Openings must not only accommodate those parts of the individual needing access, but must also be large enough to accept whatever equipment the individual must insert or remove as well. Ideally, access

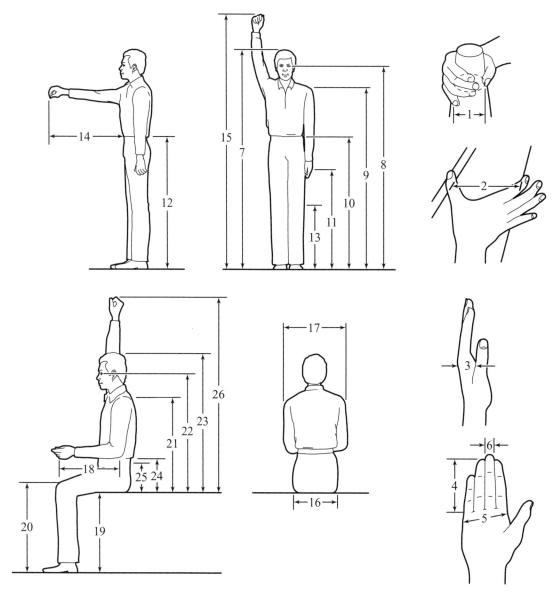

Figure 2-1: Body Dimension Illustrations

openings should provide not only room for reaching and working, but also additional space for an individual to be able to see.

Additional anthropometric and workspace information can be found in industry standards ISO 14738, EN 547-1, EN 547-2, ISO 15534-1, ISO 15534-2 and ISO 15534-3 (see details in Section 2.5), ISO 14738 being particularly useful for designers of machinery.

Table 2-1: Body Dimensions, Inch

		Men			Women		
		Percentile			Percentile		
		5th	50th	95th	5th	50th	95th
	Body Measurement	inch	inch	inch	inch	inch	inch
HAND	1. Grip Breadth (Inside Diameter)	1.65	1.89	2.05	1.57	1.69	1.81
	2. Max. Functional Hand Spread	4.80	5.59	6.38	4.29	5.00	5.71
	3. Hand Thickness	1.06	1.30	1.50	0.94	1.10	1.30
	4. Index Finger Length	2.52	2.83	3.11	2.36	2.64	2.91
	5. Hand Breadth	3.15	3.54	3.94	2.56	2.95	3.35
	6. Finger Breadth	0.75	0.83	0.91	0.63	0.71	0.79
STANDING	7. Height	64.57	69.09	73.62	59.84	63.98	68.11
	8. Eye Height	62.80	67.32	71.85	55.91	60.04	64.17
	9. Shoulder Height	52.36	56.69	61.02	48.23	52.17	56.10
	10. Elbow Height	40.16	43.50	46.85	37.20	40.16	43.11
	11. Knuckle Height	27.56	30.12	32.68	26.38	28.74	31.10
	12. Hip Height	32.87	36.02	39.17	29.92	32.87	35.83
	13. Knee Height	19.49	21.65	23.82	18.11	19.88	21.65
	14. Abdomen to Functional Pinch	19.09	24.06	29.33	no data		
	15. Functional Overhead Reach	76.77	81.89	87.01	71.06	75.79	80.51
SITTING	16. Hip Breadth	12.20	14.17	16.14	12.20	14.76	17.32
	17. Shoulder Breadth	16.73	18.50	20.28	14.17	15.75	17.32
	18. Elbow to Fist Length	12.56	14.45	16.18	no data		
	19. Seat Height (Popliteal Height)	15.55	17.52	19.49	14.17	15.94	17.72
	20. Knee Height	19.49	21.65	23.82	18.11	19.88	21.65
	21. Shoulder Height	21.46	23.62	25.79	20.08	22.24	24.41
	22. Eye Height	29.13	31.50	33.86	27.17	29.53	31.89
	23. Height	33.66	36.02	38.39	31.50	33.86	36.22
	24. Elbow Rest Height	7.68	9.65	11.61	7.28	9.25	11.22
	25. Thigh Clearance Height	5.31	6.30	7.28	4.92	6.10	7.28
	26. Functional Overhead Reach	45.47	49.41	53.35	42.13	45.67	49.21

Adapted from: Pheasant, 2002[1]

OPERATOR PHYSICAL CAPABILITIES AND LIMTATIONS: REACHING / GRASPING / MOVING / LIFTING

Ergonomics relative to individuals' physical capabilities and limitations also plays an important role in the design of machinery, not only to make the machine functional and productive, but also to assure the workspace is reasonably safe. Issues of physical movement strength and limitations, and exposure to surrounding environmental conditions must be understood during the design process.

[1] S. Pheasant, *Bodyspace, Anthropometry, Ergonomics and the Design of Work,* 2nd Ed., Taylor & Francis Inc., Philadelphia, PA, 2002

Table 2-2: Body Dimensions, Metric

		Men			Women		
		Percentile			Percentile		
		5th	50th	95th	5th	50th	95th
	Body Measurement	mm	mm	mm	mm	mm	mm
HAND	1. Grip Breadth (Inside Diameter)	42	48	52	40	43	46
	2. Max. Functional Hand Spread	122	142	162	109	127	145
	3. Hand Thickness	27	33	38	24	28	33
	4. Index Finger Length	64	72	79	60	67	74
	5. Hand Breadth	80	90	100	65	75	85
	6. Finger Breadth	19	21	23	16	18	20
STANDING	7. Height	1640	1755	1870	1520	1625	1730
	8. Eye Height	1595	1710	1825	1420	1525	1630
	9. Shoulder Height	1330	1440	1550	1225	1325	1425
	10. Elbow Height	1020	1105	1190	945	1020	1095
	11. Knuckle Height	700	765	830	670	730	790
	12. Hip Height	835	915	995	760	835	910
	13. Knee Height	495	550	605	460	505	550
	14. Abdomen to Functional Pinch	485	611	745	no data		
	15. Functional Overhead Reach	1950	2080	2210	1805	1925	2045
SITTING	16. Hip Breadth	310	360	410	310	375	440
	17. Shoulder Breadth	425	470	515	360	400	440
	18. Elbow to Fist Length	319	367	411	no data		
	19. Seat Height (Popliteal Height)	395	445	495	360	405	450
	20. Knee Height	495	550	605	460	505	550
	21. Shoulder Height	545	600	655	510	565	620
	22. Eye Height	740	800	860	690	750	810
	23. Height	855	915	975	800	860	920
	24. Elbow Rest Height	195	245	295	185	235	285
	25. Thigh Clearance Height	135	160	185	125	155	185
	26. Functional Overhead Reach	1155	1255	1355	1070	1160	1250

Adapted from: Pheasant, 2002[2]

Equipment, tools, or work areas that require reaching, grasping, or moving things must be designed to accommodate the strength capabilities and limitations of ordinary operators — being mindful of gender differences. Gender differences must be considered and planned for.

The illustrations in Figure 2-3 are examples of ergonomic information that can be found in the available literature (see citations at the end of this chapter). These types of illustrations are often accompanied with specific force (weight) limits, often representing the 5th percentile adult male (the weakest fifth percent) for the cited population group. Industry standard EN 1005-3 recommends that force limits for *professional users* correspond to the 15th *percentile* of the whole adult population (males and females) between 20 years and 65 years of age. For machines intended for *domestic users*, forces should be limited to the *1st per-*

[2] S. Pheasant, *Bodyspace, Anthropometry, Ergonomics and the Design of Work,* 2nd Ed., Taylor & Francis Inc., Philadelphia, PA, 2002

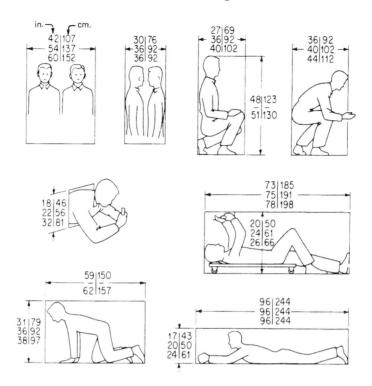

Figure 2-2: Clearances for Work Spaces

Source: Sanders and McCormick, 1993[3]

Note: These clearances may qualify as confined spaces and require a safety assessment.

centile (the weakest one percent) of the same total population. This EN 1005-3 standard also provides de-rating factors for movement velocities, movement frequencies, and application durations. For an understanding of capabilities/limits one might expect in a given situation, refer to and compare data from several sources, understanding what the information and chart numbers represent.

Among the types of movement and force limits to be considered include (but are not limited to):

- Pedal force limits
- Handwheel rotational force limits
- Hand-cranking force limits
- Finger-grip/squeeze and hand-grip/squeeze strength limits
- Hand-grasped item twisting force limits
- Weight carrying limits in various carrying modes
- Arm — up, down, in, out, pull, push — force limits
- Whole body pushing, pulling force limits
- Lifts above shoulder height weight limits

[3] Reproduced with permission of the McGraw-Hill Companies: Sanders, M. and McCormick, E. *Human Factors in Engineering and Design Seventh Edition.* McGraw-Hill: New York: 1993

Table 2-3: Access Openings

OPENING DIMENSIONS	DIMENSIONS (IN MM)		TASK
	A	B	
	, 110	120	USING COMMON SCREWDRIVER, WITH FREEDOM TO TURN HAND THROUGH 180°.
	130	115	USING PLIERS AND SIMILAR TOOLS.
	135	155	USING "T" HANDLE WRENCH, WITH FREEDOM TO TURN HAND THROUGH 180°
	270	200	USING OPEN-END WRENCH, WITH FREEDOM TO TURN WRENCH THROUGH 60°.
	120	155	USING ALLEN-TYPE WRENCH WITH FREEDOM TO TURN WRENCH THROUGH 60°.
	90	90	USING TEST PROBE.

continued on next page

Table 2-3: Access Openings (Continued)

OPENING DIMENSIONS	DIMENSIONS (IN MM)		TASK
	A	B	
	110	120	GRASPING SMALL OBJECTS (UP TO 50mm WIDE) WITH ONE HAND.
	W+45	125*	GRASPING LARGE OBJECTS (50mm OR MORE WIDE) WITH ONE HAND.
	W+75	125*	GRASPING LARGE OBJECTS WITH TWO HANDS, WITH HANDS EXTENDED THROUGH OPENINGS UP TO FINGERS.
	W+150	125*	GRASPING LARGE OBJECTS WITH TWO HANDS, WITH ARMS EXTENDED THROUGH OPENINGS UP TO WRISTS.
	W+150	125*	GRASPING LARGE OBJECTS WITH TWO HANDS, WITH ARMS EXTENDED THROUGH OPENINGS UP TO ELBOWS.

Source: MIL-HDBK-759

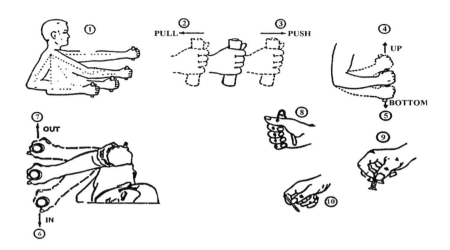

Figure 2-3: Reaching — Grasping — Moving Illustrations

Lifting is often required of machine operators while operating equipment or moving material into or out from a machine. An operator may be required to perform many lifts during a normal shift. There are limits to how much total lifting a person can do in a given workday, and it is dependent on the geometry of the lift, the weight involved, and the frequency of performance. When determining if a lifting task is reasonable, the Applications Manual for the Revised NIOSH Lifting Equation should be consulted (available online at **www.cdc.gov under "lifting**," which, at the time of this book's printing, was **http://www.cdc.gov/niosh/docs/94-110/**). It is recommended that a qualified professional review any ergonomic lifting scenarios.

Additional human physical performance information can be found in industry standards EN 1005-2, EN 1005-3, EN 1005-4 and EN 1005-5, MIL-STD-1472, and the book *Human Factors Design Handbook*, EN 1005.3 is particularly helpful for machine designers.

CRITICAL CONSIDERATIONS: Ergonomics

- Obey all applicable codes and standards when designing any machine for human interaction. See the recommended resources in Section 2.5 for more information.
- Safety and ergonomics are broad and specialized topics. Have a qualified professional evaluate all applications.

Section **MACHINE SAFETY: DESIGN PROCESS**

2.2

A properly designed machine must be reasonably safe for its intended or foreseeable use. This means that it is not just safe for operators and individuals working around the machine while it is operating and performing its primary function, but it is also safe for individuals involved with all aspects of the machine. This includes:

1. Initial assembly and set-up
2. Production job set-up
3. Machine start-up
4. Normal production operation
5. Unusual production circumstances
6. In-production adjustments, clearing and troubleshooting
7. Cleaning and routine maintenance
8. Non-routine maintenance
9. Relocation
10. Decommissioning and scrapping[4]

In addition, a machine should be reasonably safe for reasonably foreseeable misuse as well. The following foreseeable behavior of operators, maintenance personnel, and by-standers must be considered:

- Normal carelessness, inattention, or zeal (but not including deliberate and calculated misuse of the machine)
- Reflex behavior in cases of malfunction, disrupting incidents, failures, jams, etc., during use of the machine
- For some machines, particularly those foreseeably used by non-professionals, the foreseeable behavior of certain innocent, untrained, or unknowledgeable persons, such as children or disabled individuals.

To design a reasonably safe machine, the designer must understand the hazards and risks associated with the machine, and how to effectively reduce them to an acceptable level. Industry standards ISO 12100 and EN 1050 approach this task by working through two basic activities: Risk Assessment and Risk Reduction.

---

[4] Typically (ideally), the overwhelming majority of a machine's active life will be spent in process of performing its primary function, as identified as categories 2, 3, 4, 5, and 6 above. Because of this, a separate section in this chapter, Section 2.3, is devoted specifically to machine safety issues which are most important during this machine-active time, issues known as "safeguarding".

> **RECOMMENDED RESOURCES**
>
> • A list of books and industry standards is provided in Section 2.5 of this chapter.

RISK ASSESSMENT

Risk Assessment is a series of systematic steps to enable the examination of the various hazards associated with the machine. *Risk Assessment* is followed, when necessary, by *Risk Reduction*. It must be assumed that a hazard will sooner or later lead to an injury or damage to health (or property) if no Risk Reduction safety measure is taken. This process is a repeating iterative one to eliminate or reduce hazards as far as possible. Risk Assessment steps include:

1. Risk Analysis
   (a) Understand the machine, its function, requirements, and limits:
      - Understand the machine's space requirements, power requirements, operator requirements, maintenance requirements, intended use, and foreseeable misuse throughout all phases of the machine's life cycles.
   (b) Identify hazards and perform analysis:
      - Identify and describe all reasonably foreseeable hazards associated with the machine.
      - Identify all individuals who will be operating or maintaining or in the general area of the machine. Consider their skill levels and likely training.
      - Understand the frequency and duration of each hazard's exposure, as well as the relationship between exposure and likely consequences.
      - Consider all reasonably foreseeable situations and conditions.
   (c) Estimate risks:
      - Perform a systematic analysis of risk based on:
         • the **severity** of the possible harm, and
         • the **probability of occurrence**
      - Consider the individuals or property exposed, the type, frequency and duration of exposure, and the relationship between exposure and the effects.

     - Refer to recognized industry standards for risk estimation procedures, including specifically MIL-STD-882 and ANSI B11. TR3. Other standards to consider include ANIS B11-2008, ISO 12100:2010, EN 1050, ISO/TR 14121-2, ISO 14121-1, EN 1005-5, and ISO 13849-1.

2. Risk Evaluation

    (a) Based on the results from Risk Analysis, determine if Risk Reduction (or *additional* Risk Reduction) is required.

    (b) If Risk Reduction is needed, then take appropriate safety-improvement steps. These could include eliminating a hazard, providing additional safeguarding, restricting or improving access, interlocks, etc.

    (c) Repeat the Risk Assessment process until an acceptable level of risk is achieved.

There are several methods customarily used for analyzing hazards and estimating risks. Each has its own unique approach, strengths, and limitations. Although it is not imperative that any one of these methods listed should necessarily be used, some organized methodical approach should be, and the process should take place while the machine is being designed — not after. Some methods for analyzing hazards and estimating risks include:

Failure Modes, Effects, and Criticality Analysis (FMECA—sometimes known as FMEA) is a method of assessing designs or processes with respect to the various ways they can fail. Failure modes can affect safety as well as machine function. FMECA uses a worksheet (or software) to identify failure modes, their effects, risks, probabilities, and control (mitigation) methods. FMECA is covered in more detail in Chapter 12.

'What If' Analysis is a simple system of questions and answers. For machines or systems that are not too complex, "what if" questions are asked and answered in a systematic way about the machine and its operation for the purpose of evaluating the consequences of component failures, operator mistakes, or certain operational situations. The suitability of the machine, its controls, and its safeguarding and safety-related equipment are evaluated.

Fault Tree Analysis (FTA) uses deductive logic. This method starts with an unwanted event (the unwanted consequence) and works backwards to reveal the individual failures that can lead to that event. It enables individuals to find the various critical paths that can lead to the eventual unwanted event. The end event is first identified, then the various events and failures (and combinations of failures) that can lead to the final event are identified and listed, followed by estimates of probabilities of failure. A fault tree analysis can be used to determine the impact of alternative designs.

Preliminary Hazard Analysis (PHA) is an inductive analysis tool that helps identify and address machine hazards at the earliest stages of design. It is a means of creating an initial list of all hazards that may exist in every machine area and system and operation. It helps overcome the tendency to focus only on immediately obvious hazards, forcing an evaluation of potentially more serious or hidden dangers within a machine. Proposals for safety measures are the end result.

The DELPHI Technique involves providing questionnaires to a group of experts, individually, in successive steps. The results of the previous round of questions and answers, together with additional information from the others in the group, are communicated back to the participants. During the third or fourth round, the questions concentrate on those issues for which there is little or no general agreement. Because of its use of experts, this technique is notably efficient.

It should be understood that there are many methods of hazard identification and analysis. These listed are only a few. Each method has advantages for certain applications; therefore, it may be necessary to adjust or combine methods to match the situation at hand. It is important, whatever method is or methods are chosen, that hazard identification and analyses be performed early and often during the design process.

RISK REDUCTION

Risk Reduction is the process of taking sequential steps to either eliminate hazards or reduce hazards to an acceptable level. Although there are variations of lists of such steps, the following steps are commonly cited (listed in sequential order of most effective to least effective):

1. Eliminate or reduce the severity of the hazard (design the hazard out).
2. Safeguard the hazard (barrier guards or protective devices).
3. Instruct and warn the user (in manuals, and when appropriate, on the machine).
4. Describe requirements for training the user (safe work procedures training).
5. Recommend personal protective equipment (ear protection, eye protection, etc.).

The Risk Reduction process, unlike the Risk Assessment process, requires actual design creations and decisions — some take the form of designs that eliminate or provide safeguarding for hazards; others are in the form of warnings, instructions, or information (literature and manuals that provide information, instructions, and warnings are a part of the total machine). When risks have been properly assessed, and when the machine's design has satisfactorily reduced the risks, the resulting machine can be judged to have reached its safety goals. The Risk Assessment / Risk Reduction iterative cycle would then be complete. (For a detailed understanding of this process, refer to industry standards ISO 12100:2010, ANSI B11-2008, and others listed in Section 2.5.)

CRITICAL CONSIDERATIONS:
Machine Safety: Design Process

- Be sure to comply with all safety and ergonomics laws, codes, and standards. Consult the recommended resources (Section 2.5) for more information.
- Conduct Risk Assessments on all designs and processes. Conduct risk analysis early and often during the design process.
- Designing a hazard out should be the goal of any risk reduction activity. If this is not possible, other methods may be employed.
- Risk of long-term repetitive use injury should never be underestimated. This is an important factor in the design of workstations, tools, and industrial equipment.

Section

2.3

MACHINE SAFEGUARDING

Most machines have moving parts, both rotating and linear-moving, that can cause injury. Moving machine parts may be found in numerous locations on or around a machine, including:

(a) At the point of operation where work is performed on the work piece or material,

(b) In the power transmission components that transmit power throughout the machine (shafts, pulleys, belts, sprockets, chains, flywheels, couplings, spindles, gears, cams, cranks, rods, and other moving parts), or

(c) In other moving parts of the machine (machine components that move during operation, such as rotating parts, reciprocating parts, or traverse-moving parts).

All parts that move have the potential to contribute to accidents that can result in personal injury. Both rotating motion and linear movement can be dangerous.

Rotating components, even smooth rotating shafts, can grip an item of clothing or hair — even skin — and draw it into the machine. The danger of rotating components increases if they contain irregular or uneven surfaces or projecting parts, such as adjusting screws, bolts, slits, notches, or sharp edges. Rotating machine parts can also create dangerous in-running nip points with adjacent rotating or wedging components.

Vertical, horizontal, and reciprocating motion can cause injury in several ways, including causing a person to become caught between a moving machine part and some other object, shoving or knocking into a person, catching a person in a nip point, or causing injury with an unexpected movement of a sharp edge.

In addition, many machines have hydraulic or pneumatic systems that transmit power, or components that store energy. Springs, components under pressure, and elevated masses are examples of stored energy that can be released suddenly and hazardously. These power-transmitting and energy-storing components, should they fail or be released unexpectedly, pose hazards that must be considered during the machine design process. For example, a burst hydraulic line can spray oil that can be scalding hot, or can produce pin-point spray capable of penetrating skin. A whipping hose can become violent. A failed pneumatic fitting can release a potentially dangerous blast of compressed air, or release the pressure on an air cylinder causing unexpected movement and potentially dropping a load. The sudden release of a deflected spring or sudden drop of an elevated component can be very dangerous.

If a given hazard cannot be designed out of a machine (or neutralized by the nature of its remote location), then safeguarding, in some form, must be provided. Safeguarding can be categorized as either *Designed-into-the-machine* safeguarding, or *Procedural* safeguarding. (Procedural safeguarding is sometimes known as *Information for use*, or information and warnings).

Designed-into-the-machine Safeguarding can be either a guard, which is a physical barrier that prevents access to a hazardous area, or a protective device, which is a safeguarding means other than a barrier *guard* that controls access to a hazardous area. Examples of protective devices are photoelectric curtains, cable pull-backs, pressure sensitive mats, trip bars, and two-hand controls. Protective devices are often used to control access specifically to a machine's point of operation. Guards and protective devices can be (and often are) integrated together (an interlocking guard, for instance) when appropriate to enhance productivity or further enhance overall safety. Safeguarding will be detailed later in this section.

Procedural Safeguarding (Information for use) is information, instructions, procedures, and/or warnings posted on the machine, in manuals, or in training sessions instructing individuals how to operate the machine, how to avoid certain actions, or how to take certain steps to avoid a particular hazard. Procedural safeguarding is inherently less effective than barrier guards or protective devices designed into the machine because it relies on factors that are unpredictable. Procedural safeguarding relies on every individual doing everything in accordance with all instructions — 100% of the time — with no exceptions.

Safeguarding that is designed into a machine is far more effective than procedural safeguarding (information and warnings). Only when the hazard cannot be designed out of the machine and cannot be effectively safeguarded (the 1st and 2nd steps in the sequential Risk Reduction steps — see earlier) should the designer turn to those steps that fall into the 'Procedural Safeguarding' category (the 3rd, 4th, and 5th steps).

> **RECOMMENDED RESOURCES**
>
> • A list of books and industry standards are provided in Section 2.5 of this chapter.

GUARDS

When a hazard cannot be designed out of a machine, safeguarding the hazards *(guards or protective devices)* are next in priority. Physical barrier *guards* (and shields) have significant advantages over other means of safeguarding. They can provide operator protection from:

- Contact with moving parts of the machine
- Contact with splashing or misting chemicals or liquids
- Contact with thrown objects, hot chips, or other discharged objects or particles

- Contact with failed machine components (mechanical, hydraulic, pneumatic, electrical, etc.)
- Harmful contact due to human frailties and human traits, such as: curiosity, distraction, zeal, fatigue, worry, anger, illness, corner-cutting, or deliberate risk taking

Guards and shields should also be considered as protection for certain components of the machine itself. They can protect components susceptible to damage, such as electrical wires, electrical components, hydraulic lines, or small functional mechanical components that could otherwise be exposed to damage. (This damage could be due to such things as human error during machine handling, set-up, operation or maintenance and repair; machine malfunction and workpiece variations; or the machine's exposure to workplace traffic, to name a few). Guards and shields can be used to protect machine components from damage from:

- Contact with hard objects (workers' tools, workpiece material, etc.)
- Contact with a failed machine component
- Exposure to harmful liquids, mists, fumes, or chemicals
- Exposure to radiation or harmful light

Guards and shields are intended to provide protection. For the design and construction of a barrier guard to be fully effective, it should:

- Provide adequate protection from the danger zone during operation.
- Conform to applicable federal and state laws and regulations.
- Be considered as a permanent part of the machine. (If capable of being opened or removed, it should be simple and quick to close or re-install, and it should become obvious or required that it be closed or re-installed before machine operation continues.)
- Be durable and robust (able to withstand the stresses of the process and environmental conditions, able to hold up under normal wear and impact and corrosion damage, and able to withstand extensive use with minimum maintenance).
- Be easily reparable.
- Not weaken the structure of the machine.
- Be convenient — not interfering with the efficient operation of the machine, and not causing aggravation or discomfort to the operator.
- Be designed for the specific machine and for the specific danger zone.

- Have provisions for inspecting, adjusting, maintaining, and repairing the machine.
- Itself not create a hazard (having sharp edges, creating a pinch point, etc.).

Regarding guards themselves, the allowable opening in or under a machine guard changes with the distance from the point of danger. Guards close to a point of danger must restrict sizes of openings, while guards further away can allow larger openings. One standard identifying these distances and openings are cited in OSHA regulations (1910.217(c), Table O-10 — see Figure 2-4). Other industry standards provide more thorough information on this subject, including those cited in Section 2.5 of this chapter (specifically: EN 294, ISO 13852, EN 811, ISO 13853, ISO 13857, and EN 999).

There are four basic types of guards commonly used today: *fixed, adjustable, self-adjusting,* and *interlocked.* Fixed guards and interlocked guards

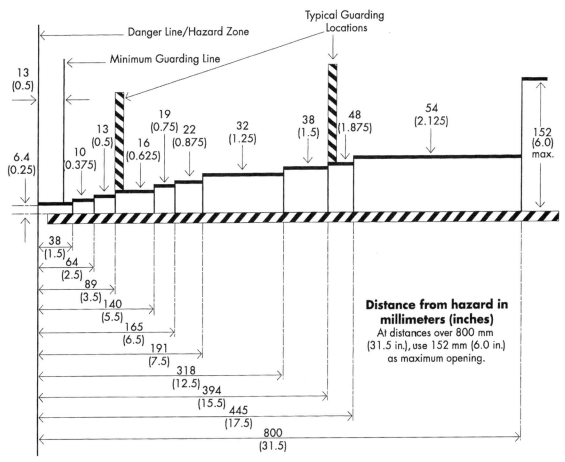

Figure 2-4: Guard Openings vs. Distances from Point of Danger

are the most common type for guarding mechanical power transmission components and other machine moving parts. For these applications, there are few reasons to employ adjustable or self-adjusting guards. For guarding a machine's point-of-operation, depending on the situation, any one of the four types of guards may be appropriate. Examples of guards can be found in the National Safety Council book *Safeguarding Concepts Illustrated* and the industry standard BSI PD 5304 Safe Use of Machinery (Section 7), as cited near the end of this chapter. For point-of-operation guarding, Table 2-4 provides a list of advantages and disadvantages for these four types of point of operation guards.

- Fixed guards, if they are designed appropriately, are the safest type of guarding. When they are in place on the machine, by their fixed design and fixed location, they provide the protection they were designed for — every time.[5] (The guarding over the top of the blade of a powered circular saw is an example of a fixed guard.)

- Adjustable guards are a type of guards that have adjustable parts to permit various machine settings and various operations. An example of an adjustable guard is a guard at the entrance to a machine's point-of-operation that can be adjusted to admit variations in work piece shapes. (Adjustable guards are often found at the point of operation of band-saws.)

- Self-adjusting guards are designed such that the whole guard or a portion of the guard is free to move and can automatically adjust (move) to accommodate movement of the machine or movement of material being processed. (Self-adjusting guards are commonly found on the underside — under the support shoe — of hand-held circular saws.)

- Interlocking guards are guards that interact with a device that is interconnected with some operational function of the machine to automatically stop (or alter) a prescribed function should the guard be out of place.[6] Interlocking guards are common on light industrial machinery where frequent, safe access to hazardous areas is required. Interlocking guards can include doors, lids, and gate guards. Gate guards are physical barrier guards that automatically close off access during the hazardous operation of the machine and open for access when the machine has finished the unsafe activity. (The top lid to a top-loading clothes washing machine is

---

[5] See industry standards BSI PD 5304 (Section 7), ANSI B11.19, ANSI B15.1, EN 953, WN 1088, ISO 14120, ISO/TR 5046, and the book *Safeguarding Concepts Illustrated*.

[6] Interlocking guards are discussed in greater detail in industry standards BSI PD 5304 (Section 9), EN 1088, ISO 14119, and in the book *Product Safety Management and Engineering* (Chapter 11). Be aware, also, that the interlocking part of interlocking guards are a part of the machine's "safety-related control system," discussed in some detail in a footnote later in this chapter.

Table 2-4: Point-of-Operation Protection: Barrier Guards
Based on U.S. Department of Labor, OSHA publication OSHA 3170-02R 2007

Type	Method of Safeguarding	Advantages	Limitations
Fixed Guard	Barrier that allows for material feeding but does not permit operator to reach the danger area.	• Can be constructed to suit many applications • Permanently encloses the point of operation or hazard area • Provides protection against machine malfunction • Allows simple, in-plant construction, with minimal maintenance	• Sometimes not practical for changing production runs involving different material or feeding methods • Machine adjustment and repair often require guard removal. • Other means of protecting maintenance personnel are often required (Lockout/Tagout).
Adjustable Guard	Barrier that adjusts for a variety of parts feeding requirements.	• Can be constructed to suit many applications • Can be adjusted to admit varying material or part sizes	• May require frequent maintenance or adjustment • Improper adjustment may make the guard ineffective.
Self-Adjusting Guard	Barrier that moves according to the size of the material or part entering point of operation. Guard is in place when machine is at rest and pushes away when stock enters the point of operation.	• Off-the-shelf guards are sometimes commercially available. • Can be adjusted to admit varying material or part sizes • Convenient for the operator	• Does not provide maximum protection • May require periodic maintenance and adjustment
Interlocking Guards	Open interlock shuts off or disengages power and prevents machine start-up when guard is out of position. Should allow for inching or jogging of machine.	• Allows quick access, in accordance with the Lockout/Tagout exception, without time-consuming removal of fixed guards	• May require periodic maintenance or adjustment • Movable sections cannot be used for manual feeding. • Some designs may be easy to defeat. • Maintenance and servicing work may be performed without interlock control circuitry.

an example of an interlocking guard, interconnected with the machine to prevent spin-cycle operation when the lid is open.)

PROTECTIVE DEVICES

For a machine's point-of-operation, a less desirable alternative to physical barrier guarding is a protective device. Point-of-operation protective devices work to prevent an operator's hands and other body parts from entering the point-of-operation danger zone during machine operation. Generally, they do not prevent discharge of hot chips, thrown particles, or liquid splash as barrier guards do. They also do not protect others in the vicinity other than the operator.

Protective devices can be categorized into four basic types: restraints, proximity detection devices, two-hand controls, and gates.

Restraints include fixed restraints that limit the operator's movement, or pull-back devices that physically pull the operator's hands out of the danger zone prior to machine cycling. Restraints protect only the operator, not others who may be in the area.

Proximity detection devices include interlocking devices, light curtains, weight-sensitive floor mats, pressure-sensitive bars or edges, contact probes, and other devices. These devices protect everyone in the area.

Two-hand controls are primarily devices placed some safe distance away from the danger zone that require both of the operator's hands to contact cycle start switches simultaneously. The distance from the danger zone is calculated such that operators cannot push the buttons and then quickly move their hands into the danger zone. These devices are also often called "two-hand no-tie-down devices". These devices protect only the operator, not others who may be in the area.

Gates are physical barriers that move to block access to the danger zone when the machine cycles. These devices protect everyone in the area.

Examples of protective devices can be found in industry standards BSI PD 5304 (including in Section 8), and in other literature cited in Section 2.5 of this chapter. Table 2-5 provides a list of advantages and disadvantages of various point-of-operation protective devices.

Table 2-5: Point-of-Operation Protection: Operator Protective Devices
Based on U.S. Department of Labor, OSHA publication OSHA 3170-02R 2007

Type	Method of Safeguarding	Advantages	Limitations
Pullback Devices	Cords are connected to the operator's wrists and linked mechanically to the machine. They automatically withdraw the hands from the point of operation **during the machine cycle.**	• Feeding of material is unrestricted • Allows the hands to enter the point of operation for feeding and removal • Provides protection even in the event of mechanical repeat	• Requires adherence to procedure • Does not protect everyone in the area • Does not protect against thrown material, either into or out of the machine • Close supervision is required to ensure proper use and adjustment. Must be inspected prior to each operator change or machine set-up • Limits operator's movement and may obstruct their work space • Improper adjustment may make device ineffective
Restraint Devices	The operator's wrists are attached to cords that are secured to a fixed anchor point. The tethers prevent the operator's hands from reaching the point of operation **at any time.**	• Feeding of material is unrestricted • Simple, few moving parts; requires little maintenance • Operator cannot reach into the danger area • Little risk of mechanical failure; provides protection even in the event of mechanical repeat	• Requires adherence to procedure • Does not protect everyone in the area • Does not protect against thrown material, either into or out of the machine • Close supervision is required to ensure proper use and adjustment. Must be inspected prior to each operator change or machine set-up • Operator must use hand tools to enter the point of operation • Limits the movement of the operator; may obstruct work space around operator

continued on next page

Table 2-5: Point-of-Operation Protection: Operator Protective Devices (Continued)

Type	Method of Safeguarding	Advantages	Limitations
Presence-Sensing Devices	These devices interlock with the machine's control system to stop operation when the sensing field (photoelectric, radio frequency, or electro-magnetic) is disturbed.	• Protects everyone in the area • Does not require adherence to procedure • Can be adjusted for feeding of different material • Allows access to load and unload the machine • Allows quick access to the guarded area for service activities	• Does not protect against thrown material, either into or out of the machine • Restricted to machines that stop operating cycle before operator can reach into danger area (e.g., machines with partial revolution clutches or brakes, or hydraulic machines) • Must be carefully maintained and adjusted • Does not protect operator in the event of a device failure • Operator may make device ineffective
Presence-Sensing Mats	These devices interlock into machine's control system to stop operation when a predetermined weight is applied to the mat. A manual reset switch must be located outside the protected zone.	• Protects everyone in the area • Does not require adherence to procedure • Full visibility and access to the work area • Can be installed as a perimeter guard or over an entire area • Configurable for many applications	• Does not protect against thrown material, either into or out of the machine • Restricted to machines that stop operating cycle before operator can reach into danger area (e.g., machines with part-revolution clutches or hydraulic machines) • Some chemicals can degrade the mats • Does not protect operator in the event of a device failure
Two-Hard Control	This device requires concurrent and continued use of both hands to operate the machine, preventing them from entering the danger area.	• Does not require adherence to procedure • Operator's hands are at a predetermined safety distance • Operator has unrestricted movement between machine cycles • Operator's hands are free to pick up new parts after completion of the dangerous part of the machine cycle • No obstruction to hand feeding	• Does not protect everyone in the area • Does not protect against thrown material, either into or out of the machine • Operator may make devices without anti-tiedown ineffective • Protects the operator only

continued on next page

Table 2-5: Point-of-Operation Protection: Operator Protective Devices (Continued)

Type	Method of Safeguarding	Advantages	Limitations
Two-Hand Trip	This device requires concurrent use of both hands to initiate a machine cycle, and prevents them from being in the danger area when the machine cycle starts.	• Does not require adherence to procedure • Operator's hands are at a predetermined safety distance • Can be adapted to multiple operations • No obstruction to hand feeding • Operator's hands are free to pick up new parts after completion of the dangerous part of the machine cycle	• Does not protect everyone in the area • Does not protect against thrown material, either into or out of the machine • Protects the operator only • Sometimes impractical because distance requirements may reduce production below acceptable level • May require adjustment with tooling changes • Requires anti-repeat feature
Type "A" Gate (moveable barrier)	These guards are applicable to mechanical power presses. They provide a barrier between the danger area and the operator (or other employees) until completion of a machine cycle.	• Protects everyone in the area • Protects against thrown material • Does not require adherence to procedure • Prevents operator from reaching into danger area during machine cycle • Provides protection from machine repeat or other malfunction	• May require frequent inspection and regular maintenance • May interfere with operator's ability to see work
Type "B" Gate (moveable barrier)	These guards are applicable to mechanical power presses and press brakes. They provide a barrier between the danger area and the operator (or other employees) during the downstroke.	• Protects everyone in the area • Protects against thrown material • Does not require adherence to procedure • Prevents operator from reaching into danger area during the dangerous part of the machine cycle • May increase production by allowing the operator to remove and feed the press on the upstroke	• Can only be used on machines with a part-revolution clutch or brake, or hydraulic machines • May require frequent inspection and regular maintenance • May interfere with the operator's ability to see work

PROCEDURAL SAFEGUARDING: INFORMATION, INSTRUCTIONS AND WARNINGS

Information and instructions, customarily in the form of machine manuals[7], are an integral part of a machine. Manuals provide information and instructions about the machine and its use. Information and instructions should be complete enough to ensure proper and safe installation, set-up, use, and maintenance. Instructions are important, but on their own, they are not enough to ensure operators will work safely.

Regarding specifically safety, manuals must also include information and warnings about hazards and risks that may not be known to users, not just when the machine is used as intended, but in reasonably foreseeable misuse situations as well. Although there is not a duty to warn of hazards that are open and obvious (the sharpness of a kitchen knife, for instance), it is important to understand that what may be obvious to a machine's designer may not be obvious to an ordinary user. *Procedural safeguarding*, the 3rd, 4th and 5th priorities[8] in the machine hazard risk reduction hierarchy list, starts with the information, instructions and warnings found in the manual. This section is intended to be only a brief introduction to the subject. The reader is encouraged to become familiar with books and industry standards on the subject, some of which are cited in Section 2.5.

Regarding warnings, when a hazard that can cause serious injury cannot be designed out of a machine (or protected by its remote location) and cannot be effectively safeguarded by a barrier guard or protective device, if the hazard is not obvious or readily known to an ordinary user, there is an obligation to provide a warning likely to be seen and understood. Although warnings provide information and instructions about hidden or unknown hazards, it must be understood that they have a limited impact on making mishaps less probable. The ultimate aim of a warning is to alter behavior, by encouraging an individual either to do something, or to avoid doing something.

Once it is determined that a warning is needed, the resulting warning must be crafted to be, at minimum, "adequate." An "adequate" warning is one

---

[7] Although customarily this information and instructions has been in manual form, increasingly, it is becoming more and more common to find electronic media used as well. The book *Writing and Designing Manuals* provides further information on writing machine manuals (cited in Section 2.5).

[8] These being: "3. Instruct and warn the user," "4. Describe requirements for training the user," and "5. Recommend personal protective equipment."

that would lead an ordinarily reasonable person to understand the hazard and take steps to avoid harm. In general, a warning can be viewed as "adequate" if:

- the warning is in a form that could reasonably be expected to catch the attention of a reasonably prudent person,
- with a message understandable to the ordinary user,
- conveying a fair indication of the nature and gravity of the harm that could result from the hazard,
- conveyed with a degree of intensity that would cause a reasonable person to exercise appropriate caution.

The development of a warning can be a simple process of providing understandable information and instructions in an effective way or, in some situations, a complex process involving product research, message development, and focus group responses. In all cases, warnings—both those included in manuals and those posted on machines—for them to be "adequate" must be clear, readable, and understandable to the ordinary user.

WARNINGS INCLUDED IN THE MANUAL

Warnings included in a machine's manual must be capable of conveying the same information as those posted on the machine itself. When writing warnings for the manual, the following should be considered:

- All warnings a user may need to know about should be included in the manual.
- From a style perspective, warnings in a manual must be reasonably consistent with each other, and they must be consistent with the warning labels posted on the machine.
- Warnings that warrant the "**DANGER**" level (DANGER being the most serious level) must be not only included in the manual but also posted on the machine itself.
- Warnings must stand out from the rest of the information in the manual.
- Never mix warnings with ordinary instructions.
- Never bury warnings in the text in such a way that it might be missed.
- If there is a section in the manual dedicated to listing all warnings and safety instructions (at the beginning of the manual, for instance), also include the warnings elsewhere in the manual text where appropriate.
- It is good practice to illustrate in the manual every warning label posted on the machine, indicating its location on the machine. Along with

these illustrations, it is good practice to provide the label's part number to simplify replacement if necessary.

It must be remembered that not all individuals who will be operating the machine will have had access to or have read the manual. And many who have read the manual will not remember certain pieces of information contained within. It is important to remember that for a warning to be effective, it must be seen, read, understood, remembered, and heeded. As a result, warnings placed only in the manual will not likely meet at least one of the requirements for it to be deemed "adequate."

WARNING LABELS POSTED ON THE MACHINE

Warnings located in the manual are simply not going to catch the attention of a machine user who has not read it or doesn't recall its contents.

Some hazards that cannot be eliminated or guarded are serious enough to warrant the posting of warning labels on the machine itself. The design of machine-posted warnings has developed over the years, based on industry experience and controlled studies. One definition of an effective warning is one that changes behavior in a way that results in a net reduction in negative consequences. It has been found that the effectiveness of machine-posted warnings improves when warnings:

- are **located near the zone of danger** itself,
- are **conspicuous** (eye-catching) and **in a location likely to be seen,**
- have a **'signal word'** of appropriate strength (such as "**CAUTION**" or "**WARNING**" or "**DANGER**"), indicating the seriousness of hazard,
- have a **hazard statement** informing the reader in a clearly stated manner what the danger is,
- have a **consequence statement** telling in a clearly understandable way of the consequence of the hazard,
- have an **instruction statement** providing clearly stated instructions on what to do to avoid the hazard.

The machine designer should have a basic understanding of warnings and what factors influence their effectiveness. Effectiveness is influenced by choice of colors, placement, 'signal words,' pictogram illustrations, wording of statements, size of letters, and durability, to name a few. The subject of warnings and what makes them effective has been studied and written about extensively. Our discussion presents only very basic information about the subject. For more information, consult industry standards such as ANSI Z535 (series), ISO 3864 (series), ISO 17398, EN 842, EN 457, EN 981 and ISO 7000, along with various books on the subject.

> **CRITICAL CONSIDERATIONS: Machine Safeguarding**
>
> - Be sure to comply with all safety and ergonomics laws, codes, and standards. Consult the recommended resources for more information.
> - Physical barrier guarding should be the first choice if a hazard cannot be designed out of the machine.
> - Procedural safeguards must be written, and training of personnel should be formalized and recorded.
> - Have a qualified safety professional evaluate all equipment, tools, and workspaces.

Section 2.4

OTHER SAFETY ISSUES

There are other machine design safety features and issues that cannot necessarily be classified as safeguarding. It is important that all aspects of machine safety be analyzed during the design process and appropriate steps and design features be considered. The following are only some of the safety-related aspects of a machine that were not included earlier in this chapter; designers must be mindful of them during the machine development process.

An excellent source for understanding a machine's safety goals and the associated design process is industry standard ISO 12100:2010. It is recommended that machine designers not only become familiar with much of the literature cited in Section 2.5, but also become familiar with specifically this standard.

> **RECOMMENDED RESOURCES**
>
> - A list of books and industry standards are provided in Section 2.5 of this chapter.

EMISSION OF AIRBORNE SUBSTANCES OR MODIFICATION OF SURROUNDING ATMOSPHERE

The machine designer must be aware of the potential for the machine to emit undesirable or potentially harmful airborne substances such as mists, vapors, fumes, particles, dust, or other contaminants, or to cause a modification of its surrounding atmosphere, such as enriched oxygen, carbon-dioxide, or nitrogen. Such airborne substance or modified atmosphere, when certain levels are exceeded, could affect not only the performance, safety, or health of individuals, but also potentially the value of surrounding property.

It is important that a machine that emits such substances or gases be designed such that such emissions can be appropriately controlled. Controlling such substance or gases can be accomplished either within the confines of the machine itself (designed into the machine), or through information and instructions provided with the machine, providing information about the emissions (nature of the substance, likely volumes of emission, and potential effects), and instructions or guidelines for their control (the type and basic capacity of emission control equipment needed).

EMISSION OF RADIATION, INTENSE LIGHT, VIBRATION, HEAT

The machine designer must consider the potential for the machine to emit radiation (X-rays, gamma rays, ultraviolet, infrared, microwave, etc.), intense light (welding beam, laser beam, etc.), substantial vibration, or substantial heat. Such emissions, if beyond an acceptable level, can affect the health and safety of individuals in the area as well as surrounding property, and must be controlled. Control can be achieved through the use of radiation filters, screens or shielding, limiting radiation power, providing remote operation, providing vibration isolators or dampers, providing cooling ventilation, etc.

EMISSION OF NOISE

All machines that have moving parts emit noise. According to U.S. OSHA requirements, employers must take steps to control noise employees are exposed to, and achieve, at minimum, levels not exceeding those cited in the regulation (Table G-16 of 29 CFR 1910.95(b)(1); see Table 2-6). Machine noise can be controlled, and designers should understand excessive noise is

Table 2-6: OSHA Noise Exposure Limits

Duration of Exposure hours	Sound Level DBA (slow response)
8	90
6	92
4	95
3	97
2	100
1.5	102
1	105
0.5	110
0.25 or less	115
Impulse Noise	140 peak

generally not desirable. Ways of reducing machine noise include increasing the mass of panel material, using sound insulation, damping or cushioning sources of impact or vibration, and reducing or muffling compressed air emissions. When noise at the operator's ears exceeds limits set in OSHA regulations, hearing protection is required, and such information and instructions must be included with the machine. Machine noise levels should be measured in accordance with ANSI B11.TR5.

HAND/ARM VIBRATION

Hand/arm vibration (HAV) is defined as the transfer of vibration from a tool to a worker's hand and arm. The amount of hand/arm vibration is characterized by the acceleration level of the tool when grasped by the worker and in use. The vibration frequencies that most affect hands and arms lie in the 5 to 1,500 Hz range. The types of machines typically associated with significant vibration include chain saws, chipping hammers, grinders, hammer drills, and powered compactors. NIOSH publication No. 89-106 "Criteria for a Recommended Standard: Occupational Exposure to Hand-Arm Vibration," 1989 (found at www.cdc.gov/niosh/89-106.html) provides helpful information.

WHOLE BODY VIBRATION

Whole body vibration (WBV) is the transfer of relatively low frequency (0.5 to 80 Hz) motion to the whole body through a broad contact area. It is most commonly transmitted through the feet when standing, or through the buttocks when sitting. Off-road unsprung vehicles are the type of machines most typically associated with the vibration, jarring, and jolting associated with WBV. Whole body vibration can jostle organs, contribute to back pain, cause fatigue, and cause a reduction in a person's work performance. NIOSH, as well as Health and Safety Executive (found at www.hse.gov.uk/), provides additional information on this subject.

MACHINE USE IN EXPLOSIVE ATMOSPHERES

If it is intended or if it is foreseeable that a machine will be used in the presence of an explosive atmosphere (gases, vapors, mists, or combustible dusts), components capable of operating in such atmospheres safely (spark-free, for instance) must be selected. The industry standards EN 1127-1, EN 50020, EN60204, as well as UL standards titled "Hazardous (Classified) Location ..." and the European Union directive ATEX95 94/9/EC provide guidance in such situations.

MOVING THE MACHINE

When it is determined moving a machine is hazardous, then provisions for jacking, lifting, and hoisting should be provided to help the moving process and enhance safety. When it is foreseeable that a machine's weight, center-of-gravity, or lift point material strength could be misunderstood by movers, moving instructions should be made available and made obvious to movers. These instructions should provide such information as the machine's weight, center-of-gravity location, lift points, loose component tie-down points, and other information necessary for a safe move. Professional riggers should be used to move extremely heavy, large, awkward, or sensitive machinery.

MACHINE STABILITY

A machine must be sufficiently stable for it to be used as intended safely. Its weight distribution and base footprint must be such that it remains stable, taking into account machine vibration, dynamic movement of components, movement of work pieces, foreseeable mishandling, accidental bumping, forces of nature, etc. If necessary, provisions should be provided for anchoring (bolting) the machine into position (to the floor, for instance).

LUBRICATION

To avoid exposing individuals to unacceptable risks when lubricating a machine, lubrication points should be located in accessible and safe-to-reach locations, when possible.

DANGER WARNING ALARM SIGNALING: AUDIBLE SIGNALS AND VISUAL SIGNALS

When the health or safety of an individual is put at risk by an unstaffed machine in a defective or dangerous state, according to accepted industry standards the machine must be equipped to transmit an appropriate audible or visual warning alarm signal indicating the danger. The warning alarm signal must be immediately and easily recognized and understood; it must have priority over all other signals (except not over *emergency* signals, which have absolute priority). The characteristics of warning signals are outlined in industry standards, including EN 981, EN 842 and EN 457.

Audible danger signals must be such that anybody who hears them recognizes them and can react immediately. The characteristics which make a danger signal effective are its sound (audibility), its ability to be recognized immediately (discriminability), and that there is absolutely no doubt as to what it refers to (unequivocability). For further details, industry standard EN 457 is recommended as a reference.

Visual danger signals must be designed such that that anybody who sees them will recognize them and respond immediately. A visual danger signal must be clearly visible, even in strong light (visibility), distinguishable from other lights and light signals (distinguishability), and understood immediately (unequivocability). The visual signals must be positioned where it reaches all of the area affected, and its message must be clearly understood — whether it refers to a machine, a group of machines, a production line, or a complete department. For further details, industry standard EN 842 is recommended as a reference.

LOCKOUT/TAGOUT REQUIREMENTS

Federal OSHA regulations (29CFR1910.147) require employers to ensure that all new machines provided to employees are capable of being locked out during service and maintenance for the purpose of preventing unexpected start-ups, the energizing of machinery, or the release of stored energy that could cause injury to employees. The lockout standard applies if (1) the employee is required to remove or bypass a guard or other safety device during service, (2) an associated danger zone exists during a machine operating cycle, or (3) the employee is required to place any body part into the machine's point-of-operation area.[9]

All new machinery and equipment needs to be designed to accept lockout devices.[10]

EMERGENCY STOP DEVICES

An emergency stop device is a manually actuated control device that requires deliberate action to bring the machine to a stop when a dangerous situation is recognized. The E-stop (emergency stop) device must be continuously operable and within easy reach. Each operator panel must contain at least one E-stop device. Additional E-stop devices must be available and readily accessible, at minimum, wherever a machine operator is intended to be or would foreseeably be during normal operation. E-stop devices should stop the machine as quickly as possible without generating additional hazards. Emergency stops must safely neutralize energy to and within that portion of the machine affected, including electrical power as well as pressurized air. E-stops are not safeguards, and are not alternatives to safeguarding.

Because emergency stop switches and circuits can remain inactive for long periods of time, it is important that they be designed with reliability in mind. In addition, instructions for maintenance requirements and periodic testing are important to assure confidence that the system will function as intended.[11]

---

[9] In manufacturing settings, tags are normally attached to lockout devices identifying the person who placed the lockout on the machine. In cases where multiple persons are working on a machine simultaneously, it is not uncommon for there to be multiple identifying tags placed on a lockout device. This prevents communication errors between work crews from jeopardizing the lockout.

[10] Plug-connected electric machinery for which exposure to hazards is controlled by unplugging the machine and by the plug and cord being under the exclusive control of the employee performing the maintenance work is exempted from OSHA's lockout/tagout standards.

[11] Emergency Stops and their associated parts are components of a part of a machine's control system identified as the machine's "safety-related control system", or SRCS. A machine's safety-related control system is that part of the machine's control system that prevents a hazardous condition from occurring, either (1) by preventing the *initiation* of a hazardous situation (e.g., two-hand controls), or (2) by detecting the onset of a hazard (e.g., emergency stop switches). Safety-related control systems are designed to perform *safety* functions. A machine's SRCS must continue to operate correctly under all foreseeable conditions, and because they perform a safety function, the components in the system must be verifiably reliable. Industry standards ISO 138491-1 "Safety Related Parts of Control Systems" (ISO 13849-1 replaces EN 849-1), EN 62061 "Safety of Machinery — Functional Safety of Safety Related Electrical, Electronic and Programmable Electronic Control Systems," and, related to control reliability of pneumatic and hydraulic systems, ANSI B 155.1 "Safety Requirements for Packaging Machinery and Packaging-Related Converting Machinery" address requirements of machine safety-related control systems and their components.

Emergency stop components and circuitry must be failsafe, and the appropriate components for E-stop use are normally clearly identified by component manufacturers. Emergency stop devices come in various forms, the most common of which are cable pulls and mushroom-type button switches. When an E-stop device is actuated, it must latch in, and it must not be possible to generate the stop command without latching in. The resetting of an E-stop device must not cause a hazardous situation. To restart the machine, a separate and deliberate action must be required.

Emergency stop device details can be found in industry standards BSI PD 5304 (section 5), EN 418, EN-13850, ISO 13852 and ISO 13850. Design and selection of components for E-stops should be performed by a qualified controls professional.

 **CRITICAL CONSIDERATIONS: Other Safety Issues**

- Comply with all laws, codes, and standards governing environmental hazards. Consult the recommended resources for more information.
- Design of an E-stop circuit and selection of related components should be conducted by a qualified controls professional.
- Lockout/Tagout procedures are required by law to isolate all sources of energy (electrical, pneumatic, hydraulic, mechanical, etc.) when maintenance activities are performed. Consult the recommended resources for Lockout/Tagout procedures and requirements.
- Have a qualified safety professional evaluate all equipment, tools, and workspaces for environmental and machine hazards.

**Section **

2.5

RECOMMENDED RESOURCES

Machinery safety and ergonomics information is found in governmental statutes, codes and regulations, industry standards, handbooks, textbooks and manuals, and scientific and technical literature found in magazines, journals and periodical articles. Safety *requirements* are found in governmental statutes. Safety *expectations* — just as important for machine safety — are found in various books, standards and periodical literature.

Society's expectations of what is acceptable relative to product safety evolves with time, and usually it is the industry standards, and the books, that most accurately reflect what society views as the 'best practices' to achieve what is viewed as reasonably safe. For this reason, although governmental statues are certainly important and necessary to be included in this literature list, the designer should understand that well-organized books and up-to-date industry standards are of paramount importance in designing safe machines and products. Because of the fluidity and time-limited exposure of periodical literature, this chapter does not attempt to cite notable magazine articles, journal papers, technical studies, government 'fact sheets' and bulletins, or other such literature, despite their importance.

GOVERNMENTAL REGULATIONS, STATUTES, CODES AND PUBLICATIONS

- **Code of Federal Regulations 29 CFR 1910:** <u>Occupational Safety and Health Standards,</u> **parts 1910.1 through 1910.399** (authorized through the Occupational Safety and Health Act—OSHA)

    This is the U.S. Department of Labor's regulation that covers general worker safety, including 1910's Subpart 'O' - *Machinery and Machine Guarding* (1910.211 through 1910.219), and Subpart 'P' - *Hand and Portable Tools and Other Hand-Held Equipment* (1910.241 through 1910.244). Parts 1910.1 through 1910.399 do not detail many of the engineering requirements for compliance. Instead, they reference other industry standards, including standards issued by or through such organizations as ANSI, ASME, ASSE, IEEE, ISO, NFPA, and UL, to name a few. OSHA standards represent the *minimum* level of regulatory compliance requirements within the United States.

- **Code of Federal Regulations 29 CFR 1926:** <u>Safety and Health Regulations for Construction,</u> **parts 1926.300, and 1926.302 through 1926.307** (authorized through the Contract Work Hours and Safety Standards Act, and the Occupational Safety and Health Act—OSHA)

    This is the U.S. Department of Labor's regulation covering certain hand tools and power tools, including those typically used in the construction industry, as found in 1926's Subpart 'I' -*Tools - Hand and Power* (1926.300 through 1926.307).

- **Code of Federal Regulations 29 CFR 1928: <u>Occupational Safety and Health Standards for Agriculture, part 1928.57</u>** (authorized through the Occupational Safety and Health Act—OSHA)

    This is the U.S. Department of Labor's regulation covering certain agricultural equipment, as included in Subpart 'D' - *Safety for Agricultural Equipment*, in section 1928.57 - *Guarding of farm field equipment, farmstead equipment, and cotton gins.*

- **Military Standards and Handbooks**

    These are often used or referred to in private industry for guidance for design, manufacturing, quality control and maintenance, relating to services, machines and equipment.

    MIL-STD-882 <u>System Safety Program Requirements</u>

    > This is a standard that addresses hazard identification and risk analysis / reduction, employing an integration of hazard identification and hazard severity applicable to machines and systems (military and non-military).

    MIL-STD-1472 <u>Human Engineering Design Criteria for Military Systems, Equipment and Facilities</u>

    > This standard is acknowledged worldwide as an authoritative source for human factors requirements and design criteria. It provides most aspects of human factors information and performance criteria / limitations helpful for the physical design and layout of machines, equipment, and facilities, including operational controls (for both military and non-military applications). This standard focuses more on task and operational performance than worker health and safety.

    MIL-HDBK-759 <u>Human Engineering Design Guidelines</u>

    > This handbook guideline provides a broad range of human factors information and performance considerations as a supplement to MIL-STD-1472.

    DOD-HDBK-743 <u>Anthropometry of U.S. Military Personnel</u>

    > This document provides body size information on military personnel of the United States, as a supplement to MIL-STD-1472.

- **NASA Reference Publication 1024 <u>Anthropometric Source Book Volume I: Anthropometry for Designers</u>** (N79-11734) National Aeronautics and Space Administration, Scientific and Technical Information Office, Lyndon B. Johnson Space Center, Houston, TX 77058 (edited by the Staff of

Anthropology Research Project, Webb Associates, Yellow Springs, Ohio, 1978)

This publication contains 550 pages of extensive and easy-to-use anthropometric information on adult men and women, including data on certain non-U.S. population groups.

BOOKS

- **Product Safety Management and Engineering,** by W. Hammer; American Society of Safety Engineers, Des Plaines, IL, 1993

  This is a thorough, well-organized, well-written text with numerous charts, checklists, and tables addressing designing safe products and machines.

- **Accident Prevention Manual for Business & Industry: Engineering & Technology,** 13th Ed., National Safety Council, 2010

  This manual is periodically published and updated by the NSC, and includes chapters on industrial safety, including machine safety.

- **Human Factors Design Handbook**, 2nd Ed., by Woodson, Tillman and Tillman; McGraw-Hill Inc., New York, NY, 1992

  This is a notably useful compilation of human factors data, including numerous guidelines, illustrations, checklists, tables, charts, diagrams, and practical examples.

- **Ergonomics A Practical Guide**, National Safety Council, 1993

  This book contains practical ergonomics information.

- **Safeguarding Concepts Illustrated**, 7th Ed., National Safety Council, 2002

  This book contains 140 pages of 300 to 400 illustrations and pictures of actual machine and equipment guards, with explanations, in well-organized chapters and groupings.

- **Human Factors in Engineering and Design**, 7th Ed., by Sanders and McCormick; McGraw-Hill Inc., New York, NY, 1993

  This book provides human factors information, emphasizing workplace locations and situations.

- **Warnings and Risk Communication**, by Wogalter, DeJoy and Laughery; Taylor & Francis Inc., Philadelphia, PA, 1999

This book provides insight into the effectiveness and ineffectiveness of warnings, based on extensive research.

- *Handbook of Warnings*, by Wogalter; Lawrence Erlbaum Associates, Mahwah, NJ, 2006

  This handbook contains extensive information on warnings as they are applied to a broad range of situations.

- *Writing and Designing Manuals*, 2nd Ed., by Schoff and Robinson; Lewis Publishers, Inc., Chelsea, MI, 1991

  This book is a guide for writing, illustrating, and organizing manuals for machine and product owners, operators, and service personnel.

- *The Measure of Man & Woman*, Dreyfuss & Associates; John Wiley & Sons, Inc., New York, NY, 2002

  This book contains extensive and useful anthropometric charts and data for U.S. civilians in the 1st, 50th, and 99th population percentiles (information beyond the more typical 5%-50%-95% numbers), as well as charts and data for children and youths.

- *Anthropometry of Infants, Children and Youths to Age 18 for Product Safety Design — SAE SP-450*, Society of Automotive Engineers, Warrenburg, PA, 1977

  This book has 627 pages of useful anthropometric information compiled in the 1970s based on 4,127 young subjects.

- *Safety and Health for Engineers*, 2nd Ed., by R. Brauer; John Wiley & Sons, Inc., Hoboken, NJ, 2006

  This book contains 740+ pages of text outlining information applicable to occupational safety and health. This is the type of reference book certified safety engineers would find helpful.

- *Bodyspace, Anthropometry, Ergonomics and the Design of Work*, 2nd Ed., by S. Pheasant; Taylor & Francis Inc., Philadelphia, PA, 2002

  This is a British publication providing some workplace ergonomics guidance.

- *Encyclopedia of Occupational Health and Safety*, *Fourth Edition (4 Volumes)*, (also available in CD Rom format), Jeanne Mager Stellman, editor; International Labour Office, 1998

  This is a set of four books containing 4,000 pages of health and safety-related information, some of which applicable to the machine designer.

INDUSTRY SAFETY STANDARDS

An industry safety standard, most of which are *voluntary* standards, is a document relating to a product, process, service, system, or personnel that is developed through a collaborative, balanced, and consensus-based approval process. Industry safety standards are developed for the purpose of identifying design or performance requirements that are viewed as necessary to achieve a basic, usually a *minimum*, level of safety, below which an individual's safety cannot be assured. A voluntary safety standard is the result of a periodic and iterative process of assessing hazards, risks, and accident data, reviewing technical developments, and balancing this information with product utility, marketplace economics, and public sentiment. Requirements contained in an industry safety standard are for the purpose of avoiding the recurrence of accidents, or avoiding the existence of hazards which are understood to be causes of accidents. Accident data, technical developments, and the threshold of the public's acceptance of a basic level of safety evolve over time. As these criteria evolve, standards can be expected to change. Any given issuance of a standard is merely a reflection of these criteria at the time of publication.

Relative to *machine* safety standards, there are a number of organizations (ANSI, CEN, ISO, BSI[12], CSA[13], ASABE[14], ASME[14], SAE[14] and UL[15], to name the most prominent ones) involved in standards development. Some are independent standards development organizations; others are engineering and technical societies. The three more prominent standards development and distribution organizations are ANSI, CEN, and ISO.

- **ANSI (American National Standards Institute)** is the prominent standards administrating organization for voluntary standards in the United States, the **ANSI** standards. Mandatory provisions of those standards (those containing the word "shall" or other mandatory language) — as well as all other (non-ANSI) standards that are cited in U.S. Federal Regulation 29 CFR 1910.6 — are adopted and incorporated as a part of the Occupational Safety and Health Act (OSHA), and thus are required by U.S. law.

[12] **BSI**, The British Standards Institution, another significant standards development and distribution organization, is the National Standards Body (NSB) of the U.K. (BSI organized the first Commonwealth Standards Conference in 1946, which led to the establishment of the International Organization for Standardization (ISO).) CEN includes BSI transpositions of EN standards.

[13] **CSA** is not a part of the Canadian government. CSA is a not-for-profit organization that represents Canada on various ISO committees.

[14] **ASABE, ASME,** and **SAE** are engineering societies that, among other things, develop safety standards applicable to the design and construction of machines associated with their industry.

[15] **UL** is an independent organization, not a part of any government, that, among other things, develops (and provides testing services for) safety standards applicable to products and machines.

- **CEN (European Community for Standardization)** is the organization that develops and manages **EN** (European) standards for the 31 participating European community countries. All 31 countries reference by law many of these EN standards, elevating them from voluntary standards to the level of legislated legal requirements.

- **ISO (International Organization for Standardization)** is the organization that develops **ISO** standards. ISO was formed to facilitate and manage international industrial standards for the broader international community (approximately 140 countries). In many countries, ISO standards, too, are referenced in laws.

In 1991, an agreement was signed by ISO and CEN to establish cooperation and coordination of European and international standards to harmonize text to create similar language in the two organizations' issued ISO and EN standards. By 2011, this process has resulted in ISO and EN standards having many of the same requirements. At the time of this book's printing, more than 30% of these standards have identical language.

Machine safety standards should be viewed as being grouped into one of three basic types: those addressing basic concepts and principles applicable to all machines, those dealing with human factors and certain types of safety devices applicable to a wide range of machines, and those offering specific requirements for specific types or classes of machines or specific industries. The following hierarchy of standards groupings has been adopted or recognized by CEN, ISO, and ANSI[16]:

- **Type 'A' standards** (fundamental safety standards): These standards give basic concepts, principles for design, and general considerations that can be applied to all machinery. Type 'A' standards provide designers and manufacturers an overall framework and guide for the design and production of machines that are safe for their intended use, including when no type 'C' standards exist.

- **Type 'B' standards** (group safety standards): These standards deal with one safety aspect, or one type of safety-related device that can be used across a wide range of machinery.

    - **Type 'B1' standards**, which address particular safety issues (e.g., safe distances, surface temperatures, noise)
    - **Type 'B2' standards**, which address safety-related devices (e.g., two-hand controls, barrier guards, interlocking devices, presence-sensing devices)

---

[16] (as observed in ANSI B11-2008)

- **Type 'C' standards** (machine-specific safety standards): These standards provide detailed safety requirements for a particular type or group of machines. (Historically, the overwhelming majority of ANSI safety standards have been machine-type-specific; thus, it would be logical to classify most of them as Type 'C' standards.)

The standards cited in this chapter are focused primarily on safety requirements applicable to the broad range of machinery in general. Because of the many different types of machines and specific industries they are used in, this chapter does not attempt to cover Type 'C' standards. (When designing a machine for which there are specific industry safety standards, it is incumbent upon the designer to become familiar with that industry's and that machine's requirements.) In addition, electrical components, although important to the machine designer, are not generally addressed (with some exceptions) in this chapter.

The following are standards with which the machine designer should be or become familiar:

- **Type 'A' Standards**

ISO 12100: 2010	*Safety of Machinery — General Principles for Design — Risk Assessment and Risk Reduction*
BSI PD 5304	*Safe Use of Machinery*
	(This Published Document from BSI, although not strictly a standard, covers practical measures and techniques to safeguard operators, maintenance personnel, and others, along with covering the safe use of machinery.)
ISO 12100-1:2009	*Safety of Machinery — Basic Concepts, General Principles for Design. Part 1: Basic Terminology, Methodology*
	(This is in process of being replaced by ISO 12100:2010)
EN 292-1	*Safety of Machinery — Basic Concepts, General Principles for Design. Part 1: Basic Terminology, Methodology*
ISO 12100-2:2009	*Safety of Machinery — Basic Concepts, General Principles for Design. Part 2: Technical Principles*
	(This is in process of being replaced by ISO 12100:2010)
EN 292-2	*Safety of Machinery — Basic Concepts, General Principles for Design. Part 2: Technical Principles and Specifications*
ANSI B 155.1	*Packaging Machinery and Packaging-Related Converting Machinery — Safety Requirements for Construction, Care, and Use*

EN 1050	*Principles for Risk Assessment*
ISO 14121-1	*Safety of Machinery — Risk Assessment — Part 1: Principles*
ISO 14121-2	*Safety of Machinery — Risk Assessment — Part 2: Practical Guidance and Examples of Methods*
EN 1070	*Safety of Machinery — Terminology*

- ## Type 'B' Standards

EN 614-1	*Ergonomic Design Principles — Terminology and General Principles*
EN 547-3	*Human Body Measurements — Anthropometric Data*
EN 1005-1	*Human Physical Performance — Terms and Definitions*
EN 1005-2	*Human Physical Performance — Manual Handling of Machinery and Component Parts of Machinery*
EN 1005-3	*Human Physical Performance — Recommended Force Limits for Machinery Operation*
EN 1005-4	*Human Physical Performance — Evaluation of Working Postures and Movements in Relation to Machinery*
EN 1005-5	*Human Physical Performance — Risk Assessment for Repetitive Handling at High Frequency*
ISO 14738	*Safety of Machinery — Anthropometric Requirements for the Design of Workstations at Machinery*
EN 349	*Minimum Gaps to Avoid Crushing of Parts of the Human Body*
ISO 13854	*Minimum Gaps to Avoid Crushing of Parts of the Human Body*
EN 294	*Safety Distances to Prevent Danger Zones being Reached by the Upper Limbs*
ISO 13852	*Safety Distances to Prevent Hazard Zones being Reached by the Upper Limbs*
EN 811	*Safety Distances to Prevent Danger Zones being Reached by the Lower Limbs*
ISO 13853	*Safety Distances to Prevent Hazard Zones being Reached by the Lower Limbs*
ISO 13857	*Safety Distances to Prevent Hazard Zones being Reached by Upper and Lower Limbs*

ISO 13855	*Safety of Machinery — Positioning of Safeguards with Respect to the Approach Speeds of Parts of the Human Body*
EN 999	*The Positioning of Protective Equipment in Respect of Approach Speeds of Parts of the Human Body*
EN 547-1	*Human Body Measurements — Principles for Determining the Dimensions Required for Openings for Whole-Body Access into Machinery*
EN 547-2	*Human Body Measurements — Principles for Determining the Dimensions Required for Access Openings*
ISO 15534 (series)	(parts 1 thru 3) *Ergonomic Design for the Safety of Machinery* (relative to access openings)
EN 563	*Temperatures of Touchable Surfaces — Ergonomics Data to Establish Temperature Limit Values for Hot Surfaces*
ISO 13732-1	*Ergonomics of the Thermal Environment — Methods for the Assessment of Human Responses to Contact with Surfaces — Part 1: Hot Surfaces*
ISO 13732-3	*Ergonomics of the Thermal Environment — Methods for the Assessment of Human Responses to Contact with Surfaces — Part 3: Cold Surfaces*
ANSI B11.19	*Performance Criteria for Safeguarding*
ISO 14120	*Safety of Machinery — Guards — General Requirements for the Design and Construction of Fixed and Movable Guards*
EN 953	*Guards — General Requirements for the Design and Construction of Fixed and Movable Guards*
CSA-Z432-04	*Safeguarding of Machines*
ANSI B15.1	*Safety Standard for Mechanical Power Transmission Apparatus* (including **ANSI B15.1 Interpretation**)
ISO 14119	*Safety of Machinery — Interlocking Devices Associated with Guards — Principles for Design and Selection* (also **Amendment 1:2007** — *Design to Minimize Defeat Possibilities*)
EN 1088	*Interlocking Devices Associated with Guards — Principles for Design and Selection*
IEC 61496-1[17]	*Safety of Machinery — Electro-Sensitive Protective Equipment — General Requirements and Tests*

[17] The IEC (International Electrotechnical Commission) is a non-governmental internationally recognized standards organization that develops and issues standards for electrical and electronic technologies and related equipment, including componentry often used in machinery.

IEC 61496-2[17]	*Safety of Machinery — Electro-Sensitive Protective Equipment — Particular Requirements for Equipment Using Active Opto-Electronic Protective Devices*
EN 982	*Safety Requirements for Fluid Power Systems and their Components — Hydraulics*
ISO 4413	*Hydraulic Fluid Power— General Rules and Safety Requirements for Systems and their Components*
EN 983	*Safety Requirements for Fluid Power Systems and their Components — Pneumatics*
ISO 4414	*Pneumatic Fluid Power— General Rules and Safety Requirements for Systems and their Components*
EN 842	*Visual Danger Signals, General Requirements, Design and Testing*
EN 457	*Audible Danger Signals, General Requirements, Design and Testing*
EN 981	*System of Auditory and Visual Danger and Information Signals*
ISO 13850	*Safety of Machinery — Emergency Stop — Principles for Design*
EN 13850	*Emergency Stop — Principles for Design*
EN 418	*Emergency Stop Equipment, Functional Aspects — Principles for Design*
ISO 13851	*Safety of Machinery — Two-Hand Control Devices — Functional Aspects and Design Principles*
EN 574	*Two-Hand Control Devices — Functional Aspects — Principles for Design*
EN 954-1	*Safety-Related Parts of Control Systems — General Principles of Design*
ISO 13849-1	*Safety-Related Parts of Control Systems (ISO 13849-1 replaces EN 954-1)*
EN 1037	*Prevention of Unexpected Start-Up*
ISO 14118	*Safety of Machinery — Prevention of Unexpected Start-Up*
ANSI/ASSE Z244.1	*Control of Hazardous Energy, Lockout/Tagout and Alternative Methods*

CSA-Z460	*Control of Hazardous Energy — Lockout and Other Methods*
ISO 14123 (series)	(parts 1 and 2) *Safety of Machinery — Reduction of Risks to Health from Hazardous Substances Emitted by Machinery*
EN 626-1	*Reduction of Risks to Health from Hazardous Substance Emitted by Machinery — Principles and Specifications for Machinery Manufacturers*
EN 1127-1	*Explosive Atmospheres — Explosion Prevention and Protection*
ANSI B11.20	*Safety Requirements for Integrated Manufacturing Systems*
ANSI/RIAR 15.06	*Safety Standard for Industrial Robots and Robot Systems*
ISO 10218-1	*Robots and Robotic Devices — Safety Requirements — Industrial Robots*
ISO 10218-2	*Robots and Robotic Devices — Safety Requirements — Industrial Robot Systems and Integration*
UL 1740	*Standard for Robots and Robotic Equipment*
ISO 11161	*Safety of Machinery — Integrated Manufacturing Systems — Basic Requirements*
CSA-Z434	*Industrial Robots and Robot Systems — General Safety Requirements*
ANSI A12.1	*Safety Requirements for Floor and Wall Openings, Railings, and Toe Boards*
ANSI/ASSE A1264.1	*Safety Requirements for Workplace Walking/Working Surfaces and their Access; Workplace, Floor, Wall and Roof Openings; Stairs and Guardrails Systems*
ISO 14122 (series)	(parts 1 thru 4) *Safety of Machinery — Permanent Means of Access to Machinery*
	(relative to working platforms, walkways, access between levels, stairs, ladders and guard-rails, etc.)
ANSI Z535.1	*Safety Color Code*
ANSI Z535.3	*Criteria for Safety Symbols*
ANSI Z535.4	*Product Safety Signs and Labels*
ANSI Z535.6	*Product Safety Information in Product Manuals, Instructions, and Other Collateral Materials*

ISO 7000	*Graphical Symbols for Use on Equipment — Index and Synopsis*
ISO 3864-2	*Graphical Symbols — Safety Colours and Safety Signs — Part 2: Design Principles for Product Safety Labels*
ISO 3864-3	*Graphical Symbols — Safety Colours and Safety Signs — Part 3: Design Principles for Graphical Symbols for Use in Safety Signs*
ISO 17398	*Safety Colours and Safety Signs — Classification, Performance and Durability of Safety Signs*
EN 894-1	*Ergonomics Requirements for the Design of Displays and Control Actuators — General*
EN 894-2	*Ergonomics Requirements for the Design of Displays and Control Actuators — Displays*
EN 894-3	*Ergonomics Requirements for the Design of Displays and Control Actuators — Control Actuators*
CSA -Z431	*Basic and Safety Principles for Man-Machine Interface, Marking and Identification — Coding Principles for Indication Devices and Actuators*
EN 60204	*Safety of Machinery — Electrical Equipment of Machines*
ANSI/NFPA 70	*U.S. National Electrical Code*
ANSI/NFPA 70E	*Standard for Electrical Safety in the Workplace*
ANSI/NFPA 79	*Electrical Standard for Industrial Machinery*
ANSI C1	*National Electrical Code*
UL 987	*Standard for Stationary and Fixed Electric Tools*
UL 745(series)	*Standard for Portable Electric Tools*
UL 1439	*Test for Sharpness of Edges on Equipment*
ANSI B11.TR1	(technical report) *Ergonomic Guidelines for the Design, Installation and Use of Machine Tools*
ANSI B11.TR3	(technical report) *Risk Assessment and Risk Reduction — A Guide to Estimate, Evaluate and Reduce Risks Associated with Machine Tools*
ANSI B11.TR5	(technical report) *Sound Level Measurement Guidelines — A Guide for Measuring, Evaluating, Documenting and Reporting Sound Levels Emitted by Machinery*
ISO/TR 14121-2	(technical report) *Safety of Machinery — Risk Assessment — Part 2: Practical Guidance and Examples of Methods*

ISO/TR 18569   (technical report) *Safety of machinery — Guidelines for the understanding and Use of safety of machinery standards*

- **Type 'C' Standards**

   ANSI, BSI, CEN, ISO, CSA, ASABE, ASME, SAE, and UL (and others) have developed many hundreds of industry-specific and machine-type-specific machinery safety standards to achieve a basic level of safety. The design of a machine should be in conformance with not only Type 'A' and Type 'B' standards, but also those standards (if they exist) that apply to that specific machine type. A list of Type 'C' standards covering all machine types and all industries is too lengthy to be included in this chapter. It is, therefore, left to the designer to learn the targeted industry and the machine's intended use so the appropriate Type 'C' standards can be obtained and used during the design process.

INTERNET WEB SITES

   In addition to this list of statutes, books, and industry standards, today's machine designer should also be aware of and take advantage of information available through the Internet. Notable sites include, but are not limited to:

- **OSHA:** www.osha.gov
   helpful link: www.osha.gov/SLTC/machineguarding/index.html
   helpful link: www.osha.gov/Publications/Mach_SafeGuard/toc.html
- **ANSI**   (American National Standards Institute):www.ansi.org
   helpful link: www.osha.gov/SLTC/machineguarding/scope98.html
- **CEN** (European Committee for Standardization): www.cen.eu/cen/pages/default.aspx
- **ISO** (International Organization for Standardization): www.iso.org/iso/home.html
- **BSI** (The British Standards Institution): www.bsigroup.com
- **CSA** (Canadian Standards Association): www.csa-international.org/about
- **SAE** (Society of Automotive Engineers): www.sae.org helpful link: http://standards.sae.org/commercial-vehicle/safety/standards/current/
- **ASABE** (American Society of Agricultural and Biological Engineers): www.asabe.org helpful link: http://asae.frymulti.com
- **ASME** (American Society of Mechanical Engineers): www.asme.org
- **UL** (Underwriters Laboratories): www.ul.com/global/eng/pages

- **National Safety Council:** www.nsc.org/Pages/Home.aspx helpful link: www.nsc.org/products_training/Products/Pages/OnlineProductCatalog. aspx
- **CDC - Workplace Safety:** http://www.cdc.gov/Workplace
  helpful link: www.cdc.gov/niosh/docs/94-110/
  helpful link: www.cdc.gov/niosh/89-106.html
- **Other Helpful Websites** www.schmersalusa.com/catalog_pdfs/<http://www. schmersalusa.com/catalog_pdfs GK1_2008.pdf (specifically, the last 65 pages) and www.sti.com/ltr2/access.php?file=pdf/807.pdf

APPLICATION OF LITERATURE TO DESIGN TOPICS

The list of codes and standards cited in Table 2-7 is not intended to be complete or exhaustive. It is intended to provide the reader with reference literature from which to learn more about safety and ergonomics in machine design.

Table 2-7: Selected Codes and Standards for Machinery Safety

Topic	Standard	Standard Title
Fundamental and Basic Machine Design Safety Principles	ISO 12100: 2010	Safety of Machinery - General Principles for Design — Risk Assessment and Risk Reduction
	BSI PD 5304	Safe Use of Machinery
	ANSI B11 - 2008	General Safety Requirements Common to ANSI B11 Machines
	ISO 12100-1 (2003)	Safety of Machinery — Basic Concepts, General Principles for Design. Part 1: Basic Terminology, Methodology
	EN 292-1	Safety of Machinery — Basic Concepts, General Principles for Design. Part 1: Basic Terminology, Methodology
	ISO 12100-2 (2003)	Safety of Machinery — Basic Concepts, General Principles for Design. Part 2: Technical Principles
	EN 292-2	Safety of Machinery — Basic Concepts, General Principles for Design. Part 2: Technical Principles and Specifications
	ANSI B 155.1	Packaging Machinery and Packaging-Related Converting Machinery — Safety Requirements for Construction, Care, and Use
	OSHA 1910.212	General Requirements for All Machines
	ISO/TR 18569 (tech report)	Safety of Machinery — Guidelines for the Understanding and Use of Safety of Machinery Standards

continued on next page

Table 2-7: Selected Codes and Standards for Machinery Safety (Continued)

Assessment of Risk	ISO 12100: 2010	Safety of Machinery — General Principles for Design — Risk Assessment and Risk Reduction
	EN 1050	Principles for Risk Assessment
	ISO/TR 14121-2 (tech report)	Safety of Machinery — Risk Assessment — Part 2: Practical Guidance and Examples of Methods
	ANSI B11 - 2008	General Safety Requirements Common to ANSI B11 Machines
	ANSI B11.TR3 (tech report)	Risk Assessment and Risk Reduction — A Guide to Estimate, Evaluate, and Reduce Risks Associated with Machine Tools
	EN 1005-5	Human Physical Performance — Risk Assessment for Repetitive Handling at High Frequency
	ANSI B 155.1	Packaging Machinery and Packaging-Related Converting Machinery — Safety Requirements for Construction, Care, and Use
	MIL-STD-882	System Safety Program Requirements
Ergonomics: Human Physical Sizes and Physical Capabilities and Limitations	MIL-STD-1472	Human Engineering Design Criteria for Military Systems, Equipment, and Facilities
	MIL-HDBK-759	Human Engineering Design Guidelines
	DOD-HDBK-743	Anthropometry of U.S. Military Personnel
	NASA Reference Publication 1024	Anthropometric Source Book Volume I: Anthropometry for Designers
	ANSI B11.TR1 (tech report)	Ergonomic Guidelines for the Design, Installation, and Use of Machine Tools
	EN 614-1	Ergonomic Design Principles — Terminology and General Principles
	EN 1005-1	Human Physical Performance — Terms and Definitions
	EN 547-3	Human Body Measurements — Anthropometric Data
	EN 1005-2	Human Physical Performance — Manual Handling of Machinery and Component Parts of Machinery
	EN 1005-3	Human Physical Performance — Recommended Force Limits for Machinery Operation
	EN 1005-4	Human Physical Performance — Evaluation of Working Postures and Movements in Relation to Machinery
	EN 1005-5	Human Physical Performance — Risk Assessment for Repetitive Handling at High Frequency
	ISO 14738	Safety of Machinery — Anthropometric Requirements for the Design of Workstations at Machinery
	EN 563	Temperatures of Touchable Surfaces — Ergonomics Data to Establish Temperature Limit Values for Hot Surfaces
	ISO 13732-1	Ergonomics of the Thermal Environment — Methods for the Assessment of Human Responses to Contact with Surfaces — Part 1: Hot Surfaces
	ISO 13732-3	Ergonomics of the Thermal Environment — Methods for the Assessment of Human Responses to Contact with Surfaces — Part 3: Cold Surfaces

continued on next page

Table 2-7: Selected Codes and Standards for Machinery Safety (Continued)

Machine Ergonomics: Ergonomic Access and Ergonomic Access Opening Sizes	MIL-STD-1472	Human Engineering Design Criteria for Military Systems, Equipment, and Facilities
	EN 547-1	Human Body Measurements — Principles for Determining the Dimensions Required for Openings for Whole-Body Access into Machinery
	EN 547-2	Human Body Measurements — Principles for Determining the Dimensions Required for Access Openings
	ISO 15534 (series)	(parts 1 thru 3) Ergonomic Design for the Safety of Machinery (relative to access openings)
	ANSI/ASSE A1264.1	Safety Requirements for Workplace Walking/Working Surfaces and Their Access; Workplace, Floor, Wall and Roof Openings; Stairs and Guardrails Systems
	ISO 14122 (series)	(parts 1 thru 4) Safety of Machinery — Permanent Means of Access to Machinery (relative to working platforms, walkways, access between levels, stairs, ladders and guard-rails, etc.)
	UL1439	Test for Sharpness of Edges on Equipment
Machine Safety: Safe Clearances, Safe Distances	OSHA 1910.217 (c)	Mechanical Power Presses Safeguarding the Point of Operation, Table O-10
	MIL-STD-1472	Human Engineering Design Criteria for Military Systems, Equipment, and Facilities
	ANSI B15.1	Safety Standard for Mechanical Power Transmission Apparatus
	ISO 13854	Minimum Gaps to Avoid Crushing of Parts of the Human Body
	EN 349	Minimum Gaps to Avoid Crushing of Parts of the Human Body
	EN 294	Safety Distances to Prevent Danger Zones being Reached by the Upper Limbs
	ISO 13852	Safety Distances to Prevent Hazard Zones being Reached by the Upper Limbs
	EN 811	Safety Distances to Prevent Danger Zones being Reached by the Lower Limbs
	ISO 13853	Safety Distances to Prevent Hazard Zones being Reached by the Lower Limbs
	ISO 13857	Safety Distances to Prevent Hazard Zones being Reached by Upper and Lower Limbs
	ISO 13855	Positioning of Safeguards with Respect to the Approach Speeds of Parts of the Human Body
	EN 999	The Positioning of Protective Equipment in Respect of Approach Speeds of Parts of the Human Body
	UL1439	Test for Sharpness of Edges on Equipment

continued on next page

Table 2-7: Selected Codes and Standards for Machinery Safety (Continued)

Guards	ANSI B11.19	Machines — Performance Criteria for Safeguarding
	ANSI B15.1	Safety Standard for Mechanical Power Transmission Apparatus
	EN 953	Guards — General Requirements for the Design and Construction of Fixed and Movable Guards
	EN 1088	Interlocking Devices Associated with Guards — Principles for Design and Selection
	EN 999	The Positioning of Protective Equipment in Respect of Approach Speeds of Parts of the Human Body
	ISO 13855	Positioning of Safeguards with Respect to the Approach Speeds of Parts of the Human Body
	ISO 14120	Safety of Machinery — Guards — General Requirements for the Design and Construction of Fixed and Movable Guards
	CSA-Z432-04	Safeguarding of Machines
	EN 294	Safety Distances to Prevent Danger Zones Being Reached by the Upper Limbs
	ISO 13852	Safety Distances to Prevent Hazard Zones Being Reached by the Upper Limbs
	EN 811	Safety Distances to Prevent Danger Zones Being Reached by the Lower Limbs
	ISO 13853	Safety Distances to Prevent Hazard Zones Being Reached by the Lower Limbs
	ISO 13857	Safety Distances to Prevent Hazard Zones Being Reached by Upper and Lower Limbs
	ISO/TR 5046	Continuous Mechanical Handling Equipment — Safety Code for Conveyors and Elevators with Chain-Elements — Examples for Guarding of Nip Points
Interlocks	ISO 14119	Safety of Machinery Interlocking Devices Associated with Guards — Principles for Design and Selection (also Amendment 1:2007 — Design to Minimize Defeat Possibilities)
	EN 1088	Interlocking Devices Associated with Guards — Principles for Design and Selection
Machine Perimeter Safeguarding	EN 294	Safety Distances to Prevent Danger Zones Being Reached by the Upper Limbs
	ISO 13852	Safety Distances to Prevent Hazard Zones Being Reached by the Upper Limbs
	EN 811	Safety Distances to Prevent Danger Zones Being Reached by the Lower Limbs
	ISO 13853	Safety Distances to Prevent Hazard Zones Being Reached by the Lower Limbs
	ISO 13857	Safety Distances to Prevent Hazard Zones Being Reached by Upper and Lower Limbs
	IEC 61496-1	Safety of machinery — Electro-Sensitive Protective Equipment — General Requirements and Tests
	IEC 61496-2	Safety of Machinery — Electro-Sensitive Protective Equipment — Particular Requirements for Equipment Using Active Opto-Electronic Protective Devices

continued on next page

Table 2-7: Selected Codes and Standards for Machinery Safety (Continued)

Walking Surfaces, Working Surfaces, Elevated Surfaces, Stairs and Ladders	OSHA 1910.23	Guarding Floor and Wall Openings and Holes
	OSHA 1910.24	Fixed Industrial Stairs
	OSHA 1910.27	Fixed Ladders
	ANSI/ASSE A1264.1	Safety Requirements for Workplace Walking/Working Surfaces and Their Access; Workplace, Floor, Wall and Roof Openings; Stairs and Guardrails Systems
	ISO 14122 (series)	(parts 1 thru 4) Safety of Machinery — Permanent Means of Access to Machinery (relative to working platforms, walkways, access between levels, stairs, ladders and guard-rails, etc.)
	SAE and ASABE standards	(for off-road, agricultural, construction and outdoor powered wheeled (and some non-wheeled) machines and equipment)
Machine Controls and Displays	MIL-HDBK-759	Human Engineering Design Guidelines
	EN 954-1	Safety-Related Parts of Control Systems — General Principles of Design
	ISO 13849 (series)	Safety of Machinery — Safety-Related Parts of Control Systems
	EN 842	Visual Danger Signals — General Requirements, Design, and Testing
	EN 894-1	Ergonomics Requirements for the Design of Displays and Control Actuators — General
	EN 894-2	Ergonomics Requirements for the Design of Displays and Control Actuators — Displays
	EN 894-3	Ergonomics Requirements for the Design of Displays and Control Actuators — Control Actuators
	CSA -Z431	Basic and Safety Principles for Man-Machine Interface, Marking and Identification — Coding Principles for Indication Devices and Actuators
	EN 457	Auditory Danger Signals — General Requirements, Design, and Testing
	EN 981	System of Auditory and Visual Danger and Information Signals
	ISO 14738	Safety of Machinery — Anthropometric Requirements for the Design of Workstations at Machinery
	EN 1005-2	Human Physical Performance — Manual Handling of Machinery and Component Parts of Machinery
	EN 1005-3	Human Physical Performance — Recommended Force Limits for Machinery Operation
	EN 1005-4	Human Physical Performance — Evaluation of Working Postures and Movements in relation to Machinery
	EN 1005-5	Human Physical Performance — Risk Assessment for Repetitive Handling at High Frequency
	ISO 13851	Safety of Machinery — Two-Hand Control Devices — Functional Aspects and Design Principles
	EN 574	Two-Hand Control Devices — Functional Aspects — Principles for Design
	ISO 7000	Graphical Symbols for Use on Equipment — Index and Synopsis
	SAE and ASABE standards	(for off-road, agricultural, construction, and outdoor powered wheeled—and some non-wheeled—machines and equipment)

continued on next page

Table 2-7: Selected Codes and Standards for Machinery Safety (Continued)

Information, Instructions and Warnings Posted on Machines	ANSI Z535.1	Safety Color Code
	ANSI Z535.3	Criteria for Safety Symbols
	ANSI Z535.4	Product Safety Signs and Labels
	ISO 3864-2	Graphical Symbols — Safety Colours and Safety Signs — Part 2: Design Principles for Product Safety Labels
	ISO 3864-3	Graphical Symbols — Safety Colours and Safety Signs — Part 3: Design Principles for Graphical Symbols for Use in Safety Signs
	ISO 17398	Safety Colours and Safety Signs — Classification, Performance, and Durability of Safety Signs
	EN 842	Visual Danger Signals — General Requirements, Design, and Testing
	EN 457	Auditory Danger Signals — General Requirements, Design, and Testing
	EN 981	System of Auditory and Visual Danger and Information Signals
	ISO 7000	Graphical Symbols for Use on Equipment — Index and Synopsis
	SAE and ASABE standards	(for off-road, agricultural, construction and outdoor powered wheeled—and some non-wheeled—machines and equipment)
Safety Information in Operator's Manuals, Service Manuals, and other accompanying literature	ANSI Z535.6	Product Safety Information in Product Manuals, Instructions, and Other Collateral Materials
Unexpected Start-Ups	EN 1037	Prevention of Unexpected Start-Up
	ISO 14118	Safety of Machinery — Prevention of Unexpected Start-Up
Emergency Stop	EN 418	Emergency Stop Equipment, Functional Aspects — Principles for Design
	EN 13850	Emergency Stop — Principles for Design
	ISO 13850	Safety of Machinery — Emergency Stop — Principles for Design
Lockout / Tagout	OSHA 1910.147	The Control of Hazardous Energy (lockout/tagout)
	ANSI/ASSE Z244.1	Control of Hazardous Energy, Lockout/Tagout, and Alternative Methods
Temperatures of Touchable Surfaces	UL 987 (Table 26.1)	Standard for Stationary and Fixed Electric Tools (Table 26.1 — Maximum Acceptable Surface Temperatures)
	EN 563	Temperatures of Touchable Surfaces — Ergonomics Data to Establish Temperature Limit Values for Hot Surfaces
	MIL-STD 1472 (pp: 5.13.4.6)	Human Engineering Design Criteria for Military Systems, Equipment, and Facilities (pp: 5.13.4.6 — Thermal Contact Hazards)
	ISO 13732-1	Ergonomics of the Thermal Environment — Methods for the Assessment of Human Responses to Contact with Surfaces — Part 1: Hot Surfaces
	ISO 13732-3	Ergonomics of the Thermal Environment — Methods for the Assessment of Human Responses to Contact with Surfaces — Part 3: Cold Surfaces

continued on next page

Table 2-7: Selected Codes and Standards for Machinery Safety (Continued)

Noise	OSHA 1910.95	Occupational Noise Exposure
	ANSI B11.TR5 (tech report)	Sound Level Measurement Guidelines — A Guide for Measuring, Evaluating, Documenting, and Reporting Sound Levels Emitted by Machinery
Emission of Hazardous Substance	EN 626-1	Reduction of Risks to Health from Hazardous Substance Emitted by Machinery — Principles and Specifications for Machinery Manufacturers
	ISO 14123 (series)	Safety of Machinery — Reduction of Risks to Health from Hazardous Substances Emitted by Machinery
	ANSI B11.TR2 (tech report)	Mist Control Considerations for the Design, Installation, and Use of Machine Tools Using Metalworking Fluids
Fluid Power Systems: Hydraulics / Pneumatics	OSHA 1910 - M	Compressed Gas and Compressed Air Equipment
	EN 982	Safety Requirements for Fluid Power Systems and Their Components — Hydraulics
	EN 983	Safety Requirements for Fluid Power Systems and Their Components — Pneumatics
	ISO 4413	Hydraulic Fluid Power — General Rules and Safety Requirements for Systems and Their Components
	ISO 4414	Pneumatic Fluid Power — General Rules and Safety Requirements for Systems and Their Components
Electrical Systems	OSHA 1910.302 thru .308	Design Safety Standards for Electrical Systems
	EN 60204	Safety of Machinery — Electrical Equipment of Machines
	UL 73	Standard for Motor Operated Appliances
	NFPA 79	Electrical Standard for Industrial Machinery
	NFPA 70E	Standard for Electrical Safety in the Workplace
	ANSI C1	National Electrical Code
Hand-Held Powered Tools	OSHA 1910 - P	Hand and Portable Powered Tools and Other Hand-Held Equipment
	UL 745 (series)	Portable Electric Tools . . . .
	UL 60745 (series)	Standard for Hand-Held Motor-Operated Electric Tools — Safety — . . . .
	ISO 11148 (series)	Safety Requirements Hand-Held Non-Electric Power Tools
Fixed Electric Machines, Equipment and Tools	UL987	Standard for Stationary and Fixed Electric Tools
Robots, Manufacturing Cells	ANSI B11.20	Safety Requirements for Integrated Manufacturing Systems
	ANSI/RIAR 15.06	Safety Standard for Industrial Robots and Robot Systems
	UL 1740	Standard for Robots and Robotic Equipment
	ISO 10218-1	Robots and Robotic Devices — Safety Requirements — Industrial Robots
	ISO 10218-2	Robots and Robotic Devices — Safety Requirements — Industrial Robot Systems and Integration
	ISO 11161	Safety of Machinery — Integrated Manufacturing Systems — Basic Requirements
	CSA-Z434	Industrial Robots and Robot Systems — General Safety Requirements

3

DIMENSIONS
AND
TOLERANCES

Contents

3.1	LIMITS, FITS, AND TOLERANCE GRADES	95
3.2	TOLERANCES ON DRAWINGS, AND GD&T	116
3.3	TOLERANCE STACK-UPS	127

Tables

3-1	Commonly Used Limits and Fits	99
3-2	Selected ANSI Tolerance Grades	100
3-3	Selected International Tolerance (IT) Grades	101
3-4	Manufacturing Process Average Tolerance Grades	102
3-5	Selected Limits of Size, Holes (Inch)	104
3-6	Selected Limits of Size, Shafts (Inch)	106
3-7	Additional Selected Limits of Size (Inch)	108
3-8	Selected Limits of Size, Holes (Metric)	110
3-9	Selected Limits of Size, Shafts (Metric)	112
3-10	Additional Selected Limits of Size (Metric)	114
3-11	Common Implied Tolerances (Inch)	117
3-12	Common Implied Tolerances (Metric)	117
3-13	ANSI and ISO Geometric Symbols	120
3-14	Area Under the Standard Normal Curve	140
3-15	Effect of Material Modifier on Stack-Up	153
3-16	Effect of Least Material Condition Modifier on Stack-Up	154
3-17	Boundary Calculations for Positioned Features of Size	163
3-18	Stack-Up Method Application Matrix	175

Section 3.1 LIMITS, FITS, AND TOLERANCE GRADES

All parts and features have some variation from ideal size, and this variation is controlled by the designer through the application of dimensional tolerances. The system of limits and fits allows the designer to quickly tolerance parts that fit together with a predetermined clearance or interference. Limits of size refer to the two tolerances or deviations applied to a dimension that set the upper and lower limits for that dimension. These tolerances are meant to be applied to nominal parts that were designed 'line to line' (without clearance when nominal size) in assemblies. The standards governing limits and fits are specific to cylindrical features and parts (holes and shafts), but these fits can also be applied to non-cylindrical features and parts like rectangular slides and pockets. When selecting fits, one must consider the loading, speed, length of engagement, temperature, and lubrication conditions of the assembly. The designer is free to use the standard fits or specify different tolerance combinations to achieve the desired result.

RECOMMENDED RESOURCES

- Oberg, Jones, Horton, Ryffel, *Machinery's Handbook*, 28th Ed., Industrial Press, New York, NY, 2008
- **ANSI B4.1:** "Preferred Limits and Fits for Cylindrical Parts"
- **ISO 286:** "ISO System of Limits and Fits"

TYPES OF FITS AND THEIR LIMITS

Fits refer to the amount of clearance or interference between mating parts. There are three basic types of fits: clearance fits, transition fits (chance of either clearance or interference), and interference fits. Each fit specifies two sets of tolerances, or limits of size: one for the hole or external feature, and one for the shaft or internal feature. These tolerances are applied to the nominal feature size when parts are designed 'line to line' (without clearance at nominal size), and can be either positive or negative. Tolerance designations are represented by a class letter followed by a tolerance grade number. The hole or internal feature's tolerance class is represented by a capital letter, and the shaft or external feature's tolerance class is represented by a lowercase letter. The larger the grade number, the wider the tolerance range. On part drawings, tolerances are given using the class letter and grade number designation, the numerical tolerances themselves, or both. For example, a hole and pin are designed nominally 'line to line' and the designer wishes to apply a press fit to the joint. The pin is a commercial item with an m6 tolerance. The designer would then need to find out what tolerance designation to apply to the hole. This will be solved in the next few paragraphs.

ANSI B4.1 governs preferred limits and fits in inch units. The ANSI system uses descriptive two-letter designations to represent fits. Each type of ANSI fit has a series of possible grades, each represented by a number. The grade indicates the degree of tightness of the fit. ANSI fit grades are not the same as tolerance grades. A graphical representation of the ANSI standard fits is shown in Figure 3-1. ISO 286 governs preferred limits and fits in metric units.

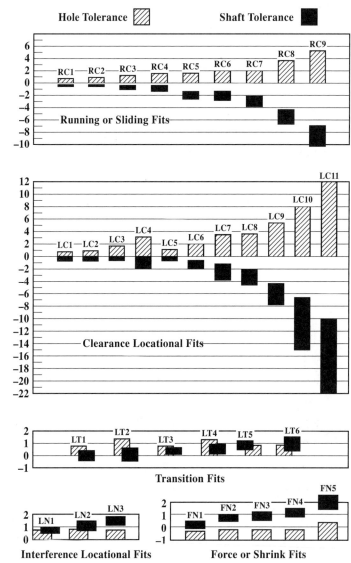

Figure 3-1:
Graphical Representation
of ANSI Standard Fits

The ISO system of fits specifies preferred tolerance combinations for each fit type. Table 3-1 lists some commonly used fits, their typical use, and their graded tolerance designations from both the ANSI and ISO standards. Revisiting the example with the m6 tolerance pin, it can be seen from Table 3-1 that a locational interference fit joint can have a tolerance combination of H7/p6. Because the pin has an m6 tolerance rather than the standard p6, some calculation will be required to adjust the tolerances. That will be discussed further in the next paragraph.

Tolerance grades, or limits of size, are graduated based on the size of the feature being toleranced. Larger features will have larger tolerance ranges. Standard ANSI tolerance grades and their numerical values are shown in Table 3-2. IT tolerance grades and their values are given in Table 3-3. The graded tolerance values corresponding to a given fit letter and grade number combination are normally obtained either through the use of charts or are built in to drafting (CAD) software. Some selections from these charts are provided in Tables 3-5 through 3-10. To use the tolerance charts, the designer must look up first the value corresponding to the letter designation. The tolerance values are then located on the letter designation chart at the intersection of the feature size row and number designation column.

Again revisiting the example with the m6 pin, the numerical limits of size for each tolerance designation can be found using Tables 3-5 through 3-10. First look up the limits of size for the diameter of the hole and pin, assuming that they are the standard locational interference fit tolerances of H7/ p6. If the pin diameter is 0.25 inch, the chart limits of size values for H7 in that size are +0.0006 / −0. For p6, the limits of size are +0.001 / +0.0006. Calculate the maximum and minimum interference between an H7 hole and a p6 pin using those limits: 0.001 / 0. For more information on performing tolerance analysis, please refer to Section 3.3. Now apply these interference values to the m6 pin and its limits of size to get the target tolerances for the locational interference fit hole. Use the tables to find the limits of size for the m6 pin with 0.25 inch diameter: +0.00059 / +0.00024. Because the least calculated interference should be 0, the upper limit for the hole to fit this oversized pin should be +0.00024. The upper limit of interference is 0.001, so the lower limit for the hole should be −0.00041.

If the designer prefers to use a tolerance designation instead of numerical values for the hole, a standard designation should be sought for limits of size of +0.00024 / −0.00041 for a diameter of 0.25 inches. Using the tables, the closest designation to those limits of size is K7 for that size of hole. Designation

K7 at 0.25 inches diameter has limits of size of +0.0001 / −0.0005. A K7 hole combined with an m6 pin in that size will result in a fit that allows between 0.00014 and 0.00109 inches of interference in the joint. The standard fit allows between 0 and 0.001, so this modified fit should work acceptably. The advantage of using the system of limits and fits is that once a fit is calculated, the designations are easily remembered and reused. A standard dowel pin has an m6 designation, so the designer can easily remember (or record) that a K7 hole will yield a satisfactory interference fit. Numerical values need not be recalled once a fit is defined. When the designer has control over the tolerances applied to both parts, using standard fit designations can speed the process.

When using force fits, the pressure required to assemble the parts can be estimated using pressure factors. The stress resulting from force fits should be calculated for a more accurate result. It is essential that the elastic limit of the parts in a force fit assembly not be exceeded because that would result in a loosening of the fit. Consult the recommended resources for calculation guidance.

Table 3-1: Commonly Used Limits and Fits

FIT TYPE	DESCRIPTION		Hole Basis		Shaft Basis
			INCH	METRIC	METRIC
Clearance Fits	**Loose Running Fit:** use with wide commercial tolerances when accuracy is not essential	RC8	H10 / c9	H11 / c11	C11 / h11
	Free Running Fit: use with high speeds when accuracy is not essential	RC7	H9 / d8	H9 / d9	D9 / h9
	Running Fit: use for moderate speeds when accuracy is not essential	RC5	H8 / e7	H8 / e8	E8 / h8
	Close Running Fit: accurate location at moderate speeds	RC4	H8 / f7	H8 / f7	F8 / h7
	Sliding Fit: moves freely, locates accurately but does not run freely	RC2	H6 / g5	H7 / g6	G7 / h6
	Locational Clearance Fit: snug but can be freely assembled and disassembled	LC2	H7 / h6	H7 / h6	H7 / h6
Transition Fits	**Locational Transition Fit:** accurate location with some interference possible	LT3	H7 / k6	H7 / k6	K7 / h6
	Locational Transition Fit: accurate location with more interference possible	LT5	H7 / n6	H7 / n6	N7 / h6
Interference Fits	**Locational Interference Fit:** accurate location with no movement under light loading	LN2	H7 / p6	H7 / p6	P7 / h6
	Medium Drive Fit: tightest interference fit for ordinary steel parts	FN2	H7 / s6	H7 / s6	S7 / h6
	Force Fit: interference fit for heavy sections and high stresses	FN4	H7 / u6	H7 / u6	U7 / h6

Table 3-2: Selected ANSI Tolerance Grades

Nominal Size, Inches		Tolerance Grade									
		Difference Between Upper and Lower Limits, Inches									
Over	To	4	5	6	7	8	9	10	11	12	13
0.00	0.12	0.00012	0.00015	0.00025	0.0004	0.0006	0.0010	0.0016	0.0025	0.004	0.006
0.12	0.24	0.00015	0.00020	0.00030	0.0005	0.0007	0.0012	0.0018	0.0030	0.005	0.007
0.24	0.40	0.00015	0.00025	0.00040	0.0006	0.0009	0.0014	0.0022	0.0035	0.006	0.009
0.40	0.71	0.00020	0.00030	0.00040	0.0007	0.0010	0.0016	0.0028	0.0040	0.007	0.010
0.71	1.19	0.00025	0.00040	0.00050	0.0008	0.0012	0.0020	0.0035	0.0050	0.008	0.012
1.19	1.97	0.00030	0.00040	0.00060	0.0010	0.0016	0.0025	0.0040	0.0060	0.010	0.016
1.97	3.15	0.00030	0.00050	0.00070	0.0012	0.0018	0.0030	0.0045	0.0070	0.012	0.018
3.15	4.73	0.00040	0.00060	0.00090	0.0014	0.0022	0.0035	0.0050	0.0090	0.014	0.022
4.73	7.09	0.00050	0.00070	0.00100	0.0016	0.0025	0.0040	0.0060	0.0100	0.016	0.025
7.09	9.85	0.00060	0.00080	0.00120	0.0018	0.0028	0.0045	0.0070	0.0120	0.018	0.028
9.85	12.41	0.00060	0.00090	0.00120	0.0020	0.0030	0.0050	0.0080	0.0120	0.020	0.030
12.41	15.75	0.00070	0.00100	0.00140	0.0022	0.0035	0.0060	0.0090	0.0140	0.022	0.035
15.75	19.69	0.00080	0.00100	0.00160	0.0025	0.0040	0.0060	0.0100	0.0160	0.025	0.040
19.69	30.09	0.00090	0.00120	0.00200	0.0030	0.0050	0.0080	0.0120	0.0200	0.030	0.050
30.09	41.49	0.00100	0.00160	0.00250	0.0040	0.0060	0.0100	0.0160	0.0250	0.040	0.060
41.49	56.19	0.00120	0.00200	0.00300	0.0050	0.0080	0.0120	0.0200	0.0300	0.050	0.080
56.19	76.39	0.00160	0.00250	0.00400	0.0060	0.0100	0.0160	0.0250	0.0400	0.060	0.100
76.39	100.90	0.00200	0.00300	0.00500	0.0080	0.0120	0.0200	0.0300	0.0500	0.080	0.125
100.90	131.90	0.00250	0.00400	0.00600	0.0100	0.0160	0.0250	0.0400	0.0600	0.100	0.160
131.90	171.90	0.00300	0.00500	0.00800	0.0120	0.0200	0.0300	0.0500	0.0800	0.125	0.200
171.90	200.00	0.00400	0.00600	0.01000	0.0160	0.0250	0.0400	0.0600	0.1000	0.160	0.250

Table 3-3: Selected International Tolerance (IT) Grades

| Nominal Feature Size, mm | | Tolerance Grade | | | | | | | | | |
Over	To	4	5	6	7	8	9	10	11	12	13
		Difference Between Upper and Lower Limits, mm									
0	3	0.003	0.004	0.006	0.010	0.014	0.025	0.040	0.600	0.100	0.140
3	6	0.004	0.005	0.008	0.012	0.018	0.030	0.048	0.750	0.120	0.180
6	10	0.004	0.006	0.009	0.015	0.022	0.036	0.058	0.900	0.150	0.220
10	18	0.005	0.008	0.011	0.018	0.027	0.043	0.070	0.110	0.180	0.270
18	30	0.006	0.009	0.013	0.021	0.033	0.052	0.084	0.130	0.210	0.330
30	50	0.007	0.011	0.016	0.025	0.039	0.062	0.100	0.160	0.250	0.390
50	80	0.008	0.013	0.019	0.030	0.046	0.074	0.120	0.190	0.300	0.460
80	120	0.010	0.015	0.022	0.035	0.054	0.087	0.140	0.220	0.350	0.540
120	180	0.012	0.018	0.025	0.040	0.063	0.100	0.160	0.250	0.400	0.630
180	250	0.014	0.020	0.029	0.046	0.072	0.115	0.185	0.290	0.460	0.720
250	315	0.016	0.023	0.032	0.052	0.081	0.130	0.210	0.320	0.520	0.810
315	400	0.018	0.025	0.036	0.057	0.089	0.140	0.230	0.360	0.570	0.890
400	500	0.020	0.027	0.040	0.063	0.097	0.155	0.250	0.400	0.630	0.970
500	630	-	-	0.044	0.070	0.110	0.175	0.280	0.440	0.700	1.100
630	800	-	-	0.050	0.080	0.125	0.200	0.320	0.500	0.800	1.250
800	1000	-	-	0.056	0.090	0.140	0.230	0.360	0.560	0.900	1.400
1000	1250	-	-	0.066	0.105	0.165	0.260	0.420	0.660	1.050	1.650
1250	1600	-	-	0.078	0.125	0.195	0.310	0.500	0.780	1.250	1.950
1600	2000	-	-	0.092	0.150	0.230	0.370	0.600	0.920	1.500	2.300
2000	2500	-	-	0.110	0.175	0.280	0.440	0.700	1.100	1.750	2.800
2500	3150	-	-	0.135	0.210	0.330	0.540	0.860	1.350	2.100	3.300

MACHINING TOLERANCES

Choice of tolerances should always take into account the manufacturing process capability as well as functional requirements. Every machining process has a tolerance capability. This can vary by machine and machinist. Some machining tolerances are given in ANSI B4.1. Table 3-4 illustrates some typical tolerance grades achieved by various machining processes. The values shown are intended only as a guide and vary depending on machine tool and operator. Tables 3-2 and 3-3 provide the numerical values for the tolerance grades as a function of part size.

Table 3-4: Manufacturing Process Average Tolerance Grades

	MACHINING OPERATION	TOLERANCE GRADES									
		4	5	6	7	8	9	10	11	12	13
This chart may be used as a general guide to determine the machining processes that will, under normal conditions, produce work within the tolerance grades indicated.	Lapping & Honing	██									
	Cylindrical Grinding		████								
	Surface Grinding		██████							
	Diamond Turning		████								
	Diamond Boring		████								
	Broaching		██████							
	Reaming			██████						
	Turning				████████			
	Boring				██████			
	Milling					██████	
	Planing & Shaping					██████	
	Drilling					██████	

LIMITS OF SIZE DATA

Tables 3-5 through 3-10 contain some of the more commonly encountered limits of size.

This page was intentionally left blank so that you may view the following tables in a more cohesive manner.

Table 3-5: Selected Limits of Size, Holes (Inch)

Basic Size Above inch	Up To & Including inch	D9 inch	D10 inch	E7 inch	E8 inch	F7 inch	F8 inch	G6 inch	G7 inch	H6 inch	H7 inch	H8 inch
0	0.12	+0.002 +0.001	+0.0026 +0.001	+0.001 +0.0006	+0.0012 +0.0006	+0.0007 +0.0003	+0.0009 +0.0003	+0.00035 +0.0001	+0.0005 +0.0001	+0.00025 0	+0.0004 0	+0.0006 0
0.12	0.24	+0.0024 +0.0012	+0.003 +0.0012	+0.0013 +0.0008	+0.0015 +0.0008	+0.0009 +0.0004	+0.0011 +0.0004	+0.00045 +0.00015	+0.00065 +0.00015	+0.0003 0	+0.0005 0	+0.0007 0
0.24	0.4	+0.003 +0.0016	+0.0038 +0.0016	+0.0016 +0.001	+0.0019 +0.001	+0.0011 +0.0005	+0.0014 +0.0005	+0.0006 +0.0002	+0.0008 +0.0002	+0.0004 0	+0.0006 0	+0.0009 0
0.4	0.71	+0.0036 +0.002	+0.0048 +0.002	+0.0019 +0.0012	+0.0022 +0.0012	+0.0013 +0.0006	+0.0016 +0.0006	+0.00065 +0.00025	+0.00095 +0.00025	+0.0004 0	+0.0007 0	+0.0010 0
0.71	1.19	+0.0045 +0.0025	+0.006 +0.0025	+0.0024 +0.0016	+0.0028 +0.0016	+0.0016 +0.0008	+0.002 +0.0008	+0.0008 +0.0003	+0.0011 +0.0003	+0.0005 0	+0.0008 0	+0.0012 0
1.19	1.97	+0.0055 +0.003	+0.007 +0.003	+0.003 +0.002	+0.0036 +0.002	+0.002 +0.001	+0.0026 +0.001	+0.001 +0.0004	+0.0014 +0.0004	+0.0006 0	+0.0010 0	+0.0016 0
1.97	3.15	+0.007 +0.004	+0.0085 +0.004	+0.0037 +0.0025	+0.0043 +0.0025	+0.0024 +0.0012	+0.003 +0.0012	+0.0011 +0.0004	+0.0016 +0.0004	+0.0007 0	+0.0012 0	+0.0018 0
3.15	4.73	+0.0085 +0.005	+0.010 +0.005	+0.0044 +0.003	+0.0052 +0.003	+0.0028 +0.0014	+0.0036 +0.0014	+0.0014 +0.0005	+0.0019 +0.0005	+0.0009 0	+0.0014 0	+0.0022 0
4.73	7.09	+0.010 +0.006	+0.012 +0.006	+0.0051 +0.0035	+0.006 +0.0035	+0.0032 +0.0016	+0.0041 +0.0016	+0.0016 +0.0006	+0.0022 +0.0006	+0.0010 0	+0.0016 0	+0.0025 0
7.09	9.85	+0.0115 +0.007	+0.014 +0.006	+0.0058 +0.004	+0.0068 +0.004	+0.0038 +0.002	+0.0048 +0.002	+0.0018 +0.0006	+0.0024 +0.0006	+0.0012 0	+0.0018 0	+0.0028 0
9.85	12.41	+0.012 +0.007	+0.015 +0.007	+0.0065 +0.0045	+0.0075 +0.0045	+0.0042 +0.0022	+0.0052 +0.0022	+0.0019 +0.0007	+0.0027 +0.0007	+0.0012 0	+0.0020 0	+0.0030 0
12.41	15.75	+0.014 +0.008	+0.017 +0.008	+0.0072 +0.005	+0.0085 +0.005	+0.0047 +0.0025	+0.006 +0.0025	+0.0021 +0.0007	+0.0029 +0.0007	+0.0014 0	+0.0022 0	+0.0035 0
15.75	19.69	+0.015 +0.009	+0.019 +0.009	+0.0075 +0.005	+0.009 +0.005	+0.0053 +0.0028	+0.0068 +0.0028	+0.0024 +0.0008	+0.0033 +0.0008	+0.0016 0	+0.0025 0	+0.0040 0

continued on next page

Table 3-5: Selected Limits of Size, Holes (Inch) (Continued)

Basic Size		H9	H10	H12	H13	H14 (Converted)	H15 (Converted)	K7	P6	P7
Above inch	Up To & Including inch	inch	inch	inch	inch	inch	inch	inch	inch	inch
0	0.12	+0.0010 / 0	+0.0016 / 0	+0.0040 / 0	+0.0060 / 0	+0.0098 / 0	+0.0157 / 0	+0 / ...	-0.00035 / -0.0006	-0.00025 / -0.00065
0.12	0.24	+0.0012 / 0	+0.0018 / 0	+0.0050 / 0	+0.0070 / 0	+0.0118 / 0	+0.0189 / 0	+0.0001 / ...	-0.0004 / -0.0007	-0.0003 / -0.0008
0.24	0.4	+0.0014 / 0	+0.0022 / 0	+0.0060 / 0	+0.0090 / 0	+0.0142 / 0	+0.0228 / 0	+0.0001 / -0.0005	-0.0005 / -0.0009	-0.0004 / -0.001
0.4	0.71	+0.0016 / 0	+0.0028 / 0	+0.0070 / 0	+0.0100 / 0	+0.0169 / 0	+0.0276 / 0	+0.0002 / -0.0005	-0.0006 / -0.001	-0.0004 / -0.0011
0.71	1.19	+0.0020 / 0	+0.0035 / 0	+0.0080 / 0	+0.0120 / 0	+0.0205 / 0	+0.0331 / 0	+0.0002 / -0.0006	-0.0007 / -0.0012	-0.0005 / -0.0013
1.19	1.97	+0.0025 / 0	+0.0040 / 0	+0.0100 / 0	+0.0160 / 0	+0.0244 / 0	+0.0394 / 0	+0.0003 / -0.0007	-0.0008 / -0.0014	-0.0006 / -0.0016
1.97	3.15	+0.0030 / 0	+0.0045 / 0	+0.0120 / 0	+0.0180 / 0	+0.0291 / 0	+0.0472 / 0	+0.0004 / -0.0008	-0.0012 / -0.0019	-0.0009 / -0.0021
3.15	4.73	+0.0035 / 0	+0.0050 / 0	+0.0140 / 0	+0.0220 / 0	+0.0343 / 0	+0.0551 / 0	+0.0004 / -0.001	-0.0013 / -0.0022	-0.0011 / -0.0025
4.73	7.09	+0.0040 / 0	+0.0060 / 0	+0.0160 / 0	+0.0250 / 0	+0.0394 / 0	+0.0630 / 0	+0.0005 / -0.0011	-0.0015 / -0.0025	-0.0012 / -0.0028
7.09	9.85	+0.0045 / 0	+0.0070 / 0	+0.0180 / 0	+0.0280 / 0	+0.0453 / 0	+0.0728 / 0	+0.0005 / -0.0013	-0.0016 / -0.0028	-0.0014 / -0.0032
9.85	12.41	+0.0050 / 0	+0.0080 / 0	+0.0200 / 0	+0.0300 / 0	+0.0512 / 0	+0.0827 / 0	+0.0006 / -0.0014	-0.0019 / -0.0031	-0.0014 / -0.0034
12.41	15.75	+0.0060 / 0	+0.0090 / 0	+0.0220 / 0	+0.0350 / 0	+0.0551 / 0	+0.0906 / 0	+0.0006 / -0.0016	-0.0021 / -0.0035	-0.0017 / -0.0039
15.75	19.69	+0.0060 / 0	+0.0100 / 0	+0.0250 / 0	+0.0400 / 0	+0.061 / 0	+0.0984 / 0	+0.0007 / -0.0018	-0.0022 / -0.0038	-0.0019 / -0.0044

Table 3-6: Selected Limits of Size, Shafts (Inch)

Basic Size		c9 inch	d8 inch	d9 inch	e7 inch	e8 inch	e9 inch	f6 inch	f7 inch	g5 inch	g6 inch	h6 inch
Above inch	Up To & Including inch											
0	0.12	-0.0025 / -0.0035	-0.0010 / -0.0016	-0.0010 / -0.0020	-0.0006 / -0.0010	-0.0006 / -0.0012	-0.0006 / -0.0016	-0.0003 / -0.00055	-0.0003 / -0.0007	-0.0001 / -0.0003	-0.0001 / -0.00035	+0 / -0.00025
0.12	0.24	-0.0028 / -0.0040	-0.0012 / -0.0019	-0.0012 / -0.0024	-0.0008 / -0.0013	-0.0008 / -0.0015	-0.0008 / -0.0020	-0.0004 / -0.0007	-0.0004 / -0.0009	-0.00015 / -0.00035	-0.00015 / -0.00045	+0 / -0.0003
0.24	0.4	-0.0030 / -0.0044	-0.0016 / -0.0025	-0.0016 / -0.0030	-0.0010 / -0.0016	-0.0010 / -0.0019	-0.0010 / -0.0024	-0.0005 / -0.0009	-0.0005 / -0.0011	-0.0002 / -0.00045	-0.0002 / -0.0006	+0 / -0.0004
0.4	0.71	-0.0035 / -0.0051	-0.0020 / -0.0030	-0.0020 / -0.0036	-0.0012 / -0.0019	-0.0012 / -0.0022	-0.0012 / -0.0028	-0.0006 / -0.0010	-0.0006 / -0.0013	-0.00025 / -0.00055	-0.00025 / -0.00065	+0 / -0.0004
0.71	1.19	-0.0045 / -0.0065	-0.0025 / -0.0037	-0.0025 / -0.0045	-0.0016 / -0.0024	-0.0016 / -0.0028	-0.0016 / -0.0036	-0.0008 / -0.0013	-0.0008 / -0.0016	-0.0003 / -0.0007	-0.0003 / -0.0008	+0 / -0.0005
1.19	1.97	-0.0050 / -0.0075	-0.0030 / -0.0046	-0.0030 / -0.0055	-0.0020 / -0.0030	-0.0020 / -0.0036	-0.0020 / -0.0045	-0.0010 / -0.0016	-0.0010 / -0.0020	-0.0004 / -0.0008	-0.0004 / -0.0010	+0 / -0.0006
1.97	3.15	-0.0060 / -0.0090	-0.0040 / -0.0058	-0.0040 / -0.0070	-0.0025 / -0.0037	-0.0025 / -0.0043	-0.0025 / -0.0055	-0.0012 / -0.0019	-0.0012 / -0.0024	-0.0004 / -0.0009	-0.0004 / -0.0011	+0 / -0.0007
3.15	4.73	-0.0070 / -0.0105	-0.0050 / -0.0072	-0.0050 / -0.0085	-0.0030 / -0.0044	-0.0030 / -0.0052	-0.0030 / -0.0065	-0.0014 / -0.0023	-0.0014 / -0.0028	-0.0005 / -0.0011	-0.0005 / -0.0014	+0 / -0.0009
4.73	7.09	-0.0080 / -0.0120	-0.0060 / -0.0085	-0.0060 / -0.0100	-0.0035 / -0.0051	-0.0035 / -0.0060	-0.0035 / -0.0075	-0.0016 / -0.0026	-0.0016 / -0.0032	-0.0006 / -0.0013	-0.0006 / -0.0016	+0 / -0.0010
7.09	9.85	-0.0100 / -0.0145	-0.0070 / -0.0098	-0.0070 / -0.0115	-0.0040 / -0.0058	-0.0040 / -0.0068	-0.0040 / -0.0085	-0.0020 / -0.0032	-0.0020 / -0.0038	-0.0006 / -0.0014	-0.0006 / -0.0018	+0 / -0.0012
9.85	12.41	-0.0120 / -0.0170	-0.0080 / -0.0110	-0.0070 / -0.0120	-0.0050 / -0.0070	-0.0050 / -0.0080	-0.0045 / -0.0095	-0.0025 / -0.0037	-0.0025 / -0.0045	-0.0008 / -0.0017	-0.0007 / -0.0019	+0 / -0.0012
12.41	15.75	-0.0140 / -0.0200	-0.0100 / -0.0135	-0.0080 / -0.0140	-0.0060 / -0.0082	-0.0060 / -0.0095	-0.0050 / -0.0110	-0.0030 / -0.0044	-0.0030 / -0.0052	-0.0010 / -0.0020	-0.0007 / -0.0021	+0 / -0.0014
15.75	19.69	-0.0160 / -0.0220	-0.0120 / -0.0160	-0.0090 / -0.0150	-0.0080 / -0.0105	-0.0080 / -0.0120	-0.0050 / -0.0110	-0.0040 / -0.0056	-0.0040 / -0.0065	-0.0012 / -0.0022	-0.0008 / -0.0024	+0 / -0.0016

continued on next page

Table 3-6: Selected Limits of Size, Shafts (Inch) (Continued)

Basic Size		h7	h8	h9	h11	k6	k7	m6	n6	p6
Above inch	**Up To & Including** inch	inch	inch	inch	inch	inch	inch	inch	inch	inch
0	0.12	+0	+0	+0	+0	+0.00024	+0.0004	+0.00031	+0.0005	+0.00065
		-0.0004	-0.0006	-0.0010	-0.0025	+0	+0	+0.00008	+0.00025	+0.0004
0.12	0.24	+0	+0	+0	+0	+0.00035	+0.0005	+0.00047	+0.0006	+0.0008
		-0.0005	-0.0007	-0.0012	-0.0030	+0.00004	+.00004	+0.00016	+0.0003	+0.0005
0.24	0.4	+0	+0	+0	+0	+0.0005	+0.0007	+0.00059	+0.0008	+0.0010
		-0.0006	-0.0009	-0.0014	-0.0035	+0.0001	+0.0001	+0.00024	+0.0004	+0.0006
0.4	0.71	+0	+0	+0	+0	+0.0005	+0.0008	+0.00071	+0.0009	+0.0011
		-0.0007	-0.0010	-0.0016	-0.0040	+0.0001	+0.0001	+0.00028	+0.0005	+0.0007
0.71	1.19	+0	+0	+0	+0	+0.0006	+0.0009	+0.00083	+0.0011	+0.0013
		-0.0008	-0.0012	-0.0020	-0.0050	+0.0001	+0.0001	+0.00031	+0.0006	+0.0008
1.19	1.97	+0	+0	+0	+0	+0.0007	+0.0011	+0.00098	+0.0013	+0.0016
		-0.0010	-0.0016	-0.0025	-0.0060	+0.0001	+0.0001	+0.00035	+0.0007	+0.0010
1.97	3.15	+0	+0	+0	+0	+0.0008	+0.0013	+0.00118	+0.0015	+0.0021
		-0.0012	-0.0018	-0.0030	-0.0070	+0.0001	+0.0001	+0.00043	+0.0008	+0.0014
3.15	4.73	+0	+0	+0	+0	+0.0010	+0.0015	+0.00138	+0.0019	+0.0025
		-0.0014	-0.0022	-0.0035	-0.0090	+0.0001	+0.0001	+0.00051	+0.0010	+0.0016
4.73	7.09	+0	+0	+0	+0	+0.0011	+0.0017	+0.00157	+0.0022	+0.0028
		-0.0016	-0.0025	-0.0040	-0.0100	+0.0001	+0.0001	+0.00059	+0.0012	+0.0018
7.09	9.85	+0	+0	+0	+0	+0.0014	+0.0020	+0.00181	+0.0026	+0.0032
		-0.0018	-0.0028	-0.0045	-0.0120	+0.0002	+0.0002	+0.00067	+0.0014	+0.0020
9.85	12.41	+0	+0	+0	+0	+0.0014	+0.0022	+0.00205	+0.0026	+0.0034
		-0.0020	-0.0030	-0.0050	-0.0120	+0.0002	+0.0002	+0.00079	+0.0014	+0.0022
12.41	15.75	+0	+0	+0	+0	+0.0016	+0.0024	+0.00224	+0.0030	+0.0039
		-0.0022	-0.0035	-0.0060	-0.0140	+0.0002	+0.0002	+0.00083	+0.0016	+0.0025
15.75	19.69	+0	+0	+0	+0	+0.0018	+0.0027	+0.00248	+0.0034	+0.0044
		-0.0025	-0.0040	-0.0060	-0.0160	+0.0002	+0.0002	+0.00091	+0.0018	+0.0028

Table 3-7: Additional Selected Limits of Size (Inch)

Basic Size		Holes		Shafts	
Above	**Up To & Including**	**S7**	**U7**	**s6**	**u6**
inch	inch	inch	inch	inch	inch
0	0.12	-0.0005 -0.0008	-0.0006 -0.001	+0.00085 +0.00060	+0.00095 +0.00070
0.12	0.24	-0.0005 -0.001	-0.0007 -0.0012	+0.0010 +0.0007	+0.0012 +0.0009
0.24	0.4	-0.0008 -0.0014	-0.001 -0.0016	+0.0014 +0.0010	+0.0016 +0.0012
0.4	0.56	-0.0009 -0.0016	-0.0011 -0.0018	+0.0016 +0.0012	+0.0018 +0.0014
0.56	0.71	-0.0009 -0.0016	-0.0011 -0.0018	+0.0016 +0.0012	+0.0018 +0.0014
0.71	0.95	-0.0011 -0.0019	-0.0013 -0.0021	+0.0019 +0.0014	+0.0021 +0.0016
0.95	1.19	-0.0011 -0.0019	-0.0015 -0.0023	+0.0019 +0.0014	+0.0023 +0.0018
1.19	1.58	-0.0014 -0.0024	-0.0021 -0.0031	+0.0024 +0.0018	+0.0031 +0.0025
1.58	1.97	-0.0014 -0.0024	-0.0024 -0.0034	+0.0024 +0.0018	+0.0034 +0.0028
1.97	2.56	-0.0015 -0.0027	-0.003 -0.0042	+0.0027 +0.0020	+0.0042 +0.0035
2.56	3.15	-0.0017 -0.0029	-0.0035 -0.0047	+0.0029 +0.0022	+0.0047 +0.0040
3.15	3.94	-0.0023 -0.0037	-0.0045 -0.0059	+0.0037 +0.0028	+0.0059 +0.0050
3.94	4.73	-0.0025 -0.0039	-0.0055 -0.0069	+0.0039 +0.0030	+0.0069 +0.0060
4.73	5.52	-0.0029 -0.0045	-0.0064 -0.008	+0.0045 +0.0035	+0.0080 +0.0070
5.52	6.3	-0.0034 -0.005	-0.0064 -0.008	+0.0050 +0.0040	+0.0080 +0.0070

continued on next page

Table 3-7: Additional Selected Limits of Size (Inch) (Continued)

Basic Size		Holes		Shafts	
Above inch	**Up To &** inch	**S7** inch	**U7** inch	**s6** inch	**u6** inch
6.3	**7.09**	-0.0039 -0.0055	-0.0074 -0.009	+0.0055 +0.0045	+0.0090 +0.0080
7.09	**7.88**	-0.0044 -0.0062	-0.0084 -0.0102	+0.0062 +0.0050	+0.0102 +0.0090
7.88	**8.86**	-0.0044 -0.0062	-0.0094 -0.0112	+0.0062 +0.0050	+0.0112 +0.0100
8.86	**9.85**	-0.0052 -0.007	-0.0114 -0.0132	+0.0072 +0.0060	+0.0132 +0.0120
9.85	**11.03**	-0.0052 -0.0072	-0.0114 -0.0134	+0.0072 +0.0060	+0.0132 +0.0120
11.03	**12.41**	-0.0062 -0.0082	-0.0132 -0.0152	+0.0082 +0.0070	+0.0152 +0.0140
12.41	**13.98**	-0.0062 -0.0084	-0.0152 -0.0174	+0.0094 +0.0080	+0.0174 +0.0160
13.98	**15.75**	-0.0072 -0.0094	-0.0172 -0.0194	+0.0094 +0.0080	+0.0194 +0.0180
15.75	**17.72**	-0.0081 -0.0106	-0.0191 -0.0216	+0.0106 +0.0090	+0.0216 +0.0200
17.72	**19.69**	-0.0091 -0.0116	-0.0211 -0.0236	+0.0116 +0.0100	+0.0236 +0.0220

Table 3-8: Selected Limits of Size, Holes (Metric)

Basic Size Above mm	Up To & Including mm	D9 mm	D10 mm	E7 mm	E8 mm	F7 mm	F8 mm	G6 mm	G7 mm	H6 mm	H7 mm	H8 mm
0	3	+0.045 / +0.020	+0.060 / +0.020	+0.024 / +0.014	+0.028 / +0.014	+0.016 / +0.006	+0.020 / +0.006	+0.008 / +0.002	+0.012 / +0.002	+0.006 / +0	+0.010 / +0	+0.014 / +0
3	6	+0.060 / +0.030	+0.078 / +0.030	+0.032 / +0.020	+0.038 / +0.020	+0.022 / +0.010	+0.028 / +0.010	+0.012 / +0.004	+0.016 / +0.004	+0.008 / +0	+0.012 / +0	+0.018 / +0
6	10	+0.076 / +0.040	+0.098 / +0.040	+0.040 / +0.025	+0.047 / +0.025	+0.028 / +0.013	+0.035 / +0.013	+0.014 / +0.005	+0.020 / +0.005	+0.009 / +0	+0.015 / +0	+0.022 / +0
10	18	+0.093 / +0.050	+0.120 / +0.050	+0.050 / +0.032	+0.059 / +0.032	+0.034 / +0.016	+0.043 / +0.016	+0.017 / +0.006	+0.024 / +0.006	+0.011 / +0	+0.018 / +0	+0.027 / +0
18	30	+0.117 / +0.065	+0.149 / +0.065	+0.061 / +0.040	+0.073 / +0.040	+0.041 / +0.020	+0.053 / +0.020	+0.020 / +0.007	+0.028 / +0.007	+0.013 / +0	+0.021 / +0	+0.033 / +0
30	50	+0.142 / +0.080	+0.180 / +0.080	+0.075 / +0.050	+0.089 / +0.050	+0.050 / +0.025	+0.064 / +0.025	+0.025 / +0.009	+0.034 / +0.009	+0.016 / +0	+0.025 / +0	+0.039 / +0
50	80	+0.174 / +0.100	+0.220 / +0.100	+0.090 / +0.060	+0.106 / +0.060	+0.060 / +0.030	+0.076 / +0.030	+0.029 / +0.010	+0.040 / +0.010	+0.019 / +0	+0.030 / +0	+0.046 / +0
80	120	+0.207 / +0.120	+0.260 / +0.120	+0.107 / +0.072	+0.125 / +0.072	+0.071 / +0.036	+0.090 / +0.036	+0.034 / +0.012	+0.047 / +0.012	+0.022 / +0	+0.035 / +0	+0.054 / +0
120	180	+0.245 / +0.145	+0.305 / +0.145	+0.125 / +0.085	+0.148 / +0.085	+0.083 / +0.043	+0.106 / +0.043	+0.039 / +0.014	+0.054 / +0.014	+0.025 / +0	+0.040 / +0	+0.063 / +0
180	250	+0.285 / +0.170	+0.355 / +0.170	+0.146 / +0.100	+0.172 / +0.100	+0.096 / +0.050	+0.122 / +0.050	+0.044 / +0.015	+0.061 / +0.015	+0.029 / +0	+0.046 / +0	+0.072 / +0
250	315	+0.320 / +0.190	+0.400 / +0.190	+0.162 / +0.110	+0.191 / +0.110	+0.108 / +0.056	+0.137 / +0.056	+0.049 / +0.017	+0.069 / +0.017	+0.032 / +0	+0.052 / +0	+0.081 / +0
315	400	+0.350 / +0.210	+0.440 / +0.210	+0.182 / +0.125	+0.214 / +0.125	+0.119 / +0.062	+0.151 / +0.062	+0.054 / +0.018	+0.075 / +0.018	+0.036 / +0	+0.057 / +0	+0.089 / +0
400	500	+0.385 / +0.230	+0.480 / +0.230	+0.198 / +0.135	+0.232 / +0.135	+0.131 / +0.068	+0.165 / +0.068	+0.060 / +0.020	+0.083 / +0.020	+0.040 / +0	+0.063 / +0	+0.097 / +0

continued on next page

Table 3-8: Selected Limits of Size, Holes (Metric) (Continued)

Basic Size Above mm	Up To & Including mm	H9 mm	H10 mm	H11 mm	H12 mm	H13 mm	H14 mm	H15 mm	K7 mm	P6 mm	P7 mm
0	3	+0.025 / +0	+0.040 / +0	+0.060 / +0	+0.100 / +0	+0.140 / +0	+0.250 / +0	+0.400 / +0	+0 / -0.010	-0.006 / -0.012	-0.006 / -0.016
3	6	+0.030 / +0	+0.048 / +0	+0.075 / +0	+0.120 / +0	+0.180 / +0	+0.300 / +0	+0.480 / +0	+0.003 / -0.009	-0.009 / -0.017	-0.008 / -0.020
6	10	+0.036 / +0	+0.058 / +0	+0.090 / +0	+0.150 / +0	+0.220 / +0	+0.360 / +0	+0.580 / +0	+0.005 / -0.010	-0.012 / -0.021	-0.009 / -0.024
10	18	+0.043 / +0	+0.070 / +0	+0.110 / +0	+0.180 / +0	+0.270 / +0	+0.430 / +0	+0.700 / +0	+0.006 / -0.012	-0.015 / -0.026	-0.011 / -0.029
18	30	+0.052 / +0	+0.084 / +0	+0.130 / +0	+0.210 / +0	+0.330 / +0	+0.520 / +0	+0.840 / +0	+0.006 / -0.015	-0.018 / -0.031	-0.014 / -0.035
30	50	+0.062 / +0	+0.100 / +0	+0.160 / +0	+0.250 / +0	+0.390 / +0	+0.620 / +0	+1.000 / +0	+0.007 / -0.018	-0.021 / -0.037	-0.017 / -0.042
50	80	+0.074 / +0	+0.120 / +0	+0.190 / +0	+0.300 / +0	+0.460 / +0	+0.740 / +0	+1.200 / +0	+0.009 / -0.021	-0.026 / -0.045	-0.021 / -0.051
80	120	+0.087 / +0	+0.140 / +0	+0.220 / +0	+0.350 / +0	+0.540 / +0	+0.870 / +0	+1.400 / +0	+0.010 / -0.025	-0.030 / -0.052	-0.024 / -0.059
120	180	+0.100 / +0	+0.160 / +0	+0.250 / +0	+0.400 / +0	+0.630 / +0	+1.000 / +0	+1.600 / +0	+0.012 / -0.028	-0.036 / -0.061	-0.028 / -0.068
180	250	+0.115 / +0	+0.185 / +0	+0.290 / +0	+0.460 / +0	+0.720 / +0	+1.150 / +0	+1.850 / +0	+0.013 / -0.033	-0.041 / -0.070	-0.033 / -0.079
250	315	+0.130 / +0	+0.210 / +0	+0.320 / +0	+0.520 / +0	+0.810 / +0	+1.300 / +0	+2.100 / +0	+0.016 / -0.036	-0.047 / -0.079	-0.036 / -0.088
315	400	+0.140 / +0	+0.230 / +0	+0.360 / +0	+0.570 / +0	+0.890 / +0	+1.400 / +0	+2.300 / +0	+0.017 / -0.040	-0.051 / -0.087	-0.041 / -0.098
400	500	+0.155 / +0	+0.250 / +0	+0.400 / +0	+0.630 / +0	+0.970 / +0	+1.550 / +0	+2.500 / +0	+0.018 / -0.045	-0.055 / -0.095	-0.045 / -0.108

Table 3-9: Selected Limits of Size, Shafts (Metric)

Basic Size Above (mm)	Up To & Including (mm)	d8 (mm)	d9 (mm)	e7 (mm)	e8 (mm)	e9 (mm)	f6 (mm)	f7 (mm)	g5 (mm)	g6 (mm)	h6 (mm)	h7 (mm)
0	3	-0.020 / -0.034	-0.020 / -0.045	-0.014 / -0.024	-0.014 / -0.028	-0.014 / -0.039	-0.006 / -0.012	-0.006 / -0.016	-0.002 / -0.006	-0.002 / -0.008	0 / -0.006	0 / -0.010
3	6	-0.030 / -0.048	-0.030 / -0.060	-0.020 / -0.032	-0.020 / -0.038	-0.020 / -0.050	-0.010 / -0.018	-0.010 / -0.022	-0.004 / -0.009	-0.004 / -0.012	0 / -0.008	0 / -0.012
6	10	-0.040 / -0.062	-0.040 / -0.076	-0.025 / -0.040	-0.025 / -0.047	-0.025 / -0.061	-0.013 / -0.022	-0.013 / -0.028	-0.005 / -0.011	-0.005 / -0.014	0 / -0.009	0 / -0.015
10	18	-0.050 / -0.077	-0.050 / -0.093	-0.032 / -0.050	-0.032 / -0.059	-0.032 / -0.075	-0.016 / -0.027	-0.016 / -0.034	-0.006 / -0.014	-0.006 / -0.017	0 / -0.011	0 / -0.018
18	30	-0.065 / -0.098	-0.065 / -0.117	-0.040 / -0.061	-0.040 / -0.073	-0.040 / -0.092	-0.020 / -0.033	-0.020 / -0.041	-0.007 / -0.016	-0.007 / -0.020	0 / -0.013	0 / -0.021
30	40	-0.080 / -0.119	-0.080 / -0.142	-0.050 / -0.075	-0.050 / -0.089	-0.050 / -0.112	-0.025 / -0.041	-0.025 / -0.050	-0.009 / -0.020	-0.009 / -0.025	0 / -0.016	0 / -0.025
40	50	-0.100 / -0.146	-0.100 / -0.174	-0.060 / -0.090	-0.060 / -0.106	-0.060 / -0.134	-0.030 / -0.049	-0.030 / -0.060	-0.010 / -0.023	-0.010 / -0.029	0 / -0.019	0 / -0.030
50	65	-0.120 / -0.174	-0.120 / -0.207	-0.072 / -0.107	-0.072 / -0.126	-0.072 / -0.159	-0.036 / -0.058	-0.036 / -0.071	-0.012 / -0.027	-0.012 / -0.034	0 / -0.022	0 / -0.035
65	80	-0.145 / -0.208	-0.145 / -0.245	-0.085 / -0.125	-0.085 / -0.148	-0.085 / -0.185	-0.043 / -0.068	-0.043 / -0.083	-0.014 / -0.032	-0.014 / -0.039	0 / -0.025	0 / -0.040
80	100	-0.170 / -0.242	-0.170 / -0.285	-0.100 / -0.146	-0.100 / -0.172	-0.100 / -0.215	-0.050 / -0.079	-0.050 / -0.096	-0.015 / -0.035	-0.015 / -0.044	0 / -0.029	0 / -0.046
100	120	-0.190 / -0.271	-0.190 / -0.320	-0.110 / -0.162	-0.110 / -0.191	-0.110 / -0.240	-0.056 / -0.088	-0.056 / -0.108	-0.017 / -0.040	-0.017 / -0.049	0 / -0.032	0 / -0.052
120	140	-0.210 / -0.299	-0.210 / -0.350	-0.125 / -0.182	-0.125 / -0.214	-0.125 / -0.265	-0.062 / -0.098	-0.062 / -0.119	-0.018 / -0.043	-0.018 / -0.054	0 / -0.036	0 / -0.057
140	160	-0.230 / -0.327	-0.230 / -0.385	-0.135 / -0.198	-0.135 / -0.232	-0.135 / -0.290	-0.068 / -0.108	-0.068 / -0.131	-0.020 / -0.047	-0.020 / -0.060	0 / -0.040	0 / -0.063

continued on next page

Table 3-9: Selected Limits of Size, Shafts (Metric) (Continued)

Basic Size Above mm	Up To & Including mm	h8 mm	h9 mm	h11 mm	k6 mm	k7 mm	m6 mm	n6 mm	p6 mm
0	3	0 / -0.014	0 / -0.025	0 / -0.060	+0.006 / 0	+0.010 / 0	+0.008 / +0.002	+0.010 / +0.004	+0.012 / +0.006
3	6	0 / -0.018	0 / -0.030	0 / -0.075	+0.009 / +0.001	+0.013 / +0.001	+0.012 / +0.004	+0.016 / +0.008	+0.020 / +0.012
6	10	0 / -0.022	0 / -0.036	0 / -0.090	+0.010 / +0.001	+0.016 / +0.001	+0.015 / +0.006	+0.019 / +0.010	+0.024 / +0.015
10	18	0 / -0.027	0 / -0.043	0 / -0.110	+0.012 / +0.001	+0.019 / +0.001	+0.018 / +0.007	+0.023 / +0.012	+0.029 / +0.018
18	30	0 / -0.033	0 / -0.052	0 / -0.130	+0.015 / +0.002	+0.023 / +0.002	+0.021 / +0.008	+0.028 / +0.015	+0.035 / +0.022
30	40	0 / -0.039	0 / -0.062	0 / -0.160	+0.018 / +0.002	+0.027 / +0.002	+0.025 / +0.009	+0.033 / +0.017	+0.042 / +0.026
40	50	0 / -0.046	0 / -0.074	0 / -0.190	+0.021 / +0.002	+0.032 / +0.002	+0.030 / +0.011	+0.039 / +0.020	+0.051 / +0.032
50	65	0 / -0.054	0 / -0.087	0 / -0.220	+0.025 / +0.003	+0.038 / +0.003	+0.035 / +0.013	+0.045 / +0.023	+0.059 / +0.037
65	80	0 / -0.063	0 / -0.100	0 / -0.250	+0.028 / +0.003	+0.043 / +0.003	+0.040 / +0.015	+0.052 / +0.027	+0.068 / +0.043
60	100	0 / -0.072	0 / -0.115	0 / -0.290	+0.033 / +0.004	+0.050 / +0.004	+0.046 / +0.017	+0.060 / +0.031	+0.079 / +0.050
100	120	0 / -0.081	0 / -0.130	0 / -0.320	+0.036 / +0.004	+0.056 / +0.004	+0.052 / +0.020	+0.066 / +0.034	+0.088 / +0.056
120	140	0 / -0.089	0 / -0.140	0 / -0.360	+0.040 / +0.004	+0.061 / +0.004	+0.057 / +0.021	+0.073 / +0.037	+0.098 / +0.062
140	160	0 / -0.097	0 / -0.155	0 / -0.400	+0.045 / +0.005	+0.068 / +0.005	+0.063 / +0.023	+0.080 / +0.040	+0.108 / +0.068

Table 3-10: Additional Selected Limits of Size (Metric)

Basic Size		Holes		Shafts		
Above	Up To & Including	S7	U7	c9	s6	u6
mm	mm	mm	mm	mm	mm	mm
0	3	-0.014 / -0.024	-0.018 / -0.028	-0.060 / -0.085	+0.020 / +0.014	+0.024 / +0.018
3	6	-0.015 / -0.027	-0.019 / -0.031	-0.070 / -0.100	+0.027 / +0.019	+0.031 / +0.023
6	10	-0.017 / -0.032	-0.022 / -0.037	-0.080 / -0.116	+0.032 / +0.023	+0.037 / +0.028
10	18	-0.021 / -0.039	-0.026 / -0.044	-0.095 / -0.138	+0.039 / +0.028	+0.044 / +0.033
18	24	-0.027 / -0.048	-0.033 / -0.054	-0.110 / -0.162	+0.048 / +0.035	+0.054 / +0.041
24	30	-0.027 / -0.048	-0.040 / -0.061	-0.110 / -0.162	+0.048 / +0.035	+0.061 / +0.048
30	40	-0.034 / -0.059	-0.051 / -0.076	-0.120 / -0.182	+0.059 / +0.043	+0.076 / +0.060
40	50	-0.034 / -0.059	-0.061 / -0.086	-0.130 / -0.192	+0.059 / +0.043	+0.086 / +0.070
50	65	-0.042 / -0.072	-0.076 / -0.106	-0.140 / -0.214	+0.072 / +0.053	+0.106 / +0.087
65	80	-0.048 / -0.078	-0.091 / -0.121	-0.150 / -0.224	+0.078 / +0.059	+0.121 / +0.102
80	100	-0.058 / -0.093	-0.111 / -0.146	-0.170 / -0.257	+0.093 / +0.071	+0.146 / +0.124
100	120	-0.066 / -0.101	-0.131 / -0.166	-0.180 / -0.267	+0.101 / +0.079	+0.166 / +0.144
120	140	-0.077 / -0.117	-0.155 / -0.195	-0.200 / -0.300	+0.177 / +0.092	+0.195 / +0.170
140	160	-0.085 / -0.125	-0.175 / -0.215	-0.210 / -0.310	+0.125 / +0.100	+0.215 / +0.190
160	180	-0.093 / -0.133	-0.195 / -0.235	-0.230 / -0.330	+0.133 / +0.108	+0.235 / +0.210
180	200	-0.105 / -0.151	-0.219 / -0.265	-0.240 / -0.355	+0.151 / +0.122	+0.265 / +0.236

continued on next page

Table 3-10: Additional Selected Limits of Size (Metric) (Continued)

Basic Size		Holes		Shafts		
Above	Up To & Including	S7	U7	c9	s6	u6
mm	mm	mm	mm	mm	mm	mm
200	225	-0.113 -0.159	-0.241 -0.287	-0.260 -0.375	+0.159 +0.130	+0.287 +0.258
225	250	-0.123 -0.169	-0.267 -0.313	-0.280 -0.395	+0.169 +0.140	+0.313 +0.284
250	280	-0.138 -0.190	-0.295 -0.347	-0.300 -0.430	+0.190 +0.158	+0.347 +0.315
280	315	-0.150 -0.202	-0.330 -0.382	-0.330 -0.460	+0.202 +0.170	+0.382 +0.350
315	355	-0.169 -0.226	-0.369 -0.426	-0.360 -0.500	+0.226 +0.190	+0.426 +0.390
355	400	-0.187 -0.244	-0.414 -0.471	-0.400 -0.540	+0.244 +0.208	+0.471 +0.435
400	450	-0.209 -0.272	-0.467 -0.530	-0.440 -0.595	+0.272 +0.232	+0.530 +0.490
450	500	-0.229 -0.292	-0.517 -0.580	-0.480 -0.635	+0.292 +0.252	+0.580 +0.540

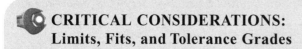

CRITICAL CONSIDERATIONS:
Limits, Fits, and Tolerance Grades

- Factors like long engagement lengths, temperature, and lubrication will affect fit.
- Standard fits are an excellent starting point, but are no substitute for careful analysis of tolerances.
- In a force fit assembly, the elastic limit of the parts must not be exceeded. Calculate the resultant stresses for all force fits to ensure proper grip.

BEST PRACTICES: Limits, Fits, and Tolerance Grades

- When choosing tolerances for holes, an H designation is preferred.
- Use the most generous grades and tolerances possible to ease manufacture.
- Standard fits can be used to calculate the total clearance or interference for a desired result. When designing around an item with given limits of size, those values can be applied to calculate the limits of size for the mating part.

TOLERANCES ON DRAWINGS, AND GD&T

Section

3.2

Choosing tolerances and representing those values on drawings are critical steps of the design process. Choosing and analyzing tolerances is addressed in Sections 3.1 and 3.3 of this chapter. Communicating the desired tolerances and design intent can be simplified using implied tolerances, as well as Geometric Dimensioning and Tolerancing (GD&T).

RECOMMENDED RESOURCES

- L. Foster, *Geo-Metrics III*, Addison-Wesley Publishing Company, New York, NY, 1994
- Oberg, Jones, Horton, Ryffel, *Machinery's Handbook*, 28th Ed., Industrial Press, New York, NY, 2008
- **ANSI Y14.5M:** "Dimensioning and Tolerancing"
- **ANSI B4.1:** "Preferred Limits and Fits for Cylindrical Parts"

IMPLIED TOLERANCES

To simplify drawings, most drawing title blocks contain a list of implied tolerance values. These tolerance values are applied to all dimensions unless otherwise specified. Tables 3-11 and 3-12 give some commonly used implied tolerances, but any tolerances may be applied to a drawing as implied tolerances. Wider tolerances generally reduce manufacturing cost, so generous implied tolerances are preferred if possible.

Table 3-11: Common Implied Tolerances (Inch)

Dimension Decimals	LINEAR TOLERANCE (inch)
X/X	±0.032
X.X	±0.015
X.XX	±0.010
X.XXX	±0.005

ANGULAR TOLERANCE
±0.5°

Table 3-12: Common Implied Tolerances (Metric)

	LINEAR TOLERANCE					
Length (mm)	0.5 - 6	6 - 30	30 - 120	120 - 400	400 - 1000	1000 - 2000
ISO Coarse	±0.2	±0.5	±0.8	±1.2	±2	±3
ISO Medium	±0.1	±0.2	±0.3	±0.5	±0.8	±1.2
ISO Fine	±0.05	±0.1	±0.15	±0.2	±0.3	±0.5

	ANGULAR TOLERANCE			
Shorter Side Length (mm)	0 - 10	10 - 50	50 - 120	120 - 400
	±1°	±0.5°	±0.32°	±0.16°

GEOMETRIC DIMENSIONING AND TOLERANCING (GD&T)

The use of Geometric Dimensioning and Tolerancing (GD&T) is the subject of many excellent books. GD&T allows the user to specify tolerances and relationships based on physical features and can yield drawings that are truer to the design intent that traditional Cartesian tolerancing. This section will serve as a basic introduction to the subject. Refer to the recommended resources for a full treatment of the proper use of GD&T and its symbols.

Datums

GD&T uses datums and a system of symbols to communicate the relationships between and tolerances of part features and surfaces. Datums are theoretically exact points, axes, or planes that are used as references to define the location and orientation of features on a part. Their typical appearance is illustrated in Figure 3-2. Although datums are usually associated with physical features, they are theoretical and have no tolerance or deviation from ideal even if the actual feature deviates. Datums should be selected to represent the function and mating relationship of a part. When selecting datums, it is helpful to use the following procedure:

1. Select datum A to be the primary constraining surface to contact the mating part when this part is placed into an assembly. This is the primary functional datum. For convenience, this surface is often a flat plane. Often this will be the bottom surface of a part, or a side surface if the part attaches on its side.

2. Select datum B to be the second constraining feature to locate the part to another part when the part is placed into an assembly. This datum is often a dowel hole or other locating feature. This is the secondary functional datum.

3. Select datum C to be the third constraining feature (if there is one) to contact the mating part when this part is placed into an assembly. This is the tertiary functional datum. This is often a slotted hole for a dowel pin. When all three datums are in use to constrain the part, it should not be able to move or rotate. It should be fully constrained.

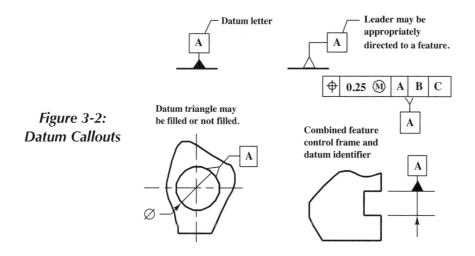

Figure 3-2:
Datum Callouts

It is possible to have more than three datums on a part, and it is also possible to have only one or two datums on a part. The typical case uses three datums. Datums can be patterns of features, such as a pair of dowel holes. In a case where datum A is a planar surface and datum B is a pair of holes, datum C is rarely needed to fully define the part because datum B can be used for location and orientation purposes. Orientation can be related to the imaginary line between the two holes in the pattern defining datum B, and location can be related to the center of either hole.

Symbols

Geometric characteristic symbols are used to specify feature and surface characteristics like orientation, location, shape, symmetry, and runout. These symbols are used as part of a feature control frame represented by a rectangle. This feature control frame contains the geometric characteristic symbol, a total tolerance, any modifiers, and the datums referenced in their order of significance. Figure 3-3 illustrates a typical feature control frame and its contents.

Table 3-13 provides geometric symbols governed by ANSI Y14.5M, which is the unified U.S. standard for both inch and metric units. For a complete coverage of ISO standards governing GD&T, the following standards are required:

- ISO 129: Technical Drawings General Principles
- ISO 2768: General Geometrical Tolerances
- ISO 8015: Fundamental Tolerance Principle
- ISO 406: Linear and Angular Dimensions
- ISO 5459: Datums and Datum Systems
- ISO 2692: Maximum Material Principle
- ISO 2692: Least Material Principle
- ISO 1101: Tolerances of Form, Orientation, Location and Run-Out
- ISO 5458: Positional Tolerancing
- ISO 3040: Cones
- ISO 1660: Profiles
- ISO 10578: Projected Tolerance Zones
- ISO 10579: Non-Rigid Parts
- ISO 7083: Symbols Proportions

When tolerances of position or profile reference datums, the two are related using "basic" dimensions. These dimensions have their value enclosed in a box. Basic dimensions are interpreted as nominal dimensions with no tolerance. The tolerance governing the feature's position(s) and orientation(s) is

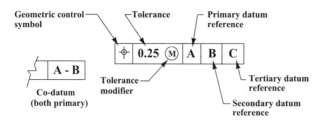

Figure 3-3: Feature Control Frame

applied through the feature control frame attached to the feature that is being positioned using the basic dimension.

The following are descriptions of the most commonly used geometric symbols (Table 3-13) and their interpretation:

<u>Straightness</u> is a specification applied mainly to cylindrical features, and less commonly to flat features like rectangular bars. In the case of a cylindrical feature, it defines a tolerance zone within which the longitudinal elements of the feature must lie. Straightness is a tolerance of form, not position, and therefore does not reference any datums. Figure 3-4 demonstrates the straightness callout and its measurement.

Table 3-13: ANSI and ISO Geometric Symbols

Symbol for	ANSI Y14.5M	ISO	Symbol for	ANSI Y14.5	ISO	Symbol for	ANSI Y14.5M	ISO
Straightness	▬	▬	Circular Runout[a]	↗	↗	Feature Control Frame	⊕ Ø0.5 Ⓜ A B C	⊕ Ø0.5 Ⓜ A B C
Flatness	▱	▱	Total Runout[a]	↗↗	↗↗	Datum Feature[a]	🔺Ａ	🔺 or 🔺Ａ
Circularity	○	○	At Maximum Material Condition	Ⓜ	Ⓜ	All Around - Profile	⊕	⊕ (proposed)
Cylindricity	⌭	⌭	At Least Material Condition	Ⓛ	Ⓛ	Conical Taper	▷	▷
Profile of a Line	⌒	⌒	Regardless of Feature Size	NONE	NONE	Slope	◿	◿
Profile of a Surface	⌓	⌓	Projected Tolerance Zone	Ⓟ	Ⓟ	Counterbore/Spotface	⌴	⌴ (proposed)
Angularity	∠	∠	Diameter	Ø	Ø	Countersink	⌵	⌵ (proposed)
Perpendicularity	⊥	⊥	Basic Dimension	50	50	Depth/Deep	↧	↧ (proposed)
Parallelism	//	//	Reference Dimension	(50)	(50)	Square (Shape)	□	□
Position	⊕	⊕	Datum Target	Ⓐ1̸Ø6	Ⓐ1̸Ø6	Dimension Not to Scale	15	15
Concentricity/Coaxiality	◎	◎	Target Point	✕	✕	Number of Times/Places	8X	8X
Symmetry	≡	≡	Dimension Origin	◑►	◑►	Arc Length	⌒105	⌒105
Radius	R	R	Spherical Radius	SR	SR	Spherical Diameter	SØ	SØ
Between[a]	◄►	None	Controlled Radius	CR	None	Statistical Tolerance	⟨ST⟩	None

[a] Arrowheads may be filled in.

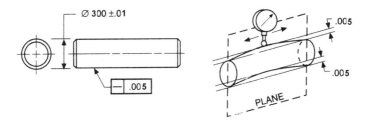

Figure 3-4: Straightness Callout and Interpretation

Flatness is a specification applied to flat surfaces. It defines a tolerance zone between two parallel planes. All elements of the actual surface must fall within these two parallel planes. Flatness is a tolerance of form, not position. As a result, it does not reference any datums, and the planes defining the tolerance zone do not need to be parallel to any datums. Figure 3-5 illustrates the flatness callout and its measurement.

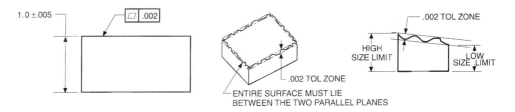

Figure 3-5: Flatness Callout and Interpretation

Circularity is a specification applied to cylindrical, conical, or spherical surfaces. On a cylindrical feature, it defines a circular tolerance zone on a plane perpendicular to the axis of the cylinder. An infinite number of planes can be assumed. The feature surface at each planar cross section must fall within the tolerance zone. This tolerance does not define the relationship between the cross sections at different planes. Circularity is a tolerance of form, not position, and therefore does not reference any datums. Figure 3-6 illustrates circularity tolerance for a conical feature.

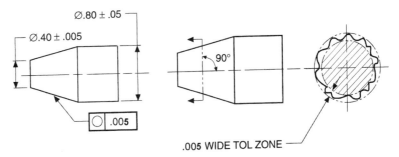

Figure 3-6: Circularity Callout and Interpretation

Cylindricity is a specification applied to a cylindrical feature. It defines a cylindrical tolerance zone with a straight axis around the cylinder, within which all points on the feature's surface must lie. Cylindricity is a tolerance of form, not position. As a result, it does not reference any datums. Figure 3-7 illustrates the cylindricity callout and meaning.

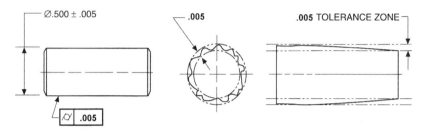

Figure 3-7: Cylindricity Callout and Interpretation

Angularity is a specification applied to a flat surface, axis, or midline of a feature. It defines a tolerance zone between two parallel planes at the specified angle from a datum. The angle will be given as a basic dimension. All points on the surface must fall between the tolerance planes. Angularity is a tolerance of orientation relative to a datum. Figure 3-8 illustrates angularity tolerance.

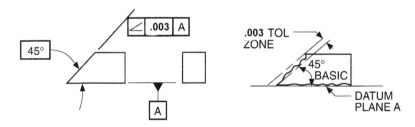

Figure 3-8: Angularity Callout and Interpretation

Perpendicularity is a specification applied to a flat surface, axis, or midline of a feature. It defines a tolerance zone between two parallel planes perpendicular to a datum. All points on the surface must fall between the tolerance planes. Perpendicularity is a tolerance of orientation relative to a datum. Figure 3-9 illustrates perpendicularity tolerance.

Parallelism is a specification applied to a flat surface, axis, or midline of a feature. It defines a tolerance zone between two parallel planes parallel to a datum. All points on the surface must fall between the tolerance planes. Parallelism is a tolerance of orientation to a datum. Figure 3-10 shows the parallelism callout and its meaning.

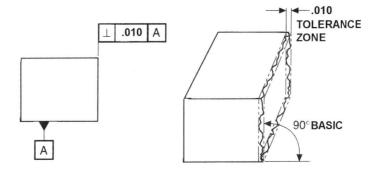

Figure 3-9: Perpendicularity Callout and Interpretation

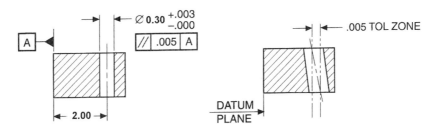

Figure 3-10: Parallelism Callout and Interpretation

Concentricity is a specification applied to cylindrical or spherical features. It defines a tolerance zone about the axis of the feature (or center point, in the case of a sphere) within which all center points of the feature must lie. Concentricity is a tolerance of location relative to a datum axis or center point. An illustration of this is shown in Figure 3-11.

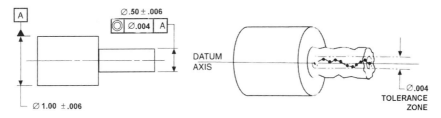

Figure 3-11: Concentricity Callout and Interpretation

Circular runout is a specification applied to cylindrical features, or features with round cross section. It defines a circular tolerance zone on a plane perpendicular to the datum axis within which all elements of the feature surface intersecting that plane must lie. An infinite number of planes can be assumed. Circular runout, unlike total runout, does not relate measurements at each section to one another. The part will be rotated about the datum in

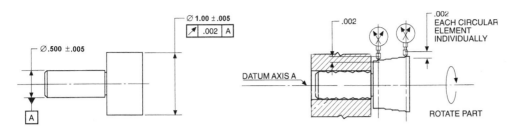

Figure 3-12: Circular Runout Callout and Interpretation

order to measure runout. Circular runout is a functional tolerance that relates the circular form at each cross section to one or more (concentric) datum axes. Figure 3-12 illustrates circular runout and its measurement.

Total runout is a specification applied to cylindrical features. It defines a cylindrical tolerance zone parallel to the datum axis within which all elements of the feature surface must lie. The part will be rotated about the datum in order to measure runout. Total runout is a functional tolerance that relates the cylindrical form to one or more (concentric) datum axes. Figure 3-13 illustrates total runout and its measurement. This specification is particularly useful when dimensioning rotating shafts that carry components like bearings or gears that are sensitive to misalignment.

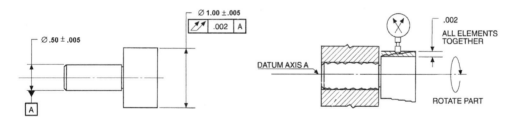

Figure 3-13: Total Runout Callout and Interpretation

Profile of a surface is a specification applied to surfaces. The surfaces can be of complex shape and must be fully defined using basic dimensions. Profile of a surface defines a tolerance zone centered on the nominal surface and following the surface shape. All points on the surface must fall within the two boundary planes. Unilateral tolerancing is possible, so consult the recommended resources for more information. Profile of a surface is a tolerance of form that may or may not reference datums. When datums are referenced, the profile form and relative position are both controlled. This specification is very powerful in that regard. Figure 3-14 illustrates this callout.

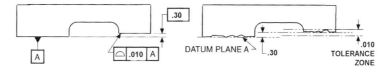

Figure 3-14: Profile of a Surface Callout and Interpretation

Position is a specification applied to center points, midplanes, or axes of features. It is most commonly applied to holes and hole patterns. Position specification defines a tolerance zone around the feature's nominal center. This tolerance zone is cylindrical in the case of hole features, and rectangular in the case of flat features. The center, midplane, or axis of the feature must lie within the tolerance zone. Position tolerance can be applied to groups of identical features and, in such a case, the tolerance zone is set up for each individual feature in the pattern. Position is a tolerance of location relative to one or more datums. Basic dimensions must be used to provide nominal location for the features bearing position tolerances. Figure 3-15 illustrates a typical position callout for holes, and Figure 3-16 shows some possible variations allowable within position tolerance

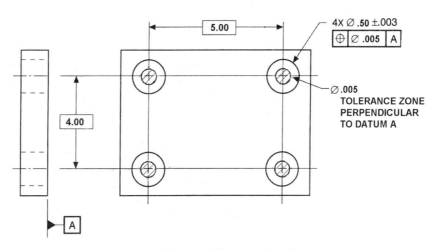

Figure 3-15: Position Callout and Tolerance Zones

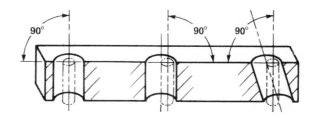

Figure 3-16: Position Tolerance Zones and Allowable Variation of a Hole Axis

zones. Compound tolerance frames are an advanced technique that may be used to control position of features within a pattern while allowing a separate tolerance to control the location of the pattern center relative to the datums. This powerful technique is illustrated in the recommended resources as well as in Chapter 4 of this book.

Modifiers

Tolerance modifiers include maximum material condition, minimum material condition, and regardless of feature size. Modifers may not be added to runout, concentricity, or symmetry specifications. Maximum material condition (MMC) is the condition in which a feature is at the limit of size corresponding to the maximum material left on the part. For a hole, MMC corresponds to the smallest hole within the stated limits of size. For an external diameter, MMC corresponds to the largest diameter within the stated limits of size. Least material condition (LMC) is the condition in which a feature is at the limit of size corresponding to the least material left on the part. LMC for a hole is the largest hole within the limits of size, and for an external diameter is the smallest diameter within the limits of size.

MMC and LMC are used to modify a tolerance or datum reference based on the size of the feature as produced, rather than its theoretical size. Regardless of feature size (RFS) indicates that the tolerance or datum reference applies to a nominal feature and not to the feature as produced. RFS is assumed in all cases unless otherwise stated. MMC is commonly used, whereas LMC is seldom used. When the MMC modifier is present, the tolerance is read as "tolerance when feature is at maximum material condition." The use of MMC permits additional tolerance when the considered feature, as produced, departs from its maximum material condition. Consult the recommended resources and Section 3.3 for guidance on the proper use of modifiers.

CRITICAL CONSIDERATIONS:
Tolerances on Drawings and GD&T

- All dimensions on a drawing must have tolerances specified, either implied or explicit.
- Dimensions and tolerances should convey design intent and functional relationships between features and surfaces.
- Analyze all critical tolerances to ensure proper fit and function.

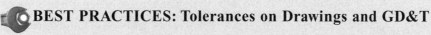

BEST PRACTICES: Tolerances on Drawings and GD&T

- Use standard symbols when applicable rather than notes. Standard symbols have clear and universal meaning, whereas notes may be misunderstood.
- Specify the loosest tolerances possible to save cost and enable a choice of manufacturing methods. This may require explicit tolerances that are looser than the drawing's implied tolerances in some cases.
- When checking drawings, a useful technique with paper drawings is to use a highlighting marker for checked dimensions and a red pen to make changes. The title block should also be checked, including any implied tolerance information.
- Apply position tolerance to all holes, and dimension their locations with basic dimensions.
- Threaded features and tapped holes are normally dimensioned with the modifier "Regardless of Feature Size (RFS)," and the tolerance is applied to the axis of the thread derived from the pitch cylinder. Exceptions to this practice must be noted on the drawing.
- When applicable, apply the modifier "Maximum Material Condition (MMC)" to allow more deviation when parts to be fit together are not produced at the maximum material condition limits of size. This can save cost by allowing more deviation while ensuring proper fit.

Section

3.3

TOLERANCE STACK-UPS
Written by Charles Gillis, P.E.

Every dimension on every feature on every mechanical component has variation. The allowable variation is specified by the designer through tolerances associated with each dimension. Understanding the effects of these variations on the assembly and assigning appropriate tolerances to dimensions requires performing tolerance stack-ups. Sometimes referred to as tolerance analysis or tolerance assignment, performing stack-ups bring together understanding of manufacturing processes and dimensioning standards (e.g. ASME Y14.5) to meet the assembly's functional requirements; they are a critical element of good design practice.

The choice of tolerance is as important as any other design choice. Tolerances must not be chosen arbitrarily, but rather with good understanding of assembly requirements, manufacturing process capabilities, and cost.

Good designs allow the largest tolerances possible to achieve the functional requirements. Manufacturing methods evolve and improve over time, and larger tolerances give manufacturers greater flexibility to choose methods: part routing, machine tool choice, setups, etc. Overly restrictive tolerances tie the hands of manufacturers and drive costs up. Excessive precision may meet the functional requirements, but is a very poor design choice.

Tolerance stack-up calculations are performed during the design phase to understand sources of variation within a physical assembly, for sensitivity analysis, and to verify that the design intent has been captured on dimensional specifications (detail drawings). Performing stack-up calculations allows designers to assign tolerances based on manufacturing capability, to determine assembly process capability, and to implement process control. This section presents the reasons and methods for conducting tolerance stack-up calculations, including several different mathematical approaches and the appropriateness of each approach. The purpose is to enable the reader to understand the effects of tolerance stack-ups on design choices, ultimately enabling better design choices.

RECOMMENDED RESOURCES

- D. Madsen and D. Madsen, *Geometric Dimensioning and Tolerancing, 8th Edition*, Goodheart-Willcox, Tinley Park, Il, 2009
- A. Newmann and S. Newmann, *GeoTol Pro - A Practical Guide to Geometric Tolerancing per ASME Y14.5-2009*, Society of Manufacturing Engineers, Dearborn, MI, 2009
- ASME Y14.5 - 2009: *Dimensioning and Tolerancing*

DESIGN PRACTICE

Good design practice involves an iterative process:

1. Determine which component dimensions contribute to critical assembly dimensions (the tolerance stack-up chain).
2. Assign preliminary tolerances.
3. Analyze the assembly tolerances.
4. Determine fitness of design, iterate as necessary.

Component dimensions are "in-specification" (in spec) if they are manufactured within their specified tolerances. The combined effect of the individual variations in an assembly may not be in spec, even if the components are. It is sometimes discovered through tolerance analysis that a concept will not achieve the assembly tolerance needed.

When a tolerance stack-up shows the assembly to be "out-of-spec," the designer reduces and/or re-distributes component dimension tolerances using some of the following methods:

1. Redesign the assembly to reduce the number of components in the stack-up chain, or utilize features that can be produced through inherently more precise processes.
2. Change the component dimensioning scheme to better represent assembly method, apply different geometric controls, or reduce the number of dimensions in the stack-up chain.
3. Eliminate fits from the stack-up chain.
4. Utilize precision locating methods (see Chapter 4) to reduce variation introduced by fits.
5. Reassign / reduce component dimension tolerances.

THE TOLERANCE STACK-UP CHAIN

The first step in a tolerance stack-up is to determine the chain of dimensions contributing to the stack-up. Consider the assembly of components in Figure 3-17. In order for the components to be assembled, it is necessary for the tab of the left component to fit within the slot of the right component. The detail drawings illustrating relevant dimensions are shown in Figure 3-18. The designer is interested in the allowable variation on the size of the lower gap of the tab / slot features, as well as the upper gap of the tab / slot features. Two stack-up calculations are required to determine that positive, non-zero gap distances exist on both the top and bottom clearances.

For chains involving only a few dimensions, it may be tempting to take shortcuts. However, an organized approach is the best approach in determining the tolerance stack-up chain. The procedure is as follows:

1. **Understand the assembly.** Analyzing unfamiliar designs may require some time to get oriented.
2. **Gather the detail component drawings**, which may be in process.
3. **Make an assembly sketch**. Hand-drawn is best, as it is often helpful to illustrate gaps and clearances, show components at extreme positions and orientations, and show other dimensions out of scale.
4. **Choose a sign convention** (e.g., positive up, positive to the right).

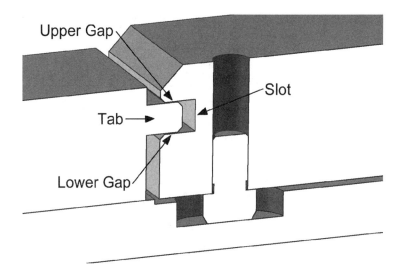

Figure 3-17: Sample Assembly for Analysis

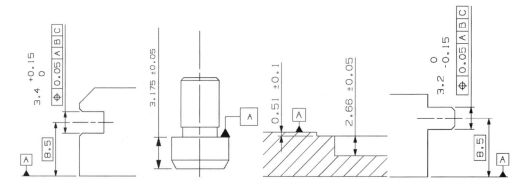

Figure 3-18: Detail Drawings of Component Parts

5. **Determine which dimension needs to be solved**. This requires a good understanding of the problem in order to convert a design concern into a specific dimension that must be discovered. The concern being addressed by the analysis may require that several stack-ups be evaluated. A stack-up to determine if a retaining ring will fit within its groove would seek to calculate the clearance gap dimension (groove width – ring thickness) and ensure it is positive. More complicated problems can be properly defined and understood with the aid of the sketch.

6. **Draw the assembly dimension and label it 'A' for assembly**. The assembly dimension is the unknown dimension you're solving for. It should be the only dimension that is not obtained from the component detail drawings.

7. **Identify contributing dimensions**. Determine which dimensions on the component detail drawings contribute to the size, position, and / or

orientation of the assembly dimension sought. These dimensions may be given with bilateral equal or unequal tolerances, unilateral tolerances, limit tolerances, title block tolerances, or geometric controls. Draw the other dimensions on the assembly sketch per the detail drawings. Label them.

8. **Make each dimension into a vector by assigning an origin and destination**. The choice of origin and destination is arbitrary, but it is important that this convention be maintained once it is chosen. The vector is illustrated with an arrowhead on the destination's extension line and an origin symbol (circle) on the origin's extension line.

9. **Chain all the dimensions together head-to-tail**. Ensure the chain of dimensions forms a complete loop with the destination of the final dimension connecting to the origin of the first. The term "loop" is used even if the vectors may all be one-dimensional (e.g.. all dimensions are up / down, or all dimensions are left / right). Although components are three-dimensional, and specifications made properly using GD&T fully define the component in 3D, tolerance analyses are generally 2D or even 1D problems. Depending on the assembly, the designer determines in steps 1 through 3 whether the stack-up problem is 1D, 2D, or 3D. The simplification to a 2D or 1D problem does not omit any information or oversimplify the analysis — it is done when the problem is, at its core, a 2D or a 1D problem. The vast majority of tolerance stack-up problems within machine design are 1D. In other fields, 2D and 3D analyses may be required more often. The rest of this section focuses on techniques for 1D stack-ups.

10. **Write the stack-up equation**. Write out the tolerance stack-up equation as the algebraic sum of dimensions, following the positive / negative sign convention. Set the sum equal to 0. Re-arrange the equation to solve for assembly dimension A.

A tolerance stack-up chain drawing clearly shows how component dimensions relate to one another and contribute to assembly dimensions. Understanding this relationship allows the design engineer to meet the functionality and cost goals. The creation of the stack-up chain drawing is the first step in analysis or assignment and is best begun when the design of the overall assembly scheme is still preliminary.

The simplified stack-up chain drawing is shown in Figure 3-19. Each dimension has a marked origin and destination, i.e., a "from" denoted by the

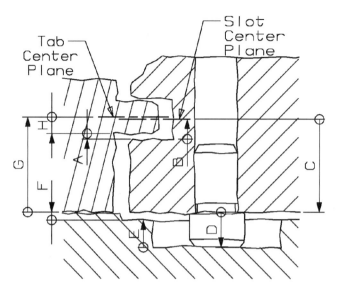

Figure 3-19: Stack-Up Chain Drawing

circular origin symbol and a "to" denoted by an arrow. Each "from" connects to the next "to." The assembly dimension connects the first "to" to the last "from." A clearance dimension is included for each fit.

The stack-up equation is first written by summing the dimensions using the sign convention ("up" arrows positive, "down" arrows negative):

$$+G - H - A + F + E + B - C - D = 0$$

Re-arranging to solve for the assembly dimension A:

$$A = B - C - D + E + F + G - H$$

The assembly dimension A contains both dimensions with positive signs (positive contributors) and negative signs (negative contributors).

The stack-up equation must be created before any calculations can be done. Some tolerance stack-up problems are very simple and can be solved with simple arithmetic without the need to write the stack-up equation. These are sometimes done mentally, or written on scrap paper that is not maintained as part of the engineering records, resulting in a loss of important engineering analysis. The step-by-step process illustrated here assures that a standard methodology is followed, ensuring a commonality of approach regardless of the complexity of the problem. This method is appropriate for both simple problems and complex stack-ups involving dozens of dimensions. It also allows checking and review by other engineers seeking to verify correctness of analysis, to issue engineering changes, or to diagnose problems encountered downstream.

PRELIMINARY TOLERANCE ASSIGNMENT

Assigning small tolerances requires producing high-precision components that increase cost. The designer must be careful not to exceed the capabilities of the manufacturing process (see Section 3.1), or to exceed cost targets. When assembly variation is sufficiently controlled by limiting allowable component variation, interchangeability is achieved. Interchangeability has many benefits to the assembly process, maintenance, and field service, but it requires careful planning.

Interchangeability may be required if a firm uses globally-sourced components and its products are installed, sold, used, or serviced globally; for example if its strategy is to be able to build-to-print from anywhere any time, and ship parts to any field location. Replacement parts and serviceable parts often need to be completely interchangeable as machining-to-fit is not an option. Interchangeable parts enable lower-skilled technicians or users to service, repair, and / or replace parts without precision assembly setup procedures. Whenever it is reasonable to assume that the component in question will be exchanged for an equivalent one, designing for interchangeability should be considered.

When interchangeability is not required, or when assembly variation cannot be adequately controlled by tolerance assignment alone, several design options are available, each requiring greater skill during the assembly process:

- **Setup at assembly**: Adjustability is designed into the assembly, assembly instructions are provided by the design engineer, and final setup dimensions and assembly tolerances are clearly indicated. Standard shop inspection tools, portable coordinate measuring machines, or custom-designed gages may be used to achieve the assembly accuracy required.
- **Assembly fixturing**: Similar to setup at assembly, though fixtures are used to hold some or all components in relationship to one another during the assembly process. Many modular fixturing systems are available, or custom fixturing may be required.

- **Assembly of matched sets**: Components are segregated by the measurements of critical dimensions. Assemblers match small components with large ones, for example — to achieve a tighter assembly variation. The stack-up guides the assembly planner in creating assembly instructions. Ball bearings are produced as matched sets of inner / outer races and balls.
- **Machining at assembly**: After the components are assembled, critical dimensions are achieved by finish machining features that were only roughed in initially. Dowel pins or other permanent fitting methods are often used to lock in the critical assembly dimensions, either by preventing disassembly or by ensuring repeatable re-assembly.

Even though the assembly's dimensional needs have been met, the task of tolerance assignment is not necessarily complete. Consider an example of seven dimensions contributing to a tolerance stack-up chain. One approach is to assign the same tolerance to each dimension in the chain, such that the total assembly tolerance is achieved. This may not be optimal because features that are simple to produce with precision could have smaller tolerances assigned, and features made using more difficult or costly operations may require larger tolerances. A second approach seeks to match each contributing dimension tolerance to the manufacturing capabilities used to produce it. Section 3.1 provides guidelines for common manufacturing processes. Good partnerships between design engineers (function, design, dimensioning) and machine shop / manufacturing personnel (process capability and cost) can prove extremely valuable. Alternative approaches to tolerance assignment incorporate manufacturing vs. tolerance cost information and sensitivity analysis to assign tolerances based on multiplicative effects of their contributions to the assembly tolerance and cost. Analysis gives just one answer, but assignment has no unique solution.

ANALYSIS AND ASSIGNMENT METHODS

Tolerance analysis and tolerance assignment are two sides of the same coin, the only difference being which dimensions are considered inputs and which are outputs. For either approach, four methods of calculation are common:

- Worst-case
- Statistical
- Combined
- Monte Carlo

Worst-Case Method

The most common method of tolerance stack-up for machine design engineers is the worst-case method. The tolerance stack-up calculation is performed twice, resulting in the maximum assembly dimension on the first pass and the minimum value on the second pass. Using this method gives the maximum possible contribution to the final result from each source of variation. Benefits to this method are its simplicity and speed. It leaves the smallest tolerances available for assignment. If the designer is more interested in reducing risk than in accurate predictions, worst-case is the method to use.

Writing all dimensions in the chain in limits-of-fit format provides simple and efficient bookkeeping. This is a shorthand way of writing what are actually two equations, and they are each evaluated. The calculation is performed twice. The maximum value of the assembly dimension, A_{max}, is obtained by adding the maximum (top value) of all positive contributors and subtracting the minimum (bottom value) of all negative contributors. A_{min} is the sum of all minimum (bottom value) positive contributors minus all maximum (top value) negative contributors.

$$\begin{pmatrix} A_{max} \\ A_{min} \end{pmatrix} = \begin{pmatrix} B_{max} \\ B_{min} \end{pmatrix} - \begin{pmatrix} C_{max} \\ C_{min} \end{pmatrix} - \begin{pmatrix} D_{max} \\ D_{min} \end{pmatrix} + \begin{pmatrix} E_{max} \\ E_{min} \end{pmatrix} + \begin{pmatrix} F_{max} \\ F_{min} \end{pmatrix} + \begin{pmatrix} G_{max} \\ G_{min} \end{pmatrix} - \begin{pmatrix} H_{max} \\ H_{min} \end{pmatrix}$$

Here is what the two equations look like when separated:

$$A_{max} = B_{max} - C_{min} - D_{min} + E_{max} + F_{max} + G_{max} - H_{min}$$
$$A_{min} = B_{min} - C_{max} - D_{max} + E_{min} + F_{min} + G_{min} - H_{max}$$

Calculating the example stack-up from Figures 3-17 through 3-19:

$$\begin{pmatrix} A_{max} \\ A_{min} \end{pmatrix} = \begin{pmatrix} 1.775 \\ 1.7 \end{pmatrix} - \begin{pmatrix} 8.525 \\ 8.475 \end{pmatrix} - \begin{pmatrix} 3.225 \\ 3.125 \end{pmatrix} + \begin{pmatrix} 2.71 \\ 2.61 \end{pmatrix} + \begin{pmatrix} 0.61 \\ 0.41 \end{pmatrix} + \begin{pmatrix} 8.525 \\ 8.475 \end{pmatrix} - \begin{pmatrix} 1.6 \\ 1.525 \end{pmatrix}$$

$$\begin{pmatrix} A_{max} \\ A_{min} \end{pmatrix} = \begin{pmatrix} 0.495 \\ -0.155 \end{pmatrix}$$

This calculation has revealed that the gap in question can obtain a negative value — meaning an interference is possible. If this design were still preliminary, the designer could address this problem by adjusting assigned tolerances.

Worst-case analysis is appropriate when the number of components produced is relatively small, and when inspection is used on every component to eliminate components whose dimensions fall outside the specification limits. When components are screened, the designer can assume that component dimensions will be produced in-spec. Worst-case analysis is also used when no quality inspection data are available to allow more conservative prediction of the produced component dimensions.

One drawback of the worst-case analysis is that it is unlikely that the actual measured assembly dimension will take on the calculated minimum or maximum value. It is more likely to be near the middle of the range, as component dimensions produced near their upper limits cancel out the effect of dimensions produced near their lower limits. This drawback becomes more pronounced when more dimensions are included in the stack-up chain.

The stack-up equation can also be used to solve a tolerance assignment problem containing a single unknown. When the assembly dimension is known, and only one dimension needs to have its dimensional limits assigned, the separated worst-case equations can be re-arranged to solve for the final tolerance to assign. For example, when the limits of dimension B are desired,

$$B_{max} = A_{max} + C_{min} + D_{min} - E_{max} - F_{max} - G_{max} + H_{min}$$
$$B_{min} = A_{min} + C_{max} + D_{max} - E_{min} - F_{min} - G_{min} + H_{max}$$

The stack-up equation used for analysis must be separated before it can be re-arranged for use in assignment. This is because of the way the analysis equation is derived: dimension A reflects the total variation of all dimensions. Using this equation for assignment may reveal that there is not enough tolerance left to give to dimension B and other assignment changes must be made. This is indicated by a meaningless result showing a minimum value for dimension B greater than its maximum value. Often, an assignment problem requires assigning tolerances to several dimensions, and the analysis equation cannot be re-arranged to solve for the missing dimension. In this case, an iterative approach is used.

Statistical Method

The statistical stack-up method shifts the focus from the possible to the probable. The resulting predictions of assembly dimensions are more realistic provided the data and/or assumptions are realistic. Statistical stack-ups also leave larger tolerances available for assignment. The analysis gives likely values and ranges depending on confidence intervals. The analysis is not a guarantee of conformance to specs because some outliers may exist. Assembly yield rates or defect rates can be predicted.

When a large number of dimensions contribute to a stack-up, the result is typically distributed according to a normal distribution. Even though individual components will be fully screened at inspection, statistical methods predict the likely assembly dimensions more accurately than the more conservative worst-case method.

Statistical stack-up calculations are appropriate when it is known or can be assumed that the contributing dimensions follow a normal distribution. When a dimension follows a normal distribution, most of the values are clustered around the mean value. It is also possible for a dimension to attain a value outside its specification limits, though with low probability. If no quality inspection reports exist for a component being designed, but data for analogous components and manufacturing processes shows a normal distribution, statistical methods may also be applied.

For large production runs, Statistical Process Control (SPC) is typically used to control the manufacturing process and ensure dimensions are under control. SPC controls the process, not the component. Sample components are inspected at intervals and used to determine whether the process is in control. Inspection is not used to determine which components fall within specifications and which do not on a part-by-part basis. Under these conditions, a normal distribution of dimensions typically results. This will typically be true of commodity hardware components like screws.

Using Variables with a Normal Statistical Distribution

The normal distribution for a given dimension is described by two parameters: the mean (μ) and standard deviation (σ). The mean indicates the average value, and the standard deviation is a measure of the variation within the sample. A larger standard deviation indicates more dispersed values of the dimension. Stacking up dimensions described by a normal distribution results in an assembly dimension with a normal distribution. This distribution can then be compared to assembly spec limits to determine the likelihood of a given assembly being out-of-spec, or the percentage of assemblies that will be produced out-of-spec (defect rate).

For the stack-up equation derived in the previous section,

$$\text{Assembly mean: } \mu_A = \mu_B - \mu_C - \mu_D + \mu_E + \mu_F + \mu_G - \mu_H$$

$$\text{Assembly standard deviation: } \sigma_A = \sqrt{\sigma_B^2 + \sigma_C^2 + \sigma_D^2 + \sigma_E^2 + \sigma_F^2 + \sigma_G^2 + \sigma_H^2}$$

Note that the mean values sum algebraically to the assembly mean whereas the standard deviations combine as the square root of the sum of the squares of the standard deviations. Note also that standard deviations all add as positive terms under the radical. A statistical stack-up using all normally-distributed dimensions is often called the Root Sum of Squares (RSS) method.

For a tolerance assignment problem where an assembly dimension is known, and all dimensions are prescribed but one, the analysis stack-up equation can be rearranged as it was in the worst-case example. If dimension B is desired,

$$\text{Dimension mean: } \mu_B = \mu_A + \mu_C + \mu_D - \mu_E - \mu_F - \mu_G + \mu_H$$

$$\text{Dimension standard deviation: } \sigma_B = \sqrt{\sigma_A^2 - \sigma_C^2 - \sigma_D^2 - \sigma_E^2 - \sigma_F^2 - \sigma_G^2 - \sigma_H^2}$$

As with the worst-cast stack-up, most assignment problems will require simultaneous assignment of tolerances to several dimensions and an iterative approach. If the calculation of the dimension standard deviation results in a negative number under the square root sign, the assignment problem can not be solved with the tolerances currently chosen. There is insufficient variation remaining to be assigned to dimension B.

Using the Standard Normal

The property of the normal curve that is most useful to machine designers is that the area under the curve, bounded between lower and upper points Z_L and Z_U on the Z-axis, represents the percentage of all Z's that will be between Z_L and Z_U. The total area under the normal distribution curve is always equal to 1, or 100%. Calculating the area under a given normal distribution curve can be tedious, so a transformation of variables is used to take advantage of tabulated values.

The standard normal is simply a normal distribution curve with $\mu = 0$ and $\sigma = 1$. The area under the standard normal curve is pictured in Figure 3-20 and tabulated in Table 3-14. The values indicate the area under the curve to the left of Z. The table is read by finding the value of Z by summing the column and row headers and locating the area at the intersection. The area under the curve to the left of $Z = -1.25$ is 0.10565, or 10.565% of the population. This percentage is found at the intersections of row "−1.2" and column "−0.05" corresponding to the value of −1.25.

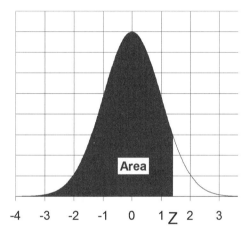

Figure 3-20: Standard Normal Curve

Because the curve is symmetric, the table only gives the area under the curve for half the curve: from the left up through $Z = 0$. To calculate the area for positive values of Z, use the identity:

$$Area(Z) = 1 - Area(-Z)$$

The following example demonstrates how to use the table to determine what percentage of assemblies fall between the upper and lower spec limits of $A_{USL} = 0.35$ and $A_{LSL} = 0$. First, calculate the assembly mean and standard deviation:

$$\mu_A = \mu_B - \mu_C - \mu_D + \mu_E + \mu_F + \mu_G - \mu_H$$
$$= 1.735 - 8.5 - 3.175 + 2.66 + 0.51 + 8.5 - 1.5625$$
$$\mu_A = 0.17$$

$$\sigma_A = \sqrt{\sigma_B^2 + \sigma_C^2 + \sigma_D^2 + \sigma_E^2 + \sigma_F^2 + \sigma_G^2 + \sigma_H^2}$$
$$\sigma_A = \sqrt{(0.0125)^2 + (0.0083)^2 + (0.0167)^2 + (0.0167)^2 + (0.0333)^2 + (0.0083)^2 + (0.0125)^2}$$
$$\sigma_A = 0.046$$

Table 3-14: Area Under the Standard Normal Curve

	0	-0.01	-0.02	-0.03	-0.04	-0.05	-0.06	-0.07	-0.08	-0.09
-3.9	0.00005	0.00005	0.00004	0.00004	0.00004	0.00004	0.00004	0.00004	0.00003	0.00003
-3.8	0.00007	0.00007	0.00007	0.00006	0.00006	0.00006	0.00006	0.00005	0.00005	0.00005
-3.7	0.00011	0.00010	0.00010	0.00010	0.00009	0.00009	0.00008	0.00008	0.00008	0.00008
-3.6	0.00016	0.00015	0.00015	0.00014	0.00014	0.00013	0.00013	0.00012	0.00012	0.00011
-3.5	0.00023	0.00022	0.00022	0.00021	0.00020	0.00019	0.00019	0.00018	0.00017	0.00017
-3.4	0.00034	0.00032	0.00031	0.00030	0.00029	0.00028	0.00027	0.00026	0.00025	0.00024
-3.3	0.00048	0.00047	0.00045	0.00043	0.00042	0.00040	0.00039	0.00038	0.00036	0.00035
-3.2	0.00069	0.00066	0.00064	0.00062	0.00060	0.00058	0.00056	0.00054	0.00052	0.00050
-3.1	0.00097	0.00094	0.00090	0.00087	0.00084	0.00082	0.00079	0.00076	0.00074	0.00071
-3	0.00135	0.00131	0.00126	0.00122	0.00118	0.00114	0.00111	0.00107	0.00104	0.00100
-2.9	0.00187	0.00181	0.00175	0.00169	0.00164	0.00159	0.00154	0.00149	0.00144	0.00139
-2.8	0.00256	0.00248	0.00240	0.00233	0.00226	0.00219	0.00212	0.00205	0.00199	0.00193
-2.7	0.00347	0.00336	0.00326	0.00317	0.00307	0.00298	0.00289	0.00280	0.00272	0.00264
-2.6	0.00466	0.00453	0.00440	0.00427	0.00415	0.00402	0.00391	0.00379	0.00368	0.00357
-2.5	0.00621	0.00604	0.00587	0.00570	0.00554	0.00539	0.00523	0.00508	0.00494	0.00480
-2.4	0.00820	0.00798	0.00776	0.00755	0.00734	0.00714	0.00695	0.00676	0.00657	0.00639
-2.3	0.01072	0.01044	0.01017	0.00990	0.00964	0.00939	0.00914	0.00889	0.00866	0.00842
-2.2	0.01390	0.01355	0.01321	0.01287	0.01255	0.01222	0.01191	0.01160	0.01130	0.01101
-2.1	0.01786	0.01743	0.01700	0.01659	0.01618	0.01578	0.01539	0.01500	0.01463	0.01426
-2	0.02275	0.02222	0.02169	0.02118	0.02068	0.02018	0.01970	0.01923	0.01876	0.01831
-1.9	0.02872	0.02807	0.02743	0.02680	0.02619	0.02559	0.02500	0.02442	0.02385	0.02330
-1.8	0.03593	0.03515	0.03438	0.03362	0.03288	0.03216	0.03144	0.03074	0.03005	0.02938
-1.7	0.04457	0.04363	0.04272	0.04182	0.04093	0.04006	0.03920	0.03836	0.03754	0.03673
-1.6	0.05480	0.05370	0.05262	0.05155	0.05050	0.04947	0.04846	0.04746	0.04648	0.04551
-1.5	0.06681	0.06552	0.06426	0.06301	0.06178	0.06057	0.05938	0.05821	0.05705	0.05592
-1.4	0.08076	0.07927	0.07780	0.07636	0.07493	0.07353	0.07215	0.07078	0.06944	0.06811
-1.3	0.09680	0.09510	0.09342	0.09176	0.09012	0.08851	0.08691	0.08534	0.08379	0.08226
-1.2	0.11507	0.11314	0.11123	0.10935	0.10749	0.10565	0.10383	0.10204	0.10027	0.09853
-1.1	0.13567	0.13350	0.13136	0.12924	0.12714	0.12507	0.12302	0.12100	0.11900	0.11702
-1	0.15866	0.15625	0.15386	0.15151	0.14917	0.14686	0.14457	0.14231	0.14007	0.13786
-0.9	0.18406	0.18141	0.17879	0.17619	0.17361	0.17106	0.16853	0.16602	0.16354	0.16109
-0.8	0.21186	0.20897	0.20611	0.20327	0.20045	0.19766	0.19489	0.19215	0.18943	0.18673
-0.7	0.24196	0.23885	0.23576	0.23270	0.22965	0.22663	0.22363	0.22065	0.21770	0.21476
-0.6	0.27425	0.27093	0.26763	0.26435	0.26109	0.25785	0.25463	0.25143	0.24825	0.24510
-0.5	0.30854	0.30503	0.30153	0.29806	0.29460	0.29116	0.28774	0.28434	0.28096	0.27760
-0.4	0.34458	0.34090	0.33724	0.33360	0.32997	0.32636	0.32276	0.31918	0.31561	0.31207
-0.3	0.38209	0.37828	0.37448	0.37070	0.36693	0.36317	0.35942	0.35569	0.35197	0.34827

To calculate the values for Z we will look up in the table:

$$\text{For } A_{USL}\text{: } Z_{USL} = \frac{A_{USL} - \mu_A}{\sigma_A} = \frac{0.35 - 0.17}{0.046} = 3.911$$

$$\text{For } A_{LSL}\text{: } Z_{LSL} = \frac{A_{LSL} - \mu_A}{\sigma_A} = \frac{0 - 0.17}{0.046} = -3.694$$

The area under the normal curve between A_{LSL} and A_{USL} is the area under the curve to the left of Z_{USL} minus the area under the curve to the left of Z_{LSL}, or:

$$= Area(3.911) - Area(-3.694)$$
$$= \left[1 - Area(-3.911)\right] - Area(-3.694)$$
$$= (1 - 0.00005) - 0.00011$$
$$= 0.99984$$

According to this sample calculation, 99.984% of assemblies will be in-spec, and 0.016% out-of-spec. The standard normal table can also be used to determine spec limits for a given desired defect rate.

Modeling the Distribution

Performing an accurate analysis using the statistical method requires knowledge of the dimensions of the produced parts. This is not always available, and assumptions must be made. Given upper and lower specification limits (USL, LSL), what will be the mean (μ) and standard deviation (σ) of the population? These values can be assumed outright when their meaning is fully understood.

Often the mean and standard deviation are assumed by way of other measures. The process capability index (C_{pk}) is used as an indicator of how well a manufacturing process is capable of producing dimensions on-target between the spec limits. The terms C_{pl} and C_{pu} are the lower and upper process capability indices, respectively.

$$C_{pk} = MIN\left(C_{pl}, C_{pu}\right)$$
$$C_{pl} = \frac{\mu - LSL}{3\sigma}$$
$$C_{pu} = \frac{USL - \mu}{3\sigma}$$

A C_{pk} of 1.0 with a mean (μ) on target (halfway between the specification limits) corresponds to a standard deviation (σ) of one-sixth of the tolerance range ($USL - LSL$), or 99.73% of dimensions in-spec. Larger values of C_{pk} indicate greater control over the process and incur greater manufacturing expense: smaller values, less control at less cost. A centered process with

C_{pk} of 2.0 represents a Six Sigma process, a goal of many corporate quality programs using SPC, and may be assumed by the design engineer purchasing components from reputable and reliable companies. A smaller assumption on C_{pk} would be more conservative.

However, knowing C_{pk} alone is not enough to select both μ and σ. When the terms C_{pl} and C_{pu} are equal, the μ is on target and centered between the spec limits. They need not be equal. A larger upper process capability index occurs when the mean is shifted closer to the LSL; a larger lower process capability index means the mean is shifted toward the USL. When C_{pl} and C_{pu} are known or can be estimated, μ and σ can be calculated:

$$\mu = \frac{LSL\dfrac{C_{pu}}{C_{pl}} + USL}{1 + \dfrac{C_{pu}}{C_{pl}}}$$

$$\sigma = \left|\frac{\mu - LSL}{3C_{pl}}\right|$$

Another method of estimating μ and σ uses the measure *target Z*, or T_z. The target Z is a measure of the off-center nature of the distribution. It is the difference between the mean and the target divided by the standard deviation. Positive T_z indicates dimension shift toward the high end of the specification and negative T_z indicates dimension shift toward the low end. Mean and standard deviation can be calculated given assumed or known T_z and C_{pk}:

$$T_z \geq 0: \ \mu = USL - 3\sigma C_{pk}, \ \sigma = \frac{USL - LSL}{2\left(T_z + 3C_{pk}\right)}$$

$$T_z \leq 0: \ \mu = LSL + 3\sigma C_{pk}, \ \sigma = \frac{LSL - USL}{2\left(T_z - 3C_{pk}\right)}$$

Root Sum of Squares Method

A statistical stack-up using all normally-distributed dimensions is often called the Root Sum of Squares (RSS) method. Like the general statistical method presented here, the Root Sum of Squares (RSS) is a tolerance analysis method used to estimate the variation of the assembly stack-up from tolerances in each of the dimensions, or to assign component tolerances given an allowable assembly variation. The Root Sum of Squares method simplifies some of the mathematics by assuming all dimensions are normally distributed and centered on their targets with $C_{pk} = 1.0$. Under these assumptions, standard deviations are calculated based on spec limits:

Standard deviation of each dimension in the stack-up: $\sigma = \dfrac{USL - LSL}{6}$

The mean and standard deviation of the assembly dimension are the same as in the previous section:

Assembly mean: $\mu_A = \mu_B - \mu_C - \mu_D + \mu_E + \mu_F + \mu_G - \mu_H$

Assembly standard deviation: $\sigma_A = \sqrt{\sigma_B^2 + \sigma_C^2 + \sigma_D^2 + \sigma_E^2 + \sigma_F^2 + \sigma_G^2 + \sigma_H^2}$

The assembly dimension is expected to fall between the limits:

$$\begin{pmatrix} A_{max} \\ A_{min} \end{pmatrix} = \mu_A \pm 3\sigma_A$$

The method's simplifying assumptions allow fast calculations at the expense of preventing the designer from incorporating known dimensional history, changing quality predictions, performing what-if scenarios, and targeting specific defect rates.

Combined Worst-Case / Statistical Method

Statistical tolerancing allows the designer to assign larger tolerances, and is therefore less costly than worst-case assignment. Statistical analysis requires that all input dimensions must fit a statistical distribution; however, this is not reasonable for all contributing dimensions. The most conservative approach would be to use worst-case analysis when any of the input dimensions are not well-represented by statistical distributions. Worst-case analysis requires that all input dimensions be modeled as worst-case, even those meeting the assumptions of statistical tolerancing. The combined method shown here is useful to incorporate the two types of tolerances in a single analysis.

When known quality inspection data are available for only some dimensions in a stack-up, these data can be combined with worst-case tolerances. Results are less conservative than worst-case, meaning generally more tolerance is available for assignment. This method is useful for assignment only, as the results are not accurate for predicting defect rates or probable stack-up distributions, and can be used when these considerations are unimportant.

To perform a combined stack-up, first determine which dimensions will be modeled as worst-case and which will be modeled using statistics. Two assembly distributions are calculated: one using the maximum stack-up of the worst-case dimensions, the other using the minimum.

The method is illustrated with an example. For the stack-up equation of the previous section,

$$A = B - C - D + E + F + G - H$$

B, D, and H are chosen to be represented as normally-distributed, each with centered $C_{pk} = 1.0$.
C, E, F, and G are chosen to be represented as worst-case.

Assembly means:
$$\mu_A |_{MAX} = \mu_B - C_{MIN} - \mu_D + E_{MAX} + F_{MAX} + G_{MAX} - \mu_H$$
$$\mu_A |_{MIN} = \mu_B - C_{MAX} - \mu_D + E_{MIN} + F_{MIN} + G_{MIN} - \mu_H$$

Assembly standard deviation: $\sigma_A = \sqrt{\sigma_B^2 + \sigma_D^2 + \sigma_H^2}$

The two distributions are plotted in Figure 3-21.

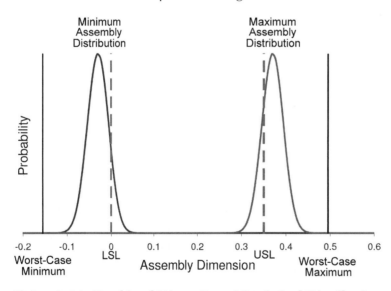

Figure 3-21: Combined Worst-Case / Statistical Distributions

It should be noted that both distributions have different means but the same standard deviation. Note that using the minimum worst-case distribution, there is nearly a 90% likelihood of the assembly dimension falling below the lower spec limit (LSL) of 0. With the maximum worst-case contributors, about 80% will exceed the upper spec limit (USL) of 0.35. These percentages are valid only when the worst-case dimensions stack together at their extremes;

the likelihood of that is a matter for the designer's judgment. Understanding the effects of the assigned tolerances allows the designer to make changes to improve the likelihood of an acceptable result, or loosen component dimensions while fully understanding the risk.

Compare the combined results to the results from the worst-case analysis and statistical analysis. The worst-case analysis predicted minimum and maximum assembly dimensions of –0.155 and 0.495, respectively. These are represented by the solid vertical lines on the graph of Figure 3-21. The combined stack-up predicts that fewer than one part per million will actually exceed those spec limits. The statistical analysis predicted that 99.984% of assembly dimensions will fall between the spec limits, with 0.016% out-of-spec. The combined analysis included some real data used for statistically-represented dimensions, whereas the statistical method used all real data. Even with only partially complete data, the combined analysis is less conservative than the worst-case analysis, though not as accurate as the full statistical analysis.

Monte Carlo Simulation

Monte Carlo simulation is a random-number-based method of tolerance stack-up that attempts to model the randomness that occurs in reality. For a single sample run, each dimension in the stack-up chain is randomly chosen to fit a preselected statistical distribution (normal, uniform, or otherwise), and the dimensions are combined according to the stack-up equation. This process is repeated thousands of times to build a probability distribution for the assembly dimension. The accuracy of a Monte Carlo simulation increases with larger sample sizes at the cost of additional computation time. Reject rates are easily calculated. Monte Carlo simulation can accurately predict the effects of nonlinear stack-up equations and allow for non-normally distributed dimensions.

Monte Carlo simulation is a very effective way to analyze tolerances that follow non-normal distributions, especially ones with highly skewed distributions. Monte Carlo simulation can also properly account for both feature modifiers and datum shift allowed through Geometric Dimensioning and Tolerancing (GD&T). It is one way to understand the variation that naturally exists in the process. Reject rates are easily calculated even though the assembly dimension may not conform to a standard statistical distribution shape. The main drawback is the complexity of modeling the simulation that is required.

So far, this section has covered the basic procedures and math involved with tolerance stack-ups. Four different mathematical approaches have been presented along with their benefits and suggestions on method choice. Readers seeking additional detail are directed to the recommended resources within this section. The statistical method introduced here is quite powerful and can be used with input dimensions with very loose or very tight process control-not necessarily only with Six-Sigma control. Root Sum of Square methods and formulas are published widely, but this is really a simple special case of the general statistical problem. Monte-Carlo simulation is also extremely powerful, can handle any input dimension probability distribution beyond just normal, can handle nonlinear (or linearized) stack-up chains, and can properly account for feature modifiers and datum shift of GD&T.

PRACTICAL APPLICATIONS

Multiple Assembly Conditions

A very common situation facing the analyst is that of the assembly with clearances between components. Clearance gaps that exist between two components involved in a stack-up need to be considered carefully. Starting with a clear understanding of the problem at hand, the analysis usually proceeds by making assumptions about which clearances exist as gaps and which are closed. It is helpful to create one stack-up chain drawing for each different set of assumptions. The particular problem at hand will dictate how many stack-up chain drawings are required. Consider the example in Figure 3-22.

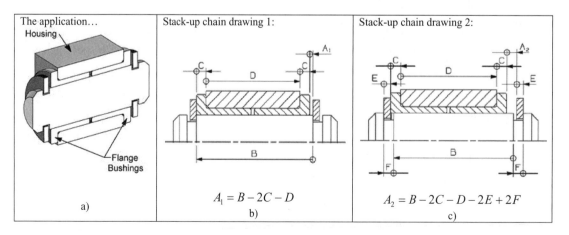

Figure 3-22: Multiple Assembly Conditions

In this assembly, the gap between the outer enclosure formed by the retaining rings and inner envelope formed by the flange bushing – housing – flange bushing sandwich is desired. Two stack-up calculations are required. Calculation 1 (Figure 3-22b) assumes both retaining rings are contacting the inner faces of their slots, and its maximum and minimum values are calculated. Calculation 2 (Figure 3-22c) assumes both retaining rings are contacting the outer faces of their slots. Its maximum and minimum values are calculated. The overall maximum and minimum gaps are given by:

$$\begin{pmatrix} A_{\max} \\ A_{\min} \end{pmatrix} = \begin{pmatrix} MAX\left(A_{1\max}, A_{2\max} \right) \\ MIN\left(A_{1\min}, A_{2\min} \right) \end{pmatrix}$$

This simple example illustrates the technique that is used on assemblies containing gaps.

Plus / Minus Tolerancing and GD&T

Every feature on a part should be fully and uniquely dimensioned and toleranced. Full specification of a feature controls the feature's form, size (if applicable), orientation, and location relative to the part's datum reference frame. Accurate tolerance stack-ups can only be done on parts dimensioned and toleranced using GD&T. Using conventional plus / minus tolerances makes an accurate stack-up impossible. Such tolerancing does not include effects of perpendicularity of holes, shape / form of surfaces, and a number of other factors that can have an immense impact. These types of variation are not explicitly controlled by the specification, and are often interpreted differently by different people or ignored entirely when performing stack-ups. There are far too many inconsistencies and assumptions required to validate parts dimensioned and toleranced using the plus / minus system alone.

Using GD&T is the first requirement for performing a tolerance stack-up. The standard provides graphical and mathematically precise definitions to account for worst cases of size, location, form, etc., and is quite comprehensive, flexible, and powerful. The part's datum reference frame is explicitly defined, derived from the actual produced part geometry. Every feature's tolerance zone is explicitly defined in 3 dimensions with respect to the part's datum reference frame. Tolerance zones are defined by basic dimensions taken from the datum reference frame, and the part's features must fall within tolerance zones. It is not enough to merely use GD&T. Controls must be applied appropriately.

This means using the right datum features and the right geometric controls to match functional requirements.

Understanding the definitions of the datum reference frame, the various geometric controls, the concepts of least and maximum material conditions, feature modifiers, datum modifiers, and the concepts of boundaries allows the design engineer to include geometric controls in stack-ups when it is appropriate to do so. It also allows the design engineer to select the right functional datum reference frame, the right geometric controls, and the right feature modifiers to control the effects of tolerances and make the right assignments.

Form Tolerance on Datum Surface

Figure 3-23 shows a part with its bottom surface controlled by a Flatness callout and designated as Datum A. From the datum definition, the ideal datum plane is defined by the highest points on the actual produced surface. Measurements of part dimensions originate at the datum plane, not the surface. In general, a tolerance stack-up chain involving this component mounted on its primary datum would include the overall ±0.3 mm size tolerance but not the 0.1 mm form tolerance.

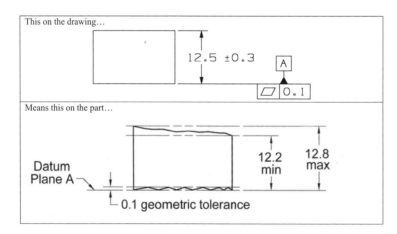

Figure 3-23: Datum Feature

Though datum surface flatness tolerances in a stack-up chain are usually not included in the calculation, there are cases where they should be. The datum derived from a basically planar surface (one designed to be planar) is defined by the highest points on the surface. Measurements are defined to originate at the datum, which is simulated in an inspection process by a granite surface plate or other device of sufficiently precise flatness that

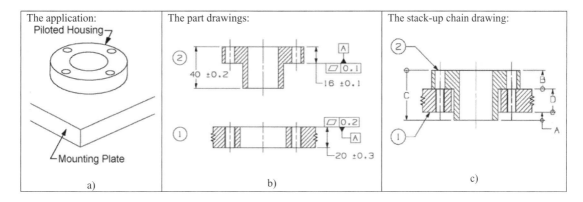

Figure 3-24: Planar Datum Features with Flatness Tolerances

it can be considered flat. When parts assemble together on the high points of their surfaces, flatness specifications are not included in the stack-up chains. Consider the application shown in Figure 3-24.

We are interested in calculating the protrusion of the pilot beyond the back of the mounting plate. Following the techniques presented in this section, the stack-up equation is:

$$A = -B + C - D$$

$$= -\begin{pmatrix} 16.1 \\ 15.9 \end{pmatrix} + \begin{pmatrix} 40.2 \\ 39.8 \end{pmatrix} - \begin{pmatrix} 20.3 \\ 19.7 \end{pmatrix}$$

$$= \begin{pmatrix} 4.6 \\ 3.4 \end{pmatrix}$$

Note that we have not included either part's datum surface flatness tolerance in the calculation. The definition of a datum from the high points on a surface is intended to simulate the mating of the part in its assembly, though the assembly will always be imperfect. In the assembly, the nearly-perfect datum simulator used in inspection is replaced by a real part with real flatness variation. It is theoretically possible, albeit unlikely, for the peaks and valleys of the part to line up perfectly with the peaks and valleys in the surface of the mating part serving as datum simulator. Figure 3-25 illustrates.

Part 1's flange thickness is 15.9 – 16.1 mm, though its local size due to the datum surface's flatness allowance could be measured anywhere from 15.8 – 16.1 mm. Part 2's thickness is 19.7 – 20.3 mm, though its local size could be measured from 19.5 – 20.3 mm. In the unlikely case that the peaks

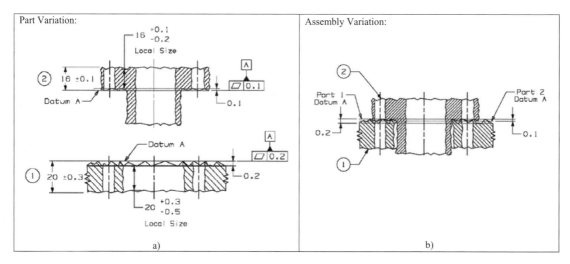

Figure 3-25: Planar Datum Feature Assembly

and valleys in the two imperfect surfaces line up with one another, the smaller flatness value (0.1 mm from part 2) should be included in the stack-up calculation. This is done by using part 2's local size of 15.8 – 16.1 mm in the stack-up calculation, rather than its actual size:

$$B' = \begin{pmatrix} 16.1 \\ 15.8 \end{pmatrix}$$

$$A = -B' + C - D$$

Smaller flatness tolerance of datum feature included:

$$= -\begin{pmatrix} 16.1 \\ 15.8 \end{pmatrix} + \begin{pmatrix} 40.2 \\ 39.8 \end{pmatrix} - \begin{pmatrix} 20.3 \\ 19.7 \end{pmatrix}$$

$$= \begin{pmatrix} 4.7 \\ 3.4 \end{pmatrix}$$

Including the smaller flatness tolerance value represents the worst-case outcome of datum surface out-of-flatness affecting an assembly stack-up, and it may be appropriate to include. A designer seeking to reduce risk would choose to include this variation or some part of it. In the case of two components assembled together having different sizes of their mating surfaces, it may be possible for even greater variation. In this example, part 1 is a very large plate. It is possible that the entire allowable range of flatness variation is used by the plate, and that the area part 1 sits on resides in a "low spot" of the datum surface. In such a situation, the flatness tolerance of the part with the larger mating surface area (part 1) should be included in the stack-up. Replace the value of D with the local size:

$$D' = \begin{pmatrix} 20.3 \\ 19.5 \end{pmatrix}$$

$$A = -B + C - D'$$

One datum feature mating surface area larger than the other:

$$= -\begin{pmatrix} 16.1 \\ 15.9 \end{pmatrix} + \begin{pmatrix} 40.2 \\ 39.8 \end{pmatrix} - \begin{pmatrix} 20.3 \\ 19.5 \end{pmatrix}$$

$$= \begin{pmatrix} 4.8 \\ 3.4 \end{pmatrix}$$

Again, whether or not this represents a likely scenario is up to the judgment of the designer, who should consider material, likely manufacturing methods, whether one or both the parts may deform under assembly loads, and other factors.

Position Tolerance

A position control placed on a feature of size located and oriented with basic dimensions specifies the allowable location and orientation of the feature's axis or center plane. The tolerance zone specified is the total size of the tolerance zone. For the center plane of a feature, the tolerance zone is the region between two parallel planes separated by the total tolerance, located symmetrically about the basic dimension. For axial features, a cylindrical tolerance zone is often specified, defining the zone as a cylindrical region whose diameter is the total tolerance and whose location and orientation are centered and aligned to the basic dimensions locating the feature to the datum reference frame.

When a position control is used, the feature in question must be located and oriented with basic dimensions to the part's datum reference frame. A tolerance stack-up problem may require including the allowable variation on the position of the feature's central axis or central plane to the part's datum reference frame. In this case, the dimension is allowed to vary from the basic dimension by one-half the total tolerance. This is how position control was incorporated in the example of Figures 3-17 through 3-19. The single dimension is shown, separated from the other dimensions of that stack-up equation:

$$\begin{pmatrix} B_{max} \\ B_{min} \end{pmatrix} = \begin{pmatrix} B_{basic} + t/2 \\ B_{basic} - t/2 \end{pmatrix}$$

Position tolerances may also need to be considered in different ways when performing a tolerance stack-up, depending on the specific problem at hand. In this example, the variation on the feature's orientation did not contribute to the total stack-up. In other cases, it may. This effect is explored later in the section.

Maximum Material Condition Modifier

The Maximum Material Condition (MMC) of a feature of size is the maximum amount of material within the stated limits of size. For example, the maximum shaft diameter or minimum hole diameter. The Least Material Condition (LMC) of a feature of size is the least amount of material within the stated limits of size. For example, the minimum shaft diameter or maximum hole diameter. Using a material modifier allows additional bonus tolerance when the feature's size departs from its MMC / LMC. The amount of this bonus is equal to difference between actual produced size and MMC size (for MMC modifier).

The example of Figures 3-17 through 3-19 is revisited. The position control on the slot has been changed to specify its tolerance of 0.05 at maximum material condition. The basic 8.5 mm dimension locates the

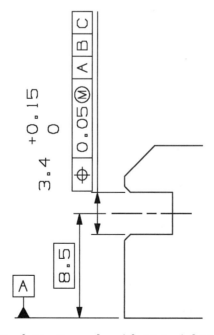

Figure 3-26: Stack-Up Example with Material Modifier

nominal center plane of the tolerance zone. The tolerance zone is the region between two parallel planes, equi-spaced about the 8.5 mm basic, whose width is the total tolerance. This total tolerance is 0.05 when the feature in question is produced at its maximum material condition of 3.4 mm. If the slot width is larger, the tolerance zone increases in size by this size difference. Table 3-15 illustrates the effect on the dimension from datum A to the slot's center plane (dimension C). Because dimension B was defined as one-half the slot width, the width of the slot is 2B. Four values spanning the allowable range of the slot width 2B are given in the leftmost column. Subsequent columns list the given print tolerance at MMC, the bonus tolerance allowed, the total resulting tolerance, the slot basic position, and the slot center plane minimum and maximum positions.

Table 3-15: Effect of Material Modifier on Stack-Up

	2B	Tol @ MMC	Bonus Tol	Total Tol	C_{basic}	C_{min}	C_{max}
MMC →	3.4	0.05	0	0.05	8.5	8.475	8.525
	3.45	0.05	0.05	0.1	8.5	8.45	8.55
	3.5	0.05	0.1	0.15	8.5	8.425	8.575
LMC →	3.55	0.05	0.15	0.2	8.5	8.4	8.6

We can account for this relationship in our stack-up chain calculation. Begin by inspecting the separated worst-case stack-up equations:

$$A_{max} = B_{max} - C_{min} - D_{min} + E_{max} + F_{max} + G_{max} - H_{min}$$
$$A_{min} = B_{min} - C_{max} - D_{max} + E_{min} + F_{min} + G_{min} - H_{max}$$

Note that dimension B_{max} occurs in the same equation with C_{min}, and C_{max} occurs with B_{min}. From the table, we note that when the slot width (2B) attains its maximum value of 3.55, C_{min} equals 8.4. The values of C used in the stack-up are therefore adjusted to reflect the bonus tolerance gained by using the material modifier:

$$C_{max} = 8.525 \text{ when } B = B_{min}$$

$$C_{min} = 8.4 \text{ when } B = B_{max}$$

$$C' = \begin{pmatrix} 8.525 \\ 8.4 \end{pmatrix}$$

Adding a maximum material modifier to both the slot (C) position, and the tab (G) position, the stack-up calculation is re-written and re-calculated:

$$\begin{pmatrix} A_{max} \\ A_{min} \end{pmatrix} = \begin{pmatrix} 1.775 \\ 1.7 \end{pmatrix} - \begin{pmatrix} 8.525 \\ 8.4 \end{pmatrix} - \begin{pmatrix} 3.225 \\ 3.125 \end{pmatrix} + \begin{pmatrix} 2.71 \\ 2.61 \end{pmatrix} + \begin{pmatrix} 0.61 \\ 0.41 \end{pmatrix} + \begin{pmatrix} 8.6 \\ 8.475 \end{pmatrix} - \begin{pmatrix} 1.6 \\ 1.525 \end{pmatrix}$$

$$\begin{pmatrix} A_{max} \\ A_{min} \end{pmatrix} = \begin{pmatrix} 0.645 \\ -0.155 \end{pmatrix}$$

Using the maximum material modifiers on the slot and tab position callouts has increased A_{max}, resulting in possible additional clearance at the slot / tab gap. The minimum gap, A_{min}, has not changed.

Least Material Condition Modifier

The same approach is used to examine the effect of specifying the slot and tab positions at least material condition. Table 3-16 shows the allowable slot position (dimension C) under various actual produced sizes of the slot width (2B). The tab's allowable position G can similarly be calculated.

Table 3-16: Effect of Least Material Condition Modifier on Stack-Up

	2B	Tol @ LMC	Bonus Tol	Total Tol	C_{basic}	C_{min}	C_{max}
MMC →	3.4	0.05	0.15	0.2	8.5	8.4	8.6
	3.45	0.05	0.1	0.15	8.5	8.425	8.575
	3.5	0.05	0.05	0.1	8.5	8.45	8.55
LMC →	3.55	0.05	0	0.05	8.5	8.475	8.525

Adding a least material modifier to both the slot (C) position, and the tab (G) position, the stack-up calculation is re-written and re-calculated:

$$\begin{pmatrix} A_{max} \\ A_{min} \end{pmatrix} = \begin{pmatrix} 1.775 \\ 1.7 \end{pmatrix} - \begin{pmatrix} 8.525 \\ 8.4 \end{pmatrix} - \begin{pmatrix} 3.225 \\ 3.125 \end{pmatrix} + \begin{pmatrix} 2.71 \\ 2.61 \end{pmatrix} + \begin{pmatrix} 0.61 \\ 0.41 \end{pmatrix} + \begin{pmatrix} 8.6 \\ 8.475 \end{pmatrix} - \begin{pmatrix} 1.6 \\ 1.525 \end{pmatrix}$$

$$\begin{pmatrix} A_{max} \\ A_{min} \end{pmatrix} = \begin{pmatrix} 0.495 \\ -0.305 \end{pmatrix}$$

Using the least material modifiers on the slot and tab position callouts has decreased A_{min}, resulting in reduced clearance at the slot / tab gap. In the current example, the minimum clearance was already negative (indicating an interference). The least material modifiers have made the interference worse. The maximum gap, A_{max}, has not changed.

Recall the functional requirement that a minimum gap of zero (0) must always be present. The maximum material modifier is the proper feature modifier to use for both slot and tab features in this application. It provides the maximum position tolerance to the machinist, increasing fabrication yield and reducing cost, while preserving the gap requirement given any condition of slot and tab size and position. When the slot's width increases beyond its maximum material condition, its allowable position tolerance increases by the same amount. In this way, slot tolerance and size play against each other to maintain stack-up control.

Two Dimensional Stack-Up: Location and Orientation Variation

Many tolerance stack-up problems are, in essence, one-dimensional (1D) chains. Two-dimensional (2D) effects do need to be considered for some problems. The details of the approach taken always begin with the stack-up chain drawing, and can vary as widely as problems themselves vary. In subsequent sections, commonly-encountered 2D stack-up effects are explored. The presentation of more complex analyses in 2D and 3D is beyond the scope of this text.

Consider the housing part in Figure 3-27. Its application is to be fastened to its base part on its bottom planar surface serving as its primary datum feature (datum A). The part mates over a precision locating pin pressed into the base part, using a hole serving as the housing's secondary datum feature (datum B). This provides location in the x and y directions. The housing's second hole mates to a diamond pin pressed into the base part, which restricts the final orientation degree of freedom, and serves as the housing's tertiary datum feature (datum C). The housing's datum reference frame is selected to match the function of the part in its assembly. The slot feature locates and guides a slide that provides a linear translating kinematic joint as part of a mechanism. The housing needs to accurately position the slide and the slide's features within the assembly.

The slot feature is specified by a size tolerance and position callout. The slot feature itself is designed as two parallel planes, though the actual produced feature will not be two parallel planes. Neither surface will be a true plane, and both will have variation in form, orientation, and position.

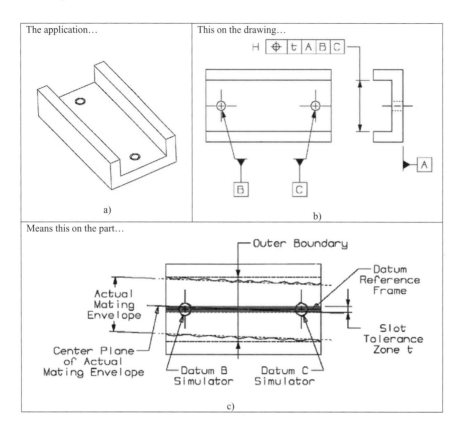

Figure 3-27: Location and Orientation

Any measurement of the slot's local size must fall within the size limits. The variation in size could result in overall taper, bow, sweep, twist of the slot, or any other variation. At the lower limit of size, the maximum material condition (MMC) of the slot, no such variation of form is permitted; and the only shape permitted is two perfectly parallel planes. Functionally, the limits of size imposed by the size specification alone dictate that a block of perfect form at the slot width's MMC fit within the slot with no interference.

The slot feature's position callout specifies a tolerance zone between a pair of parallel planes separated by the tolerance t and extending over the length of the feature. The position tolerance callout specifies that the slot's center plane must fall within the tolerance zone. The center plane of the slot feature is defined as the center plane of the slot's actual mating envelope, the most widely-separated pair of planar surfaces that could be fit within the slot contacting its sides. Variation can include being off-position along the y direction, out of orientation in the x or z directions, or any combination thereof.

When performing a tolerance stack-up, the designer needs to decide whether or not orientation effects need to be included in addition to location effects. The point of interest to the analyst is the origin of the next dimension on the stack-up chain. If we are only concerned with the permissible variation of the slot from its basic dimension, we only need to consider position, not orientation. When the point of interest is within the extent of the tolerance zone itself, the stack-up does not include any effects of orientation. Figure 3-28 illustrates.

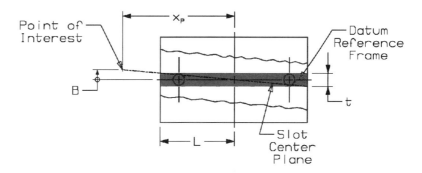

Figure 3-28: Location and Orientation

Dimension B originates on the part's datum reference frame and measures vertical position of a point on the center plane of the slot feature. The slot is basically positioned (positioned using basic dimensions) with its center plane through the secondary datum axis and tertiary datum axis. The variation on B is:

$$\begin{pmatrix} B_{max} \\ B_{min} \end{pmatrix} = f \begin{pmatrix} t/2 \\ -t/2 \end{pmatrix}$$

The term f is called the orientation factor, and its value depends on the point of interest being investigated. The dimension x_p represents the distance to the point of interest taken from an origin at the center of the slot feature's extents. The orientation factor is given by the following relationship:

$$f = MIN\left(\frac{x_p}{L}, 1\right)$$

When the orientation factor is greater than 1, orientation variation has a greater effect on the stack-up than location alone. The orientation factor will never be less than 1. In the case where it is exactly 1, orientation does not contribute any additional stack-up. The orientation factor does not attain a value greater than 1 unless the point of interest $x_p > L$. When the point of interest is within the extents of both the feature and the tolerance zone, location effects dominate over orientation effects.

Tolerances on the x_p and L dimensions used to calculate the orientation factor may be ignored, or these dimensions may be taken at their minimum and maximum values to maximize the orientation factor. In general, the tolerances present in any stack-up will be much smaller than the nominal dimensions used, and the effect of including x_p and L tolerances on the orientation factor will be negligible. Using the orientation factor assumes that only small angles result from orientation variation, allowing us to eliminate trigonometric terms and simplify the resulting nonlinear stack-up equations.

This housing part under consideration can contribute a combination of position and orientation variation when we look at the rest of the assembly and the components involved in the stack-up chain. The point of interest is often located on another component. The maximum orientation variation possible from a two-part assembly is illustrated in Figure 3-29. The orientation of the slot has a multiplying effect on the tolerance stack-up.

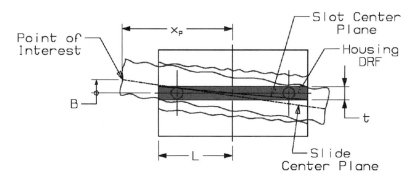

Figure 3-29: Position and Orientation Stack-Up

The orientation variation permitted due to clearance between the slot feature and mating part can increase the total orientation variation allowed at the assembly and should also be considered in the stack-up chain. Orientation is accounted for in the stack-up equation by applying the orientation factor as a multiplier to the total amount of play permitted by the slot's position tolerance and part-to-part clearance. *B* is the dimension from the housing's datum reference frame to the slot center plane, and dimension *C* is from the slot center plane to the mating part's center plane. For a worst-case stack-up:

$$A = B + C$$

$$\begin{pmatrix} A_{\max} \\ A_{\min} \end{pmatrix} = f \begin{pmatrix} t/2 \\ -t/2 \end{pmatrix} + f \begin{pmatrix} Clr/2 \\ -Clr/2 \end{pmatrix}$$

The worst-case clearance between mating parts drives the C dimension, also known as the assembly joint error. The clearance is:

$$Clr = H_{LMC} - W_{LMC}$$

The statistical analysis of a stack-up chain including orientation requires a slightly different approach. The amount of clearance available to contribute to orientation variation at the assembly is a function of a given slot width and mating part width. This gives rise to a third dimension, the angle of mis-orientation between slot and mating part, which can also vary randomly within its allowable limits. The statistical distribution on dimension C then becomes a distribution of a distribution. This breaks the statistical requirement that dimensions are independent of one another. Several options are available to the analyst:

1. Switch to the worst-case stack-up method.
2. Continue with the statistical stack-up method by one of the following methods:
   a. Split the problem into two assembly stack-up conditions (see the section on Multiple Assembly Conditions). Condition 1 assumes the mating part is oriented "up," and condition 2 assumes it is oriented "down:"

$$\mu_{C_1} = \frac{f}{2}(\mu_H - \mu_W), \sigma_{C_1} = \frac{f^2}{4}\sqrt{\mu_H^2 + \mu_W^2}$$

$$\mu_{C_2} = -\frac{f}{2}(\mu_H - \mu_W), \sigma_{C_2} = \frac{f^2}{4}\sqrt{\mu_H^2 + \mu_W^2}$$

   b. Assume a normal distribution on dimension C by assuming a mean and standard deviation. Several methods may be used according to the risk aversion of the analyst:

    i. Set the USL and LSL of a distribution for C by assuming worst-case clearance, and select a C_{pk}:

$$USL = \frac{f}{2}\left(H_{LMC} - W_{LMC}\right)$$

$$LSL = -\frac{f}{2}\left(H_{LMC} - W_{LMC}\right)$$

    ii. Set the USL and LSL of a distribution for C by selecting a confidence interval on the clearance dimension, and select a C_{pk}.

    iii. Set the USL and LSL of a distribution for C by applying the nominal clearance dimension, and select a C_{pk}:

$$USL = \frac{f}{2}\left(\mu_H - \mu_W\right)$$

$$LSL = -\frac{f}{2}\left(\mu_H - \mu_W\right)$$

3. Switch to the combined stack-up method considering the C dimension a worst-case dimension and leaving the remainder of the dimensions in the stack-up as statistical dimensions.

4. Switch to the Monte Carlo method. The Monte Carlo method can represent the orientation effects accurately.

The orientation factor is used with the feature's orientation tolerance to determine the contribution to the total stack-up. In this example, the single tolerance given for the slot's position controlled both location and orientation effects to the same degree. In other parts, additional callouts can be added to refine the control over orientation when this is needed. Refinement over orientation can be imposed by position controls, orientation control, profile control, runout control, and composite feature control frames of various types. The control permitted depends on the geometry of the feature and the needs of the application. Figure 3-30 shows the housing component with the slot feature's orientation refined through use of a composite position tolerance callout.

The orientation tolerance t_2 is less than the overall position tolerance t_1. Dimension B is calculated in this case by:

$$\begin{pmatrix} B_{max} \\ B_{min} \end{pmatrix} = \begin{pmatrix} (t_1 - t_2)/2 \\ -(t_1 - t_2)/2 \end{pmatrix} + f\begin{pmatrix} t_2/2 \\ -t_2/2 \end{pmatrix}$$

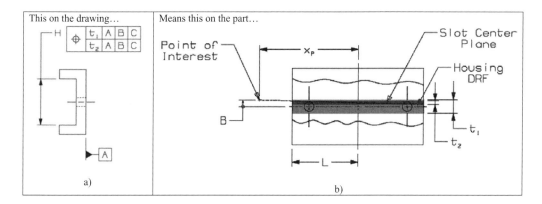

Figure 3-30: Refinement of Orientation Control

Floating Fasteners and Fixed Fasteners

The fixed and floating fastener formulas are commonly used to determine position tolerance requirements for hole features to ensure assemblability. Use of the formulas requires applying basic dimensions, position tolerances, using maximum material modifiers, and calculating the maximum material boundary for both holes and fasteners. Many key GD&T concepts are used in a very simple formula. Fixed and floating fastener applications are illustrated and explained in Chapter 4.

Boundaries

Many tolerance stack-up problems can be analyzed and solved by using the concept of boundaries. ASME Y14.5-2009 defines some terms, the maximum material boundary (MMB) and the least material boundary (LMB), that will prove useful in solving tolerance analysis and assignment problems. Each feature of size (e.g. a pair of opposed surfaces forming a slot or tab, or a cylindrical surface whose diameter can be measured) and each feature without size (e.g. a single planar surface) have these boundaries defined. In addition, each feature of size has an inner and outer boundary located and oriented basically at the feature's basic dimensions. They are mathematically ideal shapes.

The inner boundary (IB) is a worst-case boundary generated by the smallest feature minus the stated geometric tolerance and any bonus tolerance that may be applied. For an internal feature such as a cylindrical hole, the IB is also the MMB. Its shape is simply a cylinder that is guaranteed to be free of material under all allowable combinations of a feature's geometric variation including size, location, and orientation. For an external feature such as a cylindrical pin, the IB is also the LMB; its shape is a cylinder, but one within which we are guaranteed to always find material after considering all allowable variation.

The outer boundary (OB) are a worst-case boundary generated by the largest feature plus the stated geometric tolerance and any bonus tolerance that may be applied. The face of a hole will never extend beyond the OB; the OB will always be located within the material. The face of a pin will never extend beyond its OB; the OB will always be outside the material.

Figure 3-31 illustrates the concept of inner and outer boundaries. The pin's body serves as the part's primary datum feature. The head's size is controlled with a size dimension, and its location and orientation is controlled using a position control, as shown in Figure 3-31a. Figure 3-31b-c illustrate how datum A is defined, the tolerance zone basically (using basic dimensions)

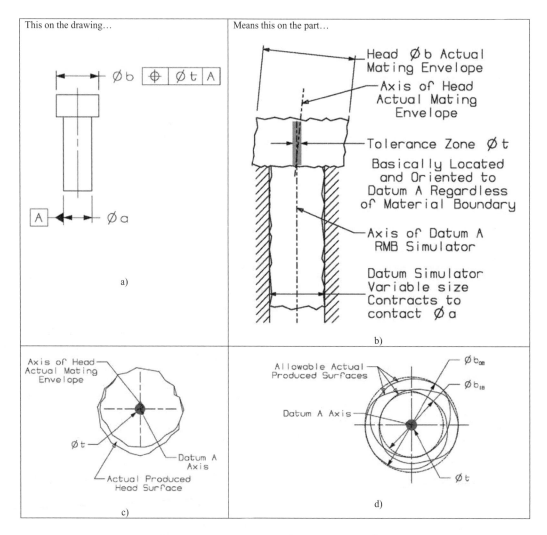

Figure 3-31: Inner and Outer Boundary

located and oriented to datum A, and the actual produced pin head feature. The pin head's actual mating envelope axis may vary from datum A in location in two directions, orientation in two directions, or any combination thereof provided the axis lies entirely within the tolerance zone. Given all the allowable variation, the outer boundary represents the worst-case limit of the pin head's surface. The inner boundary represents the innermost limit of where the pin head may be found. Note that given several allowable pin heads shown in Figure 3-31d, the IB and OB are always respected.

Maximum material boundary and least material boundary are general terms that can apply to any feature, whether it is a feature of size or not. Inner and outer boundaries are only defined for features of size. Table 3-17 lists formulas for calculating inner and outer boundaries for internal or external features of size that are located with a position callout.

We know that we gain greater control over a tolerance stack-up when there are fewer tolerances involved in the stack-up. Note in the table the inner boundary (IB) and outer boundary (OB) calculations for a hole feature positioned at maximum material condition (MMC). The IB calculation has fewer terms, thus less variation. This inner boundary is maintained regardless of the interplay between the actual produced hole size and position, and it is always preserved to be clear of material. This should remind the reader of the usefulness of this callout to position features for clearance, such as bolt holes. Note also the OB of a hole is better-controlled using a LMC modifier. The outer

Table 3-17: Boundary Calculations for Positioned Features of Size

Position Callout	Internal Feature (Hole / Slot) $d_{MMC} = d_{min}$ $d_{LMC} = d_{max}$ $IB = MMB$ $OB = LMB$	External Feature (Pin / Tab) $d_{MMC} = d_{max}$ $d_{LMC} = d_{min}$ $IB = LMB$ $OB = MMB$
⌀d ⊕ ⌀t A B C (position tolerance applies regardless of feature size)	$d_{IB} = d_{MMC} - t$ $d_{OB} = d_{LMC} + t$	$d_{IB} = d_{LMC} - t$ $d_{OB} = d_{MMC} + t$
⌀d ⊕ ⌀t Ⓜ A B C (position tolerance applies at maximum material condition)	$d_{IB} = d_{MMC} - t$ $d_{OB} = d_{LMC} + t + bonus$ $= d_{LMC} + t + (d_{LMC} - d_{MMC})$	$d_{IB} = d_{LMC} - t - bonus$ $= d_{LMC} - t - (d_{MMC} - d_{LMC})$ $d_{OB} = d_{MMC} + t$
⌀d ⊕ ⌀t Ⓛ A B C (position tolerance applies at least material condition)	$d_{IB} = d_{MMC} - t - bonus$ $= d_{MMC} - t - (d_{LMC} - d_{MMC})$ $d_{OB} = d_{LMC} + t$	$d_{IB} = d_{LMC} - t$ $d_{OB} = d_{MMC} + t + bonus$ $= d_{MMC} + t + (d_{MMC} - d_{LMC})$

boundary of a hole defines the allowable location of a mating component, which also makes sense given the LMC modifier is recommended for use positioning locational features.

Boundary Example Problem - Spring Pin

The pin and block parts shown in Figure 3-32 form an assembly, along with a spring that sits on top of the head of the pin. The function is for the pin to be guided by a running sliding fit within the block's pilot hole, and the larger diameter hole houses the spring that presses onto the pin head. The design has progressed to a complete CAD model, and detailing of components is the next step. The designer's immediate problem concerns how to specify the allowable variation on the pin's geometry and the pilot hole's features of the block that mate with the pin. Using the proper GD&T callouts allows maximum

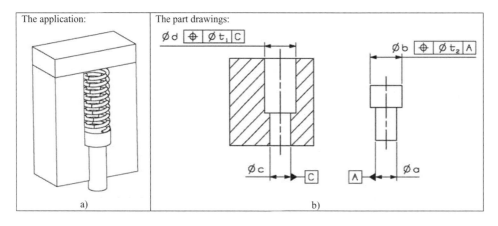

Figure 3-32: Boundary Stack-Up Example

tolerance and flexibility to the machinist making the parts, while ensuring the design intent is met in the produced parts.

There are several design objectives of the block / pin assembly, namely:

1. Establish a close sliding fit between the pin body and pilot hole.
2. The pin head and spring clearance hole must be a clearance fit.
3. The radial clearance between the pin head and spring clearance hole must not allow the spring wire to slide into the gap; the maximum radial gap size is 0.33 mm.
4. No precision machining processes should be required to make the spring clearance hole: diameter variation should be greater than 0.012 mm, and position tolerance should be greater than 0.05 mm.

In order to meet the design objectives, it is necessary to choose each of the following:

- Pin body size including tolerance: a
- Pilot hole size including tolerance: c
- Pin head size and tolerance, and position tolerance: b, t_2
- Spring clearance hole size and tolerance, and position tolerance: d, t_1

We see that each choice affects the others and the ability to satisfy the objectives. Some requirements compete with one another — e.g., the looser the fit between the pin body and pilot hole, the greater the variation in spring clearance hole / pin head gap size. Additionally, the exact machining methods used and the capabilities of those methods (hence the tolerances that are easy to hit or very difficult to hit) may be known only generally. Design tradeoffs will need to be made and an optimal balance chosen with imperfect knowledge of machining methods. These trade-offs can only be made effectively through an accurate stack-up analysis.

To determine the stack-up chain, begin with the simple case where the pin's datum A and the block's datum C are aligned. That is, the pin body is coaxial to its pilot hole. Using the knowledge of boundaries, it can be seen that the maximum radial gap between the spring clearance hole and the pin head is equal to one-half the difference between the spring clearance hole's outer boundary and the pin head's inner boundary. The minimum gap is one-half the difference between the spring clearance hole's inner boundary and the pin head's outer boundary, or:

$$Gap_{max} = \frac{1}{2}\left(d_{OB} - b_{IB}\right)$$

$$Gap_{min} = \frac{1}{2}\left(d_{IB} - b_{OB}\right)$$

The equations for calculating the boundaries are obtained from Table 3-17:

$$d_{IB} = d_{MMC} - t_1$$
$$d_{OB} = d_{LMC} + t_1$$
$$b_{IB} = b_{LMC} - t_2$$
$$b_{OB} = b_{MMC} + t_2$$

The pin head's boundaries are basically located and oriented to pin datum A (regardless of material boundary). Datum A is an axis through the pin body regardless of the pin body's size, with no datum shift allowed. The spring clearance hole's boundaries are basically located and oriented to block datum C (regardless of material boundary). Datum C is an axis through the block's pilot hole regardless of the pilot hole's size, with no datum shift allowed. The boundaries we calculate are located and oriented to their respective datum reference frames. However, we have two different parts mating together with different datum reference frames that may not be aligned to one another. Clearance between the pin body and pilot hole will allow the pin to shift location within the pilot hole, increasing the gap on one side and reducing the gap on the opposite side by the same amount.

Clearance must be maintained between the outer edge of the pin head and the inner edge of the spring clearance hole under all conditions. The location of the pin axis is determined by the pin body / pilot hole clearance and may be located anywhere within that zone at any given time; no adjustment or fixing at assembly is possible. Because the pin is allowed to move within the pilot hole, the gap calculation must be adjusted for clearance:

$$Gap_{max} = \frac{1}{2}\left(d_{OB} - b_{IB} + Clr_{max}\right)$$

$$Gap_{min} = \frac{1}{2}\left(d_{IB} - b_{OB} - Clr_{max}\right)$$

The maximum amount of slop, or clearance between the pilot hole and the pin body, is the least material condition of the hole minus the least material condition of the pin body. This represents the loosest fit or intended difference between these two mating parts:

$$Clr_{max} = c_{LMC} - a_{LMC}$$

Re-writing the stack-up equations to consolidate equations and include the design requirements, we obtain the following two relationships that must be maintained:

$$Gap_{max} = \frac{1}{2}\left(d_{LMC} + t_1 - b_{LMC} + t_2 + c_{LMC} - a_{LMC}\right) \leq 0.33$$

$$Gap_{min} = \frac{1}{2}\left(d_{MMC} - t_1 - b_{MMC} - t_2 - c_{LMC} + a_{LMC}\right) \geq 0$$

Once the stack-up equations that must be satisfied are understood, the next step is to assign preliminary tolerances. Starting with the pin body and pilot hole sizes, the design calls for a nominal Ø4 mm. Standard fits and tolerances per ISO 286-2 (See Section 3.1) are widely used to specify fits for mating parts. An H7 tolerance (+0.012 / 0 mm) for the pilot hole and g6 (–0.004 / –0.012 mm) tolerance for the pin body will produce a sliding clearance fit and are initially chosen for these features based on a long history of success using standard fits and tolerances. The maximum diametral clearance that could exist between the pin body and pilot hole is given by:

$$Clr_{max} = c_{LMC} - a_{LMC} = 4.012 - 3.988 = 0.024$$

Trial values are assigned for the remaining dimensions, and the two gap equations are checked to ensure fitness. After several iterative attempts, the following gap calculations result. All selected dimensions are shown on the completed drawings in Figure 3-33.

$$Gap_{max} = \frac{1}{2}\left(6.12 + 0.182 - 5.704 + 0.028 + 4.018 - 3.984\right) = 0.33 \le 0.33$$

$$Gap_{min} = \frac{1}{2}\left(6 - 0.182 - 5.754 - 0.028 - 4.018 + 3.984\right) = 0.001 \ge 0$$

Note some of the choices made. After recognizing the relationship between the pin body's accuracy and the pin body / pilot hole clearance, the pin body / pilot hole fit was loosened slightly to an H8/g7 fit. The spring clearance hole's size tolerance was loosened from an initial guess of Ø6 H8 all the way to Ø6 H12 (+0.12 / 0 mm), standard twist drill hole tolerance. This could be done while using a somewhat permissive Ø0.182 mm position tolerance zone for the hole, a more restrictive size tolerance on the pin head, and an Ø0.028 mm position tolerance controlling coaxiality of the pin head.

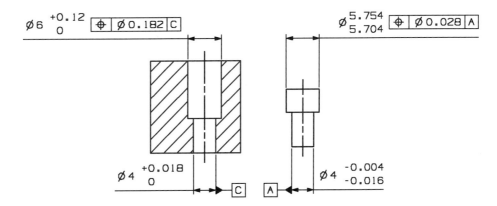

Figure 3-33: Boundary Stack-Up Results

Another designer may have made different trade-offs among the tolerances to be assigned. As an example, if after review with a skilled machinist, the designer learned that the spring clearance hole position would be a challenge, but that a pin head size of Ø5.72 – Ø5.74 mm and position tolerance of Ø0.006 mm were easily achievable, the relationships established by the stack-up equations would have shown it possible to increase the spring clearance hole position tolerance t_1 to Ø0.22 mm. Understanding the tradeoffs involved makes good tolerance assignment choices possible.

Different geometric controls could also have been applied. For example, the pin head was controlled by its size and position. A basic pin head diameter and profile control could easily have been selected to achieve the functional requirement. The profile tolerance requires the pin head's entire surface to fall between two concentric cylinders located and oriented to the pin's datum A axis. The two cylinders are separated by a radial gap centered on the basic dimension Øb. The produced pin head diameter is allowed to vary its size, form, orientation, location, or any possible combination these variations. The surface's coaxiality is no more or less important than its size, which is no more or less important than its circularity. Any of these types of variation can exist in any amount and combination so long as the surface does not go beyond the boundaries established by the profile tolerance zone. There is no one "right" answer for tolerance assignment.

Effect of Datum Shift

The concept of datum shift is an important one within GD&T. Figure 3-34 shows the housing part with the geometric controls used to specify

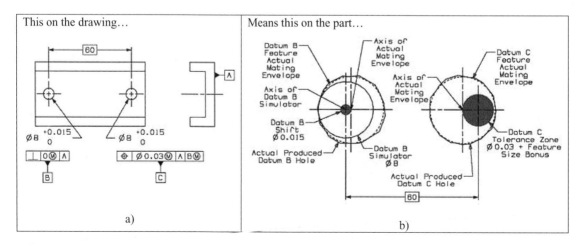

Figure 3-34: Datum Shift

the datum B and C features. Consider the process of inspecting the datum C feature. The datum reference frame used in the control of datum feature C specifies that datum B is taken *at its maximum material boundary (MMB)*. Specifying datum B at MMB specifies the size of the datum feature simulator used when inspecting the part. The datum B feature simulator used is a pin whose diameter is the feature's size at MMB, determined by the combined effects of size and geometric controls. The B datum feature's boundary is diameter 8 mm (MMC size 8 — geometric tolerance 0 applied at MMC). The datum reference frame is defined by the datum simulators, not the datum features.

The datum B modifier allows datum shift when inspecting the feature. Fixturing the part on its datums for inspection allows for some play to occur during the inspection process when the datum B feature departs from its stated maximum material condition. In this example, when the hole is larger than its maximum material boundary, the part can move on its fixturing elements when checking the feature's tolerance. The amount of datum shift is calculated the same way as bonus tolerance. Datum shift allowed to datum feature B is the actual size of the produced hole, minus the MMB. If the play allowed by the part causes the feature being inspected to come into its allowable tolerance zone, the feature passes inspection.

This allows displacement of the part by the amount of datum shift to properly locate the C datum feature. This has the effect of allowing additional position tolerance to the machinist making this precision locating feature. Functionally, applying datum shift in practice is equivalent to placing the datum B hole over its locating pin, and allowing the part to move to any degree allowed by the play between pin and hole as the C hole attempts to find its mating pin. The part shifts during the assembly process in fact allowing assembly to occur. This assembly process captured in the datum reference frame, which in this case is truly a functional datum reference frame for locating the C datum feature.

Datum shift can be used very effectively to allow additional tolerance to the machinist and inspector. Datum shift should be allowed when it helps the assembly process and does not affect the tolerance stack-up. When the feature is specified with a callout to a datum reference frame using maximum material boundary modifier, the part is allowed some wiggle on the datum simulators to get the feature into spec. The tolerance stack-up chain goes through the datum simulators, not the actual produced part features. Appropriately-chosen, functional datum reference frames allow this with ease. Difficulty

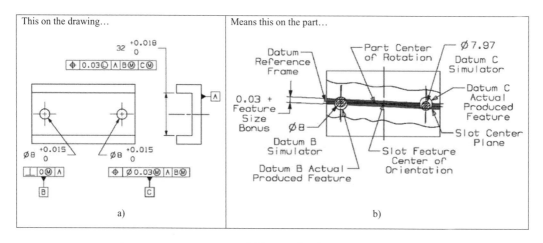

Figure 3-35: Datum Shift Applied to Slot Feature

arises, and additional variation must be included in the stack-up, when datum shift is applied inappropriately.

Figure 3-35 shows the housing part with an alternative callout used to control the slot position. The slot feature is located with respect to the part's datum reference frame, established by the flat surface and two holes at MMB. A material modifier is used to control the slot's position of 0.03 mm at least material condition. The least material condition modifier for the slot's position is appropriate for a feature of size functioning to locate a subsequent mated component.

Specifying datums B and C at maximum material boundary specifies the sizes of the datum simulators used when inspecting the part. The B datum feature's boundary is an 8 mm diameter cylinder, and the C datum feature's boundary is a 7.97 mm (MMC size 8 - geometric tolerance 0.03 applied at MMC) diameter cylinder. The datum reference frame, which locates and orients all tolerance zones, is defined by the datum simulators, not the datum features.

Datum shift must be included in tolerance stack-ups when appropriate. The way to include datum shift in a particular stack-up depends on the stack-up chain drawing. This part as specified is allowed to float on its datum simulators by the datum shift amount in order to ensure accurate location of the slot. The assembler is permitted, even expected, to use what wiggle room is physically allowed while verifying that the slot's position is where it needs to be in the assembly. The maximum datum B shift of 0.015 allows displacement of the part's datum B reference frame from the actual produced hole's position, up or down by half the datum shift amount. The maximum datum C shift of 0.018 allows further displacement of the part's datum reference frame up or down by half the datum shift amount, at that location. The combined effect is that of combined location and orientation variation. The tolerance stack-up must include a dimension between the actual produced datum feature (which connects to a feature on the previous component in the stack-up chain, possibly with gap clearance allowance) and the datum reference frame from which subsequent features are measured. The maximum mis-orientation allowed occurs about the indicated center of rotation of the part. The part may rotate about a point between the B and C datum features, here not exactly in the center due to the different datum B shift and datum C shift amounts. From this datum reference frame, the slot is allowed its locational and orientation variation specified by its position control.

In this example, the datum shift applied to the slot feature's geometric control adds to the possible assembly dimension variation, making the overall assembly less precise. In addition, the presence of datum shift in this case does not help the assembly process. The subsequent mating part is placed into the slot unconstrained by other features regardless of where the slot is. Fitting the mating part into the slot does not become more forgiving through the presence of datum shift, as it did with the datum shift allowed to the datum C feature. Functionally, the part is expected to hold its mating part in the proper location regardless of any assembly variation introduced in fitting the B and C datum features onto their functionally equivalent pins. No setup adjustment is allowed at assembly, so datum shift has been incorrectly applied to this feature control.

One might naturally ask why datum shift is used if the goal is to minimize the effects of tolerances. In general, this is a goal; but it must be balanced with goals of minimizing the cost of fabrication while maximizing production yield, inspection yield, etc. When appropriately applied, datum shift simplifies the tolerance stack-up calculation while allowing additional manufacturing tolerance and improving assembly yield. When misapplied, it has the opposite effects. A detailed stack-up calculation is not made here for misapplied datum shift; the recommendation is to apply the appropriate functional datum reference frames.

Profile Tolerances

Profile tolerance can be applied to a 3D surface or to a 2D section of a surface. Profile may or may not be related to the datum reference frame.

The more common 3D control defines a tolerance zone as the region between two surfaces offset from the basic feature being controlled. The feature can be of any shape. In a common application of profile tolerance, the feature controlled is basically related to the datum reference frame. The profile tolerance callout defines two boundaries within which the actual produced surface must lie. The boundaries are equally offset from the basic surface, and the distance between the boundaries is the total tolerance zone specified by the callout.

Profile is a very broad and general control, controlling the location, orientation, and form of the surface all to the same degree. Because the actual produced surface is permitted anywhere between the two boundaries, the amount of orientation control achieved is equal to the amount of profile control given; and the amount of form control achieved is also equal to the profile control given. When a single callout is used, all aspects of geometry are controlled to the same degree.

When dimensions in a stack-up chain originate or terminate on a surface controlled with a profile tolerance, the surface's location, orientation, and form need to be considered. What variation needs to be included in the stack-up chain depends on the application. Figure 3-36 shows an application involving a support bracket and bushing mount.

The vertical height to the bushing press-fit hole, A, is desired. The limits of the hole's centerline position are taken from the position tolerance callout

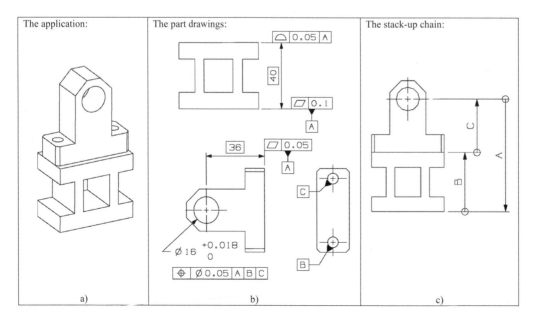

Figure 3-36: Stack-Up Example with Profile

on the housing part, assuming the part is assembled on the high points of the datum surface. The limits of the spacer place thickness, dimension *B*, are taken from the maximum and least material boundaries as specified by the profile control on that part's drawing. Mis-orientation due to allowable height difference between one side of the spacer plate's surface and the other (taper) does not contribute substantively to height variation on the bushing hole location.

$$\begin{pmatrix} A_{\max} \\ A_{\min} \end{pmatrix} = \begin{pmatrix} 40.025 \\ 39.975 \end{pmatrix} + \begin{pmatrix} 36.025 \\ 35.975 \end{pmatrix}$$

$$\begin{pmatrix} A_{\max} \\ A_{\min} \end{pmatrix} = \begin{pmatrix} 76.05 \\ 75.95 \end{pmatrix}$$

The previous discussions on orientation and form tolerances should equip the analyst with the proper decision-making tools for determining whether or not profile's effects on orientation and form need to be included in the stack-up chain and how to include them.

Composite tolerances can be built using profile controls, allowing great flexibility in the control specified. Orientation and form can be refined with respect to all datums or only those datums selected. For a fuller exploration of the controls that are possible through use of the profile control, see the recommended resources.

Orientation Tolerances

Orientation tolerances such as parallelism, perpendicularity, and angularity specify relationships between features, such as plane-to-plane, axis-to-axis, and plane-to-axis. Orientation tolerances can also be used as 2D controls, applying their control to individual elements of a surface, such as cross-sectional shapes. Orientation tolerances also control the form of features to the same degree that they specify orientation, unless additional refinement if provided by other form controls. Like all geometric tolerances, the value of an orientation tolerance is the total tolerance zone size. This is typically the total distance between parallel planes, or between parallel lines, or the diameter of a cylindrical zone used to locate an axis.

Orientation tolerances do not control location, so surfaces controlled with orientation controls need to be located through other means. For features of size, this is typically done with a position control, whereas for surfaces, profile is used. The position control or the profile control applied to the feature will generally be the only tolerance included in the stack-up chain. If orientation is used to refine the orientation control already given by a profile or position control, it may have the effect of reducing the allowable variation due to orientation.

Runout Tolerances

Runout tolerances (both circular runout and total runout) control the relationship of a surface to a datum axis. Runout can control a cylindrical surface about a coaxial datum axis, a planar surface perpendicular to a datum axis, a conical surface, or any axisymmetric surface generated as a surface of revolution. Circular runout is a 2D control defining its tolerance zone as two concentric circles—for example, two concentric circles of different diameters within which an actual produced surface must lie. Total runout defines as its tolerance zone two concentric cylindrical surfaces when controlling a cylindrical surface or two parallel planes when controlling a planar surface perpendicular to the datum axis. The datum reference frame for runout tolerances always includes at least one datum axis, and this is typically the primary datum.

Because runout controls the surface and not an axis, material modifiers are meaningless and so are not applied. Generally runout will be applied only when the surface in question is functionally important. Runout controls circularity and cylindricity (form), straightness, and coaxiality (orientation and location). It also refines the control given by a size dimension alone. Understanding the application and the control given should allow the analyst to recognize whether or not a runout tolerance needs to be included in the stack-up chain and how it is included.

SUMMARY

Table 3-18 summarizes application guidelines for which type of tolerance analysis / assignment method is suited given the problem to be solved.

Table 3-18: Stack-Up Method Application Matrix

	Worst-Case	Statistical	RSS	Combined	Monte Carlo
Inspection is used on every component to eliminate components whose dimensions fall outside the specification limits	●	⊘	⊘	○	○
Quality inspection data of contributing dimensions are available	○	●	●	●	●
Quality inspection data of contributing dimensions are unavailable	●	○	○	○	○
All contributing dimensions well-modeled by statistical distributions	⊘	●	●	⊘	●
Some contributing dimensions well-modeled by statistical distributions	○	○	○	●	○
Few dimensions contribute to stack-up chain (approx. 7 or fewer)	●	⊘	⊘	●	⊘
Many dimensions contribute to stack-up chain (approx. 7 or more)	⊘	●	●	●	●
Minimizing risk is required	●	⊘	⊘	○	⊘
Simplicity is required	●	○	●	○	⊘
Accurate prediction of assembly variation is required	⊘	●	●	⊘	●
Accurate prediction of yield / defect rate is required	⊘	●	●	⊘	●
Minimizing cost is required	⊘	●	○	○	●
Statistical process control (SPC) is used to control the variation of contributing dimensions	⊘	●	●	⊘	●
Some contributing dimensions are screened by inspection, others are controlled through SPC	○	○	○	●	○
The maximum allowable tolerance is required for assignment	⊘	●	●	○	●
The maximum possible assembly variation must be known	●	⊘	⊘	○	⊘
Some contributing dimensions controlled using material modifiers and / or datum shift	●	○	⊘	●	●
Some contributing dimensions well-modeled by non-normal statistical distribution	⊘	○	⊘	○	●

● Suitable

○ Sometimes Suitable

⊘ Not Suitable

CRITICAL CONSIDERATIONS: Tolerance Stack-Ups

- Tolerance stack-ups are the designer's responsibility.
- Clear, unambiguous part definition provided by the dimensioning and tolerancing standard is required for tolerance stack-ups to have any meaning.
- Costs can change dramatically due to small changes in tolerances.
- Tolerance stack-up rigor leads to better drawings, better designs, and better products at the right cost.
- It is sometimes discovered through tolerance analysis that a design concept will not work to achieve the assembly tolerance needed.
- Precision assembly methods can be used for minimizing (or compensating for) component variation.
- Component positions after assembly may not match the design intent. Consider possible adjustments that can be made between components during assembly and consider multiple stack-up chains when appropriate.
- Interchangeability requires strict control over tolerances and should be carefully considered as a holistic business decision.

BEST PRACTICES: Tolerance Stack-Ups

- Perform tolerance stack-up analyses and assignments for all critical assemblies.
- Follow a consistent methodology for determining the stack-up equation and the contributing dimensions.
- Hand-draw stack-up chain drawings to clearly show imperfect geometry, exaggerated effects of tolerances, and clearances between parts.
- Worst-case analysis is most appropriate when the number of components produced is relatively small and when inspection is used on every component to eliminate components whose dimensions fall outside the specification limits.
- Statistical analysis is best when contributing dimensions actually follow a normal distribution.
- Maintain calculations and sketches as part of the engineering records.
- Use tolerance stack-up analysis to root out poor design choices and poor drawing methodologies.
- Assign plus/minus tolerances for the sizes of features, not the location of features.
- Balance risk and cost using the tolerance stack-up method guidelines.
- Choose tolerances to reflect manufacturing process capability (enabled by knowledge of component manufacturing capabilities).
- Assign tolerances by beginning with functional datum reference frames, controlling part features using the appropriate geometric control and modifiers for the application, and selecting the largest permissible tolerance values.

4

PRECISION LOCATING TECHNIQUES

Written by
Charles Gillis, P.E.

Tables

4-1	Typical Hole-Making Processes	190
4-2	Two Holes / Two Pins / Two Holes Dimensions: Metric Hardware	194
4-3	Two Holes / Two Pins / Two Holes Dimensions: Inch Hardware	195
4-4	Two Holes / Two Pins / One Hole and One Slot Dimensions: Metric Hardware	202
4-5	Two Holes / Two Pins / One Hole and One Slot Dimensions: Inch Hardware	203
4-6	Two Holes / One Round Pin and One Diamond Pin / Two Holes Dimensions: Metric Hardware	206
4-7	Two Holes / One Round Pin and One Diamond Pin / Two Holes Dimensions: Inch Hardware	207
4-8	Precision Locating Application Matrix	213
4-9	Precision Locating Formula Summary	214

When mechanical design engineers design parts, they need to consider how the parts are to be located relative to one another at the assembly level. If parts are mated by bolts only, there can be significant variation, or "play," in part alignment. For example, typical clearance bolt hole drilling tolerances, combined with the precision of bolt threads, can easily result in 0.040 inch (1.0 mm) of play for bolts up to 3/8 inch (approximately 10 mm). Using some sort of precision locating features in addition to bolts can reduce the play to the neighborhood of 0.0004 inch (0.01 mm) or even zero. Precision locating features bring the additional benefits of fast, error-proof assembly without setup, and repeatable disassembly and re-assembly. These features, along with a high degree of positional accuracy, come with costs that must be balanced. The purpose of this chapter is to communicate best practices regarding the choice of precision locating techniques, including:

- Types of planar, pin / pin precision locating techniques to use
- Hardware choices
- Dimensioning techniques and geometric controls
- Hole size tolerances
- Hole positional tolerances

RECOMMENDED RESOURCES

- D. Blanding, *Exact Constraint: Machine Design Using Kinematic Principles*, ASME Press, New York, 1999
- L. Kamm, *Designing Cost-Efficient Mechanisms*, Society of Automotive Engineers, Warrendale, PA, 1993
- **ASME Y14.5 - 2009:** "Dimensioning and Tolerancing"

DESIGN REQUIREMENTS

Typically when using a precision locating technique, the designer seeks to achieve the following technical requirements:

- The assembled part must be precisely located to the base part.
- It must be possible to assemble the two parts together without binding or force-fits.
- It must be possible to disassemble the two parts without special tools.
- It must be possible to use different instances of the base and assembled parts interchangeably.
- Both parts of the assembly, as well as hardware used, should be economical to produce.

THE TWO-HOLE PROBLEM

Designers attempting to design parts that require precision locating may begin by selecting hardware. Standard components, such as dowel pins, are commonly used. Next, the required hole sizes are addressed. Typical dowel pin tables (such as Tables 5-7 and 5-5 in Chapter 5) provide "press fit" and "slip fit" dimensions. The proper sizes are selected, and the design progresses in a rather straightforward manner. A designer following this technique quickly realizes that the hole-to-hole center distance accuracy is not specified anywhere. A single "slip fit" hole will indeed slide nicely without binding onto one pin. The challenge comes when a second slip fit hole needs to slide nicely without binding onto a second pin pressed into the mating part. If the hole-to-hole center distance is not held tightly enough, successful assembly will not occur.

Figure 4-1 illustrates a typical two hole / two pin / two hole assembly. It shows the parts involved and the dimensions specifying the parts using the nomenclature that will be seen throughout this chapter. Placing one of the assembled part's holes over the first pin fixes the location of the assembled part. The second hole will only fit over the second pin in this position when the hole

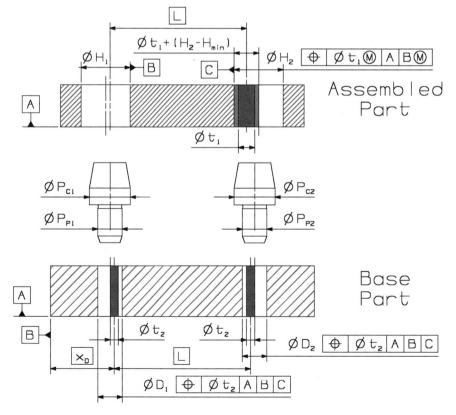

Figure 4-1: Two Holes / Two Pins / Two Holes Assembly

and pin size and locations permit it. Larger holes make assembly easier at the cost of allowing play between the parts. It can be seen that the clearance holes in the assembled part will just fit over the pair of pins pressed into the base part's holes.

Fixed And Floating Fastener Formulas

A designer attempting to solve this problem may seek out the fixed fastener and floating fastener formulas within the Geometric Dimensioning and Tolerancing (GD&T) standard, and the various books on the subject. However, these formulas may not be appropriate for reasons that will be discussed later in this section. The formulas are used to determine the required positional tolerances on the holes of fastened parts to ensure assemblability. When two parts are assembled with fasteners, such as screws or pins, and both parts have clearance fits on the fasteners, the floating fastener formula is used. Figure 4-2 shows a floating fastener application and the dimensioned part drawings.

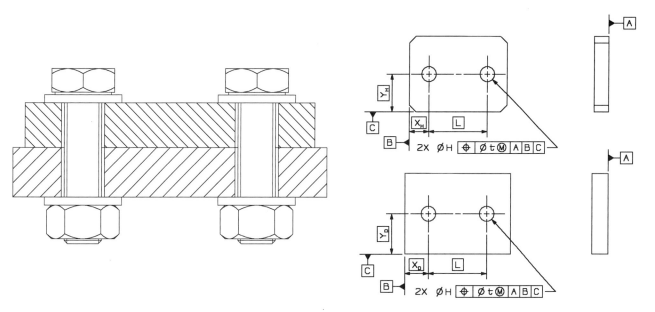

Figure 4-2: Floating Fastener Application and Part Drawings

Floating Fastener Formula: $t \leq H_{min} - P_{max}$

Here, H_{min} is the minimum hole size, P_{max} is the maximum fastener size, and t is the position tolerance at maximum material condition (MMC) that can be applied to the holes of the two parts.

When the fasteners are restrained by one of the parts, such as with tapped holes restraining screws, or press-fit holes restraining pins, the fixed fastener formula is used. Figure 4-3 shows a fixed fastener application using tapped holes in one part and the dimensioned part drawings.

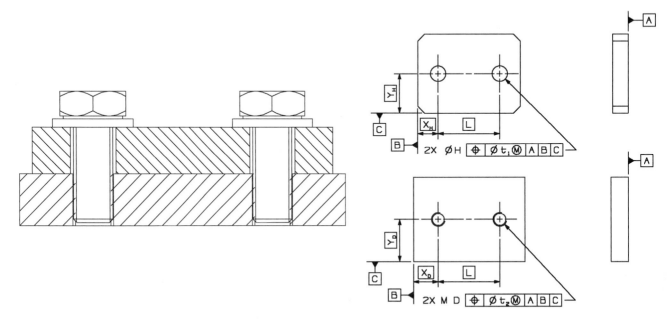

Figure 4-3: Fixed Fastener Application and Part Drawings

Fixed Fastener Formula: $t_1 + t_2 \leq H_{min} - P_{max}$

Here, t_1 and t_2 are the hole position tolerances at maximum material condition (MMC) that ensure assemblability. The two tolerances t_1 and t_2 can each be equal to one-half the total, or they can be unequally assigned to the two parts. How the values are assigned to the two parts doesn't affect assemblability as long as the total $t_1 + t_2$ is not exceeded.

When press-fit pins are used, the fixed fastener formula appears to be the right application. The problem with applying this formula to a precision locating application is that the tolerance values calculated apply only when the holes are dimensioned from the part's datum reference frame and controlled with a position callout, as shown on the part drawings in Figures 4-2 and 4-3. Both the fixed and floating fastener formulas presuppose that the part has a datum reference frame already defined elsewhere on the part and that we are seeking to position the holes with respect to this datum reference frame. However, in a precision locating application, the holes in question **are** the assembled part's datum reference frame. This makes the fixed and floating fastener method unsuitable.

For the fixed and floating fastener formulas, the holes in both parts must be positioned using the MMC modifier. The maximum material modifiers allow additional position tolerance when the actual hole size is larger than the MMC size. The additional position tolerance along with the larger hole size ensures that assemblability is maintained. This is a good tool to allow manufacturing additional tolerance while maintaining the intended function, when it is appropriate for the application. Unfortunately, this geometric control is not appropriate for a base part, whose job it is to precisely locate the press-fit pins. That application calls for a regardless-of-feature size (RFS) modifier, which is assumed if no modifier is explicitly applied.

When the holes in a part make up the part's datum reference frame, the fixed and floating fastener methods are not the best way to specify dimensions and tolerances. A different dimensioning scheme must be applied to the drawings, and different formulas must be used to determine the required tolerance. Various cases of the functional datum method used to accomplish this are detailed later in this chapter.

DATUM REFERENCE FRAME

For proper assembly of a part, its six degrees of freedom (three translational and three rotational) must be constrained against motion. A single feature will provide one or more constraints. When using Geometric Dimensioning and Tolerancing (GD&T), part specifications include the definition of a datum reference frame serving as a common reference for part definition, fabrication, and inspection. The datum reference frame is made up of several datums that are derived from their defining datum features and specifies the locations and orientations of all dimensional requirements.

Functional Datums

Good design involves specifying the part's datum features by matching how it will be fixtured / assembled in use. The datums are then functional. In the definition of the part, the datums constrain the CAD model geometry. When assembled in service, its mating parts constrain the part on its datum features. GD&T completely specifies the parts being designed. By properly specifying the datum reference frame of the part based on how it is assembled and fixtured into location, the designer gets exactly the function that was intended.

An example of a datum reference frame consists of a planar surface used as a primary datum feature and providing three constraints: one translational and two rotational. A second, non-parallel planar surface provides two more constraints (removes two more degrees of freedom): one translational and one rotational. A third, non-parallel planar surface removes the last translational degree of freedom. The part's location and orientation is fully defined in 3D space — whether inside a CAD definition or physically assembled in service.

Practical Considerations

It is important to note that using features for constraint has some limitations. In practice, what is referred to as a "constraint" or "datum feature" does not truly eliminate a degree of freedom, but rather limits this degree of freedom to some very small value of displacement (a.k.a. semi-minimum constraint design). Consider the case of a pin in a slip fit hole. Because there must be some size difference (clearance) between the pin and the hole, the two degrees of freedom are not removed entirely. Motion is restricted to a small total allowable displacement, equal to this size difference.

In practice, datum features do not always constrain all the degrees of freedom they constrain in theory. Consider again the pin that fits into the hole. A long pin's cylindrical surface would remove four degrees of freedom (two translational and two rotational), provided these same four were not already constrained by a preceding datum or competing constraint. Following a planar primary datum feature with a cylindrical secondary datum feature would overconstrain the part (provide redundant constraints); assembly would not be possible without binding or deforming the components. In contrast, a short-enough pin behaves more like a two-dimensional circle and removes only two translational degrees of freedom. We take advantage of this practicality to sidestep overconstraint by specifying short pin / hole engagement lengths as secondary and tertiary datum features.

In the preceding examples, there exists some repeatability error in the assembly. Repeatability error is a function of the precision locating technique used, the sizes of precision holes and pins, and the geometry of the mating parts. For true precision location (a.k.a. kinematic fixturing, exact constraint design, or minimum constraint design), it is necessary to permit exactly zero repeatability error. That is, when the assembly is disassembled and reassembled, the two pieces are in exactly the same position relative to one another every time. This may not be possible, practical, or necessary in any one particular application.

DESIGN PROCESS

The designer will typically begin with the problem that one part must be located accurately within an assembly. One way to achieve a suitable design is to choose one of the arrangements described later in this chapter; then choose hardware, hole dimensions, and tolerances from the table. Finally, apply the given dimensioning scheme to the parts. At this point, the design is complete.

Assemblability

Assemblability refers to the ability to successfully install a part onto its base part when both are produced to within their specifications. Interchangeability refers to the ability of any pair of base and assembled parts fabricated to the same specifications to be assembled. Designers are often required to achieve interchangeability among the parts in an assembly. That is, it must be possible to pluck a part randomly from a bin full of similar parts, and install it on its base part with 100% success. No matched sets are used, and no trial and error assembly is required.

Assemblability and interchangeability are commonly, though not always, required. Consider an assembly process with assemblers hand-picking commercial bearings from one bin and using an arbor to press them into die-cast housings selected from another bin. It may be permissible to relax the assemblability and interchangeability requirements. If these requirements are relaxed, the assemblers will be required to do some manual set-matching. In this case, if a bearing and housing in hand fit too loosely, they can quickly and easily select another housing that gives a press fit.

Consider, by contrast, a mission-critical, expensive spare part that must be shipped to a remote plant or customer site. The requirements of assemblability and interchangeability must be decided holistically. The precision locating techniques described in this chapter are presented so that assemblability and interchangeability are achieved. If assemblability and interchangeability are not required, the techniques of this chapter are unnecessary.

When assemblability and interchangeability are required, the designer's task is to select hole sizes, pins, and hole position tolerances to: a) balance machining difficulty / yield / cost, b) maintain assemblability, and c) ensure required accuracy. Solving this design problem is simplified through the use of the tables provided later in this chapter. Alternatively, the associated formulas can be used.

Using The Hardware Tables

To assist in selection of hardware and tolerancing of features, tabulated hardware, hole sizes, and tolerances are given below for both a worst-case tolerance stack-up and for a statistical tolerance stack-up. The tabulated data comes from the associated formulas. The worst-case tables provide the smallest tolerances to the engineer for assigning feature tolerances and ensure that press-fits are adequate, clearances are adequate, and assembly is possible without binding when all features are produced within their given tolerances.

The statistical values given in the hardware tables provide tolerances based on a statistical stack-up. See section 3.3 for more information on tolerance stack-up analysis and assignment methods. The statistical tables provide more tolerance to the designer by making the assumption that all hardware dimensions are randomly distributed about their target dimensions, with process capability index (c_{pk}) equal to 1.0. This represents a Three-Sigma manufacturing quality level, which may be a conservative estimate of real part and feature quality control levels. Further, successful press-fits and interchangeability are achieved with at least 99% confidence. It is up to designers to choose one set of values (either worst-case or statistical) or another, given their comfort level with these analysis assumptions and the specifics of the design. One can also conduct one's own statistical analysis using the formulas provided and different assumptions.

When hole position tolerances t_1 and t_2 are tabulated, recognize that these values do not need to be applied directly from the tables. The assemblability criterion specifies the maximum total tolerances that can be assigned $(\frac{1}{2}t_1 + t_2)$. The tabulated variables t_1 and t_2 represent suggested tolerances. Any combination of t_1 and t_2 that meet the assemblability criterion will also work.

For example, Table 4-6 suggests $t_1 = 0.033$ mm and $t_2 = 0.017$ mm for a worst-case method analysis of an 8 x 10 mm round pin and diamond pin. Because the total tolerance $\frac{1}{2}t_1 + t_2 = \frac{1}{2}(0.033) + 0.017 = 0.0335$, we could choose t_1 = 0.017 mm and t_2 = 0.025 mm, and the maximum total tolerance value required for assemblability would not be exceeded:

$$\frac{1}{2}t_1 + t_2 = \frac{1}{2}(0.017) + 0.025 = 0.0335$$

Tabulated recommendations for the two tolerances t_1 and t_2 are divided between the two in order to achieve the same center distance variation between the holes of each part. The assumption is that both parts are equally challenging to produce or are made on equipment with similar process capability. It may be advisable to permit one part more tolerance than the other. How the values are assigned to the two parts does not affect assemblability as long as the total $\frac{1}{2}t_1 + t_2$ is not exceeded.

Designing A Custom Solution

Sometimes a solution from the tables cannot be used and the engineer must design a custom approach. The typical custom design process steps are:

1. Choose one of the arrangements described in this chapter based on the summary description and application guidelines.
2. Select commercial hardware if possible. Design custom parts, if necessary.
3. Determine the press-fit hole tolerances based on the press-fit criterion.
4. Select an initial clearance hole size and tolerance based on the clearance fit criteria, if necessary.
5. Calculate the worst-case pin position tolerances required, if necessary. Assign tolerances to the two parts based on manufacturing difficulty and cost.
6. Review the design with the fabricator for additional feedback, if appropriate. Modify features or tolerances as required, maintaining the press-fit, clearance fit, and assemblability criteria.

Consideration of part geometry, manufacturing methods, and required accuracy allows the designer to choose the proper assembly type and to optimize the sizes and tolerances of the features specified to produce the assembly.

Manufacturing Considerations

Holes can be formed by punching, drilling, water jet cutting, milling, turning, boring, jig boring, electrical-discharge wire cutting (wire EDM), EDM drill, jig grinding, and even laser jet. Reaming and honing can be used as secondary operations to improve accuracy and surface finish. In general, the more accurate the process, the more expensive it is. Table 4-1 gives several hole-making processes for a single 6 mm diameter hole in steel with various tolerances. Several process options are listed, any one of which might be appropriate for a given part. Cost is for comparison purposes only and assumes $75 / hour personnel and machine rate. Actual costs will vary with machining operation used and with differing labor rates.

The appropriateness of one manufacturing method over others is a function of the part's material type and hardness, the hole size accuracy required, the hole position accuracy required, the part quantities required, the other features of the part design, and the relationship of the holes to the rest of the part's geometry. Additional considerations are the equipment available at the shop and the skill levels of the machinists. Parts should not be designed so that they can only be manufactured with one particular shop's capabilities and equipment in mind. Whenever possible, parts should be designed for the capabilities of a typical shop.

A machine shop usually has several manufacturing methods to choose from, each of which could be used to produce the holes in question. The problem then becomes choosing the technique to produce the holes that fits compatibly with all the other operations required to produce the part. Trade-offs

Table 4-1: Typical Hole-Making Processes

Hole Size / Tolerance	Process Step	Time (hour)	Cost ($)
6 ± 0.1	• Set-Up / Drill	0.2	15
6 ± 0.025	• Set-Up / Drill and Ream	0.3	22.50
6 ± 0.005	• Set-Up / Drill and Ream	0.5	37.50
	• Set-Up / Drill and Hone	1	75
	• Set-Up / Drill and Wire EDM	1.5	112.50
	• Set-Up / Drill and Jig Grind	1.5	112.50
6 ± 0.0025	• Set-Up / Drill and Ream	0.5	37.50
	• Set-Up / Drill and Hone	1	75
	• Set-Up / Drill and Wire EDM	2	150
	• Set-Up / Drill and Jig Grind	2	150

are made to minimize the time and cost to produce the part as a whole, while maximizing quality.

The choice of hole-making method may not be obvious to the designer. For example, a hole that may be easily produced by twist drilling on a drill press is instead produced by CNC milling to allow the entire part to be machined in one machine setup and avoid re-fixturing for a second process. The designer should provide the manufacturers with as much tolerance as the functionality allows. Maximizing tolerances equates to maximizing flexibility for the shop to choose the methods produce the part most economically.

Hardware

The most common hardware items used for precision locating are pins of various types. Many pin designs are available, of which cylindrical dowel pins are the most common. Locating pins are stepped with shanks of a smaller (though sometimes larger) diameter than the head. A smaller-diameter shank allows seating the pin to a controlled depth and prevents the pin from falling through if the hole is of a looser fit. A through-hole with access from the back side is required to remove a press fit pin. Tapped pins and threaded shanks are also available, as are side-mounting styles including varieties secured with set screws on flats, locating notches, or grooves.

Standard dowel pins have a small chamfer on one side and a rounded nose on the other. Careful alignment of parts is necessary as the round nose is not very forgiving of misalignment. Locating pin heads generally have a simple angled lead-in allowing smooth transition while locating. Lead-ins can be made long and round-tipped. Bullet-nose and round-nosed heads allow even smoother assembly and they allow misalignment during the locating process before the part is fully seated, and during disassembly.

PRECISION LOCATING TECHNIQUES

In the following sections, sample parts are shown to illustrate the various precision locating techniques. The drawings show only the design features relevant to the precision locating function. Real parts designed using any of these locating techniques will of course be differently shaped and contain other features as required by their function. Though designs will necessarily vary, several key points should be clear when applying these design features:

- The "assembled part" is located *with respect to* the "base part." The base part is the "from" in determining the tolerance stack-up chain, and the assembled part is the "to."
- Precision pins are used to locate the assembled part relative to the base part. The precision pins are press-fit into holes in the base part. The press-fit holes locations are dimensioned and toleranced from the base part's datum reference frame.
- The primary datum feature of the assembled part is a planar surface, which mates to a planar surface on the base part. This constrains the assembled part with respect to three degrees of freedom.
- Features (clearance holes, slots and / or faces) on the assembled part mate with the pins and form the secondary (and tertiary, if applicable) datum feature(s) of the assembled part.

Two Holes / Two Pins / Two Holes

In this technique, the two clearance holes in the assembled part are sized and positioned so that together they remove the assembled part's two remaining translational and one remaining rotational degree of freedom (see Figure 4-4). These two holes also serve collectively as the part's secondary and tertiary datum features. This technique is best when:

- Parts permit two high-precision holes to be machined easily in the same setup.
- Tight position tolerances can be held in the parts.
- High-cost manufacturing is permitted.
- Dowels can be inserted to a specific depth.
- Correct diamond pin location and orientation cannot be ensured.
- No additional steps can be taken to properly seat the part.
- Skilled technicians are not available.
- Assemblies are subject to shear loads.

It is sometimes considered poor practice to use the two hole / two pin / two hole technique with commercial precision pins because the precision of the features required to ensure successful assemblability is tighter than with any other technique. In order to ensure assemblability, clearances need to be

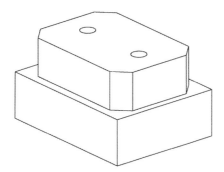 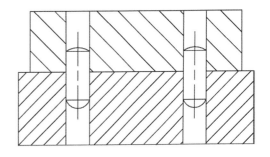

Figure 4-4: Two Holes / Two Pins / Two Holes Assembly

increased to the point that accuracy of the assembly suffers, compared to other techniques. Nevertheless, the technique is described here. Due to its simplicity, this technique serves well as a basis for understanding the overall design and analysis approach used with all the techniques.

Understanding the relationship between tolerances of hole size, pin size, hole position, and manufacturing capabilities allows the engineer to make trade-offs between precision, cost, and complexity. There can be cases where this technique is perfectly advisable. The formulas, figure, and tables that follow completely specify a base part and assembled part for precision assembly using the two holes / two pins / two holes technique (see Figure 4-5 and Tables 4-2 and 4-3).

Press-Fit Criterion: $\qquad P_{\min} - D_{\max} > 0$

Clearance Fit Criterion: $\qquad H_{\min} - P_{\max} \geq 0$

Assemblability / Interchangeability Criterion: $\qquad \dfrac{1}{2}t_1 + t_2 \leq H_{\min} - P_{\max}$

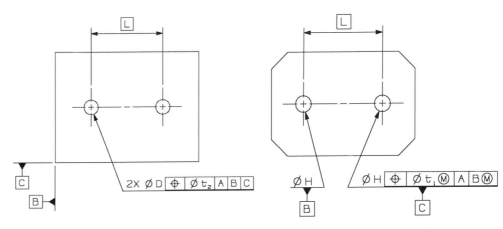

Figure 4-5: Drawings for Two Holes / Two Pins / Two Holes

Table 4-2: Two Holes / Two Pins / Two Holes Dimensions: Metric Hardware

Pin	Worst-Case Method				Statistical Method					
	D K7	t_2	H E7	t_1	D K7	t_2	H E7	t_1	Press-Fit Yield	Assembly Yield
Pin, Dowel ISO 8734 (DIN 6325 m6)-3	3	0.003	3	0.006	3	0.016	3	0.032	100.00%	99.20%
Pin, Dowel ISO 8734 (DIN 6325 m6)-4	4	0.004	4	0.008	4	0.020	4	0.040	100.00%	99.24%
Pin, Dowel ISO 8734 (DIN 6325 m6)-5	5	0.004	5	0.008	5	0.020	5	0.040	100.00%	99.24%
Pin, Dowel ISO 8734 (DIN 6325 m6)-6	6	0.004	6	0.008	6	0.020	6	0.040	100.00%	99.24%
Pin, Dowel ISO 8734 (DIN 6325 m6)-8	8	0.005	8	0.010	8	0.026	8	0.052	100.00%	99.07%
Pin, Dowel ISO 8734 (DIN 6325 m6)-10	10	0.005	10	0.010	10	0.026	10	0.052	100.00%	99.07%
Pin, Dowel ISO 8734 (DIN 6325 m6)-12	12	0.007	12	0.014	12	0.034	12	0.068	100.00%	99.07%

Table 4-3: Two Holes / Two Pins / Two Holes Dimensions: Inch Hardware

Pin	Worst-Case Method				Statistical Method					
	D -.0005 -.0010	t_2	H +.0010 +.0005	t_1	D -.0000 -.0005	t_2	H +.0010 +.0005	t_1	Press-Fit Yield	Assembly Yield
Pin, Dowel ASME B18.8.2 - 3/32	0.09325 0.09275	.0001	0.09475 0.09425	.0002	0.09375 0.09325	.0006	0.09475 0.09425	.0012	100.00%	99.36%
Pin, Dowel ASME B18.8.2 - 1/8	0.1245 0.124	.0001	0.126 0.1255	.0002	0.125 0.1245	.0006	0.126 0.1255	.0012	100.00%	99.36%
Pin, Dowel ASME B18.8.2 - 3/16	0.187 0.1865	.0001	0.1885 0.188	.0002	0.1875 0.187	.0006	0.1885 0.188	.0012	100.00%	99.36%
Pin, Dowel ASME B18.8.2 - 1/4	0.2495 0.249	.0001	0.251 0.2505	.0002	0.25 0.2495	.0006	0.251 0.2505	.0012	100.00%	99.36%
Pin, Dowel ASME B18.8.2 - 5/16	0.312 0.3115	.0001	0.3135 0.313	.0002	0.3125 0.312	.0006	0.3135 0.313	.0012	100.00%	99.36%
Pin, Dowel ASME B18.8.2 - 3/8	0.3745 0.374	.0001	0.376 0.3755	.0002	0.375 0.3745	.0006	0.376 0.3755	.0012	100.00%	99.36%
Pin, Dowel ASME B18.8.2 - 7/16	0.437 0.4365	.0001	0.4385 0.438	.0002	0.4375 0.437	.0006	0.4385 0.438	.0012	100.00%	99.36%
Pin, Dowel ASME B18.8.2 - 1/2	0.4995 0.499	.0001	0.501 0.5005	.0002	0.5 0.4995	.0006	0.501 0.5005	.0012	100.00%	99.36%
Pin, Dowel ASME B18.8.2 - 9/16	0.562 0.5615	.0001	0.5635 0.563	.0002	0.5625 0.562	.0006	0.5635 0.563	.0012	100.00%	99.36%
Pin, Dowel ASME B18.8.2 - 5/8	0.6245 0.624	.0001	0.626 0.6255	.0002	0.625 0.6245	.0006	0.626 0.6255	.0012	100.00%	99.36%

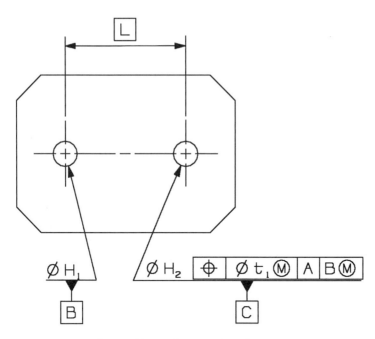

Figure 4-6: Drawing for Differently Sized Holes On Assembled Part

Variation: Differently Sized Holes On Assembled Part

This variation on the Two Holes / Two Pins / Two Holes arrangement uses two differently sized holes in the assembled part (see Figure 4-6). The design objective is to better control the assembly accuracy by reducing the repeatability error. The clearance between hole 1 (the secondary datum feature, B) and pin 1 is smaller than between hole 2 (the tertiary datum feature, C) and pin 2. Datum feature B alone constrains x and y translations, whereas tertiary datum feature C controls rotation.

Clearance Fit Criterion:

$$H_{1_{min}} - P_{max} \geq 0$$

Assemblability /
Interchangeability Criterion:

$$\frac{1}{2}t_1 + t_2 \leq \left(\frac{H_{1_{min}} + H_{2_{min}}}{2} \right) - P_{max}$$

Variation: Composite Tolerance Frame on Base Part Holes

This variation on the Two Holes / Two Pins / Two Holes arrangement uses a composite tolerance frame to control the position of the base part holes. Using composite tolerancing allows larger variation to the position of the two-hole pattern to the base part's datums while maintaining close control over hole-to-hole center distance to maintain assemblability. The position of the assembled part is allowed more variation, while still guaranteeing a center distance that allows assemblability (see Figure 4-7).

In a composite tolerance, the upper control frame specifies the larger tolerance t_3 for the two holes located and oriented to the base part's datum reference frame. These zones are designated by the larger, light grey zones. The lower frame's tolerance t_2 controls the hole-to-hole center distance. The smaller tolerance zones created by the lower frame (in dark grey) are separated by the basic dimension L but otherwise free to float up / down, left / right, and angularly within the base part's datum reference frame. Neither x_D nor y_D tie down the location of the lower frame's tolerance. Both holes must fall within both tolerance zones at their respective locations.

This on the drawing...

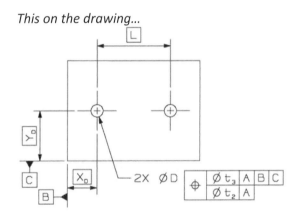

...means this on the part:

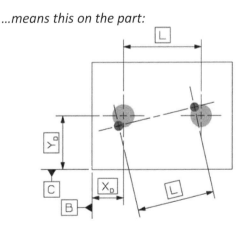

Figure 4-7: Drawing for Composite Tolerance Frame on Base Part Holes

When the holes are produced during the same machine tool setup used to create datum features B and C, it is possible to achieve the same hole-hole center distance accuracy and hole-to-datum reference frame accuracy. However, datum features B and C will often already exist in the part, having been produced on a previous machining operation in a different fixturing setup, perhaps on a different machine tool. In this case, it will be easier to control hole-to-hole center distance variation than to control the location of the hole pattern to the datum reference frame. Composite tolerancing allows this flexibility. Composite tolerancing can be used very effectively to allow the machine shop extra tolerance in the position of the hole pattern relative to the base part's datum reference frame while retaining tight control on the hole-to-hole distance.

Variation: Composite Tolerance Frame on Base Part Holes (2 Datum References)

Figure 4-8 shows another composite tolerancing application. This variation is similar to the previous one except that the orientation of the hole

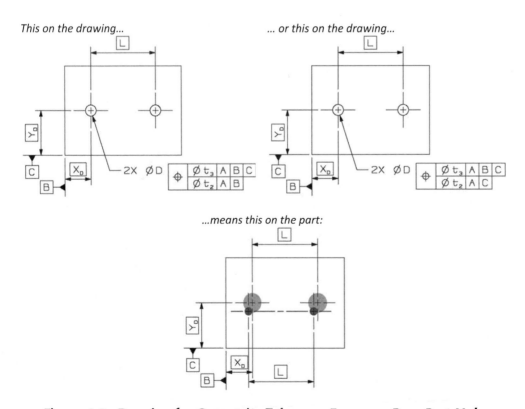

Figure 4-8: Drawing for Composite Tolerance Frame on Base Part Holes (2 Datum References)

pattern is also controlled to the part's datum reference frame. The lower tolerance zone is basically located and oriented between the features and basically oriented to the datum reference frame. The t_2 zones may move up, down, left or right within the confines of t_3, but it may not rotate. They are not tied to the x_D and y_D basic dimensions. Hole axes must fall within both zones simultaneously. Both callouts are equivalent and either one may be used because the addition of either datum B or C to the lower segment's datum reference frame locks the pattern into the orientation of that datum reference frame.

The tolerance t_2 controls the hole-to-hole distance in order to allow assemblability and interchangeability. The assembled part's location will only be accurate to within the larger zone t_3, but it will not be allowed to tilt at an angle more than t_2 allows. This method of composite tolerancing may be used when tight hole-to-hole center distance must be maintained for assemblability, and control over the orientation of the assembled part needs to be controlled to a tighter degree than position.

Variation: Dual Position Feature Control Frames

Two final variations use dual position callouts. Each callout controls the location between the holes and the location and orientation to the specified datum reference frame. These variations are used when the hole-to-hole distance is important, and the location to two of the datums is more critical than location to the third. The holes can move as a group a greater distance in one direction than the other.

The t_3 tolerance zones are basically located and oriented to the datum reference frame ABC. In Figure 4-9a, the additional t_2 tolerance zones are basically located and oriented to the datum reference frame AB with no tie to datum C. The zones can move up or down within the t_3 zone as long as the basic dimension x_D is maintained. The hole axes must fall within both zones simultaneously. Figure 4-9b's t_2 zones are not tied to datum B, and can move left-to-right within the t_3 zone remaining tied to basic dimension y_D. In both of these examples, the tolerance t_2 controls the hole-to-hole distance in order to allow assemblability. The assembled part's location will only be accurate to within the larger zone t_3 in one direction, and it will not be allowed to tilt at an angle more than t_2 allows.

Two Holes / Two Pins / One Hole and One Slot

This precision locating technique uses two precision press-fit pins to fixture the assembled part relative to the base part. The two precision pins

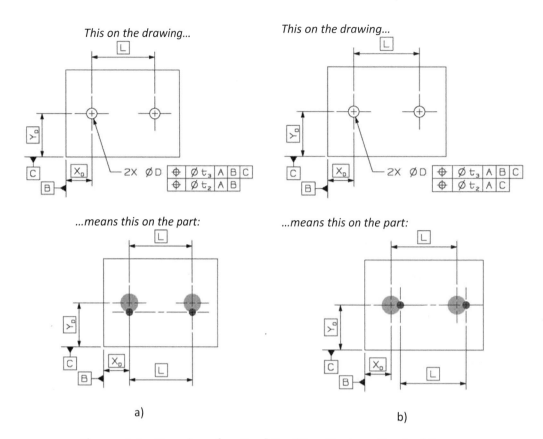

Figure 4-9: Drawings for Dual Position Feature Control Frames

are fit into one clearance hole and one clearance slot in the assembled part. The clearance hole and clearance slot are specified with the same size tolerance. The clearance hole removes the two remaining translational degrees of freedom of the assembled part. The clearance slot removes the final remaining rotational degree of freedom of the assembled part. The clearance hole serves as that part's secondary datum feature, while the clearance slot serves as that part's tertiary datum feature. In Figures 4-10 and 4-11, both open and closed slot shapes are shown. Either one may be used, depending on manufacturability of the part in question. This precision locating technique is best when:

- Assemblies are subject to shear loads (when dowel pins are used).
- Part geometry allows manufacturing of the slot.
- High-cost manufacturing is permitted.
- Correct diamond pin location and orientation cannot be ensured.
- No additional steps can be taken to properly seat the part.

- Skilled technicians are not available.
- Assemblies are subject to minimal or nonexistent shear loads (when locating pins are used).
- Very high accuracy is required.
- Accuracy in the *x* and *y* directions and accurate angular orientation is needed.

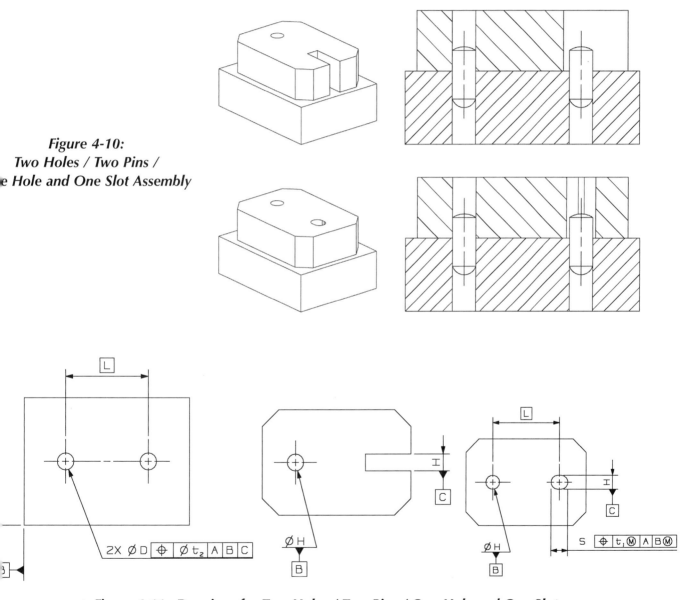

Figure 4-10:
Two Holes / Two Pins /
e Hole and One Slot Assembly

Figure 4-11: Drawings for Two Holes / Two Pins / One Hole and One Slot

Table 4-4: Two Holes / Two Pins / One Hole and One Slot Dimensions: Metric Hardware

Pin	Worst-Case Method					Statistical Method						
	D K7	t$_2$	H E7	S	t$_1$	D K7	t$_2$	H F7	S	t$_1$	Press-Fit Yield	Assembly Yield
Dowel, ISO 8734 (DIN 6325 m6)-3x6-A	3	0.05	3	4 ±0.1	0.7	3	0.05	3	4 ±0.1	1	100.00%	99.52%
Dowel, ISO 8734 (DIN 6325 m6)-4x8-A	4	0.05	4	5 ±0.1	0.7	4	0.05	4	5 ±0.1	1	100.00%	99.59%
Dowel, ISO 8734 (DIN 6325 m6)-5x10-A	5	0.05	5	6 ±0.1	0.7	5	0.05	5	6 ±0.1	1	100.00%	99.59%
Dowel, ISO 8734 (DIN 6325 m6)-6x12-A	6	0.05	6	7 ±0.2	0.6	6	0.05	6	7 ±0.2	1	100.00%	99.61%
Dowel, ISO 8734 (DIN 6325 m6)-8x16-A	8	0.05	8	9 ±0.2	0.6	8	0.05	8	9 ±0.2	1	100.00%	99.65%
Dowel, ISO 8734 (DIN 6325 m6)-10x20-A	10	0.05	10	11 ±0.2	0.6	10	0.05	10	11 ±0.2	1	100.00%	99.65%
Dowel, ISO 8734 (DIN 6325 m6)-12x24-A	12	0.05	12	13 ±0.2	0.6	12	0.05	12	13 ±0.2	1	100.00%	99.59%
Dowel, ISO 8734 (DIN 6325 m6)-16x32-A	16	0.05	16	17 ±0.2	0.6	16	0.05	16	17 ±0.2	1	100.00%	99.59%

Pin	Worst-Case Method					Statistical Method						
	D JS7	t$_2$	H H7	S	t$_1$	D H7	t$_2$	H H7	S	t$_1$	Press-Fit Yield	Assembly Yield
DIN 6321 Form B, 6x7 or 6x12	4	0.05	6	7 ±0.2	0.7	4	0.05	6	7 ±0.2	1	99.37%	99.74%
DIN 6321 Form B, 8x10 or 8x16	6	0.05	8	9 ±0.2	0.7	6	0.05	8	9 ±0.2	1	99.37%	99.74%
DIN 6321 Form B, 10x10 or 10x18	6	0.05	10	11 ±0.2	0.7	6	0.05	10	11 ±0.2	1	99.37%	99.74%
DIN 6321 Form B, 12x10 or 12x18	6	0.05	12	13 ±0.2	0.7	6	0.05	12	13 ±0.2	1	99.37%	99.76%
DIN 6321 Form B, 16x13 or 16x22	8	0.05	16	17 ±0.2	0.7	8	0.05	16	17 ±0.2	1	99.18%	99.76%
DIN 6321 Form B, 20x15 or 20x25	12	0.05	20	21 ±0.2	0.7	12	0.05	20	21 ±0.2	1	99.22%	99.76%

Table 4-5: Two Holes / Two Pins / One Hole and One Slot Dimensions: Inch Hardware

Pin	Worst-Case Method					Statistical Method					Press-Fit Yield	Assembly Yield
	D −.0005 −.0010	t_2	H +.0010 +.0005	S ±.005	t_1	D −.0000 −.0005	t_2	H +.0010 +.0005	S ±.005	t_1		
Pin, Dowel ASME B18.8.2 - 3/32	0.09325 0.09275	.002	0.09475 0.09425	.134	.030	0.09375 0.09325	.002	0.09475 0.09425	.134	.040	100.00%	99.78%
Pin, Dowel ASME B18.8.2 - 1/8	0.1245 0.124	.002	0.126 0.1255	.165	.030	0.125 0.1245	.002	0.126 0.1255	.165	.040	100.00%	99.78%
Pin, Dowel ASME B18.8.2 - 3/16	0.187 0.1865	.002	0.1885 0.188	.228	.030	0.1875 0.187	.002	0.1885 0.188	.228	.040	100.00%	99.78%
Pin, Dowel ASME B18.8.2 - 1/4	0.2495 0.249	.002	0.251 0.2505	.29	.030	0.25 0.2495	.002	0.251 0.2505	.29	.040	100.00%	99.78%
Pin, Dowel ASME B18.8.2 - 5/16	0.312 0.3115	.002	0.3135 0.313	.353	.030	0.3125 0.312	.002	0.3135 0.313	.353	.040	100.00%	99.78%
Pin, Dowel ASME B18.8.2 - 3/8	0.3745 0.374	.002	0.376 0.3755	.415	.030	0.375 0.3745	.002	0.376 0.3755	.415	.040	100.00%	99.78%
Pin, Dowel ASME B18.8.2 - 7/16	0.437 0.4365	.002	0.4385 0.438	.478	.030	0.4375 0.437	.002	0.4385 0.438	.478	.040	100.00%	99.78%
Pin, Dowel ASME B18.8.2 - 1/2	0.4995 0.499	.002	0.501 0.5005	.54	.030	0.5 0.4995	.002	0.501 0.5005	.54	.040	100.00%	99.78%
Pin, Dowel ASME B18.8.2 - 9/16	0.562 0.5615	.002	0.5635 0.563	.603	.030	0.5625 0.562	.002	0.5635 0.563	.603	.040	100.00%	99.78%
Pin, Dowel ASME B18.8.2 - 5/8	0.6245 0.624	.002	0.626 0.6255	.665	.030	0.625 0.6245	.002	0.626 0.6255	.665	.040	100.00%	99.78%

For a closed slot, length and position in the length direction must be specified on the drawing. Slots only need to be long enough to accommodate any variation in position plus pin-to-hole clearance on datum feature B. Typical slot lengths are 1 mm (0.040″) longer than their diameter. Meeting the assemblability criterion does not require precision along the slot length, and recommended slot lengths and position tolerances are given in Tables 4-4 and 4-5.

Press-Fit Criterion:

$$P_{min} - D_{max} > 0$$

Clearance Fit Criterion:

$$H_{min} - P_{max} \geq 0$$

Assemblability / Interchangeability Criterion:

$$\frac{1}{2}t_1 + t_2 \leq \left(\frac{H_{min} + S_{min}}{2}\right) - P_{max}$$

Two Holes / One Round Pin and One Diamond Pin / Two Holes

One precision round locating pin and one precision diamond pin are used to fixture the assembled part relative to the base part. This precision locating technique is best when:

- Moderate positional tolerances can be held in the parts.
- Low / moderate-cost manufacturing is desired.
- Moderate cost manufacturing is permitted.
- No additional steps can be taken to properly seat the part.
- Assemblies are subject to minimal or nonexistent shear loads.
- Very high accuracy is required.
- Accuracy in the x and y directions and accurate angular orientation is needed.

Diamond pins and their use are illustrated in Figure 4-12 (as well as Figures 4-13 and 4-14), and more information can be found in Chapter 5. The holes are shown with exaggerated clearances for illustration purposes. The round pin constrains two degrees of freedom by mating to the assembled part's secondary datum feature hole. The diamond pin must be aligned properly before it is pressed into the base part, such that its locating features constrain the assembled part against rotation by mating to the assembled part's tertiary datum feature. Care must be taken to annotate the assembly drawing to show the proper positions of the round pin and diamond pin and the orientation of the diamond pin.

Metric round head and diamond head pins are available per DIN standard 6321. Other nonstandard pins are available, even in configurable sizes. Diamond pins per DIN 6321 are the recommended hardware tabulated in

Table 4-6 (Table 4-7 shows the tabulations for inch hardware). If different pins are used, the associated formulas will need to be used.

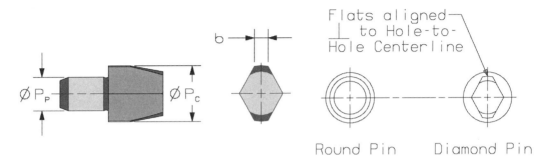

Figure 4-12: Diamond Pin and Application

Figure 4-13:
Two Holes /
One Round Pin and
One Diamond Pin /
Two Holes Assembly

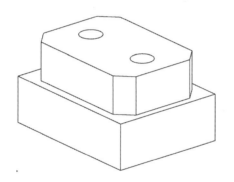
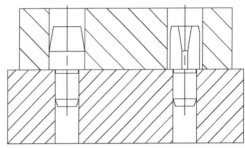

Press-Fit Criterion: $\qquad P_{p_{min}} - D_{max} > 0$

Clearance Fit Criterion: $\qquad H_{min} - P_{c_{max}} \geq 0$

Assemblability /
Interchangeability Criterion:

$$\frac{1}{2}t_1 + t_2 \leq H_{min} - P_{c_{max}} + \frac{1}{2}\sqrt{H_{min}^2 - P_{c_{max}}^2 + b_{max}^2} - \frac{b_{max}}{2}$$

Figure 4-14:
Drawings for Two Holes /
One Round Pin and
One Diamond Pin /
Two Holes

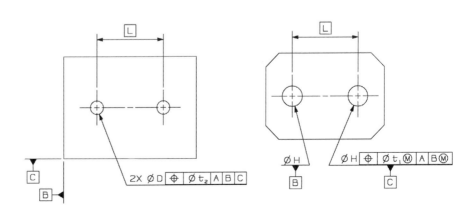

Table 4-6: Two Holes / One Round Pin and One Diamond Pin / Two Holes Dimensions: Metric Hardware

Pin	Worst-Case Method				Statistical Method					
	D JS7	t_2	H G7	t_1	D H7	t_2	H G7	t_1	Press-Fit Yield	Assembly Yield
Round Pin: DIN 6321 Form B, 6x7 or 6x12 Diamond Pin: DIN 6321 Form C, 6x7 or 6x12	4	0.015	6	0.029	4	0.062	6	0.124	99.37%	99.04%
Round Pin: DIN 6321 Form B, 8x10 or 8x16 Diamond Pin: DIN 6321 Form C, 8x10 or 8x16	6	0.017	8	0.033	6	0.066	8	0.132	99.37%	99.08%
Round Pin: DIN 6321 Form B, 10x10 or 10x18 Diamond Pin: DIN 6321 Form C, 10x10 or 10x18	6	0.015	10	0.029	6	0.057	10	0.114	99.37%	99.05%
Round Pin: DIN 6321 Form B, 12x10 or 12x18 Diamond Pin: DIN 6321 Form C, 12x10 or 12x18	6	0.020	12	0.039	6	0.078	12	0.156	99.37%	99.00%
Round Pin: DIN 6321 Form B, 16x13 or 16x22 Diamond Pin: DIN 6321 Form C, 16x13 or 16x22	8	0.019	16	0.038	8	0.075	16	0.150	99.18%	99.10%
Round Pin: DIN 6321 Form B, 20x15 or 20x25 Diamond Pin: DIN 6321 Form C, 20x15 or 20x25	12	0.018	20	0.035	12	0.068	20	0.136	99.22%	99.08%

Table 4-7: *Two Holes / One Round Pin and One Diamond Pin / Two Holes Dimensions: Inch Hardware*

Pin	Worst-Case Method				Statistical Method					
	D	t₂	H	t₁	D	t₂	H	t₁	Press-Fit Yield	Assembly Yield
Round Pin: Carr-Lane* CL-1-RP	.187	0.0018	.2505	0.0037	.187	0.0060	.2505	0.0120	100.00%	99.17%
Diamond Pin: Carr-Lane CL-1-DPX	.1867		.25		.1867		.25			
Round Pin: Carr-Lane CL-2-RP	.2495	0.0019	.313	0.0037	.2495	0.0060	.313	0.0120	100.00%	99.17%
Diamond Pin: Carr-Lane CL-2-DPX	.2492		.3125		.2492		.3125			
Round Pin: Carr-Lane CL-3-RP	.312	0.0019	.3755	0.0038	.312	0.0060	.3755	0.0120	100.00%	99.17%
Diamond Pin: Carr-Lane CL-3-DPX	.3117		.375		.3117		.375			
Round Pin: Carr-Lane CL-4-RP	.3745	0.0019	.438	0.0038	.3745	0.0060	.438	0.0120	100.00%	99.17%
Diamond Pin: Carr-Lane CL-4-DPX	.3742		.4375		.3742		.4375			
Round Pin: Carr-Lane CL-5-RP	.437	0.0019	.5005	0.0038	.437	0.0060	.5005	0.0120	100.00%	99.17%
Diamond Pin: Carr-Lane CL-5-DPX	.4367		.5		.4367		.5			
Round Pin: Carr-Lane CL-6-RP	.4995	0.0037	.563	0.0073	.4995	0.0110	.563	0.0220	100.00%	99.18%
Diamond Pin: Carr-Lane CL-6-DPX	.4992		.5625		.4992		.5625			
Round Pin: Carr-Lane CL-7-RP	.562	0.0038	.6255	0.0076	.562	0.0110	.6255	0.0220	100.00%	99.18%
Diamond Pin: Carr-Lane CL-7-DPX	.5617		.625		.5617		.625			
Round Pin: Carr-Lane CL-8-RP	.6245	0.0038	.688	0.0076	.6245	0.0110	.688	0.0220	100.00%	99.18%
Diamond Pin: Carr-Lane CL-8-DPX	.6242		.6875		.6242		.6875			
Round Pin: Carr-Lane CL-9-RP	.687	0.0038	.7505	0.0077	.687	0.0110	.7505	0.0220	100.00%	99.18%
Diamond Pin: Carr-Lane CL-9-DPX	.6867		.75		.6867		.75			
Round Pin: Carr-Lane CL-10-RP	.7495	0.0038	.813	0.0077	.7495	0.0110	.813	0.0220	100.00%	99.18%
Diamond Pin: Carr-Lane CL-10-DPX	.7492		.8125		.7492		.8125			

* Distributed by the Carr-Lane Manufacturing Co.

continued on next page

Table 4-7: Two Holes / One Round Pin and One Diamond Pin / Two Holes Dimensions: Inch Hardware (Continued)

Pin	Worst-Case Method				Statistical Method					
	D	t_2	H	t_1	D	t_2	H	t_1	Press-Fit Yield	Assembly Yield
Round Pin: Misumi** U-JPBB 0.19 - P0.19 Diamond Pin: Misumi U-JPDB 0.19 - P0.19	.187 .1865	.0005	.1885 .188	.0010	0.187 0.1865	0.0025	0.1885 0.188	0.0050	100.00%	99.20%
Round Pin: Misumi U-JPBB 0.19 - P0.25 Diamond Pin: Misumi U-JPDB 0.19 - P0.25	.187 .1865	.0006	.251 .2505	.0012	0.187 0.1865	0.0030	0.251 0.2505	0.0060	100.00%	99.10%
Round Pin: Misumi U-JPBB 0.25 - P0.25 Diamond Pin: Misumi U-JPDB 0.25 - P0.25	.2495 .249	.0005	.251 .2505	.0010	0.2495 0.249	0.0023	0.251 0.2505	0.0046	100.00%	99.14%
Round Pin: Misumi U-JPBB 0.25 - P0.31 Diamond Pin: Misumi U-JPDB 0.25 - P0.31	.2495 .249	.0005	.3135 .313	.0011	0.2495 0.249	0.0026	0.3135 0.313	0.0052	100.00%	99.15%
Round Pin: Misumi U-JPBB 0.31 - P0.31 Diamond Pin: Misumi U-JPDB 0.31 - P0.31	.312 .3115	.0005	.3135 .313	.0010	0.312 0.3115	0.0024	0.3135 0.313	0.0048	100.00%	99.25%
Round Pin: Misumi U-JPBB 0.31 - P0.38 Diamond Pin: Misumi U-JPDB 0.31 - P0.38	.312 .3115	.0006	.376 .3755	.0011	0.312 0.3115	0.0027	0.376 0.3755	0.0054	100.00%	99.14%
Round Pin: Misumi U-JPBB 0.38 - P0.50 Diamond Pin: Misumi U-JPDB 0.38 - P0.50	.3745 .374	.0006	.4385 .438	.0012	0.3745 0.374	0.0028	0.4385 0.438	0.0056	100.00%	99.06%
Round Pin: Misumi U-JPBB 0.38 - P0.56 Diamond Pin: Misumi U-JPDB 0.38 - P0.56	.3745 .374	.0006	.501 .5005	.0012	0.3745 0.374	0.0030	0.501 0.5005	0.0060	100.00%	99.22%
Round Pin: Misumi U-JPBB 0.50 - P0.56 Diamond Pin: Misumi U-JPDB 0.50 - P0.56	.4995 .499	.0006	.5635 .563	.0012	0.4995 0.499	0.0028	0.5635 0.563	0.0056	100.00%	99.07%
Round Pin: Misumi U-JPBB 0.50 - P0.63 Diamond Pin: Misumi U-JPDB 0.50 - P0.63	.4995 .499	.0006	.626 .6255	.0012	0.4995 0.499	0.0029	0.626 0.6255	0.0058	100.00%	99.20%

** Distributed by Misumi USA, Inc. Automation Components

Two Holes / Two Pins / One Slot

In this technique, two precision press-fit pins are used to fixture the assembled part relative to the base part (see Figures 4-15 and 4-16). Because a translational degree of freedom remains, it is necessary to provide additional means to ensure the assembled part is properly positioned, both during assembly and continually afterwards. The final translational degree of freedom can be removed manually at the time of assembly using setup gauges, grind

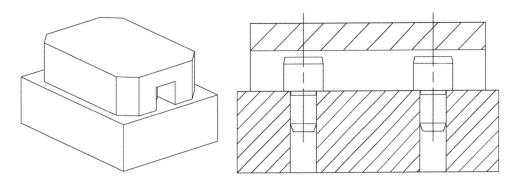

Figure 4-15: Two Holes / Two Pins / One Slot Assembly

Press-Fit Criterion: $\qquad P_{min} - D_{max} > 0$

Clearance Fit Criterion: $\qquad H_{min} - P_{max} \geq 0$

Assemblability / Interchangeability Criterion: $\qquad H_{min} - P_{max} \geq 0$

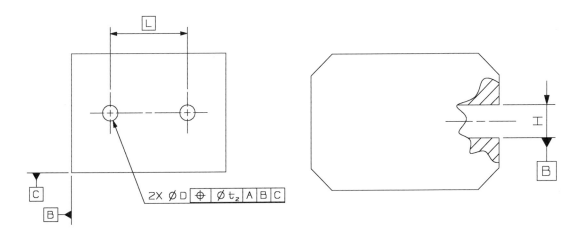

Figure 4-16: Drawings for Two Holes / Two Pins / One Slot

spacers, and the like. Alternatively, this locating technique may also be used when the part has no strict precision requirement along the slot direction and only requires one position and one orientation constraint. The feature of the assembled part that serves to remove its final degree of freedom may or may not serve as the tertiary datum feature of that part, according to the needs of the application. This precision locating technique is best when:

- Part geometry allows manufacturing of the slot.
- Moderate cost manufacturing is permitted.
- Correct diamond pin location and orientation cannot be ensured.
- Assemblies are subject to minimal or nonexistent shear loads (using step pins).
- The final degree of freedom can be constrained by other means.
- The final degree of freedom does not require precision.
- Accuracy is required in the y direction only, and accurate angular orientation is needed.
- One degree of freedom is required for final setup adjustment.

See Tables 4-4 and 4-5 for hardware and tolerance choices, which are the same for the two holes / two pins / one hole and one slot technique, (hole and slot position tolerances t_2 and t_1 given in Tables 4-4 and 4-5 may be ignored for this application).

Three Holes / Three Pins / Two Faces

In this technique, three precision press-fit pins are used to fixture the assembled part relative to the base part (see Figures 4-17 and 4-18). External faces on the assembled part form that part's datum reference frame and it is necessary to provide additional means to ensure the assembled part is contacting all three fixture pins, both during assembly and continually afterwards. When this is done, this type of assembly has zero repeatability error.

The two aligned pins remove one translational and one rotational degree of freedom of the assembled part, and the mating face serves as the secondary datum feature of that part. The third pin removes the final remaining translational degree of freedom of the assembled part, and the mating face serves as the tertiary datum feature of the part. This precision locating technique is best when:

- Additional features can be designed to provide nesting force.
- Low-cost manufacturing is desired.

- Correct diamond pin location and orientation cannot be ensured.
- Repeatability error must be exactly zero.
- Very high accuracy is required.
- Accuracy in the *x* and *y* directions and accurate angular orientation is needed.

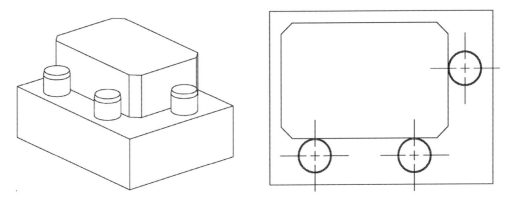

Figure 4-17: **Three Holes / Three Pins / Two Faces Assembly**

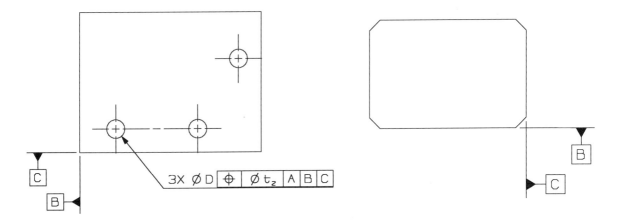

3X ∅ D | ⊕ | ∅ t₂ | A | B | C

Figure 4-18: **Drawings for Three Holes / Three Pins / Two Faces**

Press-Fit Criterion: $P_{min} - D_{max} > 0$

Clearance Fit Criterion: N/A

Assemblability /
Interchangeability Criterion: N/A

CRITICAL CONSIDERATIONS:
Precision Locating Techniques

- Be certain that precise, repeatable location between assembled components is beneficial. Unnecessary precision can be expensive.
- A good design not only works, but is achievable at the right price. Though precision assemblies of parts can be made by specifying tolerances too tight in order to "play it safe," unnecessary precision can be expensive.
- The design engineer is responsible for understanding the application thoroughly and choosing a design that is manufacturable and can accomplish the function.
- Pins must be sized appropriately to safely take any shear loads imposed on the joint during assembly and service. See Chapter 5 for more information.

BEST PRACTICES: Precision Locating Techniques

- Pin nominal diameters are generally selected to be the same as the nominal diameters of the fasteners used to secure the parts together.
- With dowel pins, specify the length of pin protrusion from the base part. This should be less than or equal to the pin nominal diameter to ensure only one or two degrees of freedom are constrained.
- Use composite position tolerancing to loosen the hole pattern position while maintaining hole-to-hole precision and to specify hole position accuracy to only those datums where it is required. It is often possible to relax tolerances in one or two directions without affecting function.
- Precision locating features should be spaced widely apart to reduce the effect of repeatability error on the assembly features requiring precise location.
- Review precision locating designs with the machine shop(s) fabricating your parts. Every part is different. Further, machinists are extremely creative and can find a way to make almost any feature. Collaborating closely is the best way to ensure that the right features are made at the right cost.
- See Chapter 5 for additional best practices when designing holes for pins.

SUMMARY

Table 4-8 summarizes application recommendations by precision locating technique and can help with selecting the appropriate technique. Table 4-9 summarizes the formulas used in this chapter.

Table 4-8: Precision Locating Application Matrix

	Application Conditions	Two Holes/ Two Pins/ Two Holes	Two Holes/ Two Pins/ Two Holes (Different Size Clearance Holes)	Two Holes/ Two Pins/ One Hole and One Slot	Two Holes/ One Round Pin and One Diamond Pin/Two Holes	Two Holes/ Two Pins/ One Slot	Three Holes/ Three Pins/ Two Faces
PART GEOMETRY	Parts permit two high-precision holes to easily be machined in the same setup	●	●	⊘	⊘	⊘	⊘
	Part geometry allows manufacturing of slot	⊘	⊘	●	⊘	●	●
	Additional features can be designed to provide nesting force	⊘	⊘	⊘	⊘	⊘	●
POSITIONAL TOLERANCE REQUIREMENT	None	⊘	⊘	⊘	⊘	●	⊘
	Loose	⊘	⊘	●	○	○	⊘
	Moderate	⊘	●	⊘	●	⊘	⊘
	Tight	●	●	⊘	●	⊘	⊘
MANUFACTURING	Low-cost manufacturing	⊘	⊘	⊘	⊘	●	●
	Moderate cost manufacturing permitted	⊘	⊘	○	●	○	●
	High-cost manufacturing permitted	●	●	●	●	●	●
ASSEMBLY PROCESS	Dowels can be inserted to a specific depth	●	●	●	⊘	●	●
	Correct diamond pin location and orientation cannot be ensured	●	●	●	⊘	●	●
	No additional steps can be taken to properly seat the part	●	●	●	●	●	⊘
	Skilled technicians not available	●	●	○	○	○	○
SERVICE LOADS	Assemblies subject to shear loads (using dowel pins)	●	●	⊘	⊘	○	○
	Assemblies subject to minimal or nonexistent shear loads	●	●	●	●	●	●
ASSEMBLY ACCURACY	Assemblies where repeatability error must be exactly zero	⊘	⊘	⊘	⊘	⊘	⊘
	Very high accuracy required	⊘	○	●	●	●	●
	The final degree of freedom can be constrained by other means	⊘	⊘	⊘	⊘	⊘	⊘
	The final degree of freedom does not require precision	⊘	⊘	⊘	⊘	●	⊘
	Accuracy in x and y directions and rotation	○	○	●	●	●	●
	Accuracy in y direction and rotation only	⊘	⊘	⊘	⊘	●	⊘
	One degree of freedom required for final setup adjustment	⊘	⊘	⊘	⊘	●	⊘

● Suitable
○ Sometimes Suitable
⊘ Not Suitable

Table 4-9: Precision Locating Formula Summary

Technique	Press-Fit Criterion*	Clearance Fit Criterion**	Assemblability / Interchangeability Criterion**
Two Holes / Two Pins / Two Holes	$P_{\min} - D_{\max} > 0$	$H_{\min} - P_{\max} \geq 0$	$\dfrac{1}{2}t_1 + t_2 \leq H_{\min} - P_{\max}$
Two Holes / Two Pins / Two Holes (Differently Sized Clearance Holes)	$P_{\min} - D_{\max} > 0$	$H_{1_{\min}} - P_{\max} \geq 0$ $H_{2_{\min}} - P_{\max} \geq 0$	$\dfrac{1}{2}t_1 + t_2 \leq \left(\dfrac{H_{1_{\min}} + H_{2_{\min}}}{2}\right) - P_{\max}$
Two Holes / Two Pins / One Hole and One Slot	$P_{\min} - D_{\max} > 0$	$H_{\min} - P_{\max} \geq 0$	Closed Slot: $\dfrac{1}{2}t_1 + t_2 \leq \left(\dfrac{H_{\min} + S_{\min}}{2}\right) - P_{\max}$ Open Slot: N/A
Two Holes / One Round Pin and One Diamond Pin / Two Holes	$P_{p_{\min}} - D_{\max} > 0$	$H_{\min} - P_{\max} \geq 0$	$\dfrac{1}{2}t_1 + t_2 \leq H_{\min} - P_{c_{\max}} + \dfrac{1}{2}\sqrt{H_{\min}^2 - P_{\max}^2 + b_{\max}^2} - \dfrac{b_{\max}}{2}$
Two Holes / Two Pins / One Slot	$P_{\min} - D_{\max} > 0$	$H_{\min} - P_{\max} \geq 0$	N/A
Three Holes / Three Pins / Two Faces	$P_{\min} - D_{\max} > 0$	N/A	N/A

* Substitute $P_{p_{\min}}$ for $P_{\min}$ when using pins with different shank (p_p) and head (p_c) diameters.

** Substitute $P_{c_{\max}}$ for $P_{\max}$ when using pins with different shank (p_p) and head (p_c) diameters.

5

PINS, KEYS, AND RETAINING RINGS

Tables

5-1	Stress Equations	218
5-2	Pin Selector	219
5-3 – 5-4	Retaining Ring Equivalency Charts	225–226
5-5 – 5-8	Dowel Pins	228–231
5-9 – 5-12	Withdrawal (Pullout) Dowel Pins	232–235
5-13	Stepped Locating and Seating Pins, Metric	237
5-14 – 5-15	Spring Pins, Coiled, Inch	238–239
5-16 – 5-17	Spring Pins, Slotted, Inch	240–241
5-18 – 5-19	Spring Pins, Coiled, Metric	242–243
5-20 – 5-22	Spring Pins, Slotted, Metric	244–247
5-23 – 5-26	Grooved Pins	248–251
5-27 – 5-30	Clevis Pins	252–255
5-31 – 5-34	Cotter Pins	256–259
5-35 – 5-36	Parallel Square Keys, Inch	260–261
5-37 – 5-38	Parallel Keys With Round Ends, Metric	262–263
5-39 – 5-40	Dimensions and Fits, Woodruff Keys, Inch	264–266
5-41 – 5-42	Axially-Assembled Internal Retaining Rings	268–270
5-43 – 5-44	Axially-Assembled External Retaining Rings	272–274
5-45 – 5-46	Radially-Assembled External (E-Ring) Retaining Rings	276–277
5-47	SolidWorks® Toolbox Components	278

Pins, keys, and retaining rings are very commonly used in machine design. Pins and keys are normally used for locating parts and transmitting shear loads. Retaining rings are used to create an artificial shoulder on a shaft or round part. This chapter provides a brief overview of stress analysis for pins and keys as well as pin, key, and retaining ring usage information and data. Consult the recommended resources and applicable standards for more information if needed.

 RECOMMENDED RESOURCES

- R. Mott, *Machine Elements in Mechanical Design*, 5th Ed., Pearson/ Prentice Hall, Upper Saddle River, NJ, 2012
- R. L. Norton, *Machine Design: An Integrated Approach*, 4th Ed., Prentice Hall, Upper Saddle River, NJ, 2011
- Oberg, Jones, Horton, Ryffel, *Machinery's Handbook*, 28th Ed., Industrial Press, New York, NY, 2008
- R. Parmley, *Standard Handbook of Fastening and Joining*, 3rd Ed., McGraw-Hill , New York, NY, 1997

PINS AND KEYS IN SHEAR

Pins and keys are normally placed in shear, and can fail by shearing or by causing bearing failure of the mating part. Dowel and locating pins are normally made of hardened steel. Keys are typically made out of ductile, low-carbon steel. Stainless steel and other materials are available for special applications. Be sure to select the proper material properties when investigating the strength of a pin, key, or mating part.

Stress equations and material properties commonly used for pins and keys are given in Table 5-1. Spring pins and grooved pins are designed for taking shear loads in joints; their shear capabilities are standardized. When more than one pin is expected to carry a shear load in a joint, consider the effects of clearances and tolerances in the assembly and their effect on load sharing. In a case where load sharing between two pins in holes must be guaranteed, the pin holes in both mating parts should be reamed for a press fit at assembly to eliminate any clearances.

To calculate the shear force on a key used in a shaft, multiply the torque by the radius at the shear point. When alternating or reversing loads are

Table 5-1: Stress Equations

A_s = shear area	SF = factor of safety	A_b = bearing area
T = torque	S_y = tensile yield strength of the material	

PINS

Direct Shear Stress

$$\tau_{AVG} \le \frac{S_y}{2(SF)}$$

Single Shear

$$\tau_{AVG} = \frac{F}{A_s}$$

Double Shear

$$\tau_{AVG} = \frac{F}{2A_s}$$

Bearing Stress

$$\sigma_b \le \frac{S_y}{SF}$$

$$\sigma_b = \frac{F}{A_b} = \frac{F}{td}$$

Single Shear

Double Shear

COMMON PIN MATERIALS

Groove and spring pin double shear capacities: Tables 5-14 – 5-26
Standard hardened ground machine dowels: highly variable
Consult manufacturers for correct material properties.

COMMON KEY MATERIALS

SAE/AISI 1018 CR steel: S_y = 53.7 kpsi (370 MPa)

SAE/AISI 1035 CR steel : S_y = 67 kpsi (462 MPa)

SAE/AISI 1045 CR steel: S_y = 77 kpsi (531 MPa)

Consult manufacturers for correct material properties.

PARALLEL KEYS

Direct Shear Stress

$$\tau_{AVG} = \frac{F}{A_s} = \frac{2T}{DWL}$$

$$\tau_{AVG} \le \frac{S_y}{2(SF)}$$

Bearing Stress

$$\sigma_b = \frac{F}{A_b} = \frac{4T}{DLH}$$

$$\sigma_b \le \frac{S_y}{SF}$$

Shear plane
Shear area = WL
Force of shaft on key
Reaction of hub on key
$F = \frac{T}{D/2}$
Shaft
Force F distributed over bearing area, $L(H/2) = A_s$
Hub

present, investigate the possibility of fatigue failure. When significant loads are involved, consider the stress concentrations in the shaft caused by the keyway. Consult the recommended resources for calculation guidance. Always use appropriate safety factors when performing calculations and sizing components. A safety factor of 3 is often recommended for keys.

PINS

Pins are an inexpensive way to prevent relative movement between parts. Dowel and locating pins are the weapons of choice for precision locating. Taper pins are recommended for very precise location of parts, or for accurate location of parts that must be frequently disassembled. This book does not cover taper pins, but they are discussed in the recommended resources. For more information on the use of pins for precision locating, see Chapter 4 of this book.

The chart in Table 5-2 can help with the selection of pins based on performance characteristics. What follows are descriptions of some of the most commonly used types of pins. Dimensions and fits are given for the most common types and sizes. For more selection and sizes, please consult the relevant standards and recommended resources.

Table 5-2: Pin Selector

COMPONENT	LOCATING ACCURATELY	FREQUENT DISASSEMBLY	SHEAR LOADS	NEW PARTS	USED PARTS	INSTALL WITHOUT PRESS	RESISTS LOOSENING
Dowel Pin	●	⊘	●	●	⊘	⊘	⊘
Oversize Dowel Pin	●	⊘	●	⊘	●	⊘	⊘
Withdrawal Dowel	●	⊘	●	●	⊘	⊘	⊘
Taper Pin	●	●	●	●	●	⊘	⊘
Stepped Locating Pin	●	⊘	⊘	●	⊘	⊘	⊘
Spring Pin	⊘	⊘	●	●	●	●	●
Groove Pin	⊘	⊘	●	●	●	●	●
Clevis Pin	O	●	●	●	●	●	⊘
Cotter Pin	⊘	●	⊘	●	●	●	●

● SUITABLE
O SOMETIMES SUITABLE
⊘ NOT SUITABLE

Dowel pins come in a variety of types. The most commonly used type is the standard hardened ground machine dowel pin, which is slightly oversized. ASME inch dowels are oversized by 0.0002 inch, and ISO metric dowels are oversized to an m6 tolerance to ensure a tight press fit. Soft dowels, production dowels, and straight pins are less commonly used. Pins called "oversized" dowels run 0.001 inch oversize and are meant to be used as replacement pins after disassembly. These are less commonly used and are not covered in this book. In special cases, dowel pins can be custom fitted at assembly to eliminate clearances. For this method, a pilot hole is drilled at the dowel location on one part, and a precision dowel hole is manufactured on the other part. The two parts are positioned at assembly. At that point the pilot hole is drilled and reamed using the other part as a guide, and a dowel is placed. The dimensions and recommended hole tolerances of some common dowel sizes are given in Tables 5-5 – 5-8. The applicable standards are:

ASME B18.8.2: "Taper Pins, Dowel Pins, Straight Pins, Grooved Pins, and Spring Pins (Inch Series)"

ISO 8734: "Parallel pins, of hardened steel and martensitic stainless steel (Dowel pins)"

Type A ISO dowels are through hardened, whereas type B are case hardened. The diameter of this pin has an m6 tolerance, as do DIN EN ISO 8734 and DIN 6325 dowels.

Withdrawal, or pullout dowel pins have an internal thread and are used when a dowel hole must be blind or if the assembly makes normal dowel removal impractical. Pullout dowels are available with spiral or flat vents for use in blind holes. This is essential for allowing air to escape during insertion. The dimensions and recommended hole tolerances of some common sizes are shown in Tables 5-9 – 5-12. The applicable standards are:

ASME B18.8.2: "Taper Pins, Dowel Pins, Straight Pins, Grooved Pins, and Spring Pins (Inch Series)"

ISO 8735: "Parallel pins with internal thread, of hardened steel and martensitic stainless steel"

Stepped locating pins can be used to locate parts as accurately as dowel pins. Locating pins of form A are often called seating pins because their heads are flattened and held to a precise height. Form B pins have round heads that are larger than the pin shank. Locating pins of form C are often called diamond pins because of their diamond-shaped head. Diamond pins are normally paired with a standard round locating pin (form B) or dowel and provide location in only one direction. An advantage of this type of pin is its stepped shape which prevents pins from falling through a hole due to vibration or improper press fit. A serious disadvantage of the stepped locating pin is its low shear strength compared to standard dowels. Another consideration is the higher cost of these pins compared to standard dowels. There is no U.S. (inch) standard, though parts are commercially available. The dimensions of some common metric sizes and recommended hole tolerances are shown in Table 5-13. The applicable metric standard is:

DIN 6321:   "Locating and Seating Pins"

Spring pins come in two basic types: coiled and slotted. They are sometimes called roll pins. Both types resist loosening due to shock and vibration, though coiled pins have a better ability to compress after installation if needed. Stresses are more uniform through coiled pins than slotted pins. Both are ideal for easy installation and can be reused. An alternative to spring pins are grooved pins, which have much greater shear strength due to their solid cross section. When sizing spring pins for shafts, do not cross-drill shafts with holes greater than 1/3 of the shaft diameter. Heavy duty spring pins are recommended when the shaft is hardened or when the pin is 1/4 of the shaft diameter or greater. A chamfer must be added at the hole entrance to prevent tearing of the pin during installation. The dimensions and hole tolerances for some common types and sizes are shown in Tables 5-14 – 5-22. It is important to note that the metric ASME standard slotted pins have a higher shear strength and lower insertion force than the same diameter ISO pins. The applicable standards are:

ASME B18.8.2:   "Taper Pins, Dowel Pins, Straight Pins, Grooved Pins, and Spring Pins (Inch Series)"
ASME B18.8.4M: "Spring Pins — Slotted (Metric Series)"
ISO 8752:   "Spring-type straight pins — Slotted, heavy duty"
ISO 8750:   "Spring-type straight pins — Coiled, standard duty"
ISO 8748:   "Spring-type straight pins — Coiled, heavy duty"

Grooved pins are solid pins with axial grooves along their sides. The burrs created by the grooves cause an interference fit with the sides of a hole. These pins resist loosening as a result of the interference. An alternative to grooved pins are spring pins, with grooved pins having significantly greater shear strength due to their solid cross section. The three most commonly used grooved pins are taper, center grooved, and half grooved with neck. Taper grooved pins are fully embedded in a part or assembly. Center grooved pins are designed to engage at the midpoint and can have both ends exposed or used in joints. Grooved pins with neck are especially suitable for anchoring extension springs. When inserting a grooved pin into a part that is harder than the pin, add a chamfer to the hole entrance to prevent shearing of the groove burrs during insertion. The dimensions and recommended hole tolerances of some common types and sizes are shown in Tables 5-23 – 5-26. The applicable standards are:

ASME B18.8.2:	"Taper Pins, Dowel Pins, Straight Pins, Grooved Pins, and Spring Pins (Inch Series)"
ISO 8744:	"Grooved pins — Full-length taper grooved"
ISO 8742:	"Grooved pins — One-third-length centre grooved"
DIN 1469:	"Grooved pins — half length grooved with gorge"

Clevis pins are used in double shear to link parts together, normally part of a "shackle" arrangement. They are available with or without a head, though headed types are most commonly used in machine design. The hole through the shaft is for installation of a cotter pin. Many clevis pins are available with multiple cotter pin holes, making them of adjustable length. The dimensions of some common sizes and their recommended hole tolerances are shown in Tables 5-27 – 5-30. The applicable standards are:

ASME B18.8.1:	"Clevis Pins and Cotter Pins (Inch Series)"
ISO 2341:	"Clevis pins with head"

Cotter pins, or split pins, are made to be inserted through a hole and the tines bent to prevent the pin from backing out. Cotter pins are often used with clevis pins, but can also be inserted into cross-drilled bolts, shafts, and other parts. Cotter pins should not be reused. The dimensions of some common sizes are shown in Tables 5-31 – 5-34. The applicable standards are:

ASME B18.8.1:	"Clevis Pins and Cotter Pins (Inch Series)"
ISO 1234:	"Split pins"

KEYS

Keys are normally used to prevent relative movement and transmit torque between hubs and shafts. They are sometimes used in other types of joints to provide precision locating in one plane when heavy shear loads are present. Keys are held to close tolerances and are suitable for precision locating. When used with shafts, keys are usually sized based on standard guidelines related to shaft size. Keys are normally placed in shear, sometimes with reversing loads. The subject of keys in shear was covered earlier in this chapter. Keys are typically made of ductile material and are normally designed to fail before the keyways in the parts fail.

Key length on shafts should be limited to less than 1.5 times the shaft diameter to avoid torsional effects. However, key length should always be long enough to carry the required loads without distorting the keyways. If insufficient length is available to carry the required load, two keys can be placed 90° apart to share the load. Key shear stress and bearing stress should always be calculated, especially when loads are significant or when dealing with soft materials. The equations in Table 5-1 can be used to analyze stresses and determine the required key length.

The most common types of keys used in modern machine design are parallel, Woodruff, and tapered. What follows are brief descriptions of these types of keys.

Parallel keys are the most common type of key used. They must be secured axially through the use of a set screw, blind slot, or some other method. Set screws in the hub are often used to axially secure parallel keys. Most metric parallel keys come standard with radii on their ends. These are sometimes called "feather keys." These types of keys can be axially captured on hubs or shafts using milled blind slots. These are the parallel keys of choice when working with metric machinery. Feather keys are also available in inch sizes, though not standardized.

There are two commonly used fits for U.S. standard parallel keys specified by ASME B17.1: Class 1 (clearance) fit, and Class 2 (transition) fit. Class 1 fits use an undersized key, and Class 2 fits use an oversized key. Metric keys and keyways are most commonly used with clearance and transition fits as well. Tight fits are part of the standard, but are not often used. Use a clearance fit for unidirectional loading. In applications with variable loads, the set screw that secures the key axially acts to stabilize the joint. For reversing loading, use a transition fit, preferably with a clamping hub to eliminate backlash. For heavy reversing loading, a tight fit may be required with a clamping or interference fit hub. Be sure to perform appropriate calculations when sizing and selecting fits for keys.

According to the U.S. standard, square keys are preferred over rectangular for shafts less than 6.5 inches in diameter. Metric keys according to DIN 6885 ("feather keys") are square for shafts up to 22 mm in diameter. The dimensions of some of the most common parallel key types and sizes are shown in Tables 5-35 – 5-38 with shaft sizes, recommended set screws, and common keyway fits. The applicable standards are:

> ASME B17.1:   "Keys and Keyseats"
> DIN 6885:   "Drive Type Fastenings without Taper Action; Parallel Keys, Keyways, Deep Pattern"

Woodruff keys are made to fit in a semicircular pocket in a shaft, and are recommended when used near shaft shoulders to reduce stress concentrations in the shaft. They are axially captive. Stress calculations for Woodruff keys can be complex due to their shape. Consult the recommended resources for more information. Fits for U.S. standard Woodruff keys are defined as the loosest fit that will still cause the key to stick in the shaft keyseat. Metric standards specify two fits: sliding and tight. The dimensions of some common key sizes are shown in Tables 5-39 – 5-40 with shaft sizes and keyway fits. The applicable standards are:

> ASME B17.2:   "Woodruff Keys and Keyseats"
> DIN 6888:   "Drive Type Fastenings without Taper Action; Woodruff Keys, Dimensions and Application"

Tapered keys are less commonly used, and are driven axially into a tapered slot on the hub to lock the hub onto the shaft with no backlash. Tapered keys are appropriate for heavy loads and especially good for reversing loads. The most common tapered key has a gib head to facilitate removal. Dimensions of these keys and keyseats are not provided in this text. See the recommended resources for more information on tapered keys and their use.

> ASME B17.1:   "Keys and Keyseats"
> DIN 6887:   "Stressed-type fastenings with taper action — Taper keys with gib heads, keyways — Dimensions and application"

RETAINING RINGS

Retaining rings are used to create an artificial shoulder in a housing bore (internal retaining ring) or on a shaft (external retaining ring.) Retaining rings are not designed to accommodate heavy thrust loads or impacts. Retaining rings have rated thrust load capacities that are dependent on ring material. Consult the manufacturer for thrust load data. Retaining rings come in a variety of materials and finishes. Avoid material combinations that could result in galvanic corrosion.

Retaining rings are used to save cost, weight, or space. Low profile retaining rings are available for tight installation spaces. E-rings are radially-assembled retaining rings and therefore can be placed anywhere along a shaft. These rings are also available in smaller sizes than standard external retaining rings. Use these rings to retain shafts in light duty assemblies instead of other fasteners to save cost, size, and weight.

Retaining rings are installed in tightly controlled grooves. Where possible, chamfers should be avoided in combination with retaining ring grooves or abutments where chamfers would reduce the retaining ring groove walls or contact faces abutting the ring. Reduction of contact area will reduce the thrust load bearing capacity of the assembly.

Among inch size retaining rings, the three most common types are designated NA1, NA2, and NA3. NA1 is an axially-assembled internal retaining ring. NA2 is an axially-assembled external retaining ring. NA3 is a radially-assembled external retaining ring, often called an E-ring due to its shape. These retaining rings are governed by ASME B18.27.1, which is no longer available. These rings are made by a number of companies who hold to the standard dimensions. Table 5-3 shows four major manufacturers and their equivalent standard products.

Metric retaining rings are governed by DIN standards. Axially-assembled internal retaining rings are governed by DIN 472. Axially-assembled

Table 5-3: Retaining Ring Equivalency Chart, Inch

Type	ASME Designation	Brand Name Catalog Number			
		Rotor Clip	Arcon	Waldes	I.R.R
Internal	NA1	SH	1400	5100	3100
External	NA2	HO	N1300	N5000	3000
E-Ring	NA3	E	1500	5133	1000

external retaining rings are governed by DIN 471. E-rings are governed by DIN 6799. Table 5-4 shows four major manufacturers and their equivalent standard products.

Table 5-4: Retaining Ring Equivalency Chart, Metric

Type	DIN Standard	Brand Name Catalog Number			
		Rotor Clip	Arcon	Waldes	Seeger
Internal	472	DHO	D1300	D1300	J
External	471	DSH	D1400	D1400	A
E-Ring	6799	DE	D1500	D1500	RA

The dimensions of some common inch and metric retaining ring sizes are shown in Tables 5-41 – 5-46 with shaft and groove sizes.

CRITICAL CONSIDERATIONS: Pins, Keys, and Retaining Rings

- The accurate location of two parts normally requires no more than two locating components in addition to the plane of contact in the joint. The use of too many locating components in any joint can lead to unwanted stresses or installation problems. See Chapter 4 for more information on accurate location methods.
- Select pin and key sizes appropriate to the loading conditions. Calculate shear and bearing stresses and use appropriate factors of safety.
- When sizing holes for dowel pins, hole fits and tolerances may vary with material, temperature, and application.
- Retaining rings are not designed to accommodate heavy thrust loads or impacts.
- When loads are significant, consider the stress concentrations caused in the shaft by a keyway.
- Keys are inexpensive sacrificial elements. Always design such that the key fails before anything else does in case of an overload.

BEST PRACTICES: Pins, Keys, and Retaining Rings

- Pins are normally installed into press fit holes in one part, and fit into slightly looser (locational transition or slip fit) holes or slots in another part. Press fit holes in both parts are not recommended unless needed for load sharing between pins. Pins should be pressed in to a depth of 1.5 – 2 times the pin diameter for good locating accuracy.
- Do not press fit solid pins into cylindrical blind holes unless they are relieved to allow air to escape during installation. If a blind hole must be used, a vented withdrawal (pullout) dowel is recommended.
- Provision for removal of press fit pins should be made, usually through the use of a through hole. The through hole can be either bigger or smaller than the pin hole itself. If a through hole cannot be accommodated, a withdrawal (pullout) type dowel can be used.
- Avoid specifying long holes for pins that must be held to tight tolerances. Hold the pin hole to its required tolerances only where the pin is captivated. Complete the through hole with either a larger or smaller drill from the other side. This is more cost effective than holding the entire through hole to tight tolerances.
- Design for the loosest key fit possible for the application to ease installation.
- Use retaining rings to retain shafts or rods to save space and weight when there are no significant thrust loads.

COMPONENT DATA

Dimensional information and fits for some of the most commonly used pins, keys, and retaining rings are provided in the following pages in Tables 5-5 – 5-46. For more sizes and types, please consult the relevant standards and recommended resources. Fits shown are recommendations only. Proper tolerance analysis should be undertaken in applications where fits are critical. A chart showing standard hardware items available in Solidworks® Toolbox can be found in Table 5-47. This chart can be particularly useful in determining what items are most commonly used and which standard applies to each part.

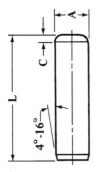

Table 5-5: Dowel Pins, Inch
(Diameters Up to 1 inch)
Part Complies With ASME B18.8.2 – 1995

Nominal Diameter inch		Tip Crown Length inch	Diameter Tolerance inch	Press Fit Hole Tolerance inch	Tight Press Fit Hole, Soft Material Tolerance inch	Slip Fit Hole Tolerance inch
1/16	0.0625	0.020				
5/64	0.0781	0.026				
3/32	0.0938	0.031				
1/8	0.1250	0.041				
5/32	0.1562	0.052				
3/16	0.1875	0.062	+0.0003	0	-0.0005	+0.0010
1/4	0.2500	0.083	+0.0001	-0.0005	-0.0010	+0.0005
5/16	0.3125	0.104				
3/8	0.3750	0.125				
7/16	0.4375	0.146				
1/2	0.5000	0.167				
5/8	0.6250	0.208				
3/4	0.7500	0.250				
7/8	0.8750	0.293				
1	1.0000	0.333				

Shaded = non-preferred sizes

Table 5-6: *Lengths, Dowel Pins, Inch*
(Diameters up to 1 inch and 3 inches long)
Part Complies With ASME B18.8.2 – 1995

Nominal Diameter (inch)		Common Lengths Up To 3 Inches ±0.010 (inch)																				
		0.188	0.25	0.313	0.375	0.438	0.5	0.563	0.625	0.75	0.875	1	1.125	1.25	1.375	1.5	1.75	2	2.25	2.5	2.75	3
1/16	0.0625	0.188	0.25	0.313	0.375	0.438	0.5	0.563	0.625	0.75	0.875	1		1.25		1.5						
5/64	0.0781	0.188	0.25	0.313	0.375	0.438	0.5	0.563	0.625	0.75	0.875	1										
3/32	0.0938	0.188	0.25	0.313	0.375	0.438	0.5	0.563	0.625	0.75	0.875	1	1.125	1.25	1.375	1.5	1.75					3
1/8	0.1250	0.188	0.25	0.313	0.375	0.438	0.5	0.563	0.625	0.75	0.875	1	1.125	1.25	1.375	1.5	1.75	2	2.25	2.5		
5/32	0.1562	0.188	0.25	0.313	0.375	0.438	0.5	0.563	0.625	0.75	0.875	1	1.125	1.25	1.375	1.5	1.75	2				
3/16	0.1875		0.25	0.313	0.375	0.438	0.5	0.563	0.625	0.75	0.875	1	1.125	1.25	1.375	1.5	1.75	2	2.25	2.5		3
1/4	0.2500		0.25	0.313	0.375	0.438	0.5	0.563	0.625	0.75	0.875	1	1.125	1.25	1.375	1.5	1.75	2	2.25	2.5	2.75	3
5/16	0.3125				0.375		0.5		0.625	0.75	0.875	1		1.25	1.375	1.5	1.75	2	2.25	2.5	2.75	3
3/8	0.3750				0.375		0.5		0.625	0.75	0.875	1		1.25	1.375	1.5	1.75	2	2.25	2.5	2.75	3
7/16	0.4375						0.5			0.75	0.875	1		1.25	1.375	1.5	1.75	2	2.25	2.5	2.75	3
1/2	0.5000						0.5			0.75	0.875	1		1.25		1.5	1.75	2	2.25	2.5	2.75	3
5/8	0.6250										0.875	1		1.25		1.5	1.75	2	2.25	2.5	2.75	3
3/4	0.7500											1		1.25		1.5	1.75	2	2.25	2.5	2.75	3
7/8	0.8750																	2	2.25	2.5		3
1	1.0000																	2	2.25	2.5		3

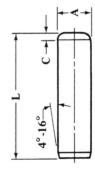

Shaded = non-preferred sizes

Table 5-7: Dowel Pins, Metric
(Diameters Up to 20 mm)
Part Complies With ISO 8734: 1997

Nominal Diameter mm	Tip Chamfer Length mm	Diameter Tolerance (m6) mm	Press Fit Hole Tolerance (K7) mm	Tight Press Fit Hole Tolerance Soft Material (P6) mm	Slip Fit Hole Tolerance (E7) mm
1	0.200				
1.5	0.300	+0.008	0	-0.006	+0.024
2	0.350	+0.002	-0.010	-0.012	+0.014
2.5	0.400				
3	0.500				
4	0.630	+0.012	+0.003	-0.009	+0.032
5	0.800	+0.004	-0.009	-0.017	+0.020
6	1.200				
8	1.600	+0.015	+0.005	-0.012	+0.040
10	2.000	+0.006	-0.010	-0.021	+0.025
12	2.500	+0.018	+0.006	-0.015	+0.050
16	3.000	+0.007	-0.012	-0.026	+0.032
20	3.500	0.021 / 0.008	+0.006 / -0.015	-0.018 / -0.031	+0.061 / +0.040

Table 5-8: Lengths, Dowel Pins, Metric
(Diameters up to 20 mm and 100 mm long)
Part Complies With ISO 8734: 1997

Nominal Diameter	Common Lengths Up To 100mm (mm)																														
	Length Tolerance																														
mm	±0.25						±0.5															±0.75									
	3	4	5	6	8	10	12	14	16	18	20	22	24	26	28	30	32	35	40	45	50	55	60	65	70	75	80	85	90	95	100
1	3	4	5	6	8	10																									
1.5		4	5	6	8	10	12	14	16																						
2			5	6	8	10	12	14	16	18	20																				
2.5				6	8	10	12	14	16	18	20	22	24																		
3					8	10	12	14	16	18	20	22	24	26	28	30															
4						10	12	14	16	18	20	22	24	26	28	30	32	35	40												
5							12	14	16	18	20	22	24	26	28	30	32	35	40	45	50										
6								14	16	18	20	22	24	26	28	30	32	35	40	45	50	55	60								
8										18	20	22	24	26	28	30	32	35	40	45	50	55	60	65	70	75	80				
10												22	24	26	28	30	32	35	40	45	50	55	60	65	70	75	80	85			
12														26	28	30	32	35	40	45	50	55	60	65	70	75	80	85	90		
16																			40	45	50	55	60	65	70	75	80	85	90	95	100
20																					50	55	60	65	70	75	80	85	90	95	100

Table 5-9: Withdrawal (Pullout) Dowel Pins, Inch
(Diameters Up to 1 inch)
Part Complies With ASME B18.8.2 – 1995

Nominal Diameter inch		Thread Size	Tip Crown Length inch	Diameter Tolerance inch	Press Fit Hole Tolerance inch	Tight Press Fit Hole, Soft Material Tolerance inch	Slip Fit Hole Tolerance inch
1/4	0.2500	#8	0.083	+0.0003	0	-0.0005	+0.0010
5/16	0.3125	#10	0.104	+0.0001	-0.0005	-0.001	+0.0005
3/8	0.3750	#10	0.125				
7/16	0.4375	1/4	0.146				
1/2	0.5000	1/4	0.167				
5/8	0.6250	1/4	0.208				
3/4	0.7500	5/16	0.250				
1	1.0000	5/16	0.333				

Table 5-10: *Lengths, Withdrawal (Pullout) Dowel Pins, Inch*
(Diameters up to 1 inch and 3 inches long)
Part Complies With ASME B18.8.2 – 1995

Nominal Diameter inch		Common Lengths Up To 3 Inches ±0.010 inch									
1/4	0.2500	0.5	0.75	1	1.25	1.5	1.75	2	2.25	2.5	
5/16	0.3125		0.75	1	1.25	1.5	1.75	2	2.25	2.5	3
3/8	0.3750		0.75	1	1.25	1.5	1.75	2	2.25	2.5	3
7/16	0.4375			1		1.5		2			
1/2	0.5000		0.75	1	1.25	1.5	1.75	2	2.25	2.5	3
5/8	0.6250			1	1.25	1.5	1.75	2	2.25	2.5	3
3/4	0.7500			1	1.25	1.5	1.75	2		2.5	3
1	1.0000							2		2.5	3

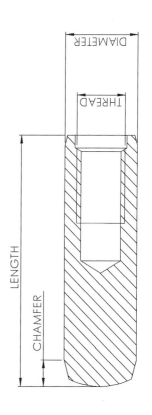

Table 5-11: *Withdrawal (Pullout) Dowel Pins, Metric*
(Diameters Up to 25 mm)
Part Complies With ISO 8735: 1997

Nominal Diameter mm	Thread Size	Tip Chamfer Length mm	Diameter Tolerance (m6) mm	Press Fit Hole Tolerance (K7) mm	Tight Press Fit Hole, Soft Material Tolerance (P6) mm	Slip Fit Hole Tolerance (E7) mm
6	M4	2.1	+0.012 +0.004	+0.003 -0.009	-0.009 -0.017	+0.032 +0.020
8	M5	2.6	+0.015 +0.006	+0.005 -0.010	-0.012 -0.021	+0.040 +0.025
10	M6	3				
12	M6	3.8	+0.018 +0.007	+0.006 -0.012	-0.015 -0.026	+0.050 +0.032
16	M8	4.6				
20	M10	6	+0.021 +0.008	+0.006 -0.015	-0.018 -0.031	+0.061 +0.040
25	M16	6				

Table 5-12: *Lengths, Withdrawal (Pullout) Dowel Pins, Metric*
(Diameters up to 25 mm and 100 mm long)
Part Complies With ISO 8735: 1997

Nominal Diameter mm	Common Lengths Up To 100 mm — mm — Length Tolerance																						
	±0.50													±0.75									
	16	18	20	22	24	26	28	30	32	35	40	45	50	55	60	65	70	75	80	85	90	95	100
6	16	18	20	22	24	26	28	30	32	35	40	45	50	55	60								
8		18	20	22	24	26	28	30	32	35	40	45	50	55	60	65	70	75	80				
10				22	24	26	28	30	32	35	40	45	50	55	60	65	70	75	80	85	90	95	100
12						26	28	30	32	35	40	45	50	55	60	65	70	75	80	85	90	95	100
16									32	35	40	45	50	55	60	65	70	75	80	85	90	95	100
20											40	45	50	55	60	65	70	75	80	85	90	95	100
25													50	55	60	65	70	75	80	85	90	95	100

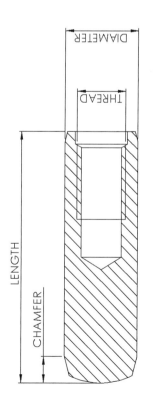

This page was intentionally left blank so that you may view the following tables in a more cohesive manner.

Table 5-13: Stepped Locating and Seating Pins, Metric
Part Complies With DIN 6321

Head Nominal Diameter d1 mm	Shank Nominal Diameter d2 mm	Head Height		Shank Length L2 mm	Head Chamfer Length (Forms B & C) L4 mm	d1 Head Diameter Tolerance (g6) mm	d1 Slip Fit Hole Tolerance (H7) mm	d2 Shank Diameter Tolerance (n6) mm	d2 Press Fit Hole Tolerance (H7) mm	d2 Tight Press Fit Hole Tolerance (K7) mm
		Seating Form A L1 (h9) mm	Long Forms B & C L1 mm							
6	4	5	12	6	4	-0.004 / -0.012	+0.012 / 0	+0.016 / +0.008	+0.012 / 0	+0.003 / -0.009
8	6	-	16	9	6	-0.005 / -0.014	+0.015 / 0			
10	6	6	18	9	6		+0.018 / 0			
12	6	-	18	9	6	-0.006 / -0.017				
16	8	8	22	12	8			+0.019 / +0.010	+0.015 / 0	+0.005 / -0.010
20	12	-	25	18	9	-0.007 / -0.020	+0.021 / 0	+0.023 / +0.012	+0.018 / 0	+0.006 / -0.012
25	12	10	25	18	9					

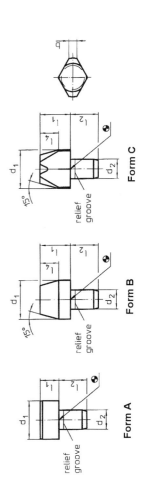

Form A

Form B

Form C

relief groove

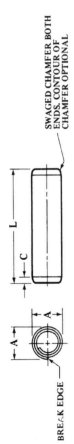

SWAGED CHAMFER BOTH ENDS. CONTOUR OF CHAMFER OPTIONAL

BREAK EDGE

Table 5-14: *Spring Pins, Coiled, Inch*
(Diameters up to ¾ inch)

Part Complies With ASME B18.8.2 - 2000

Nominal Diameter		Diameter Tolerance		Hole Tolerance	End Chamfer Length	SAE Material Number			
						Double Shear Load, Min, lbf			
		Standard Duty	Heavy Duty			Standard Duty		Heavy Duty	
size	inch	inch	inch	inch	inch	1070-1095, 51420 and 6150H	30302 and 30304	1070-1095, 51420 and 6150H	30302 and 30304
1/32	0.031	+0.004 +0.002	-	+0.001 0	0.024	90	65	-	-
-	0.039	+0.005 +0.002	-	+0.001 -0.001		135	100	-	-
3/64	0.047		-			190	145	-	-
-	0.052		-			250	190	-	-
1/16	0.062	+0.010 +0.005	+0.008 +0.004	+0.003 -0.001	0.028	330	265	475	360
5/64	0.078	+0.011 +0.005	+0.009 +0.004		0.032	550	425	800	575
3/32	0.094	+0.013 +0.006	+0.011 +0.005	+0.004 -0.001	0.038	775	600	1150	825
7/64	0.109	+0.015 +0.007	+0.012 +0.005			1050	825	1500	1150
1/8	0.125	+0.017 +0.008	+0.014 +0.006	+0.004 -0.003	0.044	1400	1100	2000	1700
5/32	0.156	+0.019 +0.009	+0.016 +0.007		0.048	2200	1700	3100	2400
3/16	0.188	+0.021 +0.01	+0.018 +0.008	+0.005 -0.002	0.055	3150	2400	4500	3500
7/32	0.219	+0.025 +0.012	+0.022 +0.010	+0.006 -0.003	0.065	4200	3300	5900	4600
1/4	0.250					5500	4300	7800	6200
5/16	0.312	+0.028 +0.013	+0.025 +0.011	+0.007 -0.004	0.08	8700	6700	12000	9300
3/8	0.375	+0.031 +0.014	+0.028 +0.012	+0.008 -0.005	0.095	12600	9600	18000	14000
7/16	0.438					17000	13300	23500	18000
1/2	0.500	+0.035 +0.016	+0.032 +0.014	+0.008 -0.007	0.110	22500	17500	32000	25000
5/8	0.625	+0.036 +0.017	+0.033 +0.015	+0.010 -0.007	0.125	35000	-	48000	-
3/4	0.750	+0.037 +0.018	+0.034 +0.016		0.150	50000	-	70000	-

Table 5-15: Lengths, Spring Pins, Coiled, Inch
(Diameters up to ¾ inch and 3 inches long)
Part Complies With ASME B18.8.2 - 2000

Common Lengths Up To 3 Inches — ±0.025 Worst Case, inch

Nominal Diameter (size)	inch	0.125	0.188	0.25	0.312	0.375	0.438	0.5	0.563	0.625	0.688	0.75	0.813	0.875	0.938	1	1.125	1.25	1.375	1.5	1.625	1.75	1.875	2	2.25	2.5	2.75	3
1/32	0.031 (a)	0.125	0.188	0.25	0.312	0.375	0.438	0.5																				
–	0.039 (a)	0.125	0.188	0.25	0.312	0.375	0.438	0.5																				
3/64	0.047 (a)	0.125	0.188	0.25	0.312	0.375	0.438	0.5																				
–	0.052 (a)	0.125	0.188	0.25	0.312	0.375	0.438	0.5																				
1/16	0.062		0.188	0.25	0.312	0.375	0.438	0.5	0.563	0.625																		
5/64	0.078			0.25	0.312	0.375	0.438	0.5	0.563	0.625	0.688	0.75																
3/32	0.094			0.25	0.312	0.375	0.438	0.5	0.563	0.625	0.688	0.75	0.813	0.875	0.938	1												
7/64	0.109			0.25	0.312	0.375	0.438	0.5	0.563	0.625	0.688	0.75	0.813	0.875	0.938	1												
1/8	0.125				0.312	0.375	0.438	0.5	0.563	0.625	0.688	0.75	0.813	0.875	0.938	1	1.125	1.25										
5/32	0.156						0.438	0.5	0.563	0.625	0.688	0.75	0.813	0.875	0.938	1	1.125	1.25	1.375	1.5	1.625	1.75						
3/16	0.188							0.5	0.563	0.625	0.688	0.75	0.813	0.875	0.938	1	1.125	1.25	1.375	1.5	1.625	1.75						
7/32	0.219							0.5	0.563	0.625	0.688	0.75	0.813	0.875	0.938	1	1.125	1.25	1.375	1.5	1.625	1.75						
1/4	0.25							0.5	0.563	0.625	0.688	0.75	0.813	0.875	0.938	1	1.125	1.25	1.375	1.5	1.625	1.75						
5/16	0.312											0.75	0.813	0.875	0.938	1	1.125	1.25	1.375	1.5	1.625	1.75	1.875	2	2.25	2.5	2.75	3
3/8	0.375											0.75	0.813	0.875	0.938	1	1.125	1.25	1.375	1.5	1.625	1.75	1.875	2	2.25	2.5	2.75	3
7/16	0.438															1	1.125	1.25	1.375	1.5	1.625	1.75	1.875	2	2.25	2.5	2.75	3
1/2	0.5																	1.25	1.375	1.5	1.625	1.75	1.875	2	2.25	2.5	2.75	3
5/8	0.625																							2	2.25	2.5	2.75	3
3/4	0.75																							2	2.25	2.5	2.75	3

a) Standard Duty Only

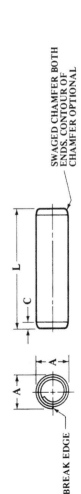

BREAK EDGE

SWAGED CHAMFER BOTH ENDS. CONTOUR OF CHAMFER OPTIONAL

CHAMFER BOTH ENDS, CONTOUR OF CHAMFER OPTIONAL

BREAK EDGE

40° MIN

Table 5-16: Spring Pins, Slotted, Inch
(Diameters up to 3/4 inch)
Part Complies With ASME B18.8.2 - 2000

Nominal Diameter		Diameter Tolerance	Hole Tolerance	End Chamfer Length	SAE Material Number		
					Double Shear Load, Min, lbf		
size	inch	inch	inch	inch	1070-1095, 51420, and 6150H	30302 and 30304	Beryllium Copper
1/16	0.062	+0.007 +0.004	+0.003 0	0.028 0.007	430	250	270
5/64	0.078	+0.008 +0.005	+0.003 0	0.032 0.008	800	460	500
3/32	0.094	+0.009 +0.005	+0.003 0	0.038 0.008	1150	670	710
1/8	0.125	+0.010 +0.006	+0.004 0	0.044 0.008	1875	1090	1170
9/64	0.141	+0.008 +0.004	+0.003 -0.001	0.044 0.008	2175	1260	1350
5/32	0.156	+0.011 +0.006	+0.004 0	0.048 0.010	2750	1600	1725
3/16	0.188	+0.011 +0.006	+0.004 -0.001	0.055 0.011	4150	2425	2600
7/32	0.219	+0.013 +0.007	+0.005 0	0.065 0.011	5850	3400	3650
1/4	0.250	+0.014 +0.008	+0.006 0	0.065 0.012	7050	4100	4400
5/16	0.312	+0.018 +0.009	+0.006 0	0.080 0.014	10800	6300	6750
3/8	0.375	+0.020 +0.010	+0.007 0	0.095 0.016	16300	9500	10200
7/16	0.438	+0.021 +0.01	+0.007 -0.001	0.095 0.017	19800	11500	12300
1/2	0.500	+0.024 +0.013	+0.010 0	0.110 0.025	27100	15800	17000
5/8	0.625	+0.028 +0.015	+0.011 0	0.125 0.030	46000	18800	-
3/4	0.750	+0.034 +0.019	+0.014 0	0.150 0.030	66000	23200	-

Table 5-17: Lengths, Spring Pins, Slotted, Inch
(Diameters up to ¾ inch and 3 inches long)
Part Complies With ASME B18.8.2 - 2000

Nominal Diameter		Common Lengths Up To 3 Inches (inch)																									
size	inch	0.1875	0.25	0.3125	0.375	0.4375	0.5	0.5625	0.625	0.6875	0.75	0.8125	0.875	0.938	1	1.125	1.25	1.375	1.5	1.625	1.75	1.875	2	2.25	2.5	2.75	3
1/16	0.062	0.1875	0.25	0.3125	0.375	0.4375	0.5	0.5625	0.625	0.6875	0.75	0.8125	0.875	0.938	1												
5/64	0.078	0.1875	0.25	0.3125	0.375	0.4375	0.5	0.5625	0.625	0.6875	0.75	0.8125	0.875	0.938	1	1.125	1.25	1.375	1.5								
3/32	0.094	0.1875	0.25	0.3125	0.375	0.4375	0.5	0.5625	0.625	0.6875	0.75	0.8125	0.875	0.938	1	1.125	1.25	1.375	1.5								
1/8	0.125			0.3125	0.375	0.4375	0.5	0.5625	0.625	0.6875	0.75	0.8125	0.875	0.938	1	1.125	1.25	1.375	1.5	1.625	1.75	1.875	2				
9/64	0.141																										
5/32	0.156					0.4375	0.5	0.5625	0.625	0.6875	0.75	0.8125	0.875	0.938	1	1.125	1.25	1.375	1.5	1.625	1.75	1.875	2	2.25	2.5		
3/16	0.188						0.5	0.5625	0.625	0.6875	0.75	0.8125	0.875	0.938	1	1.125	1.25	1.375	1.5	1.625	1.75	1.875	2	2.25	2.5		
7/32	0.219						0.5	0.5625	0.625	0.6875	0.75	0.8125	0.875	0.938	1	1.125	1.25	1.375	1.5	1.625	1.75	1.875	2	2.25	2.5	2.75	3
1/4	0.25						0.5	0.5625	0.625	0.6875	0.75	0.8125	0.875	0.938	1	1.125	1.25	1.375	1.5	1.625	1.75	1.875	2	2.25	2.5	2.75	3
5/16	0.312										0.75	0.8125	0.875	0.938	1	1.125	1.25	1.375	1.5	1.625	1.75	1.875	2	2.25	2.5	2.75	3
3/8	0.375										0.75		0.875		1		1.25		1.5		1.75		2	2.25	2.5	2.75	3
7/16	0.438														1		1.25		1.5		1.75		2	2.25	2.5	2.75	3
1/2	0.5																1.25		1.5		1.75		2	2.25	2.5	2.75	3
5/8	0.625																						2	2.25	2.5	2.75	3
3/4	0.75																						2	2.25	2.5	2.75	3

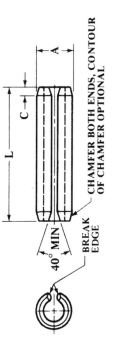

40° MIN
BREAK EDGE
CHAMFER BOTH ENDS, CONTOUR OF CHAMFER OPTIONAL

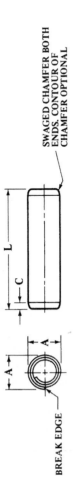

BREAK EDGE

SWAGED CHAMFER BOTH ENDS. CONTOUR OF CHAMFER OPTIONAL

Table 5-18: Spring Pins, Coiled, Metric
(Diameters up to 20 mm)
Part Complies With ISO 8750: 2007 (Heavy Duty)
ISO 8748: 2007 (Standard Duty)

Nominal Diameter mm	Diameter Tolerance — Standard Duty mm	Diameter Tolerance — Heavy Duty mm	Hole Tolerance (H12) mm	End Chamfer Length mm	Material — Double Shear Load, Min, kN — Standard Duty — Steel and Martensitic Alloys	Material — Double Shear Load, Min, kN — Standard Duty — Austenitic Stainless Steel	Material — Double Shear Load, Min, kN — Heavy Duty — Steel and Martensitic Alloys	Material — Double Shear Load, Min, kN — Heavy Duty — Austenitic Stainless Steel
0.8	+0.11 / +0.05	-	+0.100 / 0	0.3	0.4	0.3	-	-
1	+0.15 / +0.05	-		0.3	0.6	0.45	-	-
1.2	+0.15 / +0.05	-		0.4	0.9	0.65	-	-
1.5	+0.23 / +0.12	+0.21 / +0.11		0.5	1.45	1.05	1.9	1.45
2	+0.25 / +0.13	+0.23 / +0.12		0.7	2.5	1.9	3.5	2.5
2.5	+0.28 / +0.15	+0.25 / +0.12		0.7	3.9	2.9	5.5	3.8
3	+0.30 / +0.15	+0.29 / +0.14		0.9	5.5	4.2	7.6	5.7
3.5	+0.34 / +0.17	+0.30 / +0.15	+0.120 / 0	1.0	7.5	5.7	10	7.6
4	+0.40 / +0.20	+0.35 / +0.15		1.1	9.6	7.6	13.5	10
5	+0.50 / +0.25	+0.40 / +0.18		1.3	15	11.5	20	15.5
6	+0.55 / +0.25	+0.40 / +0.18		1.5	22	16.8	30	23
8	+0.63 / +0.30	+0.55 / +0.25	+0.150 / 0	2	39	30	53	41
10	+0.80 / +0.35	+0.65 / +0.30		2.5	62	48	84	64
12	+0.85 / +0.40	+0.75 / +0.35	+0.180 / 0	3	89	67	120	91
14	+0.95 / +0.45	+0.85 / +0.40		3.5	120	-	165	-
16	+1.00 / +0.45	+0.90 / +0.40		4	155	-	210	-
20	+1.10 / +0.40	+1.00 / +0.40	+0.210 / 0	4.5	250	-	340	-

Table 5-19: Lengths, Spring Pins, Coiled, Metric
(Diameters up to 20 mm and 100 mm long)
Part Complies With ISO 8750: 2007 (Heavy Duty)
ISO 8748: 2007 (Standard Duty)

Nominal Diameter mm	Common Lengths Up To 100 mm ±0.5 mm																													
	4	5	6	8	10	12	14	16	18	20	22	24	26	28	30	32	35	40	45	50	55	60	65	70	75	80	85	90	95	100
0.8 (a)	4	5		8	10	12																								
1 (a)	4	5		8	10	12																								
1.2 (a)	4	5		8	10	12																								
1.5	4	5		8	10	12	14	16																						
2			6	8	10	12	14	16	18	20																				
2.5				8	10	12	14	16	18	20	22	24	26																	
3				8	10	12	14	16	18	20	22	24	26	28	30															
3.5				8	10	12	14	16	18	20	22	24	26	28	30	32	35													
4				8	10	12	14	16	18	20	22	24	26	28	30	32	35	40												
5						12	14	16	18	20	22	24	26	28	30	32	35	40	45											
6						12	14	16	18	20	22	24	26	28	30	32	35	40	45	50	55									
8								16	18	20	22	24	26	28	30	32	35	40	45	50	55	60	65	70	75					
10										20	22	24	26	28	30	32	35	40	45	50	55	60	65	70	75	80				
12												24	26	28	30	32	35	40	45	50	55	60	65	70	75	80	85			
14																														
16																	35	40	45	50	55	60	65	70	75	80	85	90	95	100
20																		45	50	55	60	65	70	75	80	85	90	95	100	

a) Standard Duty Only

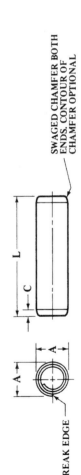

BREAK EDGE

SWAGED CHAMFER BOTH ENDS. CONTOUR OF CHAMFER OPTIONAL

Table 5-20: *Spring Pins, Slotted, Metric*
(Diameters up to 20 mm)
Part Complies With ASME B18.8.4M – 2000 (R2005) Type B

				Material		
			End Chamfer Length	Carbon Steel & 400 Series Stainless Steel	300 Series Stainless Steel	Beryllium Copper
Nominal Diameter	Diameter Tolerance	Hole Tolerance				
mm	mm	mm	mm	Double Shear Load, Min, kN		
1.5	+0.16 +0.08	+0.06 0	0.7 0.2	1.8	1	1.1
2	+0.19 +0.10	+0.07 0	0.8 0.2	3.5	2	2.2
2.5	+0.22 +0.12	+0.08 0	0.9 0.2	5.5	3.2	3.5
3	+0.25 +0.14	+0.10 0	1.0 0.2	7.8	4.5	4.9
4	+0.30 +0.16	+0.12 0	1.2 0.3	12.3	7.2	7.7
5	+0.33 +0.17	+0.12 0	1.4 0.3	19.6	11.4	12.3
6	+0.36 +0.18	+0.12 0	1.6 0.4	28.5	16.6	17.8
8	+0.45 +0.22	+0.15 0	2.0 0.4	48.8	28.4	30.5
10	+0.51 +0.25	+0.15 0	2.4 0.5	79.1	46.1	49.4
12	+0.55 +0.28	+0.18 0	2.8 0.6	104.1	60.7	65

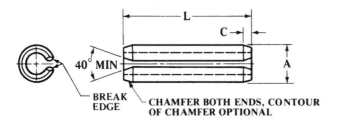

Table 5-21: Spring Pins, Slotted, Metric
(Diameters up to 20 mm)
Part Complies With ISO 8752: 2009 (Heavy Duty)

Nominal Diameter mm	Diameter Tolerance mm	Hole Tolerance (H12) mm	End Chamfer Length mm	Material Carbon Steel & 400 Series Stainless Steel Double Shear Load, Min, kN
1	+0.30 +0.20		0.35 0.15	0.7
1.5	+0.30 +0.20		0.45 0.25	1.58
2	+0.40 +0.30	+0.100 0	0.55 0.35	2.82
2.5	+0.40 +0.30		0.60 0.40	4.38
3	+0.50 +0.30		0.70 0.50	6.32
3.5	+0.50 +0.30		0.80 0.60	9.06
4	+0.60 +0.40		0.85 0.65	11.24
4.5	+0.60 +0.40	+0.120 0	1.00 0.80	15.36
5	+0.60 +0.40		1.10 0.90	17.54
6	+0.70 +0.40		1.40 1.20	26.04
8	+0.80 +0.50	+0.150 0	2.00 1.60	42.76
10	+0.80 +0.50		2.40 2.00	70.16
12	+0.80 +0.50		2.40 2.00	104.1
13	+0.80 +0.50		2.40 2.00	115.1
14	+0.80 +0.50	+0.180 0	2.40 2.00	144.7
16	+0.80 +0.50		2.40 2.00	171
18	+0.90 +0.50		2.40 2.00	222.5
20	+0.90 +0.50	+0.210 0	3.40 3.00	280.6

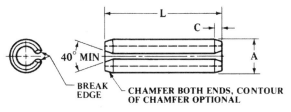

40° MIN
BREAK EDGE
CHAMFER BOTH ENDS, CONTOUR OF CHAMFER OPTIONAL
L
C
A

This page was intentionally left blank so that you may view the following tables in a more cohesive manner.

Table 5-22: Lengths, Spring Pins, Slotted, Metric
(Diameters up to 20 mm and 100 mm long)
Part Complies With ASME B18.8.4M – 2000 (R2005) Type B
ISO 8752: 2009 (Heavy Duty)

Nominal Diameter mm	Common Lengths Up To 100 mm (mm) — Length Tolerance																													
	±0.50 Worst Case																				±0.75 Worst Case									
	4	5	6	8	10	12	14	16	18	20	22	24	26	28	30	32	35	40	45	50	55	60	65	70	75	80	85	90	95	100
1																														
1.5	4	5	6	8	10	12	14	16	18	20	22	24	26	28																
2	4	5	6	8	10	12	14	16	18	20	22	24	26	28	30	32	35	40												
2.5		5 (a)	6	8	10	12	14	16	18	20	22	24	26	28	30	32	35	40	45 (b)	50 (b)										
3			6	8	10	12	14	16	18	20	22	24	26	28	30	32	35	40	45	50 (b)										
3.5 (b)				8	10	12	14	16	18	20	22	24	26	28	30	32	35	40	45	50										
4				8	10	12	14	16	18	20	22	24	26	28	30	32	35	40	45	50	55	60 (b)								
4.5																														
5					10	12	14	16	18	20	22	24	26	28	30	32	35	40	45	50	55	60 (b)								
6						12	14	16	18	20	22	24	26	28	30	32	35	40	45	50	55	60	65	70	75					
8								16	18	20	22	24	26	28	30	32	35	40	45	50	55	60	65	70	75	80	85	90	95	100
10										20	22	24	26	28	30	32	35	40	45	50	55	60	65	70	75	80	85	90	95	100
12											22 (a)	24	26	28	30	32	35	40	45	50	55	60	65	70	75	80	85	90	95	100
13 (b)																														
14 (b)																														
16 (b)																	35	40	45	50	55	60	65	70	75	80	85	90	95	100
18 (b)																														
20 (b)																		40	45	50	55	60	65	70	75	80	85	90	95	100

a) ASME Standard Pins Only
b) ISO Standard Pins Only

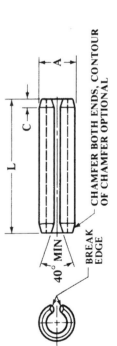

40° MIN

BREAK EDGE

CHAMFER BOTH ENDS, CONTOUR OF CHAMFER OPTIONAL

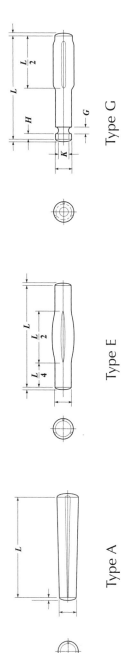

Type A

Type E

Type G

Table 5-23: Grooved Pins, Inch
(Diameters up to ½ inch)
Part Complies With ASME B18.8.2 - 2000

Nominal Diameter Size	Nominal Diameter inch	Diameter Tolerance inch	Hole Tolerance inch	End Radius Length inch	Neck Diameter K inch	Neck Width G inch	Groove Distance From Crown H inch	Double Shear Load, Minimum — Low Carbon Steel (lbf)	Alloy Steel (lbf)	Corrosion Resistant Steel (lbf)
1/16	0.0625	+0.000 / -0.001	+0.0015 / 0	0.0115 / 0.0015	–	–	–	410	720	540
3/32	0.0938	+0.000 / -0.001	+0.0018 / 0	0.0141 / 0.0041	0.067 / 0.057	0.038 / 0.028		890	1600	1240
1/8	0.1250	+0.000 / -0.002	+0.0021 / 0	0.018 / 0.008	0.088 / 0.078		0.041 / 0.031	1600	2820	2200
5/32	0.1563	+0.000 / -0.002	+0.0024 / 0	0.022 / 0.012	0.109 / 0.099	0.069 / 0.059		2300	4520	3310
3/16	0.1875	+0.000 / -0.002	+0.0028 / 0	0.023 / 0.013	0.130 / 0.120		0.057 / 0.047	3310	6440	4760
7/32	0.2188	+0.000 / -0.002	+0.0031 / 0	0.027 / 0.017	0.151 / 0.141	0.101 / 0.091		4510	8770	6480
1/4	0.2500	+0.000 / -0.002	+0.0034 / 0	0.031 / 0.021	0.172 / 0.162		0.072 / 0.062	5880	11500	8460
5/16	0.3125	+0.000 / -0.002	+0.0041 / 0	0.039 / 0.029	0.214 / 0.204	0.132 / 0.122		7660	17900	12700
3/8	0.3750	+0.000 / -0.002	+0.0047 / 0	0.044 / 0.034	0.255 / 0.245		0.104 / 0.094	11000	26000	18200
7/16	0.4375	+0.000 / -0.002	+0.0053 / 0	0.052 / 0.042	0.298 / 0.288	0.195 / 0.185		15000	35200	24800
1/2	0.5000	+0.000 / -0.002	+0.006 / 0	0.057 / 0.047	0.317 / 0.307		0.135 / 0.125	19600	46000	32400

Table 5-24: *Lengths, Grooved Pins, Inch*
(Diameters up to ½ inch and Lengths to 3 inches)
Part Complies With ASME B18.8.2 - 2000

Nominal Diameter		Common Lengths Up To 3 Inches ±0.01 inch										
Size	inch											
1/16	0.0625 (a)	0.25	0.375	0.5	0.625	0.75						
3/32	0.0938		0.375	0.5	0.625	0.75	0.875	1				
1/8	0.125			0.5	0.625	0.75	0.875	1	1.25			
5/32	0.1563			0.5	0.625	0.75	0.875	1	1.25	1.5	1.75	2
3/16	0.1875			0.5		0.75	0.875	1	1.25	1.5	1.75	2	2.25
7/32	0.2188											
1/4	0.25					0.75		1	1.25	1.5	1.75	2	2.25
5/16	0.3125					0.75		1	1.25	1.25			
3/8	0.375							1	1.25	1.5	1.75	2	2.5
7/16	0.4375											
1/2	0.5											

a) Size not available with neck

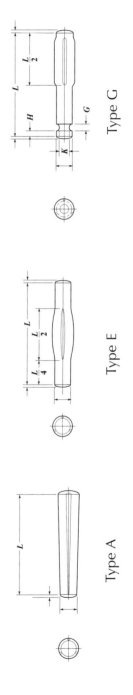

Type G

Type E

Type A

ISO 8744

ISO 8742

DIN 1469 Type C

Table 5-25: Grooved Pins, Metric
Straight Parts Comply With ISO 8744: 1997

ISO 8742: 1997

Necked Part Complies With DIN 1469: 1978 Type C

Nominal Diameter mm	Diameter Tolerance mm	Hole Tolerance (H11) mm	End Radius Length mm	Neck Diameter K mm	Neck Width G mm	Groove Distance From Crown H mm	Double Shear Load, Minimum Material: Steel Straight Grooved Pin kN	Grooved Pin With Neck kN
1.5 (a)	0 / -0.025	+0.060 / 0	0.2	-	-	-	1.6	-
2	0 / -0.025	+0.060 / 0	0.25	1.0	0.8	0.8	2.84	2.85
2.5	0 / -0.025	+0.060 / 0	0.3	1.2	0.8	0.8	4.4	4.25
3	0 / -0.025	+0.060 / 0	0.4	1.5	1.0	1.0	6.4	6.15
4	0 / -0.075	+0.075 / 0	0.5	2.4	1.4	1.4	11.3	10.6
5	0 / -0.075	+0.075 / 0	0.6	2.8	1.6	1.6	17.6	16.5
6	0 / -0.075	+0.075 / 0	0.8	3.8	1.6	1.6	25.4	22.8
8	0 / -0.090	+0.090 / 0	1.0	5.0	2.0	2.0	45.2	40.5
10	0 / -0.090	+0.090 / 0	1.2	6.8	2.6	2.6	70.4	63.2
12	0 / -0.110	+0.110 / 0	1.6	8.2	3.0	3.0	101.8	91
14 (b)	0 / -0.110	+0.110 / 0		9.6	3.0	3.0	-	124
16	0 / -0.130	+0.130 / 0	2.0	11.0	4.0	4.0	181	156.8
20	0 / -0.130	+0.130 / 0	2.5	14.0	5.0	5.0	283	236.5
25	0 / -0.130	+0.130 / 0	3.0	18.0	6.0	6.0	444	370.1

a) Not available with neck
b) Not available in tapered or center groove style

Table 5-26 Lengths, Grooved Pins, Metric
(Diameters up to 25 mm and Lengths to 100 mm)

Straight Parts Comply With ISO 8744: 1997 and ISO 8742: 1997

Necked Part Complies With DIN 1469: 1978 Type C

Diameter mm	8	10	12	14	16	18	20	22	24	25	26	28	30	32	35	40	45	50	55	60	65	70	75	80	85	90	95	100
1.5 (a)																												
2	6 (b)	8 (c)	10 (c)	12	14 (a)	16	18 (a)	20	22 (a)	24 (a)	25 (b)	26 (a)	28 (a)	30														
2.5	6 (b)	8 (c)	10 (c)	12	14 (a)	16	18 (a)	20	22 (a)	24 (a)	25 (b)	26 (a)	28 (a)	30														
3	6 (b)	8 (c)	10 (c)	12	14 (a)	16	18 (a)	20	22 (a)	24 (a)	25 (b)	26 (a)	28 (a)	30	32 (a)	35	40											
4	8 (a,c)	10 (c)	12 (c)	14 (a,c)	16 (c)	18 (a)	20	22 (a)	24 (a)	25 (b)	26 (a)	28 (a)	30	32 (a)	35	40	45	50	55	60								
5	8 (a,c)	10 (c)	12 (c)	14 (a,c)	16 (c)	18 (a)	20	22 (a)	24 (a)	25 (b)	26 (a)	28 (a)	30	32 (a)	35	40	45	50	55	60								
6		10 (c)	12 (c)	14 (a,c)	16 (c)	18 (a)	20	22 (a)	24 (a)	25 (b)	26 (a)	28 (a)	30	32 (a)	35	40	45	50	55	60	65	70	75	80				
8			12 (a,c)	14 (a,c)	16 (c)	18 (a,c)	20 (c)	22 (a,c)	24 (a,c)	25 (b)	26 (a)	28 (a)	30	32 (a)	35	40	45	50	55	60	65	70	75	80	85	90	95	100
10				14 (a,c)	16 (c)	18 (a,c)	20 (c)	22 (a,c)	24 (a,c)	25 (b)	26 (a,c)	28 (a,c)	30 (c)	32 (a)	35	40	45	50	55	60	65	70	75	80	85	90	95	100
12				14 (a,c)	16 (c)	18 (a,c)	20 (c)	22 (a,c)	24 (a,c)	25 (b)	26 (a,c)	28 (a,c)	30 (c)	32 (a,c)	35 (c)	40	45	50	55	60	65	70	75	80	85	90	95	100
14 (b)							20			25			30		35	40	45	50	55	60	65	70	75	80	85	90	95	100
16									24 (a,c)		26 (a,c)	28 (a,c)	30 (c)	32 (a,c)	35 (c)	40 (c)	45	50	55	60	65	70	75	80	85	90	95	100
20											26 (a,c)	28 (a,c)	30 (c)	32 (a,c)	35 (c)	40 (c)	45	50	55	60	65	70	75	80	85	90	95	100
25											26 (a,c)	28 (a,c)	30 (c)	32 (a,c)	35 (c)	40 (c)	45	50	55	60	65	70	75	80	85	90	95	100

Common Lengths Up To 100mm — mm

a) Size not available with neck
b) Size only available with neck
c) Size not available in center groove style

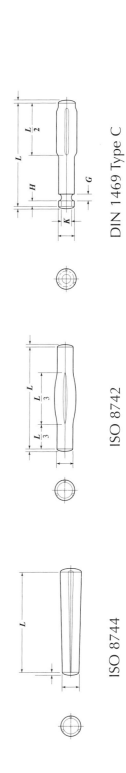

DIN 1469 Type C

ISO 8742

ISO 8744

Table 5-27: Clevis Pins, Inch
(Diameters up to 1 inch)
Part Complies With ASME B18.8.1 - 1994

Pin Nominal Diameter		Diameter Tolerance	Head Diameter	Head Height	Radius Under Head	Distance from End to Hole Center	Hole Diameter	Recommended Cotter Pin	
Size	inch	inch	inch	inch	inch	inch	inch	Size	inch
3/16	0.188	-0.002 / -0.007	0.32 / 0.30	0.07 / 0.05		0.09	0.088 / 0.073	1/16	0.062
1/4	0.250	-0.002 / -0.007	0.38 / 0.36	0.10 / 0.08		0.09	0.088 / 0.073	1/16	0.062
5/16	0.312	-0.001 / -0.006	0.44 / 0.42	0.10 / 0.08		0.12	0.119 / 0.104	3/32	0.093
3/8	0.375	-0.002 / -0.007	0.51 / 0.49	0.13 / 0.11	0.015 / 0.005	0.12	0.119 / 0.104	3/32	0.093
7/16	0.438	-0.002 / -0.007	0.57 / 0.55	0.16 / 0.14		0.12	0.119 / 0.104	3/32	0.093
1/2	0.500		0.63 / 0.61	0.16 / 0.14		0.15	0.151 / 0.136	1/8	0.125
5/8	0.625		0.82 / 0.80	0.21 / 0.19		0.15	0.151 / 0.136	1/8	0.125
3/4	0.750	-0.004 / -0.009	0.94 / 0.92	0.26 / 0.24		0.18	0.182 / 0.167	5/32	0.156
7/8	0.875		1.04 / 1.02	0.32 / 0.30		0.18	0.182 / 0.167	5/32	0.156
1	1.000		1.19 / 1.17	0.35 / 0.33		0.18	0.182 / 0.167		

L

Hole

Table 5-28 Lengths, Clevis Pins, Inch
(Diameters up to 1 inch and Lengths to 5 inches)
Part Complies With ASME B18.8.1 - 1994

Pin Nominal Diameter			Common Overall Lengths Up To 5 Inches +0.02 / -0 inch																					
Size	inch		0.500	0.750	0.875	1.000	1.125	1.250	1.375	1.500	1.625	1.750	1.875	2.000	2.125	2.250	2.500	2.750	3.000	3.250	3.500	4.000	4.500	5.000
3/16	0.188	Overall		0.750		1.000		1.250											3.000					
		Effective		0.616		0.866		1.116											2.866					
1/4	0.250	Overall	0.500	0.750		1.000		1.250	1.375	1.500	1.625			2.000			2.500		3.000	3.250	3.500			
		Effective	0.366	0.616		0.866		1.116	1.241	1.366	1.491			1.866			2.366		2.866	3.116	3.366			
5/16	0.312	Overall		0.750	0.875	1.000		1.250	1.375	1.500	1.625		1.875	2.000										
		Effective		0.571	0.696	0.821		1.071	1.196	1.321	1.446		1.696	1.821										
3/8	0.375	Overall				1.000		1.250		1.500				2.000	2.125						3.500			
		Effective				0.821		1.071		1.321				1.821	1.946						3.321			
7/16	0.438	Overall								1.500														
		Effective								1.321														
1/2	0.500	Overall				1.000	1.125	1.250	1.375	1.500		1.750		2.000		2.250	2.500	2.750	3.000		3.500	4.000	4.500	
		Effective				0.775	0.900	1.025	1.150	1.275		1.525		1.775		2.025	2.275	2.525	2.775		3.275	3.775	4.275	
5/8	0.625	Overall								1.500		1.750		2.000		2.250			3.000		3.500			5.000
		Effective								1.275		1.525		1.775		2.025			2.775		3.275			4.775
3/4	0.750	Overall																2.750	3.000		3.500			
		Effective																2.479	2.729		3.229			
7/8	0.875	Overall																			3.500	4.000		
		Effective																			3.229	3.729		
1	1.000	Overall												2.000				2.750				4.000		
		Effective												1.729				2.479				3.729		

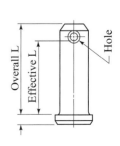

Overall L
Effective L
Hole

Table 5-29: Clevis Pins, Metric
(Diameters up to 24 mm)

Part Complies With ISO 2341: 1986 Type B

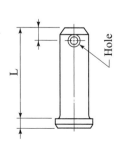

Pin Nominal Diameter mm	Diameter Tolerance (h11) mm	Head Nominal Diameter mm	Head Tolerance (h14) mm	Head Nominal Height mm	Head Height Tolerance (js14) mm	Radius Under Head mm	Distance from End to Hole Center Minimum mm	Hole Nominal Diameter mm	Hole Tolerance (H13) mm	Recommended Cotter Pin Size (a) mm
3	0 -0.060	5	0 -0.30	1.0	+0.125 -0.125	0.6	1.6	0.8	+0.140 0	0.8
4	0 -0.075	6	0 -0.36	1.6			2.2	1		1.0
5		8		2.0			2.9	1.2		1.2
6		10		3.0			3.2	1.6		1.6
8	0 -0.090	14	0 -0.43	4.0			3.5	2		2.0
10		18			+0.150 -0.150		4.5	3.2		3.2
12	0 -0.110	20	0 -0.52				5.5			
14		22		4.5			6	4	+0.180 0	4.0
16		25		5.0		1.0	7			
18		28					8	5		5.0
20	0 -0.130	30	0 -0.62	5.5						
22		33								
24		36		6.0			9	6.3	+0.220 0	6.3

a) When reversing loads are expected, use the next size up.

Table 5-30 Lengths, Clevis Pins, Metric
(Diameters up to 24 mm and Lengths to 100 mm)
Part Complies With ISO 2341: 1986 Type B

Common Overall Lengths Up To 100 mm — mm — Length Tolerance

Pin Nominal Diameter mm		±0.5								±0.75									
		16.00	20.00	25.00	30.00	35.00	40.00	45.00	50.00	55.00	60.00	65.00	70.00	75.00	80.00	85.00	90.00	95.00	100.00
3	Overall	16.00	20.00	25.00	30.00														
	Effective	13.93	17.93	22.93	27.93														
4	Overall	16.00	20.00	25.00	30.00	35.00	40.00												
	Effective	13.23	17.23	22.23	27.23	32.23	37.23												
5	Overall	16.00	20.00	25.00	30.00	35.00	40.00	45.00	50.00										
	Effective	12.43	16.43	21.43	26.43	31.43	36.43	41.43	46.43										
6	Overall	16.00	20.00	25.00	30.00	35.00	40.00	45.00	50.00	55.00	60.00								
	Effective	11.93	15.93	20.93	25.93	30.93	35.93	40.93	45.93	50.93	55.93								
8	Overall	16.00	20.00	25.00	30.00	35.00	40.00	45.00	50.00	55.00	60.00	65.00	70.00	75.00	80.00				
	Effective	11.43	15.43	20.43	25.43	30.43	35.43	40.43	45.43	50.43	55.43	60.43	65.43	70.43	75.43				
10	Overall		20.00	25.00	30.00	35.00	40.00	45.00	50.00	55.00	60.00	65.00	70.00	75.00	80.00	85.00	90.00	95.00	100.00
	Effective		13.81	18.81	23.81	28.81	33.81	38.81	43.81	48.81	53.81	58.81	63.81	68.81	73.81	78.81	83.81	88.81	93.81
12	Overall				30.00	35.00	40.00	45.00	50.00	55.00	60.00	65.00	70.00	75.00	80.00	85.00	90.00	95.00	100.00
	Effective				22.81	27.81	32.81	37.81	42.81	47.81	52.81	57.81	62.81	67.81	72.81	77.81	82.81	87.81	92.81
14	Overall				30.00	35.00	40.00	45.00	50.00	55.00	60.00	65.00	70.00	75.00	80.00	85.00	90.00	95.00	100.00
	Effective				21.91	26.91	31.91	36.91	41.91	46.91	51.91	56.91	61.91	66.91	71.91	76.91	81.91	86.91	91.91
16	Overall					35.00	40.00	45.00	50.00	55.00	60.00	65.00	70.00	75.00	80.00	85.00	90.00	95.00	100.00
	Effective					26.91	31.91	36.91	41.91	46.91	51.91	56.91	61.91	66.91	71.91	76.91	81.91	86.91	91.91
18	Overall					35.00	40.00	45.00	50.00	55.00	60.00	65.00	70.00	75.00	80.00	85.00	90.00	95.00	100.00
	Effective					25.41	30.41	35.41	40.41	45.41	50.41	55.41	60.41	65.41	70.41	75.41	80.41	85.41	90.41
20	Overall						40.00	45.00	50.00	55.00	60.00	65.00	70.00	75.00	80.00	85.00	90.00	95.00	100.00
	Effective						29.41	34.41	39.41	44.41	49.41	54.41	59.41	64.41	69.41	74.41	79.41	84.41	89.41
22	Overall							45.00	50.00	55.00	60.00	65.00	70.00	75.00	80.00	85.00	90.00	95.00	100.00
	Effective							34.41	39.41	44.41	49.41	54.41	59.41	64.41	69.41	74.41	79.41	84.41	89.41
24	Overall								50.00	55.00	60.00	65.00	70.00	75.00	80.00	85.00	90.00	95.00	100.00
	Effective								37.74	42.74	47.74	52.74	57.74	62.74	67.74	72.74	77.74	82.74	87.74

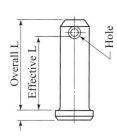

Overall L
Effective L
Hole

Table 5-31: Cotter Pins, Inch
(Diameters up to 5/32 inch)

Part Complies With ASME B18.8.1 – 1994

Square Cut Type

Plane of Contact with Gage

Pin Diameter		Max. Wire Width	Minimum Head Diameter	Minimum Prong Extension	Gage Hole Diameter ±0.001	Clevis Hole Diameter
Nominal Size	Max. inch	inch	inch	inch	inch	inch
1/32	0.032	0.032	0.06	0.01	0.047	0.057 (a) / 0.042 (a)
3/64	0.048	0.048	0.09	0.02	0.062	0.072 (a) / 0.057 (a)
1/16	0.060	0.060	0.12	0.03	0.078	0.088 / 0.073
5/64	0.076	0.076	0.16	0.04	0.094	0.104 (a) / 0.089 (a)
3/32	0.090	0.090	0.19	0.04	0.109	0.119 / 0.104
7/64	0.104	0.104	0.22	0.05	0.125	0.135 (a) / 0.120 (a)
1/8	0.120	0.120	0.25	0.06	0.141	0.151 / 0.136
9/64	0.134	0.134	0.28	0.06	0.156	0.166 (a) / 0.151 (a)
5/32	0.150	0.150	0.31	0.07	0.172	0.182 / 0.167

a) Theoretical hole size. No clevis exists for this size.

Table 5-32 Lengths, Cotter Pins, Inch
(Diameters up to 5/32 inch and Lengths to 3 inches)
Part Complies With ASME B18.8.1 – 1994
Square Cut Type

Pin Diameter		Common Lengths Up To 3 Inches																	
Nominal Size	Max. inch	inch																	
		0.25	0.3125	0.375	0.4375	0.5	0.625	0.75	0.875	1	1.125	1.25	1.5	1.75	2	2.25	2.5	2.75	3
1/32	0.032	0.25	0.3125	0.375		0.5	0.625	0.75		1		1.25	1.5	1.75	2				
3/64	0.048	0.25	0.3125	0.375	0.4375	0.5	0.625	0.75	0.875	1		1.25	1.5	1.75	2				
1/16	0.060	0.25	0.3125	0.375	0.4375	0.5	0.625	0.75	0.875	1		1.25	1.5	1.75	2	2.25	2.5		3
5/64	0.076			0.375		0.5	0.625	0.75		1		1.25	1.5	1.75	2	2.25	2.5		
3/32	0.090			0.375		0.5	0.625	0.75	0.875	1	1.125	1.25	1.5	1.75	2	2.25	2.5		3
7/64	0.104					0.5	0.625	0.75		1		1.25	1.5	1.75	2				3
1/8	0.120			0.375		0.5	0.625	0.75	0.875	1	1.125	1.25	1.5	1.75	2	2.25	2.5	2.75	3
9/64	0.134																		
5/32	0.150					0.5		0.75		1	1.125	1.25	1.5	1.75	2	2.25	2.5		3

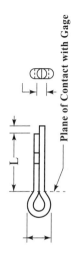

Plane of Contact with Gage

Table 5-33: Cotter Pins, Metric
(Diameters up to 6.3 mm)
Part Complies With ISO 1234: 1997 (DIN 94)

Pin Diameter		Max. Wire Width	Maximum Head Diameter	Maximum Prong Extension	Recommended Hole Diameter		
Nominal Size	Max. mm	mm	mm	mm	Nominal mm	Tolerance (H13) mm	
0.6	0.5	0.5	1.0	1.6	0.6	+0.140 0	
0.8	0.7	0.7	1.4		0.8		
1.0	0.9	0.9	1.8		1.0		
1.2	1.0	1	2.0		1.2		
1.6	1.4	1.4	2.8	2.5	1.6	+0.250 0	
2.0	1.8	1.8	3.6		2.0		
2.5	2.3	2.3	4.6		2.5		
3.2	2.9	2.9	5.8	3.2	3.2	+0.300 0	
4.0	3.7	3.7	7.4	4.0	4.0		
5.0	4.6	4.6	9.2		5.0		
6.3	5.9	5.9	11.8		6.3	+0.360 0	

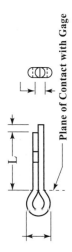

L

Plane of Contact with Gage

Table 5-34 Lengths, Cotter Pins, Metric
(Diameters up to 6.3 mm and Lengths to 100 mm)
Part Complies With ISO 1234: 1997 (DIN 94)

Common Lengths Up To 100 mm — mm — Length Tolerance

Pin Diameter Nominal mm	±0.5			±1.0							±1.5								±2.0		
	6	8	10	12	16	18	20	22	25	28	32	36	40	45	50	56	63	71	80	90	100
0.6																					
0.8																					
1.0	6	8	10	12	16	18	20														
1.2		8	10	12	16	18	20	22	25												
1.6		8	10	12	16	18	20	22	25		32	36									
2.0			10	12	16	18	20	22	25	28	32	36	40								
2.5		8	10	12	16	18	20	22	25	28	32	36	40	45	50						
3.2		8	10	12	16	18	20	22	25	28	32	36	40	45	50						
4.0					16	18	20	22	25	28	32	36	40	45	50	56	63	71			
5.0								22	25		32	36	40	45	50		63	71	80	90	
6.3												36	40	45	50		63	71	80	90	100

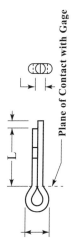

L

Plane of Contact with Gage

Table 5-35: *Parallel Square Keys, Inch*
(Widths up to ½ inch)
Part Complies With ASME B17.1 – 1967 (R2003)

Shaft Diameter (inch)	Set Screw Size	Key Width Nominal (inch)	Key Width Tol. Class 1	Key Width Tol. Class 2	Key Height Nominal (inch)	Key Height Tol. Class 1	Key Height Tol. Class 2	SHAFT Keyseat Width Nominal	SHAFT KW Tol. Class 1 Fit	SHAFT KW Tol. Class 2 Fit	SHAFT Keyseat Depth Nominal	SHAFT KD Tol. Class 1 & 2	HUB Keyseat Width Nominal	HUB KW Tol. Class 1 Fit	HUB KW Tol. Class 2 Fit	HUB Keyseat Depth Nominal	HUB KD Tol. Class 1 & 2
Up to and Including 0.3125	#6	0.0625	0 / -0.002	+0.001 / 0	0.0625	0 / -0.002	+0.001 / 0	0.0625	+0.002 / 0	+0.002 / 0	0.03125	0 / -0.015	0.0625	+0.002 / 0	+0.002 / 0	0.03125	+0.010 / 0
Over 0.3125 Including 0.4375	#10	0.09375	0 / -0.002	+0.001 / 0	0.09375	0 / -0.002	+0.001 / 0	0.09375	+0.002 / 0	+0.002 / 0	0.046875	0 / -0.015	0.09375	+0.002 / 0	+0.002 / 0	0.046875	+0.010 / 0
Over 0.4375 Including 0.5625	#10	0.125	0 / -0.002	+0.001 / 0	0.125	0 / -0.002	+0.001 / 0	0.125	+0.002 / 0	+0.002 / 0	0.0625	0 / -0.015	0.125	+0.002 / 0	+0.002 / 0	0.0625	+0.010 / 0
Over 0.5625 Including 0.875	1/4	0.1875	0 / -0.002	+0.001 / 0	0.1875	0 / -0.002	+0.001 / 0	0.1875	+0.002 / 0	+0.002 / 0	0.09375	0 / -0.015	0.1875	+0.002 / 0	+0.002 / 0	0.09375	+0.010 / 0
Over 0.875 Including 1.25	5/16	0.25	0 / -0.002	+0.001 / 0	0.25	0 / -0.002	+0.001 / 0	0.25	+0.002 / 0	+0.002 / 0	0.125	0 / -0.015	0.25	+0.002 / 0	+0.002 / 0	0.125	+0.010 / 0
Over 1.25 Including 1.375	3/8	0.3125	0 / -0.002	+0.001 / 0	0.3125	0 / -0.002	+0.001 / 0	0.3125	+0.002 / 0	+0.002 / 0	0.15625	0 / -0.015	0.3125	+0.002 / 0	+0.002 / 0	0.15625	+0.010 / 0
Over 1.375 Including 1.75	3/8	0.375	0 / -0.002	+0.001 / 0	0.375	0 / -0.002	+0.001 / 0	0.375	+0.002 / 0	+0.002 / 0	0.1875	0 / -0.015	0.375	+0.002 / 0	+0.002 / 0	0.1875	+0.010 / 0
Over 1.75 Including 2.25	1/2	0.5	0 / -0.002	+0.001 / 0	0.5	0 / -0.002	+0.001 / 0	0.5	+0.002 / 0	+0.002 / 0	0.25	0 / -0.015	0.5	+0.002 / 0	+0.002 / 0	0.25	+0.010 / 0

Hub
Key
Shaft

W
H

Table 5-36 *Lengths, Parallel Square Keys, Inch*
(Widths up to ½ inch and Lengths to 3 inches)
Part Complies With ASME B17.1 – 1967 (R2003)

Key Size (Square)	Nominal Key Width	Common Key Lengths Up To 3 Inches inch											
1/16	0.0625												
3/32	0.09375												
1/8	0.125	0.5 (b)	0.75	1	1.25	1.5		1.75	2				
3/16	0.1875	0.5	0.75	1	1.25	1.5		1.75	2	2.25	2.5	2.75	3
1/4	0.25		0.75	1	1.25	1.5		1.75	2	2.25	2.5	2.75	3
5/16	0.3125		0.75	1	1.25	1.5		1.75	2	2.25	2.5	2.75	3
3/8	0.375		0.75	1	1.25	1.5		1.75	2	2.25	2.5	2.75	3
1/2	0.5			1	1.25	1.5	1.625 (b)	1.75	2	2.25	2.5	2.75	3

a) Undersized Only (Class 1)
b) Oversized Only (Class 2)

Table 5-37: *Parallel Keys With Round Ends, Metric*
(Widths up to 16 mm)
Part Complies With DIN 6885: 1968 Type A

Shaft Diameter mm	Set Screw Size	Key Width Nominal mm	Key Width Tolerance (h9) mm	Key Height Nominal mm	SHAFT Keyseat Width Nominal mm	SHAFT Keyseat Width Sliding Fit (H9) mm	SHAFT Keyseat Width Transition Fit (N9) mm	SHAFT Keyseat Depth Nominal mm	SHAFT Keyseat Depth Tolerance mm	HUB Keyseat Width Nominal mm	HUB Keyseat Width Sliding Fit (D10) mm	HUB Keyseat Width Transition Fit (JS9) mm	HUB Keyseat Depth Nominal mm	HUB Keyseat Depth Tolerance mm
Over 6 Including 8	M1.6	2	0 / -0.025	2	2	+0.025 / 0	-0.004 / -0.029	1.2	+0.1 / 0	2	+0.060 / +0.020	+0.0125 / -0.0125	1	+0.1 / 0
Over 8 Including 10	M2	3	0 / -0.025	3	3	+0.025 / 0	-0.004 / -0.029	1.8	+0.1 / 0	3	+0.060 / +0.020	+0.0125 / -0.0125	1.4	+0.1 / 0
Over 10 Including 12	M3	4	0 / -0.030	4	4	+0.030 / 0	0 / -0.030	2.5	+0.1 / 0	4	+0.078 / +0.030	+0.015 / -0.015	1.8	+0.1 / 0
Over 12 Including 17	M4	5	0 / -0.030	5	5	+0.030 / 0	0 / -0.030	3	+0.1 / 0	5	+0.078 / +0.030	+0.015 / -0.015	2.3	+0.1 / 0
Over 17 Including 22	M5	6	0 / -0.030	6	6	+0.030 / 0	0 / -0.030	3.5	+0.1 / 0	6	+0.078 / +0.030	+0.015 / -0.015	2.8	+0.1 / 0
Over 22 Including 30	M6	8	0 / -0.036	7	8	+0.036 / 0	0 / -0.036	4	+0.2 / 0	8	+0.098 / +0.040	+0.018 / -0.018	3.3	+0.2 / 0
Over 30 Including 38	M8	10	0 / -0.036	8	10	+0.036 / 0	0 / -0.036	5	+0.2 / 0	10	+0.098 / +0.040	+0.018 / -0.018	3.3	+0.2 / 0
Over 38 Including 44	M10	12	0 / -0.043	8	12	+0.043 / 0	0 / -0.043	5	+0.2 / 0	12	+0.120 / +0.050	+0.0215 / -0.0215	3.3	+0.2 / 0
Over 44 Including 50	M12	14	0 / -0.043	9	14	+0.043 / 0	0 / -0.043	5.5	+0.2 / 0	14	+0.120 / +0.050	+0.0215 / -0.0215	3.8	+0.2 / 0
Over 50 Including 58	M12	16	0 / -0.043	10	16	+0.043 / 0	0 / -0.043	6	+0.2 / 0	16	+0.120 / +0.050	+0.0215 / -0.0215	4.3	+0.2 / 0

Note: With broached keyways, IT grade 8 should be used instead of 9 for width tolerances

Table 5-38 Lengths, Parallel Keys With Round Ends, Metric
(Widths up to 16 mm and Lengths to 100 mm)
Part Complies With DIN 6885: 1968 Type A

Common Key Lengths Up To 100 mm (mm) — Length Tolerance

Nominal Key Width mm	Key: 0/−0.2; Keyway: +0.2/0							Key: 0/−0.3; Keyway: +0.3/0									Key: 0/−0.5; Keyway: +0.5/0	
2	10		15			20												
3	10		15			20	25											
4	10	12	15	16	18	20	25	30	35	40	45	50						
5	10	12	15	16	18	20	25	30	35	40	45	50						
6	10	12	15	16	18	20	25	30	35	40	45	50	55	60				
8			15	16	18	20	25	30	35	40	45	50	55	60	70	80		
10			15	16	18	20	25	30	35	40	45	50	55	60	70	80		
12						20	25	30	35	40	45	50	55	60	70	80		
14								30	35	40	45	50	55	60	70	80	90	
16										40	45	50	55	60	70	80	90	100

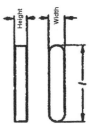

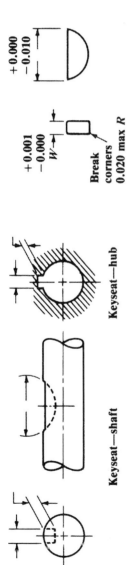

Keyseat—shaft

Keyseat—hub

+0.000
−0.010

+0.001
−0.000
W

Break
corners
0.020 max *R*

Table 5-39: Woodruff Keys, Inch
(Widths up to ½ inch)
Part Complies With ASME B17.2 – 1967 (R2003)

Shaft Diameter inch	Key Number	Key Width +0.001 / 0 inch Nominal	SHAFT Keyseat Width Nominal inch	SHAFT Keyseat Width Tolerance inch	SHAFT Keyseat Depth +0.005 / −0 inch	Seat Diameter Nominal inch	Seat Diameter TOLERANCE − inch	Seat Diameter TOLERANCE + inch	HUB Keyseat Width +0.002 / −0 inch	HUB Keyseat Depth +0.005 / −0 inch
Up to and Including 0.3125	202	0.0625	0.0625	+0.0005 / −0.0010	0.0728	0.25	0	0.018	0.0635	0.0372
	202.5				0.1038	0.312				
	203				0.1358	0.375				
	204				0.1668	0.5				
Over 0.3125 Including 0.4375	302.5	0.09375	0.09375	+0.00055 / −0.00095	0.0882	0.312	0	0.018	0.0948	0.0529
	303				0.1202	0.375				
	304				0.1511	0.5				
	305				0.1981	0.625				
Over 0.4375 Including 0.5625	403	0.125	0.125	+0.0005 / −0.0010	0.1045	0.375	0	0.018	0.1260	0.0685
	404				0.1355	0.5				
	405				0.1825	0.625				
	406				0.2455	0.75				
Over 0.5625 Including 0.875	605	0.1875	0.1875	+0.0005 / −0.0012	0.1513	0.625	0	0.018	0.1885	0.0997
	606				0.2143	0.75	0			
	607				0.2763	0.875	0			
	608				0.3393	1	0	0.02		
	609				0.3853	1.125	0			
	610				0.4483	1.25	0	0.023		
	617-1				0.3073	2.125	0	0.035		
	617				0.4323	2.125	0			

continued on next page

Table 5-39: Woodruff Keys, Inch (Continued)

Range									
				806	0.1830	0.75	0	0.018	
				807	0.2450	0.875	0	0.02	
				808	0.3080	1			
Over 0.875				809	0.3540	1.125			0.1310
Including 1.25	0.25	0.25	+0.0005 / −0.0013	810	0.4170	1.25	0	0.023	0.2510
				811	0.4640	1.375			
				812	0.5110	1.5	0	0.035	
				817-1	0.2760	2.125			
				817	0.4010	2.125			
				822-1	0.4640	2.75			
				822	0.6200	2.75			
				1008	0.2768	1	0	0.02	
				1009	0.3228	1.125			
				1010	0.3858	1.25		0.023	0.1622
Over 1.25				1011	0.4328	1.375	0		
Including 1.375	0.3125	0.3125	+0.0005 / −0.0014	1012	0.4798	1.5			0.3135
				1017-1	0.2448	2.125			
				1017	0.3698	2.125	0	0.035	
				1022-1	0.4328	2.75			
				1022	0.5888	2.75			
				1208	0.2455	1	0	0.02	
				1210	0.3545	1.25			
				1211	0.4015	1.375	0	0.023	0.1935
Over 1.375				1212	0.4485	1.5			
Including 1.75	0.375	0.375	+0.0005 / −0.0015	1217-1	0.2135	2.125			0.3760
				1217	0.3385	2.125			
				1222-1	0.4015	2.75	0	0.035	
				1222	0.5575	2.75			
				1228	0.7455	3.5			
Over 1.75				1622-1	0.3390	2.75			0.2560
Including 2.25	0.5	0.5	+0.0005 / −0.0015	1622	0.4950	2.75	0	0.035	0.5010
				1628	0.6830	3.5			

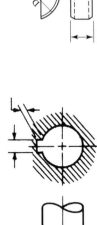

Keyseat—hub

Keyseat—shaft

Sharp edges removed
0.16 Minimum

Table 5-40 Woodruff Keys, Metric
(Widths up to 10 mm)
Part Complies With DIN 6888: 1956

Allocation I: Where torque is transmitted by the key

Shaft Diameter mm	Key Width Nominal mm	Key Width Tolerance (h9) mm	Keyseat Width Nominal mm (SHAFT)	Keyseat Width Sliding Fit (N9) mm	Keyseat Width Tight Fit (P9) mm	Keyseat Depth Nominal mm (SHAFT)	Keyseat Depth Tolerance mm	Seat Diameter Nominal mm	Seat Diameter Tolerance mm	Keyseat Width Nominal mm (HUB)	Keyseat Width Sliding Fit (JS9) mm	Keyseat Width Tight Fit (P9) mm	Keyseat Depth Nominal mm (HUB)	Keyseat Depth Tolerance mm
Over 3 Including 4	1	0 / −0.025	1	−0.004 / −0.029	−0.006 / −0.031	1	+0.1 / 0	4	+0.5 / 0	1	+0.0125 / −0.0125	−0.006 / −0.031	0.6	+0.1 / 0
Over 4 Including 6	1.5	0 / −0.025	1.5	−0.004 / −0.029	−0.006 / −0.031	2	+0.1 / 0	7	+0.5 / 0	1.5	+0.0125 / −0.0125	−0.006 / −0.031	0.8	+0.1 / 0
Over 6 Including 8	2	0 / −0.025	2	−0.004 / −0.029	−0.006 / −0.031	1.8	+0.1 / 0	7	+0.5 / 0	2	+0.0125 / −0.0125	−0.006 / −0.031	1	+0.1 / 0
						2.9	+0.1 / 0	10	+0.5 / 0					
Over 8 Including 10	2.5	0 / −0.025	2.5	−0.004 / −0.029	−0.006 / −0.031	2.9	+0.1 / 0	10	+0.5 / 0	2.5	+0.0125 / −0.0125	−0.006 / −0.031	1	+0.1 / 0
	3	0 / −0.025	3	−0.004 / −0.029	−0.006 / −0.031	2.5	+0.1 / 0	10	+0.5 / 0	3	+0.0125 / −0.0125	−0.006 / −0.031	1.4	+0.1 / 0
						3.8	+0.1 / 0	13	+0.5 / 0					
						5.3	+0.1 / 0	16	+0.5 / 0					
Over 10 Including 12	4	0 / −0.030	4	0 / −0.030	−0.012 / −0.042	3.5	+0.1 / 0	13	+0.5 / 0	4	+0.015 / −0.015	−0.012 / −0.042	1.7	+0.1 / 0
						5	+0.1 / 0	16	+0.5 / 0					
						6	+0.1 / 0	19	+0.5 / 0					
Over 12 Including 17	5	0 / −0.030	5	0 / −0.030	−0.012 / −0.042	4.5	+0.1 / 0	16	+0.5 / 0	5	+0.015 / −0.015	−0.012 / −0.042	2.2	+0.1 / 0
						5.5	+0.1 / 0	19	+0.5 / 0					
						7	+0.2 / 0	22	+0.5 / 0					
Over 17 Including 22	6	0 / −0.030	6	0 / −0.030	−0.012 / −0.042	5.1	+0.1 / 0	19	+0.5 / 0	6	+0.015 / −0.015	−0.012 / −0.042	2.6	+0.1 / 0
						6.6	+0.2 / 0	22	+0.5 / 0					
						8.6	+0.2 / 0	28	+0.5 / 0					
Over 22 Including 30	8	0 / −0.036	8	0 / −0.036	−0.015 / −0.051	6.2	+0.2 / 0	22	+0.5 / 0	8	+0.018 / −0.018	−0.015 / −0.051	3	+0.1 / 0
						8.2	+0.2 / 0	28	+0.5 / 0					
						10.2	+0.2 / 0	32	+0.5 / 0					
Over 30 Including 38	10	0 / −0.036	10	0 / −0.036	−0.015 / −0.051	7.8	+0.2 / 0	28	+0.5 / 0	10	+0.018 / −0.018	−0.015 / −0.051	3.4	+0.2 / 0
						9.8	+0.2 / 0	32	+0.5 / 0					
						12.8	+0.2 / 0	45	+0.5 / 0					

Note: With broached keyways, IT grade 8 should be used instead of 9 for width tolerances

This page was intentionally left blank so that you may view the following tables in a more cohesive manner.

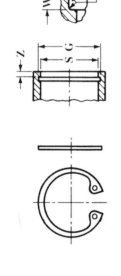

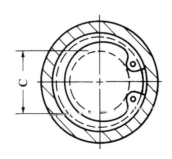

Table 5-41 Axially-Assembled Internal Retaining Rings, Inch
(Up to 2 inch bore) Part complies with NA1

Housing Bore Diameter (S) inch	Groove Diameter (G) inch	Groove Tolerance inch	Groove Width (W) inch	Width Tolerance inch	Max Allowable Radii in Groove (R) inch	Groove Distance From End (Z) inch	Unplated Ring Thickness inch	Thickness Tolerance inch	Installation Inner Clearance Diameter inch	Installed Inner Clearance Diameter (C) inch
0.250	0.268	+0.001	0.020	+.002	0.005	0.027	0.015		0.115	0.133
0.312	0.330	-0.001	0.020	-0	0.005	0.027	0.015		0.173	0.191
0.375	0.397		0.029		0.005	0.033	0.025		0.204	0.226
0.438	0.461		0.029		0.005	0.036	0.025		0.230	0.254
0.453	0.477	+0.002	0.029		0.005	0.036	0.025		0.250	0.274
0.500	0.530	-0.002	0.039		0.005	0.045	0.035		0.260	0.290
0.512	0.542		0.039		0.005	0.045	0.035		0.270	0.300
0.562	0.596		0.039	+0.003	0.005	0.051	0.035		0.275	0.305
0.625	0.665		0.039	-0	0.005	0.060	0.035	+0.002	0.340	0.380
0.688	0.732		0.039		0.005	0.066	0.035	-0.002	0.400	0.440
0.750	0.796		0.039		0.005	0.069	0.035		0.450	0.490
0.777	0.825		0.046		0.005	0.072	0.042		0.475	0.520
0.812	0.862		0.046		0.005	0.075	0.042		0.490	0.540
0.866	0.920		0.046		0.005	0.081	0.042		0.540	0.590
0.875	0.931	+0.003	0.046		0.005	0.084	0.042		0.545	0.600
0.901	0.959	-0.003	0.046		0.005	0.087	0.042		0.565	0.620
0.938	1.000		0.046		0.005	0.093	0.042		0.610	0.670
1.000	1.066		0.046		0.005	0.099	0.042		0.665	0.730
1.023	1.091		0.046		0.010	0.102	0.042		0.690	0.755

continued on next page

Table 5-41 Axially-Assembled Internal Retaining Rings, Inch (Continued)

1.062	1.130		0.056		0.010	0.102	0.050		0.685	0.750
1.125	1.197		0.056		0.010	0.108	0.050		0.745	0.815
1.181	1.255		0.056		0.010	0.111	0.050		0.790	0.860
1.188	1.262		0.056		0.010	0.111	0.050		0.800	0.870
1.250	1.330		0.056		0.010	0.120	0.050		0.875	0.955
1.259	1.339	+0.004	0.056		0.010	0.120	0.050	+0.002	0.885	0.965
1.312	1.396	−0.004	0.056		0.010	0.126	0.050	−0.002	0.930	1.010
1.375	1.461		0.056	+.004	0.010	0.129	0.050		0.990	1.070
1.378	1.464		0.056	−0	0.010	0.129	0.050		0.990	1.070
1.438	1.528		0.056		0.010	0.135	0.050		1.060	1.150
1.456	1.548		0.056		0.010	0.138	0.050		1.080	1.170
1.500	1.594		0.056		0.010	0.141	0.050		1.120	1.210
1.562	1.658		0.068		0.010	0.144	0.062		1.140	1.230
1.575	1.671		0.068		0.010	0.144	0.062		1.150	1.240
1.625	1.725		0.068		0.010	0.150	0.062		1.150	1.250
1.653	1.755		0.068		0.010	0.153	0.062		1.170	1.270
1.688	1.792	+0.005	0.068		0.010	0.156	0.062	+0.003	1.230	1.330
1.750	1.858	−0.005	0.068		0.010	0.162	0.062	−0.003	1.260	1.360
1.812	1.922		0.068		0.010	0.165	0.062		1.340	1.380
1.850	1.962		0.068		0.010	0.168	0.062		1.350	1.460
1.875	1.989		0.068		0.010	0.171	0.062		1.370	1.480
1.938	2.056		0.068		0.010	0.177	0.062		1.460	1.580
2.000	2.122		0.068		0.010	0.183	0.062		1.520	1.640

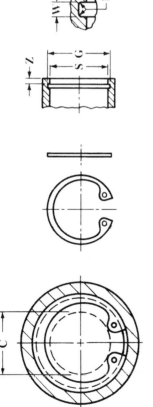

Table 5-42 *Axially-Assembled Internal Retaining Rings, Metric*
(Up to 50 mm bore) Part Complies With DIN 472: 1981

Housing Bore Diameter (S) mm	Groove Diameter (G) mm	Groove Tolerance mm	Min. Groove Width (W) mm	Width Tolerance (H13) mm	Max Allowable Radii in Groove (R) mm	Min. Groove Distance From End (Z) mm	Unplated Ring Thickness mm	Thickness Tolerance mm	Installation Inner Clearance Diameter (C) mm
8.0	8.4	+0.09 / -0	0.90	+0.14 / -0	0.08	0.60	0.80	+0 / -0.05	3.00
9.0	9.4	+0.09 / -0	0.90	+0.14 / -0	0.08	0.60	0.80	+0 / -0.05	3.70
10.0	10.4	+0.11 / -0	1.10	+0.14 / -0	0.10	0.60	1.00	+0 / -0.06	3.30
11.0	11.4	+0.11 / -0	1.10	+0.14 / -0	0.10	0.60	1.00	+0 / -0.06	4.10
12.0	12.5	+0.11 / -0	1.10	+0.14 / -0	0.10	0.80	1.00	+0 / -0.06	4.90
13.0	13.6	+0.11 / -0	1.10	+0.14 / -0	0.10	0.90	1.00	+0 / -0.06	5.40
14.0	14.6	+0.11 / -0	1.10	+0.14 / -0	0.10	0.90	1.00	+0 / -0.06	6.20
15.0	15.7	+0.11 / -0	1.10	+0.14 / -0	0.10	1.10	1.00	+0 / -0.06	7.20
16.0	16.8	+0.11 / -0	1.10	+0.14 / -0	0.10	1.20	1.00	+0 / -0.06	8.00
17.0	17.8	+0.11 / -0	1.10	+0.14 / -0	0.10	1.20	1.00	+0 / -0.06	8.80
18.0	19.0	+0.11 / -0	1.10	+0.14 / -0	0.10	1.50	1.00	+0 / -0.06	9.40
19.0	20.0	+0.13 / -0	1.10	+0.14 / -0	0.10	1.50	1.00	+0 / -0.06	10.40
20.0	21.0	+0.13 / -0	1.10	+0.14 / -0	0.10	1.50	1.00	+0 / -0.06	11.20
21.0	22.0	+0.13 / -0	1.10	+0.14 / -0	0.10	1.50	1.00	+0 / -0.06	12.20
22.0	23.0	+0.13 / -0	1.10	+0.14 / -0	0.10	1.50	1.00	+0 / -0.06	13.20
24.0	25.2	+0.21 / -0	1.30	+0.14 / -0	0.12	1.80	1.20	+0 / -0.06	14.80
25.0	26.2	+0.21 / -0	1.30	+0.14 / -0	0.12	1.80	1.20	+0 / -0.06	15.50
26.0	27.2	+0.21 / -0	1.30	+0.14 / -0	0.12	1.80	1.20	+0 / -0.06	16.10
28.0	29.4	+0.21 / -0	1.30	+0.14 / -0	0.12	2.10	1.20	+0 / -0.06	17.90

continued on next page

Table 5-42 Axially-Assembled Internal Retaining Rings, Metric (Continued)

30.0	31.4		1.30		0.12	2.10	1.20		19.90
31.0	32.7		1.30		0.12	2.60	1.20		20.00
32.0	33.7		1.30		0.12	2.60	1.20		20.60
34.0	35.7		1.60		0.15	2.60	1.50		22.60
35.0	37.0		1.60		0.15	3.00	1.50		23.60
36.0	38.0	+0.25	1.60	+0.14	0.15	3.00	1.50	+0	24.60
37.0	39.0	-0	1.60	-0	0.15	3.00	1.50	-0.06	25.40
38.0	40.0		1.60		0.15	3.00	1.50		26.40
40.0	42.5		1.85		0.18	3.80	1.75		27.80
42.0	44.5		1.85		0.18	3.80	1.75		29.60
45.0	47.5		1.85		0.18	3.80	1.75		32.00
47.0	49.5		1.85		0.18	3.80	1.75		33.50
48.0	50.5		1.85		0.18	3.80	1.75		34.50
50.0	53.0	+0.30 -0	2.15		0.20	4.50	2.00	+0 -0.07	36.30

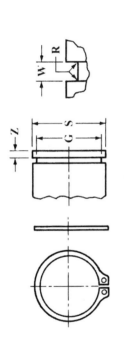

Table 5-43 Axially-Assembled External Retaining Rings, Inch
(Up to 2 inch shaft)
Part complies with NA2

Shaft Diameter (S) inch	Groove Diameter (G) inch	Groove Tolerance inch	Groove Width (W) inch	Width Tolerance inch	Max Allowable Radii in Groove (R) inch	Groove Distance From End (Z) inch	Unplated Ring Thickness inch	Thickness Tolerance inch	Installation Outer Clearance Diameter inch	Installed Outer Clearance Diameter (C) inch
0.125	0.117		0.012		0.0000	0.012	0.010	+0.001	0.222	0.214
0.156	0.146		0.012		0.0000	0.015	0.010	−0.001	0.270	0.260
0.188	0.175	+0.0015	0.018		0.0000	0.018	0.015		0.298	0.286
0.197	0.185	−0.0015	0.018	+0.002	0.0000	0.018	0.015		0.319	0.307
0.219	0.205		0.018	−0	0.0000	0.021	0.015		0.338	0.324
0.236	0.222		0.029		0.0000	0.021	0.015		0.355	0.341
0.250	0.230		0.029		0.0030	0.030	0.025		0.450	0.430
0.276	0.255		0.029		0.0030	0.031	0.025		0.480	0.460
0.281	0.261		0.029		0.0030	0.030	0.025		0.490	0.470
0.312	0.290		0.029		0.0030	0.033	0.025		0.540	0.520
0.344	0.321		0.029		0.0030	0.033	0.025		0.570	0.550
0.354	0.330		0.029		0.0030	0.036	0.025		0.590	0.570
0.375	0.352	+0.002	0.029	+0.003	0.0050	0.036	0.025	+0.002	0.610	0.590
0.394	0.369	−0.002	0.029	−0	0.0050	0.037	0.025	−0.002	0.620	0.600
0.406	0.382		0.029		0.0050	0.036	0.025		0.630	0.610
0.438	0.412		0.029		0.0050	0.039	0.025		0.660	0.640
0.461	0.435		0.029		0.0050	0.039	0.025		0.680	0.660
0.469	0.443		0.029		0.0050	0.039	0.025		0.680	0.660
0.500	0.468		0.029		0.0050	0.048	0.035		0.770	0.740
0.551	0.519		0.039		0.0050	0.048	0.035		0.810	0.780
0.562	0.530		0.039		0.0050	0.048	0.035		0.820	0.790

continued on next page

Table 5-43 Axially-Assembled External Retaining Rings, Inch (Continued)

0.594		0.039		0.0050	0.052	0.035		0.860	0.830
0.625		0.039		0.0050	0.055	0.035		0.900	0.870
0.669		0.039		0.0050	0.060	0.035		0.930	0.890
0.672		0.039		0.0050	0.060	0.035		0.930	0.890
0.688		0.046		0.0050	0.063	0.042		1.010	0.970
0.704	+0.003	0.046		0.0050	0.069	0.042		1.090	1.050
0.733	-0.003	0.046	+0.004	0.0050	0.072	0.042		1.120	1.080
0.762		0.046	-0	0.0050	0.075	0.042		1.150	1.100
0.791		0.046		0.0050	0.078	0.042		1.180	1.130
0.821		0.046		0.0050	0.081	0.042	+0.002	1.210	1.160
0.882		0.046		0.0050	0.084	0.042	-0.002	1.340	1.290
0.926		0.046		0.0050	0.087	0.042		1.390	1.340
0.940		0.046		0.0100	0.090	0.042		1.410	1.350
0.961		0.046		0.0100	0.093	0.042		1.430	1.370
0.998		0.056		0.0100	0.096	0.050		1.500	1.440
1.059		0.056		0.0100	0.099	0.050		1.550	1.490
1.118		0.056		0.0100	0.105	0.050		1.610	1.540
1.176	+0.004	0.056		0.0100	0.111	0.050		1.690	1.620
1.232	-0.004	0.056		0.0100	0.120	0.050		1.750	1.670
1.291		0.056		0.0100	0.126	0.050		1.800	1.720
1.350		0.056		0.0100	0.132	0.050		1.870	1.790
1.406		0.056	+0.004	0.0100	0.141	0.050		1.990	1.900
1.468		0.068	-0	0.0100	0.141	0.062		2.010	2.010
1.529		0.068		0.0100	0.144	0.062		2.100	2.080
1.589		0.068		0.0100	0.148	0.062		2.170	2.160
1.650	+0.005	0.068		0.0100	0.150	0.062		2.240	2.210
1.669	-0.005	0.068		0.0100	0.154	0.062	+0.003	2.310	2.230
1.708		0.068		0.0100	0.156	0.062	-0.003	2.330	2.280
1.769		0.068		0.0100	0.159	0.062		2.380	2.340
1.857		0.068		0.0100	0.168	0.062		2.440	2.460
1.886		0.068		0.0100	0.171	0.062		2.570	2.490
2.000								2.600	

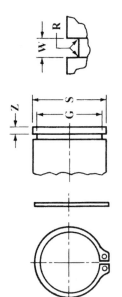

Table 5-44 *Axially-Assembled External Retaining Rings, Metric*
(Up to 50 mm shaft)
Part Complies With DIN 471: 1981

Shaft Diameter (S) mm	Groove Diameter (G) mm	Groove Tolerance mm	Groove Width (W) mm	Width Tolerance (H13) mm	Max Allowable Radii in Groove (R) mm	Min. Groove Distance From End (Z) mm	Unplated Ring Thickness mm	Thickness Tolerance mm	Installation Outer Clearance Diameter (C) mm
3.0	2.8	+0 −0.04	0.50	+0.14 −0	0.040	0.30	0.40	+0 −0.05	7.00
4.0	3.8		0.50		0.040	0.30	0.40		8.60
5.0	4.8	+0 −0.048	0.70		0.060	0.30	0.60		10.30
6.0	5.7		0.80		0.070	0.50	0.70		11.70
7.0	6.7		0.90		0.080	0.50	0.80		13.50
8.0	7.6	+0 −0.06	0.90		0.080	0.60	0.80		14.70
9.0	8.6		1.10		0.100	0.60	1.00		16.00
10.0	9.6		1.10		0.100	0.60	1.00		17.00
11.0	10.5		1.10		0.100	0.80	1.00	+0 −0.06	18.00
12.0	11.5		1.10		0.100	0.80	1.00		19.00
13.0	12.4	+0 −0.11	1.10		0.100	0.90	1.00		20.20
14.0	13.4		1.10		0.100	0.90	1.00		21.40
15.0	14.3		1.10		0.100	1.10	1.00		22.60
16.0	15.2		1.10		0.100	1.20	1.00		23.80
17.0	16.2		1.10		0.100	1.20	1.00		25.00
18.0	17.0		1.30		0.120	1.50	1.20		26.20
19.0	18.0		1.30		0.120	1.50	1.20		27.20
20.0	19.0	+0 −0.13	1.30		0.120	1.50	1.20		28.40
21.0	20.0		1.30		0.120	1.50	1.20		29.60
22.0	21.0		1.30		0.120	1.50	1.20		30.80

continued on next page

Table 5-44 Axially-Assembled External Retaining Rings, Metric (Continued)

24.0	22.9		1.30		0.120	1.70	1.20		33.20
25.0	23.9		1.30		0.120	1.70	1.20		34.20
26.0	24.9	+0 / −0.21	1.30		0.120	1.70	1.20		35.50
28.0	26.6		1.60		0.150	2.10	1.50		37.90
29.0	27.6		1.60	+0.14 / −0	0.150	2.10	1.50		39.10
30.0	28.6		1.60		0.150	2.10	1.50		40.50
32.0	30.3		1.60		0.150	2.60	1.50	+0 / −0.06	43.00
34.0	32.3		1.60		0.150	2.60	1.50		45.40
35.0	33.0		1.60		0.150	3.00	1.50		46.80
36.0	34.0		1.85		0.175	3.00	1.75		47.80
38.0	36.0	+0 / −0.25	1.85		0.175	3.00	1.75		50.20
40.0	37.5		1.85		0.175	3.80	1.75		52.60
42.0	39.5		1.85		0.175	3.80	1.75		55.70
45.0	42.5		1.85		0.175	3.80	1.75		59.10
48.0	45.5		1.85		0.175	3.80	1.75		62.50
50.0	47.0		2.15		0.200	4.50	2.00	+0 / −0.07	64.50

Table 5-45 Radially-Assembled External (E-Ring) Retaining Rings, Inch
Part complies with NA3

Shaft Diameter (S) inch	Groove Diameter (G) inch	Groove Tolerance inch	Groove Width (W) inch	Width Tolerance inch	Max Allowable Radii in Groove (R) inch	Groove Distance From End (Z) inch	Unplated Ring Thickness inch	Thickness Tolerance inch	Installed Outer Clearance Diameter (Y) inch
0.040	0.026	+0.002	0.012	+0.002	0.000	0.014	0.010	+0.001	0.090
0.062	0.052	-0	0.012	-0	0.000	0.010	0.010	-0.001	0.165
0.094	0.074		0.020		0.005	0.020	0.015		0.200
0.125	0.095		0.020		0.005	0.030	0.015		0.240
0.140	0.105		0.029		0.005	0.034	0.025		0.285
0.156	0.116		0.029		0.005	0.040	0.025		0.295
0.188	0.147		0.029		0.005	0.040	0.025	+0.002	0.350
0.250	0.210	+0.003	0.029	+0.003	0.005	0.040	0.025	-0.002	0.540
0.375	0.303	-0	0.039	-0	0.010	0.072	0.035		0.680
0.438	0.343		0.039		0.010	0.094	0.035		0.710
0.500	0.396		0.046		0.015	0.104	0.042		0.820
0.625	0.485		0.046		0.015	0.140	0.042		0.960
0.750	0.580		0.056		0.015	0.170	0.050		1.140
0.875	0.675		0.056		0.015	0.200	0.050		1.320

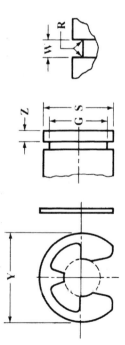

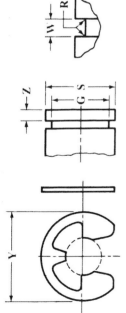

Table 5-46 Radially-Assembled External (E-Ring) Retaining Rings, Metric

Part Complies With DIN 6799: 1981

Shaft Diameter (S) FROM mm	TO mm	Groove Diameter (G) mm	Groove Tolerance mm	Groove Width (W) mm	Width Tolerance mm	Max Allowable Radii in Groove (R) mm	Min. Groove Distance From End (Z) mm	Unplated Ring Thickness mm	Thickness Tolerance mm	Installed Outer Clearance Diameter (Y) mm
1.0	1.4	0.8	+0 / -0.04	0.24	+0.04 / -0	0.020	0.4	0.20	+0.02 / -0.02	2.25
1.4	2.0	1.2		0.34		0.030	0.6	0.30		3.25
2.0	2.5	1.5	+0 / -0.06	0.44		0.040	0.8	0.40		4.25
2.5	3.0	1.9		0.54	+0.05 / -0	0.050	1.0	0.50		4.80
3.0	4.0	2.3		0.64		0.060	1.0	0.60		6.30
4.0	5.0	3.2	+0 / -0.075	0.64		0.060	1.0	0.60		7.30
5.0	7.0	4.0		0.74		0.070	1.2	0.70		9.30
6.0	8.0	5.0		0.74		0.070	1.2	0.70		11.30
7.0	9.0	6.0		0.74		0.070	1.2	0.70		12.30
8.0	11.0	7.0	+0 / -0.09	0.94		0.090	1.5	0.90		14.30
9.0	12.0	8.0		1.05		0.100	1.8	1.00		16.30
10.0	14.0	9.0		1.15	+0.08 / -0	0.110	2.0	1.10	+0.03 / -0.03	18.80
11.0	15.0	10.0		1.25		0.120	2.0	1.20		20.40
13.0	18.0	12.0	+0 / -0.11	1.35		0.130	2.5	1.30		23.40
16.0	24.0	15.0		1.55		0.150	3.0	1.50		29.40
20.0	31.0	19.0		1.80		0.175	3.5	1.75		37.60
25.0	38.0	24.0	+0 / -0.13	2.05		0.200	4.0	2.00		44.60
32.0	42.0	30.0		2.55		0.250	4.5	2.50		52.60

Table 5-47: SolidWorks® 2010 Toolbox Components
Selections Covered in This Text

SOLIDWORKS® TOOLBOX PATH			SOLIDWORKS® PART NAME	STANDARD OR SOURCE OF DATA
ANSI INCH	Pins	All Pins	Dowel Pin	ASME B18.8.2
			Spring Pin Slotted	ASME B18.8.2
			Clevis Pin	ASME B18.8.2
	Keys	Woodruff Keys	Woodruff Keys (B17.2)	ASME B17.2
	Retaining Rings	External	Basic NA1-ANSI B27.1	ASME B18.27.1
			E-type NA3-ANSI B27.1	ASME B18.27.1
		Internal	Basic NA2-ANSI B27.1	ASME B18.27.1
ISO	Pins	Parallel	Parallel Pin (Hard) ISO -8734	ISO 8734
			Parallel Pin (Hard) ISO - 8735	ISO 8735
			Split Pin ISO - 1234	ISO 1234
		Spring	Spring Pin (Coiled HD) ISO - 8748	ISO 8748
			Spring Pin (Coiled SD) ISO - 8750	ISO 8750
			Spring Pin (Slotted HD) ISO - 8752	ISO 8752
		Clevis	Clevis Pin (Headed) ISO - 2341	ISO 2341 Form B
		Grooved	Groove Pin (Half Length, Center)	ISO 8743
	Keys	Parallel Keys	Square Key	ISO 2491
		Woodruff Keys	Woodruff Key	ISO 3912
DIN	Keys	Parallel Keys	Parallel Key DIN 6885	DIN 6885
	Retaining Rings	Internal	Circlip - normal - DIN 472	DIN 472
		External	Circlip - normal - DIN 471	DIN 471
			Lock Washer - DIN 6799	DIN 6799

6

PIPE THREADS, THREADED FASTENERS AND WASHERS

Contents

6.1	PIPE AND PORT THREADS	281
6.2	THREADED FASTENERS AND WASHERS	285

Tables

6-1	NPT Dimensions, Sizes Up To 3	283
6-2	ISO 7/1 (Tapered) Dimensions, Sizes Up To 3	284
6-3	ISO 228/1 (Parallel) Dimensions, Sizes Up To 3	285
6-4	UNS Threads Up To 1 Inch	287
6-5	ISO Threads Up To 24 mm	287
6-6 – 6-7	Tap Drill Sizes	289–292
6-8 – 6-9	Recommended Tap Drill Pilot Depths	293
6-10	Fastener Selector	294
6-11 – 6-12	Approximate Proof Loads for Bolts	296
6-13	Bolt Load and Stress Equations	297
6-14	Fastener Torque Approximations	299
6-15 – 6-16	Approximate Tightening Torques for Bolts	300–301
6-17	Set Screw Selector	302
6-18	SolidWorks® Toolbox 2010 Fasteners	309
6-19 – 6-22	Hex Socket Head Cap Screws	310–313
6-23 – 6-26	Low Head Hex Socket Cap Screws	314–317
6-27 – 6-34	Hex Socket Set Screws	318–325
6-35 – 6-38	Hex Socket Head Shoulder Screws	326–329
6-39 – 6-42	Hex Socket Button Head Cap Screws	330–333
6-43 – 6-46	Hex Socket Flat Countersunk Head Cap Screws	334–337
6-47 – 6-50	Hex Head Cap Screws	338–341
6-51 – 6-53	Hex Nuts	342–344
6-54 – 6-55	Hex Lock Nuts, Prevailing Torque, Nonmetallic Insert	344–345
6-56 – 6-59	Washers	346–349
6-60 – 6-61	High Collar Helical Spring Lock Washers	350–351
6-62 – 6-65	Helical Coil Threaded Inserts	352–355

Section

6.1

PIPE AND PORT THREADS

Pipe and port threads are threads designed to provide a pressure and fluid tight seal. Pipe threads are male, and port threads are female. There are three common types of pipe and port threads used in the USA: NPT, BSP, and ISO. Connectors with pipe and port threads are often called fittings.

RECOMMENDED RESOURCES

- Oberg, Jones, Horton, Ryffel, **Machinery's Handbook,** 28th Ed., Industrial Press, New York, NY, 2008
- **ANSI/ASME B1.20.1:** "Pipe Threads, General Purpose (Inch)"
- **ISO 7/1:** "Pipe threads where pressure-tight joints are made on the threads -- Part 1: Dimensions, tolerances and designation" (Metric)
- **ISO 228/1:** "Pipe threads where pressure-tight joints are not made on the threads -- Part 1: Dimensions, tolerances and designation" (Metric)
- **ANSI B1.20.3:** "Dryseal Pipe Threads (Inch)"

STANDARDS

The American national standard pipe thread (designated NPT) is a tapered thread designed to provide a pressure tight seal using thread engagement. This thread is governed by ANSI/ASME B1.20.1. NPT fittings are used in combination with sealing compound or PTFE (Teflon®[1]) tape to provide a pressure tight seal. A less common thread, called Dry-seal or NPTF, can produce a pressure-tight seal without the use of sealing compound or tape. This thread is governed by ANSI/ASME B1.20.3. The NPTF thread is used in refrigeration systems and systems where tape or sealing compound cannot be used.

Metric tapered pipe and port threads are not interchangeable with NPT, but are somewhat interchangeable with each other. BSPT is the British standard for tapered pipe thread, and it is equivalent to the ISO tapered thread governed by ISO 7/1. ISO tapered male threads are designed to provide an effective seal in either a parallel or tapered threaded port when sealing compound or PTFE tape are applied.

---

[1] *Teflon® is a registered trademark of DuPont*

Metric parallel pipe and port threads are governed by ISO 228/1. These are identical to the British BSPP thread. Parallel pipe threads are meant to engage in parallel port threads. In this case, the seal is provided by a gasket at the port entrance since the thread engagement will not seal against fluid effectively. Parallel port threads can also be used with a tapered pipe thread, in which case the fluid seal is accomplished due to thread engagement.

CRITICAL CONSIDERATIONS: Pipe and Port Threads

- NPT and ISO tapered threads must be used with sealing compound or tape to ensure a pressure-tight seal.
- Heat or high pressure may cause sealing compounds or tape to fail.
- PTFE tape and sealing compounds are sometimes not allowed.
- For high pressure applications, longer thread engagement is needed. See the recommended resources for more information.

BEST PRACTICES: Pipe and Port Threads

- Avoid using both NPT and ISO tapered threads on the same part, and avoid mixing the fittings. They are difficult to tell apart visually, but are not interchangeable.
- When specifying pipe or port threads, use the following formats:
- Inch: Male or female 1/8 NPT: 1/8-27 NPT

    Omitting the threads per inch is also common: 1/8 NPT

Metric:

    Male metric tapered thread: ISO 7/1 – R 1/8
    Female metric tapered thread: ISO 7/1 – Rc 1/8
    Male metric parallel thread: ISO 228/1 – G 1/8
    Female metric parallel thread: ISO 228/1 – G 1/8

PIPE AND PORT THREAD DIMENSIONS

Tables 6-1 through 6-3 contain dimensions of the most common standard pipe and port threads.

Table 6-1: NPT Dimensions, Sizes Up To 3
Complies with ANSI/ASME B1.20.1 - 1983 (R2006)

Size	Threads per Inch	Pitch Diameter At Gauge Plane (E_1) inch	Gauge Length (L_1) inch	External			Internal		Joint
				Pipe Outer Diameter (D) inch	Max. Length of Useful Thread (L_2) inch	Tap Drill Diameter inch	Depth of Thread ($L_1 + L_3$) inch		Handtight Engagement (L_1) inch
1/16	27	0.28118	0.1600	0.3125	0.2611	0.242	0.2711		0.1600
1/8	27	0.37360	0.1615	0.4050	0.2639	0.332	0.2726		0.1615
1/4	18	0.49163	0.2278	0.5400	0.4018	0.438	0.3945		0.2278
3/8	18	0.62701	0.2400	0.6750	0.4078	0.562	0.4067		0.2400
1/2	14	0.77843	0.3200	0.8400	0.5337	0.703	0.5343		0.3200
3/4	14	0.98887	0.3390	1.0500	0.5457	0.906	0.5533		0.3390
1	11.5	1.23863	0.4000	1.3150	0.6828	1.141	0.6609		0.4000
1 1/4	11.5	1.58338	0.4200	1.6600	0.7068	1.484	0.6809		0.4200
1 1/2	11.5	1.82234	0.4200	1.9000	0.7235	1.719	0.6809		0.4200
2	11.5	2.29627	0.4360	2.3750	0.7565	2.188	0.6969		0.4360
2 1/2	8	2.76216	0.6820	2.8750	1.1375	2.609	0.9320		0.6820
3	8	3.38850	0.7660	3.5000	1.2000	3.234	1.0160		0.7660

Angle of taper from center line is 1.783°
Angle between thread faces is 60°

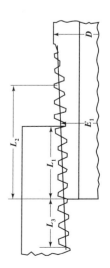

Table 6-2: ISO 7/1 (Tapered) Dimensions, Sizes Up To 3
Complies with ISO 7/1: 1994

Size	Threads per Inch	Pitch Diameter At Gauge Plane (E_1) mm	Gauge Length (L_1) mm	External		Internal			Joint
				British Std. Pipe OD (D, Converted from Inch) mm	Max. Length of Useful Thread (L_2) mm	Tap Drill Diameter mm	Depth of Thread ($L_1 + L_3$) mm		Handtight Engagement (L_1) mm
1/16	28	7.723	4.0	7.94	7.4	6.40	6.5		5.1
1/8	28	9.728	4.0	10.21	7.4	8.40	6.5		5.1
1/4	19	13.157	6.0	13.49	11	11.20	9.7		7.7
3/8	19	16.662	6.4	17.20	11.4	14.75	10.1		8.1
1/2	14	20.955	8.2	21.31	15	18.25	13.2		10.5
3/4	14	26.441	9.5	26.90	16.3	23.75	14.5		11.8
1	11	33.249	10.4	33.71	19.1	30.00	16.8		13.3
1 1/4	11	41.910	12.7	42.39	21.4	38.50	19.1		15.6
1 1/2	11	47.803	12.7	48.31	21.4	44.50	19.1		15.6
2	11	59.614	15.9	60.33	25.7	56.00	23.4		18.8
2 1/2	11	75.184	17.5	76.10	30.2	71.00	26.7		21
3	11	87.884	20.6	88.90	33.3	85.50	29.8		24.1

Angle of taper from center line is 1.783°
Angle between thread faces is 55°

Table 6-3: ISO 228/1 (Parallel) Dimensions, Sizes Up To 3
Complies with ISO 228/1: 2000

Size	Threads per Inch	Major Diameter mm	Minor Diameter mm
1/16	28	7.723	6.561
1/8	28	9.728	8.566
1/4	19	13.157	11.445
3/8	19	16.662	14.950
1/2	14	20.955	18.631
5/8	14	22.911	20.587
3/4	14	26.441	24.117
7/8	14	30.201	27.877
1	11	33.249	30.291
1 1/8	11	37.897	34.939
1 1/4	11	41.910	38.952
1 1/2	11	47.803	44.845
1 3/4	11	53.746	50.788
2	11	59.614	56.656
2 1/4	11	65.710	62.752
2 1/2	11	75.184	72.226
2 3/4	11	81.534	78.576
3	11	87.884	84.926

Angle of taper from center line is 0°
Angle between thread faces is 55°

Section 6.2 **THREADED FASTENERS AND WASHERS**

This section covers topics related to fastener threads, threaded fasteners, washers, threaded inserts, and tap drills. Consult the recommended resources and the relevant standards for complete information if needed.

RECOMMENDED RESOURCES

- Juvinall, Marshek, *Fundamentals of Machine Component Design,* 2nd Ed., John Wiley & Sons, Inc., New York, NY, 1991
- R. L. Norton, *Machine Design: An Integrated Approach,* 4th Ed., Prentice Hall, Upper Saddle River, NJ, 2011
- Oberg, Jones, Horton, Ryffel, *Machinery's Handbook,* 28th Ed., Industrial Press, New York, NY, 2008
- R. Parmley, *Standard Handbook of Fastening and Joining,* 3rd Ed., McGraw-Hill , New York, NY, 1997
- **Bolt Science Web Site:** www.boltscience.com

FASTENER THREADS

In the United States, fasteners typically have Unified National Standard (UNS) threads. Metric threads are typically ISO threads. The anatomy of a fastener thread is shown in Figure 6-1. Internal and external threads share the same pitch diameter. Pitch diameter is used when evaluating thread shear, but not used when specifying the thread. The thread callout will be close to, though always slightly greater than, the major diameter.

Fastener threads are available in coarse and fine series. UNS threads include UNC (coarse), UNF (fine), and UNEF (extra fine) pitches. ISO offers coarse, fine, and extra fine pitches as well. Coarse series threads are economical and are the most commonly used for general applications. They are also least likely to cross thread or strip. Fine threads are resistant to loosening due to vibration, but are not recommended for use with soft materials. Extra fine threads are not commonly used. Tables 6-4 and 6-5 show commonly used inch and metric standard threads for both coarse and fine series. It is important to note that UNS threads are described using the number of threads per inch, where ISO threads are described using their actual pitch measurement. UNS threads are normally called out using their nominal thread size and threads per inch as follows: **#6-32** or **¼-20**

The difference between ANSI metric and ISO metric threads is only in the way the threads are called out. These threads are functionally the same. For ANSI metric threads, the callout includes both the nominal diameter and pitch. In contrast, the ISO standard callout includes only the nominal diameter for coarse threads and diameter x pitch for fine threads.

ANSI Metric:	(Coarse) **M8x1.25**	(Fine) **M8x1**
ISO:	(Coarse) **M8**	(Fine) **M8x1**

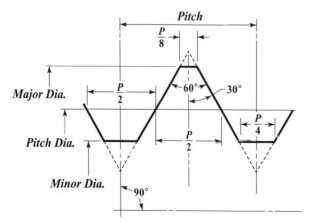

Figure 6-1 Anatomy of a Fastener Thread

Table 6-4: UNS Threads
Up To 1 Inch

Nominal Thread Size	Decimal Size inch	Threads Per Inch	
		Coarse	Fine
#0	0.060	-	80
#1	0.073	64	72
#2	0.086	56	64
#3	0.099	48	56
#4	0.112	40	48
#5	0.125	40	44
#6	0.138	32	40
#8	0.164	32	36
#10	0.190	24	32
#12	0.216	24	28
1/4	0.250	20	28
5/16	0.313	18	24
3/8	0.375	16	24
7/16	0.438	14	20
1/2	0.500	13	20
9/16	0.563	12	18
5/8	0.625	11	18
3/4	0.750	10	16
7/8	0.875	9	14
1	1.000	8	12

Table 6-5: ISO Threads
Up To 24 mm

Nominal Thread Size	Pitch (mm)	
	Coarse	Fine
1	0.25	-
1.2	0.25	-
1.6	0.35	-
2	0.40	-
2.5	0.45	-
3	0.50	-
4	0.70	-
5	0.80	-
6	1.00	-
8	1.25	1.00
10	1.50	1.25
12	1.75	1.50
16	2.00	1.50
20	2.50	2.00
24	3.00	2.00

Thread classes for UNS threads correspond to the amount of tolerance and allowance in the thread. Allowance is the gap between the theoretical and actual thread surfaces. UNS threads have classes 1A, 2A, and 3A for external threads and 1B, 2B, and 3B for internal threads. 2A and 2B are normally used for general applications, and these classes are implied if omitted from the thread calllout. 1A and 1B provide greater tolerance and allowance (looser fit) and are recommended when threads are expected to be dirty or damaged. An example of an internal inch coarse thread callout including class is: **¼-20 UNC-2B**.

ISO thread tolerances are specified for the pitch diameter and the major diameter. A major diameter tolerance of 6g is normally used on bolts for general applications. A tolerance of 6H is normally used on nuts and female threads for general applications. For long thread engagements, a tolerance of 6G may be necessary on the female thread to prevent seizing. An example of a metric thread callout for an external thread that includes these tolerances is: **M12 × 1 – 5g6g**. In this example, tolerance 5g refers to the pitch diameter and tolerance 6g refers to the major diameter. The tolerances are often omitted from the callout and 5g6g (external) or 5H6H (internal) is implied.

TAP DRILLS

A tap drill creates a hole appropriately sized for tapping. For commercial work in Aluminum or other soft materials, a target of 75% of thread is typical. Higher percentages of thread are recommended for thin materials. 50% of thread is sometimes recommended for steel and like materials. Table 6-6 is a list of inch tap drill sizes, and Table 6-7 is a list of metric tap drill sizes. Shaded entries in the tables are those most commonly used.

The pilot hole depth for blind tapped holes is the subject of much debate and depends on many variables, such as the part material and the type of process used to manufacture the threads. One conservative method of calculating pilot hole depth allows clearance for a pointed tap tip equal to one minor diameter of the tap. This method also allows clearance of 6 pitches for plug taps (3 pitches for bottoming taps) for the imperfect threads. There is an additional allowance of 1.5 pitches clearance for chips to accumulate in the hole. This method assumes a high strength material with long chip formation. The values calculated using this method are shown in Tables 6-8 and 6-9. The formulas for this method are as follows:

PLUG TAP: Pilot Depth = Thread Depth + Minor Dia +7.5*Pitch
BOTTOMING TAP: Pilot Depth = Thread Depth + Minor Dia + 4.5*Pitch

If shorter pilot holes are needed, they may be achieved through different taps, machining methods, and increased labor. A recommended minimum pilot hole depth is normally considered to be 5 pitches for plug taps and 3 pitches for bottoming taps. To save manufacturing cost, provide generous pilot hole depths whenever such an allowance is possible. Pilot hole depths for various sized fasteners can be determined using Tables 6-8 and 6-9. Add the values from the chart to the required thread depth to get the pilot hole depth. Several methods of calculation are presented in the chart. It is up to designers to determine which depth calculation method is best for their application.

Table 6-6: Tap Drill Sizes, Inch

Tap	Pitch	Tap Drill	Decimal Equivalent inch	Probable Hole Size inch	Percent of Thread
#0	80	56	0.0465	0.0480	74
		3/64	0.0469	0.0484	71
#1	64	54	0.0550	0.0565	81
		53	0.0595	0.0610	59
	72	53	0.0595	0.0610	67
		1/16	0.0625	0.0640	50
#2	56	51	0.0670	0.0687	74
		50	0.0700	0.0717	62
		49	0.0730	0.0747	49
	64	50	0.0700	0.0717	70
		49	0.0730	0.0747	56
#3	48	48	0.0760	0.0779	78
		5/64	0.0781	0.0800	70
		47	0.0785	0.0804	69
		46	0.0810	0.0829	60
		45	0.0820	0.0839	56
	56	46	0.0810	0.0829	69
		45	0.0820	0.0839	65
		44	0.0860	0.0879	48
#4	40	44	0.0860	0.0880	74
		43	0.0890	0.0910	65
		42	0.0935	0.0955	51
		3/32	0.0938	0.0958	50
	48	42	0.0935	0.0955	61
		3/32	0.0938	0.0958	60
		41	0.0960	0.0980	52
#5	40	40	0.0980	0.1003	76
		39	0.0995	0.1018	71
		38	0.1015	0.1038	65
		37	0.1040	0.1063	58
	44	38	0.1015	0.1038	72
		37	0.1040	0.1063	63
		36	0.1065	0.1088	55
#6	32	37	0.1040	0.1063	78
		36	0.1065	0.1091	71
		7/64	0.1094	0.1120	64
		35	0.1100	0.1126	63
		34	0.1110	0.1136	60
		33	0.1130	0.1156	55
	40	34	0.1110	0.1136	75
		33	0.1130	0.1156	69
		32	0.1160	0.1186	60

continued on next page

Table 6-6: Tap Drill Sizes, Inch (Continued)

Tap	Pitch	Tap Drill	Decimal Equivalent inch	Probable Hole Size inch	Percent of Thread
#8	32	29	0.1360	0.1389	62
		28	0.1405	0.1434	51
	36	29	0.1360	0.1389	70
		28	0.1405	0.1434	57
		9/64	0.1406	0.1435	57
#10	24	27	0.1440	0.1472	79
		26	0.1470	0.1502	74
		25	0.1495	0.1527	69
		24	0.1520	0.1552	64
		23	0.1540	0.1572	61
		5/32	0.1563	0.1595	56
		22	0.1570	0.1602	55
	32	5/32	0.1563	0.1595	75
		22	0.1570	0.1602	73
		21	0.1590	0.1622	68
		20	0.1610	0.1642	64
		19	0.1660	0.1692	51
#12	24	11/64	0.1719	0.1754	75
		17	0.1730	0.1765	73
		16	0.1770	0.1805	66
		15	0.1800	0.1835	60
		14	0.1820	0.1855	56
	28	16	0.1770	0.1805	77
		15	0.1800	0.1835	70
		14	0.1820	0.1855	66
		13	0.1850	0.1885	59
		3/16	0.1875	0.1910	54
1/4	20	9	0.1960	0.1998	77
		8	0.1990	0.2028	73
		7	0.2010	0.2048	70
		13/64	0.2031	0.2069	66
		6	0.2040	0.2078	65
		5	0.2055	0.2093	63
		4	0.2090	0.2128	57
	28	3	0.2130	0.2168	72
		7/32	0.2188	0.2226	59
		2	0.2210	0.2248	55
5/16	18	F	0.2570	0.2608	72
		G	0.2610	0.2651	66
		17/64	0.2656	0.2697	59
		H	0.2660	0.2701	59
	24	H	0.2660	0.2710	78
		I	0.2720	0.2761	67
		J	0.2770	0.2811	58

continued on next page

Table 6-6: Tap Drill Sizes, Inch (Continued)

Tap	Pitch	Tap Drill	Decimal Equivalent inch	Probable Hole Size inch	Percent of Thread
3/8	16	5/16	0.3125	0.3169	72
		O	0.3160	0.3204	68
		P	0.3230	0.3274	59
	24	21/64	0.3281	0.3325	79
		Q	0.3320	0.3364	71
		R	0.3390	0.3434	58
7/16	14	T	0.3580	0.3626	81
		23/64	0.3594	0.3640	79
		U	0.3680	0.3726	70
		3/8	0.3750	0.3796	62
		V	0.3770	0.3816	60
	20	W	0.3860	0.3906	72
		25/64	0.3906	0.3952	65
		X	0.3970	0.4016	55
1/2	13	27/64	0.4219	0.4266	73
		7/16	0.4375	0.4422	58
	20	29/64	0.4531	0.4578	65
9/16	12	15/32	0.4688	0.4736	82
		31/64	0.4844	0.4892	68
	18	1/2	0.5000	0.5048	80
		33/64	0.5156	0.5204	58
5/8	11	17/32	0.5313	0.5362	75
		35/64	0.5469	0.5518	62
	18	9/16	0.5625	0.5674	80
		37/64	0.5781	0.5831	58
3/4	10	41/64	0.6406	0.6456	80
		21/32	0.6563	0.6613	68
	16	11/16	0.6875	0.6925	71
7/8	9	49/64	0.7656	0.7708	72
		25/32	0.7812	0.7864	61
	14	51/64	0.7969	0.8021	79
		13/16	0.8125	0.8177	62
1	8	55/64	0.8594	0.8653	83
		7/8	0.8750	0.8809	73
		57/64	0.8906	0.8965	64
		29/32	0.9063	0.9122	54
	12	29/32	0.9063	0.9123	81
		59/64	0.9219	0.9279	67
		15/16	0.9375	0.9435	52
	14	59/64	0.9219	0.9279	78
		15/16	0.9375	0.9435	61

Table 6-7: Tap Drill Sizes, Metric

Tap	Pitch	Tap Drill mm	Closest Inch Drill Size	Approx. Percent of Thread
M1.6	0.35	1.25	55	75
		1.35	54	50
	0.20	1.40	54	75
M2	0.40	1.60	52	75
		1.75	50	50
	0.25	1.75	50	75
M2.5	0.45	2.05	46	75
		2.20	44	50
	0.35	2.15	44	75
M3	0.50	2.50	39	75
		2.70	36	50
	0.35	2.65	37	75
M3.5	0.60	2.90	32	75
		3.10	31	50
	0.35	3.15	1/8	75
M4	0.70	3.30	30	75
		3.50	28	50
	0.50	3.50	29	75
M5	0.80	4.20	19	75
		4.50	16	50
	0.50	4.50	16	75
M6	1.00	5.00	8	75
		5.40	4	50
	0.75	5.25	5	75
M8	1.25	6.80	H	75
		7.20	J	50
	1.00	7.00	J	75
M10	1.50	8.50	R	75
		9.00	T	50
	1.25	8.80	11/32	75
M12	1.75	10.30	13/32	75
		10.90	27/64	50
	1.25	10.80	27/64	75
M14	2.00	12.10	15/32	75
		12.70	1/2	50
	1.50	12.50	1/2	75
M16	2.00	14.00	35/64	75
		14.75	37/64	50
	1.50	14.50	37/64	75
M18	2.50	15.50	39/64	75
		16.50	21/32	50
	1.50	16.50	21/32	75
M20	2.50	17.50	11/16	75
		18.50	47/64	50
	1.50	18.50	47/64	75
M22	2.50	19.50	49/64	75
	1.50	20.50	13/16	75
M24	3.00	21.00	53/64	75
	2.00	22.00	7/8	75

Table 6-8: Recommended Tap Drill Pilot Depths, Inch

Screw Size and Coarse Threads Per Inch	Coarse Pitch in	Minor Diameter d in	Distance To Extend Pilot Hole Beyond Threads in				
			Plug Tap	Chip Extracting Plug Tap	Bottoming Tap	5 Pitches	3 Pitches
#1-64	0.0156	0.0527	0.170	0.146	0.100	0.078	0.047
#2-56	0.0179	0.0628	0.197	0.170	0.116	0.089	0.054
#3-48	0.0208	0.0719	0.228	0.197	0.134	0.104	0.063
#4-40	0.0250	0.0795	0.267	0.230	0.155	0.125	0.075
#5-40	0.0250	0.0925	0.280	0.243	0.168	0.125	0.075
#6-32	0.0313	0.0974	0.332	0.285	0.191	0.156	0.094
#8-32	0.0313	0.1234	0.358	0.311	0.217	0.156	0.094
#10-24	0.0417	0.1359	0.448	0.386	0.261	0.208	0.125
1/4 -20	0.0500	0.1850	0.560	0.485	0.335	0.250	0.150
5/16-18	0.0556	0.2403	0.657	0.574	0.407	0.278	0.167
3/8 -16	0.0625	0.2938	0.763	0.669	0.481	0.313	0.188
7/16-14	0.0714	0.3447	0.880	0.773	0.559	0.357	0.214
1/2 -13	0.0769	0.4001	0.977	0.862	0.631	0.385	0.231
5/8 -11	0.0909	0.5069	1.189	1.052	0.780	0.455	0.273
3/4 -10	0.1000	0.6201	1.370	1.220	0.920	0.500	0.300
7/8 -9	0.1111	0.7307	1.564	1.397	1.064	0.556	0.333
1-18	0.1250	0.8376	1.775	1.588	1.213	0.625	0.375

Table 6-9: Recommended Tap Drill Pilot Depths, Metric

Screw Size	Coarse Pitch mm	Minor Diameter d mm	Distance To Extend Pilot Hole Beyond Threads mm				
			Plug Tap	Chip Extracting Plug Tap	Bottoming Tap	5 Pitches	3 Pitches
3	0.50	2.39	6.1	5.4	3.9	2.5	1.5
4	0.70	3.14	8.4	7.3	5.2	3.5	2.1
5	0.80	4.02	10.0	8.8	6.4	4.0	2.4
6	1.00	4.77	12.3	10.8	7.8	5.0	3.0
8	1.25	6.47	15.8	14.0	10.2	6.3	3.8
10	1.50	8.16	19.4	17.2	12.7	7.5	4.5
12	1.75	9.85	23.0	20.4	15.1	8.8	5.3
14	2.00	11.55	26.6	23.6	17.6	10.0	6.0
16	2.00	13.55	28.6	25.6	19.6	10.0	6.0
20	2.50	16.93	35.7	31.9	24.4	12.5	7.5
24	3.00	18.93	41.4	36.9	27.9	15.0	9.0

FASTENER TYPES, MATERIALS, AND SELECTION

Bolts are generally meant to be used with a nut, whereas screws can be threaded directly into tapped holes and tightened from the head. For further discussion of the differences, consult the recommended resources. Hex socket head cap screws are one of the most commonly used type of threaded fastener in industrial machine design. This head type requires no radial wrench clearance and is suitable for general use. Low head socket cap screws, flat head socket cap screws, and button head socket cap screws have smaller wrench sockets and are not suitable for high tightening torque applications. Hex head screws and bolts are also commonly used in machine design, and are ideal for radial access areas or high strength applications. The chart in Table 6-10 can help with the selection of fasteners based on performance characteristics. Many of the most common fasteners and related items are detailed later in this section.

Table 6-10: Fastener Selector

COMPONENT	GENERAL USE	VERY HIGH LOADS	ACCURATE POSITIONING	RADIAL SHAFT CLAMPING	RESISTS LOOSENING
Socket Head Cap Screw	●	○	⊘	⊘	⊘
Socket Head Set Screw	●	⊘	⊘	●	⊘
Socket Head Shoulder Screw	●	○	○	⊘	⊘
Button Head Cap Screw	○	⊘	⊘	⊘	⊘
Machine Screw	○	⊘	⊘	⊘	⊘
Flat Head Cap Screw	○	⊘	○	⊘	⊘
Hex Head Screw	●	○	⊘	⊘	⊘
Hex Bolt	●	●	⊘	⊘	⊘
Hex Nut	●	●	⊘	⊘	⊘
Lock Nut	●	●	⊘	⊘	●
Washer	●	●	○	⊘	⊘
Lock Washer	●	●	⊘	⊘	●

● SUITABLE
○ SOMETIMES SUITABLE
⊘ NOT SUITABLE

Threaded fasteners come in many materials and finishes. Plain steel fasteners are common, but have no corrosion protection. Black oxide coated steel fasteners are also commonly used, and have some corrosion protection and lubricity. A third common finish for steel fasteners is Zinc plating, which offers good corrosion protection. Stainless steel fasteners are commonly used in special applications where material compatibility is critical, but they carry a lower load rating than regular steel fasteners. Plastic fasteners and other materials are available for special applications.

GRADES AND STRENGTH OF FASTENERS

SAE grade specifications for UNS fasteners correspond to the minimum mechanical properties required of that fastener. Higher grade screws or bolts should be specified in high strength applications. Metric screws have class numbers that designate their material properties. For typical machine design applications, SAE grades 5 and 8 are commonly used, as are metric classes 8.8, 10.9, and 12.9. Lower grades are seen when very small fasteners are involved. Always use appropriate safety factors when performing calculations and sizing components.

Stripping of threads occurs when threads fail in shear. The best way to guard against stripping is to increase the length of engagement. Standard UNS and ISO nuts less than one inch in diameter are designed such that their stripping strength exceeds the tensile strength of the screw. When designing a joint with a tapped hole, a thread engagement length equal to the screw diameter is sufficient for same material combinations. For steel screws in Aluminum, a thread engagement of twice the screw diameter is recommended to prevent stripping. See the recommended resources for more information on calculating shear stress of threads.

Proof load is the maximum axial load that can be applied to a bolt without causing permanent deformation. Proof loads of common coarse thread bolts are given in Table 6-11 and 6-12. To prevent loosening or joint separation and increase resistance to fatigue, bolts and screws must be preloaded in tension. For statically loaded joints, bolts can be preloaded to an initial tension as high as 90% of proof load. The bolts in dynamically loaded joints are usually preloaded to 75% of proof load or more.

The stress in a bolt in a joint is a function of the stiffness of the joint, the bolt preload, and any external forces on the joint. To calculate the tensile load on a bolt, use the equations in Table 6-13. The tensile stress in a bolt must never exceed its yield strength. In other words, the tensile force on the bolt (including preload and external forces) must never exceed its proof load. Fatigue failure of bolts can be of concern with fluctuating loading of a joint. Fatigue safety factors for bolts can be calculated using the equations in Table 6-13. Always use appropriate factors of safety when evaluating stress in fasteners.

Table 6-11: Approximate Proof Loads, Inch Bolts

Screw Size	Decimal Diameter	Threads per Inch	K_f Fatigue Stress Factor	A_t Tensile Stress Area	Strength Grade	
					5	8
					Proof Load	
	in			in^2	lbf	
#1	0.0730	64	5.7497	0.0026	221	312
#2	0.0860	56	5.7586	0.0037	315	444
#3	0.0990	48	5.7674	0.0049	417	588
#4	0.1120	40	5.7763	0.0060	510	720
#5	0.1250	40	5.7852	0.0080	680	960
#6	0.1380	32	5.7940	0.0091	774	1092
#8	0.1640	32	5.8117	0.0140	1190	1680
#10	0.1900	24	5.8294	0.0175	1488	2100
1/4	0.2500	20	5.8703	0.0318	2703	3816
5/16	0.3125	18	5.9129	0.0524	4454	6288
3/8	0.3750	16	5.9555	0.0775	6588	9300
7/16	0.4375	14	5.9980	0.1063	9036	12756
1/2	0.5000	13	6.0406	0.1419	12062	17028
5/8	0.6250	11	6.1258	0.2260	19210	27120
3/4	0.7500	10	6.2109	0.3345	28433	40140
7/8	0.8750	9	6.2961	0.4617	39245	55404
1	1.0000	8	6.3812	0.6057	51485	72684
S_{ut} = Ultimate Tensile Strength (kpsi)					120	150

Table 6-12: Approximate Proof Loads, Metric Bolts

Screw Size	Pitch	K_f Fatigue Stress Factor	A_t Tensile Stress Area	Property Class		
				8.8	10.9	12.9
				Proof Load		
	mm		mm^2	N		
3	0.50	5.78	5.03	2917	4175	4879
4	0.70	5.81	8.78	5092	7287	8517
5	0.80	5.83	14.18	8224	11769	13755
6	1.00	5.86	20.12	11670	16700	19516
8	1.25	5.91	36.61	21234	30386	35512
10	1.50	5.97	57.99	33634	48132	56250
12	1.75	6.02	84.27	48877	69944	81742
14	2.00	6.08	115.44	66955	95815	111977
16	2.00	6.13	156.67	90869	130036	151970
20	2.50	6.24	244.79	146874	203176	237446
24	3.00	6.34	352.50	211500	292575	341925
S_{ut} = Ultimate Tensile Strength (Mpa)				830	1040	1220

Table 6-13: Bolt Load and Stress Equations

F = tensile load in a bolt	F_i = initial tension (preload)
F_{EXT} = externally applied axial force (positive = tension)	
A_t = tensile stress area of the bolt (in^2 or mm^2) (See Tables 6-11 and 6-12)	
S_{ut} = ultimate tensile strength of bolt (See Tables 6-11 and 6-12)	

TENSILE LOAD IN A BOLT k = stiffness constants (See recommended resources)	HARD JOINTS (Assumes $k_c = 3k_b$) $$F = F_i + \frac{F_{EXT}}{4}$$ GENERAL CASE: $$F = F_i + \frac{k_b}{k_b + k_c} F_{EXT}$$
INITIAL TENSION (BOLT PRELOAD)	STATICALLY LOADED JOINTS: $$F_i \leq 0.9 F_P$$ DYNAMICALLY LOADED JOINTS: $$F_i \geq 0.75 F_P$$
PROOF LOAD F_p = proof load (lbf or N) S_p = proof strength of the bolt (psi or Mpa)	$$F_p = A_t S_p$$
FATIGUE SAFETY FACTOR	$$SF_f = \frac{S_e(S_{ut} - \sigma_i)}{S_e(\sigma_m - \sigma_i) + S_{ut}\sigma_a}$$
INITIAL TENSILE STRESS	$$\sigma_i = \frac{F_i}{A_t}$$
MEAN AND ALTERNATING STRESS K_f = see Tables 6-11 and 6-12 $K_{fm} \cong 1$ for most preloaded bolts See Section 11.1 for calculation method	$$\sigma_a = K_f \frac{F - F_i}{2A_t} \qquad \sigma_m = K_{fm} \frac{F + F_i}{2A_t}$$
ENDURANCE STRENGTH Estimated endurance limit and correction factors for steel bolts	$S_e = C_{load}C_{size}C_{surf}C_{temp}C_{reliab}S_e'$ $S_e' \cong 0.5 S_{ut}$ when $S_{ut} < 200$ kpsi (1400 MPa) $C_{load} = 1$ for bending $C_{size} = 1$ for $d \leq 0.3$ in (8 mm) $C_{size} = 0.869 d^{-0.097}$ for 0.3 in $< d \leq 10$ in (254 mm) $C_{surf} = 2.7(S_{ut})^{-0.265}$ for machined or cold rolled surface (S_{ut} in kpsi) $C_{temp} = 1$ for $T \leq 840°F$ (450°C) $C_{reliab} = 0.868$ for 95% reliability $C_{reliab} = 0.620$ for 99.9999% reliability

TORQUE OF FASTENERS

The most common and practical way to apply a preload to a threaded fastener is by applying torque to the head for a screw or the nut for a bolt. Two methods of approximating bolt torque for a given preload are given in Table 6-14. Please see the recommended resources for more information on calculating tightening torque. Tables 6-15 and 6-16 provide some approximate torque values for preloading coarse thread steel bolts. The tabulated values were calculated using the equation $T_i = 0.21 F_i d$

CRITICAL CONSIDERATIONS: Threaded Fasteners and Washers

- Always select screws that are sized appropriately for the loading conditions. Use appropriate factors of safety when evaluating and sizing components.
- Fasteners can fail in several ways. They can strip, shear, fail axially, or fail in torsion. They can also loosen or experience fatigue failure due to cyclic loading. Please see the recommended resources for complete calculation methods for all of these failure modes.
- Proper preloading of fasteners is critical for rated performance. Consult the recommended resources for more information on calculating torque, tension, and preload.
- Bolt or screw loosening is usually caused by vibration or transverse cyclic loading of the joint. Proper preloading can help prevent loosening, but the best method of prevention is a locking nut or washer. Prevailing torque lock nuts with nonmetallic inserts are commonly used, as are thread locking compounds.
- Torque wrenches are typically only accurate within ±25%.
- Stress relaxation and subsequent loss of preload is a concern with elevated temperatures.
- Corrosion can cause premature joint failure. Protect against both environmental and galvanic corrosion effects through the use of coatings, treatments, and corrosion-resistant materials.
- Do not reuse nuts. Deformation of the threads can occur during normal use.
- Do not reuse screws or bolts that have been previously preloaded to 90% of proof load.

Table 6-14: Fastener Torque Approximations

T_i = initial torque (lbf-ft or N-m) F_i = initial axial preload (lbf or N) (See Table 6-13) d = major diameter (ft or m) μ = coefficient of friction l = thread lead (distance per pitch) in ft or m	
TORQUE FOR LUBRICATED STEEL BOLTS Coefficient of friction is assumed to be 0.15	$T_i = 0.21 F_i d$
TORQUE FOR BOLTS SUBJECT TO OTHER COEFFICIENTS OF FRICTION Assumes the thread and bearing coefficients of friction are the same. $\mu \cong 0.15$ for lubricated steel on steel $\mu \cong 0.8$ for clean dry steel on steel	$T_i = F_i(0.159l + 1.156\mu d)$

BEST PRACTICES: Threaded Fasteners and Washers

- Consider socket head cap screws for most industrial machine design applications.
- Use coarse threads whenever possible to save cost and ease assembly. Coarse threads are especially recommended for soft materials like Aluminum.
- Use a washer when fastening soft materials or for use with oversized holes or slots. Washers can be placed under the heads of screws and under nuts to protect the parts being fastened and provide additional bearing surface.
- Specify the same size screws wherever possible in an assembly to save cost and ease the assembly process.
- Add a chamfer to the entrance of tapped holes to prevent cross threading and eliminate burrs that could interfere with proper pre-loading of the assembly.
- Washers can be used under a screw or bolt head in cases where the hole in the part may interfere with the fillet under the fastener head. This may be more cost effective than chamfering the hole.
- Favor through tapped holes over blind tapped holes unless the material is very thick.
- When specifying a blind tapped hole, allow as much pilot hole depth as possible.
- Beware of counterfeit and non-standard fasteners. These fasteners may not meet proof strength requirements.

Table 6-15: Approximate Tightening Torques, Inch Bolts
Assumes coefficient of friction = 0.15 (lubricated steel on steel)

	Screw Size	Threads Per Inch	Major Diameter in	Strength Grade	
				5	8
				Tightening Torque lbf-ft	
75% Proof Load	#1	64	0.0730	0.2	0.3
	#2	56	0.0860	0.4	0.5
	#3	48	0.0990	0.5	0.8
	#4	40	0.1120	0.7	1.1
	#5	40	0.1250	1.1	1.6
	#6	32	0.1380	1.4	2.0
	#8	32	0.1640	2.6	3.6
	#10	24	0.1900	3.7	5.2
	1/4	20	0.2500	8.9	12.5
	5/16	18	0.3125	18.3	25.8
	3/8	16	0.3750	32.4	45.8
	7/16	14	0.4375	51.9	73.2
	1/2	13	0.5000	79.2	111.7
	5/8	11	0.6250	157.6	222.5
	3/4	10	0.7500	279.9	395.1
	7/8	9	0.8750	450.7	636.3
	1	8	1.0000	675.7	954.0
90% Proof Load	#1	64	0.0730	0.3	0.4
	#2	56	0.0860	0.4	0.6
	#3	48	0.0990	0.7	0.9
	#4	40	0.1120	0.9	1.3
	#5	40	0.1250	1.3	1.9
	#6	32	0.1380	1.7	2.4
	#8	32	0.1640	3.1	4.3
	#10	24	0.1900	4.5	6.3
	1/4	20	0.2500	10.6	15.0
	5/16	18	0.3125	21.9	30.9
	3/8	16	0.3750	38.9	54.9
	7/16	14	0.4375	62.3	87.9
	1/2	13	0.5000	95.0	134.1
	5/8	11	0.6250	189.1	267.0
	3/4	10	0.7500	335.9	474.2
	7/8	9	0.8750	540.8	763.5
	1	8	1.0000	810.9	1144.8

Table 6-16: Approximate Tightening Torques, Metric Bolts
Assumes coefficient of friction = 0.15 (lubricated steel on steel)

Screw Size	Pitch	Property Class		
		8.8	10.9	12.9
		Tightening Torque		
	mm	N-m		
75% Proof Load				
3	0.5	1.4	2.0	2.3
4	0.7	3.2	4.6	5.4
5	0.8	6.5	9.3	10.8
6	1	11.0	15.8	18.4
8	1.25	26.8	38.3	44.7
10	1.5	53.0	75.8	88.6
12	1.75	92.4	132.2	154.5
14	2	147.6	211.3	246.9
16	2	229.0	327.7	383.0
20	2.5	462.7	640.0	748.0
24	3	799.5	1105.9	1292.5
90% Proof Load				
3	0.5	1.7	2.4	2.8
4	0.7	3.8	5.5	6.4
5	0.8	7.8	11.1	13.0
6	1	13.2	18.9	22.1
8	1.25	32.1	45.9	53.7
10	1.5	63.6	91.0	106.3
12	1.75	110.9	158.6	185.4
14	2	177.2	253.5	296.3
16	2	274.8	393.2	459.6
20	2.5	555.2	768.0	897.5
24	3	959.4	1327.1	1551.0

COMPONENT INFORMATION

What follows are some of the most commonly used types of fasteners and related hardware, like washers and threaded inserts. Dimensions are given for the most common sizes. For more options and sizes, please consult the relevant standards or a manufacturer. A chart showing standard hardware items available in Solidworks® Toolbox can be found in Table 6-18. This chart can be particularly useful in determining what items are most commonly used and which standard applies to each part.

Hex socket head cap screws, normally called socket head cap screws, are excellent for general use. They require no radial wrench clearance and can be used with or without a nut. The screw neck has a fillet that can interfere with the entrance of a clearance or threaded hole. It is good practice to add a 60° chamfer to the entrance of the clearance hole to clear this fillet, especially with close fit holes and hardened parts. The dimensions of some common sizes and their clearance hole sizes are shown in Tables 6-19 – 6-22. The applicable standards are:

ASME B18.3:   "Socket Cap, Shoulder, and Set Screws, Hex and Spline Keys (Inch Series)"

ISO 4762:   "Hexagon socket head cap screws"

Low head hex socket cap screws are available for use in tight spaces. These cap screws are not suitable for high tightening torque or high load applications. Their shallow hex socket has a tendency to strip when over-tightened. The screw neck has a fillet that can interfere with the entrance of a clearance or threaded hole. It is good practice to add a 60° chamfer to the entrance of the clearance hole to clear this fillet, especially with close fit holes and hardened parts. The dimensions of some common sizes and their clearance hole sizes are shown in Tables 6-23 – 6-26. The applicable standards are:

ASME B18.3:   "Socket Cap, Shoulder, and Set Screws, Hex and Spline Keys (Inch Series)"

DIN 7984:   "Socket Low Head Cap Screws"

Hex socket set screws have a variety of clamping and locating uses, but are most often installed radially to secure a hub to a shaft. They are best used in light loading conditions. Set screws come in several tip types, including

Table 6-17: Set Screw Selector

Socket Set Screw Tip Type	NON-MARRING	EXTRA FIRM HOLD	ALLOWS SHAFT ROTATION	ECCENTRIC SHAFT CLAMPING
Cup Point	⊘	●	⊘	⊘
Flat Point	O	⊘	⊘	⊘
Cone Point	⊘	●	⊘	●
Dog Point	●	⊘	●	⊘
Oval Point	●	⊘	⊘	⊘

● SUITABLE

O SOMETIMES SUITABLE

⊘ NOT SUITABLE

cone point, cup point, dog point, and flat point. Other tip shapes like oval are less commonly used but available. Table 6-17 can be helpful in selecting a set screw type. Cup and cone point set screws will mar the surface they are tightened against, so they are not recommended for applications that require adjustment. For true non-marring performance, use a brass tipped set screw. It is common practice to machine a flat on a shaft where the set screw will clamp so as not to damage the shafting surface with the set screw. To retain a shaft with a groove without marring the surface and allow shaft rotation, use a dog point set screw. It is possible to use two set screws in the same hole to provide locking. This arrangement is better than using a long set screw and nut from a balance perspective. The dimensions of some common sizes are shown in Tables 6-27 – 6-34. The applicable standards are:

ASME B18.3:	"Socket Cap, Shoulder, and Set Screws, Hex and Spline Keys (Inch Series)"
ISO 4027:	"Hexagon socket set screws with cone point"
ISO 4029:	"Hexagon socket set screws with cup point"
ISO 4028:	"Hexagon socket set screws with dog point"
ISO 4026:	"Hexagon socket set screws with flat point"

Hex socket head shoulder screws are designed with a smooth shaft which steps down to a threaded end. They are sometimes called shoulder bolts or stripper bolts. Shoulder screws are commonly used to provide precision location between parts or act as a pivot shaft. These fasteners are best used with a nut rather than threaded into a hole, though both applications are commonly seen. Because of the undercut between the thread and shoulder, these fasteners have a lower tightening torque and shear strength than pins or screws of the same size. When piloting the shoulder into a tight fitting hole, provide a clearance hole for the threaded portion of the shoulder screw and use a nut to fasten it. Use great care when piloting the shoulder into a tight fitting hole when the threaded portion is used with a tapped hole in the part. This can cause over-constraint, which generates stresses and can cause installation problems if not carefully toleranced. A chamfer should be added to any shoulder hole smaller than the head fillet outer diameter. The dimensions of some common sizes are shown in Tables 6-35 – 6-38. The applicable standards are:

ASME B18.3:	"Socket Cap, Shoulder, and Set Screws, Hex and Spline Keys (Inch Series)"
ISO 7379:	"Hexagon socket head shoulder screws"

Hex socket button head cap screws are meant for lighter duty applications where cosmetic appearance is important or where sharp corners and protrusions must be minimized. These screws are not meant for high tightening torque applications, and have smaller hex sockets than standard socket head cap screws. The dimensions of some common sizes and their clearance hole sizes are shown in Tables 6-39 – 6-42. The applicable standards are:

ASME B18.3:	"Socket Cap, Shoulder, and Set Screws, Hex and Spline Keys (Inch Series)"
ISO 7380:	"Hexagon socket button head screws"

Hex socket flat countersunk head cap screws are intended to be installed in a countersunk hole so that the head is completely below the part surface. They are often called flat head cap screws. The countersunk head can be used to provide some self-centering capability. With their smaller hex sockets, these screws are not intended for high tightening torque applications. The dimensions of some common sizes and their clearance hole sizes are shown in Tables 6-43 – 6-46. The applicable standards are:

ASME B18.3:	"Socket Cap, Shoulder, and Set Screws, Hex and Spline Keys (Inch Series)"
ISO 10642:	"Hexagon socket countersunk head screws"

Hex head cap screws are also known as finished hex bolts. Hex bolts and screws are similar, but hex screws have finished surfaces under the head and their dimensions are more tightly controlled so that they can be used as screws. Heavy hex screws are available for high load applications. Always provide radial clearance for tightening with a wrench or socket tool. Grade A metric hex head cap screws are most commonly used in light industrial machinery. The dimensions of some common sizes and their clearance hole sizes are shown in Tables 6-47 – 6-50. The applicable standards are:

ASME B18.2.1:	"Square and Hex Bolts and Screws, Inch Series"
ISO 4017:	"Hexagon head screws -- Product grades A and B"
DIN 933:	"Hexagon head screws thread up to the head - Product grades A and B"
ISO 4014:	"Hexagon head bolts -- Product grades A and B"

Hex nuts are available in two standard thicknesses: regular and thin. Thin hex nuts are often called jam nuts. Jam nuts are often used in pairs. When tightened against each other, one nut prevents the other from turning. Heavy hex nuts are available for high strength applications. Hex nuts are designed to be used with bolts or screws, and are meant to be turned while holding the fastener still. The dimensions of some common sizes are shown in Tables 6-51 – 6-53. The applicable standards are:

ASME B18.2.2:	"Nuts for General Applications: Machine Screw Nuts, Hex, Square, Hex Flange, and Coupling Nuts (Inch Series)"
ASME B18.6.3:	"Machine Screws and Machine Screw Nuts"
ISO 4032:	"Hexagon nuts, style 1 -- Product grades A and B"
ISO 4035:	"Hexagon thin nuts (chamfered) -- Product grades A and B"

Prevailing torque hex lock nuts are one of the most common types of hex lock nuts used in modern machinery. One of the most commonly used configurations of this nut typically has a plastic insert, often nylon, which conforms tightly to the screw threads and prevents loosening due to shock or vibration. Prevailing torque hex lock nuts with nonmetallic inserts can be easily removed but should be replaced after disassembly. They do not deform or damage the screw. Ensure that the threaded fastener protrudes at least one or two pitches past the nut. The locking element engages near the nut outlet. The dimensions of some common sizes are shown in Tables 6-54 – 6-55. There are no inch standards, though the parts are commonly available. The applicable metric standards are:

ISO 7040:	"Prevailing torque type hexagon nuts (with nonmetallic insert), style 1 -- Property classes 5, 8 and 10
DIN 985:	"Prevailing torque type hexagon nut with nylon insert"

Standard washers are designed to distribute loads over a wider area than the head of the fastener securing it. Regular, narrow, and wide (fender) versions are available, as are hardened and heavy duty washers. Washers are recommended for use with soft materials or elongated holes or slots to prevent deformation or marring of the part being fastened. Consider using a washer under the head of screws or bolts that require heavy torque. Large washers are sometimes used as spring retainers when a spring rides a shaft such as a shoulder screw. When used over slots or oversized holes, standard washers may deflect under heavy loading, so hardened or heavy duty washers are recommended. The dimensions of some common sizes of standard washers are shown in Tables 6-56 – 6-59. The applicable standards are:

ASME B18.22.1:   "Plain Washers"
ISO 7089:   "Plain washers -- Normal series -- Product grade A"
ISO 7093:   "Plain washers -- Large series -- Part 1: Product grade A"

Lock washers are sometimes used under nuts and screws to prevent loosening due to shock, vibration, or wear. Other methods of thread locking, such as lock nuts and/or thread locking compounds are generally preferable, but lock washers can be useful in cases where low fastener torque is expected or required. When normal tightening torque is used, lock washers are not needed and may in fact cause problems. There are many types of lock washers, but the most commonly used lock washer type in machine design is the high collar, helical spring lock washer. These washers act like a heavy spring between the fastener and the part being fastened, and act to maintain fastener tension under variable loading conditions. High collar lock washers are particularly suitable for use under socket head cap screws; they are smaller in diameter than other types of lock washers. Due to their split configuration, spring lock washers can mar the surface of a part, so it is not unusual to add a regular flat washer underneath the lock washer. The dimensions of some common sizes of standard washers are shown in Tables 6-60 – 6-61. The applicable standards are:

ASME B18.21.1:   "Washers: Helical Spring-Lock, Tooth Lock, and Plain Washers (Inch Series)"
DIN 7980:   "Spring Lock Washers With Square Ends For Cheese Head Screws"

Threaded inserts are used to strengthen and protect internal threads in soft materials. Threaded inserts are best used in soft materials where frequent tightening and loosening of screws is expected or when high tightening torque is required. Threaded inserts can be used to minimize galvanic corrosion between materials, as with steel screws in Aluminum parts. Another common application is the repair of threads in a hole. The most commonly used threaded inserts are helical coils of diamond cross section wire. Helical coil threaded inserts are installed into tapped holes to form a smaller internal thread compliant with standard thread dimensions. Threaded inserts are offered with a variety of surface treatment options, and screw-locking inserts are available. Tables 6-62 – 6-65 contain dimensions of standard helical coil threaded inserts and their appropriate hole dimensions. When calling out a threaded insert installation on a drawing, a common format to follow is:

DRILL AND TAP FOR [brand or standard] HELICAL COIL THREADED INSERT [catalog # or size and length]. The procedure for installing helical coil inserts is as follows:

1. Drill the pilot hole to the specified depth.
2. Tap the pilot hole using a special threaded insert tap to the specified depth. These taps can be purchased wherever you purchase the inserts. They are identified by the final installed insert thread designation.
3. Install the insert to the specified depth using an installation tool.
4. Break off the installation tang (if applicable) with a special tool.

Best practices for use of Helical coil threaded inserts can be found on the next page.

HELICAL COIL THREADED INSERT BEST PRACTICES

- It is good practice to add a 120° included angle countersink to the installation hole opening to help guide the insert and to eliminate burrs.
- The length of engagement with a threaded insert will affect load carrying capacity. The rule of thumb with screw engagement is to aim for 1.5 times the screw diameter or more.
- Threaded inserts must be purchased and installed, and will add cost to a part. Be conservative in using threaded inserts to save cost.
- If material thickness allows, drill to a depth for a plug tap rather than a bottoming tap.
- An excellent source for helical coil inserts online is www.emhart. com. The catalog for their line of screw thread inserts contains complete engineering information.
- Threaded inserts are of a larger diameter than the threaded hole they create. Be sure to provide enough material around the hole to accommodate the insert.

COMPONENT DATA: See the following pages for Tables 6-18 – 6-65.

Table 6-18: SolidWorks® Toolbox 2010 Fasteners
Selections covered in this book

SOLIDWORKS® TOOLBOX PATH			SOLIDWORKS® PART NAME	STANDARD OR SOURCE OF DATA
ANSI INCH	Bolts and Screws	Set Screws (Socket)	Socket Set Screw Flat Point	ASME B18.3
			Socket Set Screw Cup Point	
			Socket Set Screw Cone Point	
			Socket Set Screw Half Dog Point	
		Socket Head Screws	Socket Head Cap Screw	
			Socket Head Shoulder Screw	
			Socket Button Head Cap Screw	
			Socket Countersunk Head Cap Screw	
		Hex Head	Hex Finished Bolt	ASME B18.2.1
	Nuts	Hex Nuts	Hex Nut	ASME B18.2.2
			Hex Jam Nut	
		Machine Screw	Machine Screw Nut Hex	ASME B18.6.3
	Washers	Plain Washers Type B	Regular Flat Washer Type B	ASME B18.22.1
			Wide Flat Washer Type B	
		Spring Lock Washers	Hi-Collar Spring Lock Washer	ASME B18.21.1
ISO	Bolts and Screws	Hexagon Socket Head Screws	Hex Socket Head ISO 4762	ISO 4762
			Hex Socket Head Shoulder ISO 7379	ISO 7379
			Hex Socket Button Head ISO 7380	ISO 7380
			Hex Socket CTSK Head ISO 10642	ISO 10642
		Set Screws - Socket	Socket Set Screw Cone Point ISO 4027	ISO 4027
			Socket Set Screw Cup Point ISO 4029	ISO 4029
			Socket Set Screw Dog Point ISO 4028	ISO 4028
			Socket Set Screw Flat Point ISO 4026	ISO 4026
		Hex Bolts and Screws	Hex Screw Grade AB ISO 4017	ISO 4017
			Hex Screw Grade AB ISO 4014	ISO 4014
	Nuts	Hex Nuts	Hex Nut Style 1 ISO - 4032	ISO 4032
			Hex Thin Nut Grade AB ISO - 4035	ISO 4035
		Hex Nuts - Prevailing Torque	Prevailing Torque Nut Style 1 - ISO 7040	ISO 7040
	Washers	Plain Washers	Washer - ISO 7089 Normal Grade A	ISO 7089
			Washer - ISO 7093 Large Grade A and C	ISO 7093
DIN	Bolts and Screws	Hexagon Socket Head Screws	Hex Socket Head DIN 7984	DIN 7984

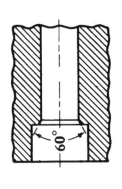

FILLET

Table 6-19: Hex Socket Head Cap Screw, Inch
(Up to 1 inch diameter)
Part Complies With ASME B18.3-2003

Size	Decimal Size inch	Nominal Clearance Hole Diameter					Counterbore Diameter (H14) inch	Max Head Diameter inch	Maximum Head Height inch	Hex Width Across Flats inch	60° Countersink Diameter inch
		Normal (H13) inch	Normal Tolerance inch	Loose (H14) inch	Close (H12) inch						
#0	0.0600	0.076		0.094	0.067	0.125	0.096	0.060	0.05	0.074
#1	0.0730	0.089	+0.006	0.104	0.081	0.156	0.118	0.073	1/16	0.087
#2	0.0860	0.102	0	0.116	0.094	0.188	0.140	0.086	5/64	0.102
#3	0.0990	0.116		0.128	0.106	0.219	0.161	0.099	5/64	0.115
#4	0.1120	0.128		0.144	0.120	0.219	0.183	0.112	3/32	0.13
#5	0.1250	0.156	+0.007	0.172	0.141	0.250	0.205	0.125	3/32	0.145
#6	0.1380	0.170	0	0.185	0.154	0.281	0.226	0.138	7/64	0.158
#8	0.1640	0.196		0.213	0.180	0.313	0.270	0.164	9/64	0.188
#10	0.1900	0.221		0.238	0.206	0.375	0.312	0.190	5/32	0.218
1/4	0.2500	0.281	+0.009	0.297	0.266	0.438	0.375	0.250	3/16	0.278
5/16	0.3125	0.344	0	0.359	0.328	0.531	0.469	0.312	1/4	0.346
3/8	0.3750	0.406		0.422	0.391	0.625	0.562	0.375	5/16	0.415
7/16	0.4375	0.469	+0.011	0.484	0.453	0.719	0.656	0.438	3/8	0.483
1/2	0.5000	0.562	0	0.609	0.531	0.813	0.750	0.500	3/8	0.552
5/8	0.6250	0.688		0.734	0.656	1.000	0.938	0.625	1/2	0.689
3/4	0.7500	0.812		0.906	0.781	1.188	1.125	0.750	5/8	0.828
7/8	0.8750	0.938	+0.013	1.031	0.906	1.375	1.312	0.875	3/4	0.963
1	1.0000	1.094	0	1.156	1.031	1.625	1.500	1.000	3/4	1.1

Table 6-20: Lengths, Hex Socket Head Cap Screw, Inch
(Up to 1 inch diameter and Lengths to 4 inches) Part
Complies With ASME B18.3-2003

Common Lengths Up To 4 Inches (inch)

Size	Coarse Threads Per Inch	0.125	0.188	0.25	0.313	0.375	0.5	0.625	0.75	0.875	1	1.25	1.5	1.75	2	2.25	2.5	2.75	3	3.25	3.5	3.75	4
#0	(80)	0.125	0.188	0.25	0.313	0.375	0.5	0.625	0.75		1												
#1	64		0.188	0.25	0.313	0.375	0.5	0.625	0.75		1												
#2	56	0.125	0.188	0.25	0.313	0.375	0.5	0.625	0.75	0.875	1	1.25	1.5										
#3	48	0.125	0.188	0.25	0.313	0.375	0.5	0.625	0.75	0.875	1	1.25	1.5										
#4	40	0.125	0.188	0.25	0.313	0.375	0.5	0.625	0.75	0.875	1	1.25	1.5	1.75	2								
#5	40	0.125	0.188	0.25	0.313	0.375	0.5	0.625	0.75	0.875	1	1.25	1.5	1.75	2								
#6	32	0.125	0.188	0.25	0.313	0.375	0.5	0.625	0.75	0.875	1	1.25	1.5	1.75	2	2.25	2.5	2.75	3				
#8	32	0.125	0.188	0.25	0.313	0.375	0.5	0.625	0.75	0.875	1	1.25	1.5	1.75	2	2.25	2.5	2.75	3	3.25	3.5		
#10	24			0.25	0.313	0.375	0.5	0.625	0.75	0.875	1	1.25	1.5	1.75	2	2.25	2.5	2.75	3	3.25	3.5	3.75	4
1/4	20			0.25	0.313	0.375	0.5	0.625	0.75	0.875	1	1.25	1.5	1.75	2	2.25	2.5	2.75	3	3.25	3.5	3.75	4
5/16	18			0.25	0.313	0.375	0.5	0.625	0.75	0.875	1	1.25	1.5	1.75	2	2.25	2.5	2.75	3	3.25	3.5	3.75	4
3/8	16					0.375	0.5	0.625	0.75	0.875	1	1.25	1.5	1.75	2	2.25	2.5	2.75	3	3.25	3.5	3.75	4
7/16	14						0.5	0.625	0.75	0.875	1	1.25	1.5	1.75	2	2.25	2.5	2.75	3	3.25	3.5	3.75	4
1/2	13						0.5	0.625	0.75	0.875	1	1.25	1.5	1.75	2	2.25	2.5	2.75	3	3.25	3.5	3.75	4
5/8	11						0.5	0.625	0.75	0.875	1	1.25	1.5	1.75	2	2.25	2.5	2.75	3	3.25	3.5	3.75	4
3/4	10								0.75	0.875	1	1.25	1.5	1.75	2	2.25	2.5	2.75	3	3.25	3.5	3.75	4
7/8	9											1.25	1.5	1.75	2	2.25	2.5	2.75	3	3.25	3.5	3.75	4
1	8											1.25	1.5	1.75	2	2.25	2.5	2.75	3	3.25	3.5	3.75	4

Note: Some lengths may be partially threaded. Check with the manufacturer.

( ) = Fine pitch only commonly available in this size.

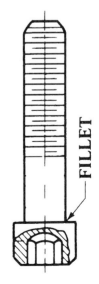

FILLET

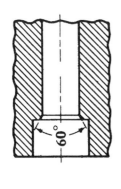

FILLET

Table 6-21: Hex Socket Head Cap Screw, Metric
(Up to M24)

Part Complies With ISO 4762: 2004

Size	Nominal Clearance Hole Diameter				Counterbore Diameter (H14) mm	Max Head Diameter mm	Maximum Head Height mm	Hex Width Across Flats mm	60° Countersink Min. Diameter mm
	Normal (H13) mm	Normal Tolerance mm	Loose (H14) mm	Close (H12) mm					
M1.6	1.8	+0.140	2.0	1.7	3.5	3	1.6	1.5	1.8
M2	2.4	0	2.6	2.2	4.4	3.8	2.0	1.5	2.2
M2.5	2.9		3.1	2.7	4.4	4.5	2.5	2.0	2.7
M3	3.4	+0.180	3.6	3.2	6.5	5.5	3.0	2.5	3.2
M4	4.5	0	4.8	4.3	8.0	7.0	4.0	3.0	4.4
M5	5.5		5.8	5.3	10.0	8.5	5.0	4.0	5.4
M6	6.6	+0.220	7.0	6.4	11.0	10.0	6.0	5.0	6.5
M8	9.0	0	10.0	8.4	15.0	13.0	8.0	6.0	8.8
M10	11.0		12.0	10.5	18.0	16.0	10.0	8.0	10.8
M12	13.5	+0.270	14.5	13.0	20.0	18.0	12.0	10.0	13.2
M14	15.5	0	16.5	15.0	24.0	21.0	14.0	12.0	15.2
M16	17.5		18.5	17.0	26.0	24.0	16.0	14.0	17.2
M20	22.0	+0.330	24.0	21.0	33.0	30.0	20.0	17.0	21.6
M24	26.0	0	28.0	25.0	40.0	36.0	24.0	19.0	25.6

Table 6-22: Lengths, Hex Socket Head Cap Screw, Metric
(Up to M24 and Lengths to 100 mm)
Part Complies With ISO 4762: 2004

Size	Coarse Pitch mm	Common Lengths Up To 100 mm mm																		
M1,6	0.35	3	4	5	6															
M2	0.4	3	4	5	6	8	10	12												
M2.5	0.45		4	5	6	8	10	12	16	20	25									
M3	0.5		4	5	6	8	10	12	16	20	25	30	35	40	45	50				
M4	0.7			5	6	8	10	12	16	20	25	30	35	40	45	50	55	60	70	80
M5	0.8				6	8	10	12	16	20	25	30	35	40	45	50	55	60	65	70	75	80	100
M6	1				6	8	10	12	16	20	25	30	35	40	45	50	55	60	65	70	75	80	100
M8	1.25						10	12	16	20	25	30	35	40	45	50	55	60	65	70	75	80	100
M10	1.5						10	12	16	20	25	30	35	40	45	50	55	60	65	70	75	80	100
M12	1.75							12	16	20	25	30	35	40	45	50	55	60	65	70	75	80	100
M14	2									20	25	30	35	40	45	50	55	60	65	70	75	80	100
M16	2									20	25	30	35	40	45	50	55	60	65	70	75	80	100
M20	2.5											30		40	45	50	55	60	65	70	75	80	100
M24	3													40	45	50		60		70		80	100

Note: Some lengths may be partially threaded. Check with manufacturer.

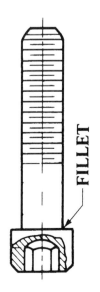

FILLET

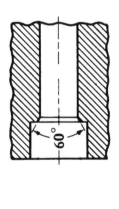

FILLET

Table 6-23: Low Head Hex Socket Cap Screw, Inch
(Up to 1 inch diameter)
Part Complies With ASME B18.3 - 2003

| Size | Decimal Size inch | Nominal Clearance Hole Diameter | | | | Counterbore Diameter (H14) inch | Max Head Diameter inch | Maximum Head Height inch | Hex Width Across Flats inch | 60° Countersink Diameter inch |
		Normal (H13) inch	Normal Tolerance inch	Loose (H14) inch	Close (H12) inch					
#4	0.1120	0.128	+0.007	0.144	0.120	0.219	0.183	0.059	0.05	0.13
#5	0.1250	0.156	0	0.172	0.141	0.250	0.205	0.065	1/16	0.145
#6	0.1380	0.170		0.185	0.154	0.281	0.226	0.072	1/16	0.158
#8	0.1640	0.196		0.213	0.180	0.313	0.270	0.085	5/64	0.188
#10	0.1900	0.221	+0.009	0.238	0.206	0.375	0.312	0.098	3/32	0.218
1/4	0.2500	0.281	0	0.297	0.266	0.438	0.375	0.127	1/8	0.278
5/16	0.3125	0.344		0.359	0.328	0.531	0.437	0.158	5/32	0.346
3/8	0.3750	0.406	+0.011	0.422	0.391	0.625	0.562	0.192	3/16	0.415
7/16	0.4375	0.469	0	0.484	0.453	0.719	0.625	0.223	7/32	0.483
1/2	0.5000	0.562		0.609	0.531	0.813	0.750	0.254	1/4	0.552
5/8	0.6250	0.688		0.734	0.656	1.000	0.875	0.316	5/16	0.689

Table 6-24: Lengths, Low Head Hex Socket Cap Screw, Inch
(Up to 1 inch diameter and Lengths to 3 inches)
Part Complies With ASME B18.3 - 2003

Size	Coarse Threads Per Inch	Common Lengths Up To 3 Inches inch							
#4	40								
#5	40								
#6	32	0.25	0.375	0.5	0.625	0.75	1		
#8	32	0.25	0.375	0.5	0.625	0.75	1	1.25	1.5
#10	24	0.25	0.375	0.5	0.625	0.75	1	1.25	1.5
1/4	20	0.25	0.375	0.5	0.625	0.75	1	1.25	1.5	1.75	2
5/16	18		0.375	0.5	0.625	0.75	1	1.25	1.5	1.75	2
3/8	16			0.5	0.625	0.75	1	1.25	1.5	1.75	2
7/16	14								
1/2	13					0.75	1	1.25	1.5	1.75	2
5/8	11								

Note: Some lengths may be partially threaded. Check with the manufacturer.

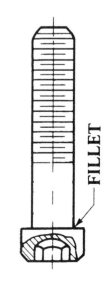

FILLET

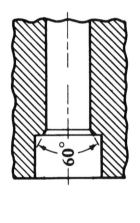

FILLET

Table 6-25: Low Head Hex Socket Cap Screw, Metric
(Up to M24)
Part Complies With DIN 7984

Size	Nominal Clearance Hole Diameter				Counterbore Diameter (H14) mm	Max Head Diameter mm	Maximum Head Height mm	Hex Width Across Flats mm	60° Countersink Min. Diameter mm
	Normal (H13) mm	Normal Tolerance mm	Loose (H14) mm	Close (H12) mm					
M3	3.4	+0.180 0	3.6	3.2	6.5	5.5	2.0	2.0	3.2
M4	4.5		4.8	4.3	8.0	7.0	2.8	2.5	4.4
M5	5.5		5.8	5.3	10.0	8.5	3.5	3.0	5.4
M6	6.6	+0.220 0	7.0	6.4	11.0	10.0	4.0	4.0	6.5
M8	9.0		10.0	8.4	15.0	13.0	5.0	5.0	8.8
M10	11.0	+0.270 0	12.0	10.5	18.0	16.0	6.0	7	10.8
M12	13.5		14.5	13.0	20.0	18.0	7.0	8	13.2
M16	17.5		18.5	17.0	26.0	24.0	9.0	12	17.2
M20	22.0	+0.330 0	24.0	21.0	33.0	30.0	11.0	14	21.6

Table 6-26: Lengths, Low Head Hex Socket Cap Screw, Metric
(Up to M24 and Lengths to 100 mm)
Part Complies With DIN 7984

Size	Coarse Pitch mm	Common Lengths Up To 100 mm mm															
		5	6	8	10	12	16	20	25	30	35	40	45	50	60	70	80
M3	0.5	5	6	8	10	12	16	20	25	30	35	40	45	50			
M4	0.7		6	8	10	12	16	20	25	30	35	40					
M5	0.8		6	8	10	12	16	20	25	30	35	40	45	50			
M6	1			8	10	12	16	20	25	30	35	40	45	50			
M8	1.25				10	12	16	20	25	30	35	40	45	50	60		
M10	1.5					12	16	20	25	30	35	40	45	50	60	70	80
M12	1.75							20	25	30	35	40	45	50	60	70	80
M16	2									30	35	40	45	50	60	70	80
M20	2.5											40		50	60	70	80

NOTE: Some lengths may be partially threaded. Check with the manufacturer.

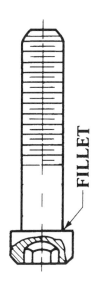

FILLET

Table 6-27: Cone Point Hex Socket Set Screw, Inch
(Up to 1 inch diameter and Lengths to 2 inches)
Part Complies With ASME B18.3 - 2003

Size	Coarse Threads Per Inch	Decimal Size (inch)	Hex Width Across Flats (inch)	Common Lengths Up To 2 Inches (inch)													
				0.0625	0.125	0.1875	0.25	0.3125	0.375	0.5	0.625	0.75	1	1.25	1.5	1.75	2
#0	(80)	0.0600	0.028														
#1	64	0.0730	0.035														
#2	56	0.0860	0.035	0.0625	0.125	0.1875	0.25	0.3125	0.375	0.5							
#3	48	0.0990	0.050														
#4	40	0.1120	0.050		0.125	0.1875	0.25	0.3125	0.375	0.5							
#5	40	0.1250	1/16		0.125	0.1875	0.25	0.3125	0.375	0.5							
#6	32	0.1380	1/16		0.125	0.1875	0.25	0.3125	0.375	0.5	0.625	0.75	1				
#8	32	0.1640	5/64		0.125	0.1875	0.25	0.3125	0.375	0.5	0.625	0.75	1				
#10	24	0.1900	3/32		0.125	0.1875	0.25	0.3125	0.375	0.5	0.625	0.75	1	1.25			
1/4	20	0.2500	1/8		0.125	0.1875	0.25	0.3125	0.375	0.5	0.625	0.75	1	1.25	1.5	1.75	2
5/16	18	0.3125	5/32			0.1875	0.25	0.3125	0.375	0.5	0.625	0.75	1	1.25	1.5		
3/8	16	0.3750	3/16				0.25	0.3125	0.375	0.5	0.625	0.75	1	1.25	1.5		
7/16	14	0.4375	7/32					0.3125	0.375	0.5	0.625	0.75	1	1.25	1.5		
1/2	13	0.5000	1/4						0.375	0.5	0.625	0.75	1	1.25	1.5	1.75	2
5/8	11	0.6250	5/16							0.5	0.625	0.75	1	1.25	1.5	1.75	2
3/4	10	0.7500	3/8								0.625	0.75	1	1.25	1.5	1.75	2
7/8	9	0.8750	1/2									0.75	1	1.25	1.5	1.75	2
1	8	1.0000	9/16														

( ) = Fine thread only available in this size.

Table 6-28: Cup Point Hex Socket Set Screw, Inch
(Up to 1 inch diameter and Lengths to 2 inches)
Part Complies With ASME B18.3 - 2003

Length columns below fall under **Common Lengths Up To 2 Inches** (inch).

Size	Coarse Threads Per Inch	Decimal Size (inch)	Hex Width Across Flats (inch)	Cup Point Max Diameter (inch)	0.0625	0.125	0.1875	0.25	0.3125	0.375	0.5	0.625	0.75	1	1.25	1.5	1.75	2
#0	(80)	0.0600	0.028	0.033	0.0625	0.125	0.1875	0.25	0.3125	0.375								
#1	64	0.0730	0.035	0.040		0.125	0.1875	0.25										
#2	56	0.0860	0.035	0.047	0.0625	0.125	0.1875	0.25	0.3125	0.375	0.5	0.625	0.75					
#3	48	0.0990	0.050	0.054	0.0625	0.125	0.1875	0.25	0.3125	0.375	0.5	0.625						
#4	40	0.1120	0.050	0.061	0.0625	0.125	0.1875	0.25	0.3125	0.375	0.5	0.625	0.75	1	1.25			
#5	40	0.1250	1/16	0.067		0.125	0.1875	0.25	0.3125	0.375	0.5	0.625	0.75	1				
#6	32	0.1380	1/16	0.074		0.125	0.1875	0.25	0.3125	0.375	0.5	0.625	0.75	1	1.25	1.5	1.75	2
#8	32	0.1640	5/64	0.087		0.125	0.1875	0.25	0.3125	0.375	0.5	0.625	0.75	1	1.25	1.5	1.75	2
#10	24	0.1900	3/32	0.102		0.125	0.1875	0.25	0.3125	0.375	0.5	0.625	0.75	1	1.25	1.5	1.75	2
1/4	20	0.2500	1/8	0.132		0.125	0.1875	0.25	0.3125	0.375	0.5	0.625	0.75	1	1.25	1.5	1.75	2
5/16	18	0.3125	5/32	0.172			0.1875	0.25	0.3125	0.375	0.5	0.625	0.75	1	1.25	1.5	1.75	2
3/8	16	0.3750	3/16	0.212				0.25	0.3125	0.375	0.5	0.625	0.75	1	1.25	1.5	1.75	2
7/16	14	0.4375	7/32	0.252				0.25	0.3125	0.375	0.5	0.625	0.75	1	1.25	1.5	1.75	2
1/2	13	0.5000	1/4	0.291				0.25	0.3125	0.375	0.5	0.625	0.75	1	1.25	1.5	1.75	2
5/8	11	0.6250	5/16	0.371						0.375	0.5	0.625	0.75	1	1.25	1.5	1.75	2
3/4	10	0.7500	3/8	0.450						0.375	0.5	0.625	0.75	1	1.25	1.5	1.75	2
7/8	9	0.8750	1/2	0.530							0.5	0.625	0.75	1	1.25	1.5	1.75	2
1	8	1.0000	9/16	0.609									0.75	1	1.25	1.5	1.75	2

( ) = Fine thread only available in this size.

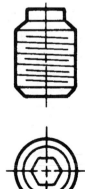

Table 6-29: Half Dog Point Hex Socket Set Screw, Inch
(Up to 1 inch diameter and Lengths to 2 inches)
Part Complies With ASME B18.3 - 2003

Size	Coarse Threads Per Inch	Decimal Size inch	Hex Width Across Flats inch	Dog Point Max Diameter inch	Dog Point Max Length inch	Common Lengths Up To 2 Inches inch
#0	(80)	0.0600	0.028	0.040	0.017	
#1	64	0.0730	0.035	0.049	0.021	
#2	56	0.0860	0.035	0.057	0.024	
#3	48	0.0990	0.050	0.066	0.027	0.125
#4	40	0.1120	0.050	0.075	0.030	0.125, 0.1875, 0.25, 0.3125, 0.375, 0.5
#5	40	0.1250	1/16	0.083	0.033	0.125, 0.1875, 0.25, 0.375
#6	32	0.1380	1/16	0.092	0.038	0.125, 0.1875, 0.25, 0.3125, 0.375, 0.625, 0.75, 1
#8	32	0.1640	5/64	0.109	0.043	0.125, 0.1875, 0.25, 0.3125, 0.375, 0.625, 0.75, 1
#10	24	0.1900	3/32	0.127	0.049	0.125, 0.1875, 0.25, 0.3125, 0.375, 0.5, 0.625, 0.75, 1, 1.25, 1.5
1/4	20	0.2500	1/8	0.156	0.067	0.25, 0.3125, 0.375, 0.5, 0.625, 0.75, 1, 1.25, 1.5
5/16	18	0.3125	5/32	0.203	0.082	0.3125, 0.375, 0.5, 0.625, 0.75, 1, 1.25, 1.5, 1.75, 2
3/8	16	0.3750	3/16	0.250	0.099	0.3125, 0.375, 0.5, 0.625, 0.75, 1, 1.25, 1.5, 1.75, 2
7/16	14	0.4375	7/32	0.297	0.114	0.5
1/2	13	0.5000	1/4	0.344	0.130	0.5, 0.625, 0.75, 1, 1.25, 1.5, 1.75, 2
5/8	11	0.6250	5/16	0.469	0.164	0.625, 0.75, 1, 1.25, 1.5, 1.75, 2
3/4	10	0.7500	3/8	0.562	0.196	0.75, 1, 1.25, 1.5, 1.75, 2
7/8	9	0.8750	1/2	0.656	0.227	
1	8	1.0000	9/16	0.750	0.260	

( ) = Fine thread only available in this size.

Table 6-30: Flat Point Hex Socket Set Screw, Inch
(Up to 1 inch diameter and Lengths to 2 inches)
Part Complies With ASME B18.3 - 2003

Size	Coarse Threads Per Inch	Decimal Size inch	Hex Width Across Flats inch	Flat Point Max Diameter inch	Common Lengths Up To 2 Inches (inch)													
					0.0625	0.125	0.1875	0.25	0.3125	0.375	0.5	0.625	0.75	1	1.25	1.5	1.75	2
#0	(80)	0.0600	0.028	0.033														
#1	64	0.0730	0.035	0.040														
#2	56	0.0860	0.035	0.047		0.125	0.1875	0.25										
#3	48	0.0990	0.050	0.054		0.125	0.1875	0.25		0.375								
#4	40	0.1120	0.050	0.061	0.0625	0.125	0.1875	0.25	0.3125	0.375	0.5	0.625						
#5	40	0.1250	1/16	0.067		0.125	0.1875	0.25	0.3125	0.375	0.5							
#6	32	0.1380	1/16	0.074		0.125	0.1875	0.25	0.3125	0.375	0.5	0.625	0.75					
#8	32	0.1640	5/64	0.087		0.125	0.1875	0.25	0.3125	0.375	0.5	0.625	0.75	1				
#10	24	0.1900	3/32	0.102		0.125	0.1875	0.25	0.3125	0.375	0.5	0.625	0.75	1	1.25	1.5		
1/4	20	0.2500	1/8	0.132			0.1875	0.25	0.3125	0.375	0.5	0.625	0.75	1	1.25	1.5	1.75	2
5/16	18	0.3125	5/32	0.172				0.25	0.3125	0.375	0.5	0.625	0.75	1	1.25	1.5	1.75	2
3/8	16	0.3750	3/16	0.212				0.25	0.3125	0.375	0.5	0.625	0.75	1	1.25	1.5	1.75	2
7/16	14	0.4375	7/32	0.252						0.375	0.5	0.625	0.75	1				
1/2	13	0.5000	1/4	0.291						0.375	0.5	0.625	0.75	1	1.25	1.5	1.75	2
5/8	11	0.6250	5/16	0.371						0.375	0.5	0.625	0.75	1	1.25	1.5	1.75	2
3/4	10	0.7500	3/8	0.450							0.5	0.625	0.75	1	1.25	1.5	1.75	2
7/8	9	0.8750	1/2	0.530									0.75	1	1.25	1.5		
1	8	1.0000	9/16	0.609														

( ) = Fine thread only available in this size.

Table 6-31: Cone Point Hex Socket Set Screw, Metric
(Up to M24 and Lengths to 50 mm)
Part Complies With ISO 4027: 2003

Size	Coarse Pitch mm	Hex Width Across Flats mm	Cone Tip Max Diameter mm	Common Lengths Up To 50 mm — mm														
				3	4	5	6	8	10	12	16	20	25	30	35	40	45	50
M1.6	0.35	0.7	0.40															
M2	0.4	0.9	0.50	3	4	5			10									
M3	0.5	1.5	0.75	3	4	5	6	8	10	12	16	20						
M4	0.7	2.0	1.00		4	5	6	8	10	12	16	20	25		35	40		
M5	0.8	2.5	1.25			5	6	8	10	12	16	20	25	30	35			
M6	1	3.0	1.50			5	6	8	10	12	16	20	25	30	35			
M8	1.25	4.0	2.00				6	8	10	12	16	20	25	30	35	40		50
M10	1.5	5.0	2.50					8	10	12	16	20	25	30	35	40		50
M12	1.75	6.0	3.00							12	16	20	25	30	35	40	45	50
M16	2	8.0	4.00								16		25	30	35	40		50
M20	2.5	10.0	5.00										25		35			50
M24	3	12.0	6.00										25					

Table 6-32: Cup Point Hex Socket Set Screw, Metric
(Up to M24 and Lengths to 50 mm)
Part Complies With ISO 4029: 2003

Size	Coarse Pitch mm	Hex Width Across Flats mm	Cup Point Max Diameter mm	Common Lengths Up To 50 mm (mm)															
M1.6	0.35	0.7	0.8	2.5	3	4	5												
M2	0.4	0.9	1.0	2.5	3	4	5	6	8	10								
M3	0.5	1.5	1.4		3	4	5	6	8	10	12	16	20	25				
M4	0.7	2.0	2.0			4	5	6	8	10	12	16	20	25	30	35	40		
M5	0.8	2.5	2.5				5	6	8	10	12	16	20	25	30	35	40		
M6	1	3.0	3.0				5	6	8	10	12	16	20	25	30	35	40	45	50
M8	1.25	4.0	5.0					6	8	10	12	16	20	25	30	35	40	45	50
M10	1.5	5.0	6.0					6	8	10	12	16	20	25	30	35	40	45	50
M12	1.75	6.0	8.0							10	12	16	20	25	30	35	40	45	50
M16	2	8.0	10.0								12	16	20	25	30	35	40	45	50
M20	2.5	10.0	14.0									16	20	25	30	35	40	45	50
M24	3	12.0	16.0											25	30	35	40	45	50

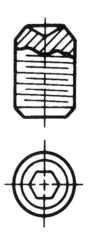

Table 6-33: Dog Point Hex Socket Set Screw, Metric
(Up to M24 and Lengths to 50 mm)
Part Complies With ISO 4028: 2003

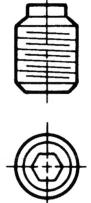

Size	Coarse Pitch mm	Hex Width Across Flats mm	Dog Point Max Diameter mm	Dog Point Max Length (Short) mm	Dog Point Max Length (Long) mm	Common Lengths Up To 50mm mm
M1.6	0.35	0.7	0.8	(0.65)	1.05	
M2	0.4	0.9	1.0	(0.75)	1.25	
M3	0.5	1.5	2.0	(1)	1.75	(4) (5) 6 8 10 12
M4	0.7	2.0	2.5	(1.25)	2.25	(5) (6) 8 10 12 16 20 25 30 40
M5	0.8	2.5	3.5	(1.5)	2.75	(6) 8 10 12 16 20 25 30 35 40
M6	1	3.0	4.0	(1.75)	3.25	(6) (8) 10 12 16 20 25 30 35 40 50
M8	1.25	4.0	5.5	(2.25)	4.3	(8) (10) 12 16 20 25 30 35 40 45 50
M10	1.5	5.0	7.0	(2.75)	5.3	(10) (12) 16 20 25 30 35 40 45 50
M12	1.75	6.0	8.5	(3.25)	6.3	(12) (16) 20 25 30 35 40 45 50
M16	2	8.0	12.0	(4.3)	8.36	(20) 25 30 35 40 45 50
M20	2.5	10.0	15.0	(5.3)	10.36	(20) (25) 30 35 40 45 50
M24	3	12.0	18.0	(6.3)	12.43	(30)

Table 6-34: *Flat Point Hex Socket Set Screw, Metric*
(Up to M24 and Lengths to 50 mm)
Part Complies With ISO 4026: 2003

Size	Coarse Pitch (mm)	Hex Width Across Flats (mm)	Flat Point Max Diameter (mm)	Common Lengths Up To 50 mm (mm)														
				3	4	5	6	8	10	12	16	20	25	30	35	40	45	50
M1.6	0.35	0.7	0.8															
M2	0.4	0.9	1.0	3	4	5	6	8										
M3	0.5	1.5	2.0	3	4	5	6	8	10	12	16							
M4	0.7	2.0	2.5	3	4	5	6	8	10	12	16	20	25					
M5	0.8	2.5	3.5		4	5	6	8	10	12	16	20	25	30	35	40		
M6	1	3.0	4.0		4	5	6	8	10	12	16	20	25	30	35	40		
M8	1.25	4.0	5.5			5	6	8	10	12	16	20	25	30	35	40	45	50
M10	1.5	5.0	7.0				6	8	10	12	16	20	25	30	35	40	45	50
M12	1.75	6.0	8.5					8	10	12	16	20	25	30	35	40	45	50
M16	2	8.0	12.0								16	20	25	30	35	40	45	50
M20	2.5	10.0	15.0									20	25	30	35	40	45	50
M24	3	12.0	18.0										25	30	35	40	45	50

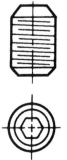

Table 6-35: Hex Socket Head Shoulder Screw, Inch
(Up to 1 inch shoulder diameter)
Part Complies With ASME B18.3 - 2003

Shoulder Size	Nominal Shoulder Diameter inch	Shoulder Tolerance inch	Under Head Fillet Outer Diameter inch	Thread Size	Thread Length inch	Max Head Diameter inch	Max Head Height inch	Hex Socket Size inch
1/4	0.2500		0.2780	#10-24	0.375	0.375	0.188	1/8
5/16	0.3125	-0.002	0.3465	1/4 - 20	0.438	0.438	0.219	5/32
3/8	0.3750		0.4150	5/16 - 18	0.500	0.562	0.250	3/16
1/2	0.5000	-0.004	0.5520	3/8 - 16	0.625	0.750	0.312	1/4
5/8	0.6250		0.6890	1/2 - 13	0.750	0.875	0.375	5/16
3/4	0.7500		0.8280	5/8 - 11	0.875	1.000	0.500	3/8
1	1.0000		1.1000	3/4 - 10	1.000	1.312	0.625	1/2

Table 6-36: Lengths, Hex Socket Head Shoulder Screw, Inch
(Up to 1 inch shoulder diameter and Lengths to 4 inches)
Part Complies With ASME B18.3 - 2003

Shoulder Size	Thread Size	Common Shoulder Lengths Up To 4 Inches (inch)																				
		0.125	0.188	0.25	0.313	0.375	0.5	0.625	0.75	1	1.3	1.5	1.8	2	2.3	2.5	2.8	3	3.3	3.5	3.8	4
1/4	#10-24	0.125	0.188	0.25	0.313	0.375	0.5	0.625	0.75	1	1.3	1.5	1.8	2	2.3	2.5	2.8	3	3.3	3.5	3.8	4
5/16	1/4 - 20			0.25	0.313	0.375	0.5	0.625	0.75	1	1.3	1.5	1.8	2	2.3	2.5	2.8	3	3.3	3.5	3.8	4
3/8	5/16 - 18			0.25	0.313	0.375	0.5	0.625	0.75	1	1.3	1.5	1.8	2	2.3	2.5	2.8	3	3.3	3.5	3.8	4
1/2	3/8 - 16					0.375	0.5	0.625	0.75	1	1.3	1.5	1.8	2	2.3	2.5	2.8	3	3.3	3.5	3.8	4
5/8	1/2 - 13					0.375	0.5	0.625	0.75	1	1.3	1.5	1.8	2	2.3	2.5	2.8	3	3.3	3.5	3.8	4
3/4	5/8 - 11						0.5	0.625	0.75	1	1.3	1.5	1.8	2	2.3	2.5	2.8	3	3.3	3.5	3.8	4
1	3/4 - 10									1	1.3	1.5	1.8	2	2.3	2.5	2.8	3	3.3	3.5		4

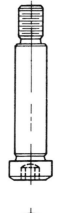

Table 6-37: Hex Socket Head Shoulder Screw, Metric
(Up to 25 mm shoulder diameter)
Part Complies With ISO 7379: 1983

Nominal Shoulder Diameter mm	Shoulder Tolerance (f9) mm	Under Head Fillet Max Outer Diameter mm	Thread Size	Max Thread Length mm	Max Head Diameter mm	Max Head Height mm	Hex Socket Distance Between Flats mm
(6)	-0.010 -0.040	7	M5	9.75	10	4.5	3
6.5	-0.013 -0.049	7.5	M5	9.75	10	4.5	3
8		9.2	M6	11.25	13	5.5	4
10		11.2	M8	13.25	16	7.0	5
(12)	-0.016 -0.059	14.2	M10	16.40	18	9.0	6
13		15.2	M10	16.40	18	9.0	6
16		18.2	M12	18.40	24	11	8
20	-0.020 -0.072	22.4	M16	22.40	30	14	10
25		27.4	M20	27.40	36	16	12

( ) = Size not ISO standard but commonly available in the USA.

Table 6-38: *Lengths, Hex Socket Head Shoulder Screw, Metric*
(Up to 25 mm shoulder diameter and Lengths to 100 mm)
Part Complies With ISO 7379: 1983

Nominal Shoulder Diameter mm	Thread Size	Common Shoulder Lengths Up To 100mm +0.25 -0 mm											
(6)	M5	6	10	12	16	20	25	30	40				50
6.5	M5												
8	M6		10	12	16	20	25	30	40	45			50
10	M8												
(12)	M10			12	16	20	25	30	40	45			50
13	M10												
16	M12					20	25	30	35	40	45		50
20	M16							30	35	40	45		50
25	M20												

( ) = Size not ISO standard but commonly available in the USA.

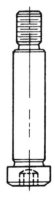

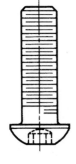

Table 6-39: Hex Socket Button Head Cap Screw, Inch
(Up to 1 inch diameter)
Part Complies With ASME B18.3 - 2003

| Size | Decimal Size inch | Nominal Clearance Hole Diameter | | | | Max Head Diameter inch | Max Head Height inch | Hex Width Across Flats inch |
		Normal (H13) inch	Normal Tolerance inch	Loose (H14) inch	Close (H12) inch			
#0	0.0600	0.076	+0.006	0.094	0.067	0.114	0.032	0.035
#1	0.0730	0.089	0	0.104	0.081	0.139	0.039	0.050
#2	0.0860	0.102		0.116	0.094	0.164	0.046	0.050
#3	0.0990	0.116		0.128	0.106	0.188	0.052	1/16
#4	0.1120	0.128	+0.007	0.144	0.120	0.213	0.059	1/16
#5	0.1250	0.156	0	0.172	0.141	0.238	0.066	5/64
#6	0.1380	0.170		0.185	0.154	0.262	0.073	5/64
#8	0.1640	0.196		0.213	0.180	0.312	0.087	3/32
#10	0.1900	0.221	+0.009	0.238	0.206	0.361	0.101	1/8
1/4	0.2500	0.281	0	0.297	0.266	0.437	0.132	5/32
5/16	0.3125	0.344		0.359	0.328	0.547	0.166	3/16
3/8	0.3750	0.406	+0.011	0.422	0.391	0.656	0.199	7/32
1/2	0.5000	0.562	0	0.609	0.531	0.875	0.265	5/16
5/8	0.6250	0.688		0.734	0.656	1.000	0.331	3/8

Table 6-40: *Lengths, Hex Socket Button Head Cap Screw, Inch*
(Up to 1 inch diameter and Lengths to 4 inches)
Part Complies With ASME B18.3 - 2003

Size	Coarse Threads Per Inch	Common Lengths Up To 4 Inches (inch)																		
#0	(80)	0.125	0.1875	0.25	0.3125	0.375	0.5	0.625	0.75	1										
#1	(72)	0.125	0.1875	0.25	0.3125	0.375	0.5	0.625	0.75	1										
#2	56	0.125	0.1875	0.25	0.3125	0.375	0.5	0.625	0.75	1										
#3	48	0.125	0.1875	0.25	0.3125	0.375	0.5	0.625	0.75	1										
#4	40	0.125	0.1875	0.25	0.3125	0.375	0.5	0.625	0.75	1	1.25									
#5	40	0.125	0.1875	0.25		0.375	0.5	0.625	0.75	1	1.25	1.5								
#6	32	0.125	0.1875	0.25	0.3125	0.375	0.5	0.625	0.75	1	1.25	1.5	1.75	2		2.5		3		
#8	32	0.125	0.1875	0.25	0.3125	0.375	0.5	0.625	0.75	1	1.25	1.5	1.75	2		2.5		3		
#10	24		0.1875	0.25	0.3125	0.375	0.5	0.625	0.75	1	1.25	1.5	1.75	2	2.25	2.5		3		
1/4	20			0.25	0.3125	0.375	0.5	0.625	0.75	1	1.25	1.5	1.75	2	2.25	2.5		3		
5/16	18					0.375	0.5	0.625	0.75	1	1.25	1.5	1.75	2	2.25	2.5	2.75	3	3.5	4
3/8	16					0.375	0.5	0.625	0.75	1	1.25	1.5	1.75	2	2.25	2.5	2.75	3	3.5	4
1/2	14								0.75	1	1.25	1.5	1.75	2	2.25	2.5	2.75	3	3.5	4
5/8	13									1	1.25	1.5	1.75	2	2.25	2.5		3	3.5	4

( ) = Fine thread only available in this size.

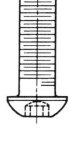

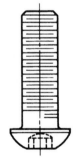

Table 6-41: Hex Socket Button Head Cap Screw, Metric
(Up to M24)
Part Complies With ISO 7380: 2004

| Size | Nominal Clearance Hole Diameter | | | | Max Head Diameter mm | Max Head Height mm | Hex Width Across Flats mm |
	Normal (H13) mm	Normal Tolerance mm	Loose (H14) mm	Close (H12) mm			
M3	3.4	+0.180	3.6	3.2	5.7	1.65	2.0
M4	4.5	0	4.8	4.3	7.6	2.20	2.5
M5	5.5		5.8	5.3	9.5	2.75	3.0
M6	6.6	+0.220	7.0	6.4	10.5	3.30	4.0
M8	9.0	0	10.0	8.4	14.0	4.40	5.0
M10	11.0		12.0	10.5	17.5	5.50	6.0
M12	13.5	+0.270	14.5	13.0	21.0	6.60	8.0
M16	17.5	0	18.5	17.0	28.0	8.80	10.0

Table 6-42: Lengths, Hex Socket Button Head Cap Screw, Metric
(Up to M24 and Lengths to 100 mm)
Part Complies With ISO 7380: 2004

Size	Coarse Pitch mm	Common Lengths Up To 100 mm mm												
M3	0.50	6	8	10	12	16	20	25						
M4	0.70	6	8	10	12	16	20	25	30	35	40			
M5	0.80	6	8	10	12	16	20	25	30	35	40			
M6	1.00	6	8	10	12	16	20	25	30	35	40	45	50	
M8	1.25			10	12	16	20	25	30	35	40	45	50	
M10	1.50				12	16	20	25	30	35	40	45	50	
M12	1.75					16	20	25	30	35	40	45	50	
M16	2.00									35				

Note: Some lengths may be partially threaded. Check with the manufacturer.

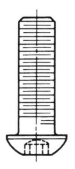

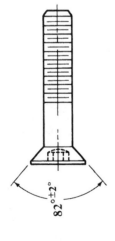

82°±2°

Table 6-43: Hex Socket Flat Countersunk Head Cap Screw, Inch
(Up to 1 inch diameter)
Part Complies With ASME B18.3 - 2003

Size	Decimal Size inch	Nominal Clearance Hole Diameter				Max Head Diameter (To Theoretical Sharps) inch	Head Height (Approximate) inch	Hex Socket Distance Between Flats inch
		Normal (H13) inch	Normal Tolerance inch	Loose (H14) inch	Close (H12) inch			
#0	0.0600	0.076		0.094	0.067	0.138	0.044	0.035
#1	0.0730	0.089	+0.006	0.104	0.081	0.168	0.054	0.050
#2	0.0860	0.102	0	0.116	0.094	0.197	0.064	0.050
#3	0.0990	0.116		0.128	0.106	0.226	0.073	1/16
#4	0.1120	0.128		0.144	0.120	0.255	0.083	1/16
#5	0.1250	0.156	+0.007	0.172	0.141	0.281	0.090	5/64
#6	0.1380	0.170	0	0.185	0.154	0.307	0.097	5/64
#8	0.1640	0.196		0.213	0.180	0.359	0.112	3/32
#10	0.1900	0.221	+0.009	0.238	0.206	0.411	0.127	1/8
1/4	0.2500	0.281	0	0.297	0.266	0.531	0.161	5/32
5/16	0.3125	0.344		0.359	0.328	0.656	0.198	3/16
3/8	0.3750	0.406	+0.011	0.422	0.391	0.781	0.234	7/32
7/16	0.4375	0.469	0	0.484	0.453	0.844	0.234	1/4
1/2	0.5000	0.562		0.609	0.531	0.938	0.251	5/16
5/8	0.6250	0.688		0.734	0.656	1.188	0.324	3/8
3/4	0.7500	0.812	+0.013	0.906	0.781	1.438	0.396	1/2
7/8	0.8750	0.938	0	1.031	0.906	1.688	0.468	9/16
1	1.0000	1.094		1.156	1.031	1.938	0.540	5/8

Table 6-44: Lengths, Hex Socket Flat Countersunk Head Cap Screw, Inch

(Up to 1 inch diameter and Lengths to 4 inches)
Part Complies With ASME B18.3 - 2003

Common Lengths Up To 4 Inches (Length Measured from Top of Head) — inch

Size	Coarse Threads Per Inch	0.125	0.1875	0.25	0.3125	0.375	0.5	0.625	0.75	1	1.25	1.5	1.75	2	2.25	2.5	2.75	3	3.5	4
#0	(80)	0.125	0.1875	0.25	0.3125	0.375	0.5	0.625	0.75	1										
#1	(72)	0.125	0.1875	0.25	0.3125	0.375	0.5	0.625	0.75	1										
#2	56	0.125	0.1875	0.25	0.3125	0.375	0.5	0.625	0.75	1										
#3	48		0.1875	0.25	0.3125	0.375	0.5	0.625	0.75	1										
#4	40		0.1875	0.25	0.3125	0.375	0.5	0.625	0.75	1	1.25	1.5								
#5	40		0.1875	0.25	0.3125	0.375	0.5	0.625	0.75	1	1.25	1.5								
#6	32		0.1875	0.25	0.3125	0.375	0.5	0.625	0.75	1	1.25	1.5	1.75	2	2.25	2.5		3		
#8	32			0.25	0.3125	0.375	0.5	0.625	0.75	1	1.25	1.5	1.75	2	2.25	2.5		3		
#10	24			0.25	0.3125	0.375	0.5	0.625	0.75	1	1.25	1.5	1.75	2	2.25	2.5		3	3.5	4
1/4	20			0.25		0.375	0.5	0.625	0.75	1	1.25	1.5	1.75	2	2.25	2.5	2.75	3	3.5	4
5/16	18					0.375	0.5	0.625	0.75	1	1.25	1.5	1.75	2	2.25	2.5	2.75	3	3.5	4
3/8	16						0.5	0.625	0.75	1	1.25	1.5	1.75	2	2.25	2.5	2.75	3	3.5	4
7/16	14							0.625	0.75	1	1.25	1.5	1.75	2	2.25	2.5	2.75	3	3.5	4
1/2	13								0.75	1	1.25	1.5	1.75	2	2.25	2.5	2.75	3	3.5	4
5/8	11								0.75	1	1.25	1.5	1.75	2	2.25	2.5	2.75	3	3.5	4
3/4	10								0.75	1	1.25	1.5	1.75	2	2.25	2.5	2.75	3	3.5	4
7/8	9													2		2.5		3	3.5	4
1	8													2	2.25	2.5	2.75	3	3.5	4

Note: Some lengths may be partially threaded. Check with the manufacturer.

( ) = Fine pitch only commonly available in this size.

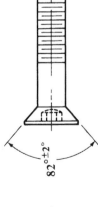

82° ±2°

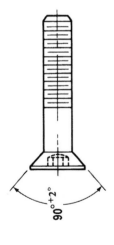

Table 6-45: Hex Socket Flat Countersunk Head Cap Screw, Metric
(Up to M24)
Part Complies With ISO 10642: 2004

Size	Nominal Clearance Hole Diameter					Max Head Diameter (To Theoretical Sharps)	Head Height (Approximate)	Hex Socket Distance Between Flats
	Normal (H13)	Normal Tolerance	Loose (H14)	Close (H12)				
	mm	mm	mm	mm		mm	mm	mm
M3	3.4	+0.180	3.6	3.2		6.72	1.86	2.0
M4	4.5	0	4.8	4.3		8.96	2.48	2.5
M5	5.5		5.8	5.3		11.20	3.10	3.0
M6	6.6	+0.220	7.0	6.4		13.44	3.72	4.0
M8	9.0	0	10.0	8.4		17.92	4.96	5.0
M10	11.0		12.0	10.5		22.40	6.20	6.0
M12	13.5	+0.270	14.5	13.0		26.88	7.44	8.0
M14	15.5	0	16.5	15.0		30.80	8.40	10.0
M16	17.5		18.5	17.0		33.60	8.80	10.0
M20	22.0	+0.330 0	24.0	21.0		40.32	10.16	12.0

Table 6-46: Lengths, Hex Socket Flat Countersunk Head Cap Screw, Metric
(Up to M24 and Lengths to 100 mm)
Part Complies With ISO 10642: 2004

Size	Coarse Pitch mm	Common Lengths Up To 100 mm (Length Measured from Top of Head) mm																		
		5	6	8	10	12	16	20	25	30	35	40	45	50	55	60	70	80	90	100
M3	0.50	5	6	8	10	12	16	20	25	30	35	40								
M4	0.70		6	8	10	12	16	20	25	30	35	40	45	50		60				
M5	0.80	5	6	8	10	12	16	20	25	30	35	40	45	50		60				
M6	1.00		6	8	10	12	16	20	25	30	35	40	45	50	55	60	70	80		
M8	1.25				10	12	16	20	25	30	35	40	45	50	55	60	70	80	90	100
M10	1.50					12	16	20	25	30	35	40	45	50	55	60	70	80	90	100
M12	1.75							20	25	30	35	40	45	50	55	60	70	80	90	100
M14	2.00								25	30	35	40	45	50	55	60	70	80		
M16	2.00								25	30	35	40	45	50	55	60	70	80	90	100
M20	2.50										35	40	45	50	55	60	70	80	90	100

NOTE: Some lengths may be partially threaded. Check with the manufacturer.

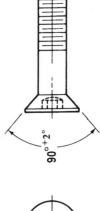

90°+2°

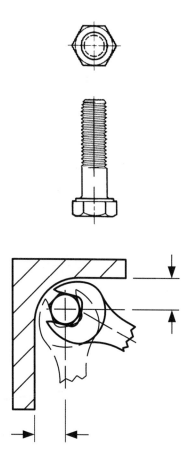

Table 6-47: Hex Head Cap Screw, Inch
(Up to 1 inch diameter)
Part Complies With ASME B18.2.1 - 1996

| Size | Max Body Diameter inch | Nominal Clearance Hole Diameter | | | | Max Head Height inch | Max Head Distance Between Corners inch | Max Head Distance Between Flats inch | Open End Wrench Clearance Radius inch | Minimum Socket Drive Clearance Diameter inch |
		Normal (H13) inch	Normal Tolerance inch	Loose (H14) inch	Close (H12) inch					
1/4	0.2500	0.281	+0.009	0.297	0.266	0.163	0.505	0.438	0.470	0.713
5/16	0.3125	0.344	0	0.359	0.328	0.211	0.577	0.500	0.520	0.727
3/8	0.3750	0.406		0.422	0.391	0.243	0.650	0.562	0.590	0.808
7/16	0.4375	0.469	+0.011	0.484	0.453	0.291	0.722	0.625	0.640	0.920
1/2	0.5000	0.562	0	0.609	0.531	0.323	0.866	0.750	0.770	1.140
5/8	0.6250	0.688		0.734	0.656	0.403	1.083	0.938	0.970	1.340
3/4	0.7500	0.812	+0.013	0.906	0.781	0.483	1.299	1.125	1.140	1.570
7/8	0.8750	0.938	0	1.031	0.906	0.563	1.516	1.312	1.390	1.850
1	1.0000	1.094		1.156	1.031	0.627	1.732	1.500	1.470	2.055

Table 6-48: Lengths, Hex Head Cap Screw, Inch
(Up to 1 inch diameter and Lengths to 4 inches)
Part Complies With ASME B18.2.1 - 1996

Size	Coarse Threads Per Inch	Common Lengths Up To 4 Inches inch																	
		0.375	0.5	0.625	0.75	0.875	1	1.25	1.5	1.75	2	2.25	2.5	2.75	3	3.25	3.5	3.75	4
1/4	20	0.375	0.5	0.625	0.75	0.875	1	1.25	1.5	1.75	2	2.25	2.5	2.75	3	3.25	3.5	3.75	4
5/16	18		0.5	0.625	0.75	0.875	1	1.25	1.5	1.75	2	2.25	2.5	2.75	3	3.25	3.5	3.75	4
3/8	16		0.5	0.625	0.75	0.875	1	1.25	1.5	1.75	2	2.25	2.5	2.75	3	3.25	3.5	3.75	4
7/16	14				0.75	0.875	1	1.25	1.5	1.75	2	2.25	2.5	2.75	3	3.25	3.5	3.75	4
1/2	13				0.75	0.875	1	1.25	1.5	1.75	2	2.25	2.5	2.75	3	3.25	3.5	3.75	4
5/8	11						1	1.25	1.5	1.75	2	2.25	2.5	2.75	3	3.25	3.5	3.75	4
3/4	10						1	1.25	1.5	1.75	2	2.25	2.5	2.75	3	3.25	3.5	3.75	4
7/8	9								1.5	1.75	2	2.25	2.5	2.75	3	3.25	3.5	3.75	4
1	8								1.5	1.75	2	2.25	2.5	2.75	3	3.25	3.5	3.75	4

Note: Some lengths may be partially threaded. Check with the manufacturer.

Table 6-49: Hex Head Cap Screw, Metric
(Up to M24) Part Complies With ISO 4017: 1999 Grade A and DIN 933 and ISO 4014: 1999 Grade A

| Screw Size | Nominal Clearance Hole Diameter | | | | Max Head Height | Min. Head Distance Between Corners | Max. Head Distance Between Flats | Minimum Box Wrench Clearance Radius | Minimum Socket Drive Clearance Diameter |
| | Normal (H13) | Normal Tolerance | Loose (H14) | Close (H12) | | | | | |
	mm	mm	mm	mm	mm	mm	mm	mm	mm
M1.6	1.8	+0.140	2.0	1.7	1.225	3.410	3.2	4.56	9.0
M2	2.4	0	2.6	2.2	1.525	4.320	4.0	4.56	9.0
M2.5	2.9		3.1	2.7	1.825	5.450	5.0	5.26	10.0
M3	3.4	+0.180	3.6	3.2	2.125	6.010	5.5	6.66	11.0
M4	4.5	0	4.8	4.3	2.925	7.660	7.0	7.91	13.0
M5	5.5		5.8	5.3	3.650	8.790	8.0	8.26	15.0
M6	6.6	+0.220	7.0	6.4	4.150	11.050	10.0	10.16	18.0
M8	9.0	0	10.0	8.4	5.450	14.380	13.0	12.31	24.0
M10	11.0		12.0	10.5	6.580	17.770	16.0	14.26	28.0
M12	13.5	+0.270	14.5	13.0	7.680	20.030	18.0	15.41	33.0
M14	15.5	0	16.5	15.0	8.980	23.360	21.0	17.66	36.0
M16	17.5		18.5	17.0	10.180	26.750	24.0	19.81	40.0
M20	22.0	+0.330	24.0	21.0	12.715	33.530	30.0	24.51	46.0
M24	26.0	0	28.0	25.0	15.215	39.980	36.0	28.81	58.0

Table 6-50: Lengths, Hex Head Cap Screw, Metric
(Up to M24 and Lengths to 100 mm)
Part Complies With ISO 4017: 1999 Grade A and DIN 933
and ISO 4014: 1999 Grade A

Screw Size	Coarse Pitch mm	Common Lengths Up To 100 mm mm																		
M1.6	0.35																			
M2	0.40																			
M2.5	0.45																			
M3	0.50	5	6	8	10	12	16	20	25	30	35									
M4	0.70		6	8	10	12	16	20	25	30	35	40	45	50						
M5	0.80		6	8	10	12	16	20	25	30	35	40	45	50	55	60	65	70	80	90
M6	1.00			8	10	12	16	20	25	30	35	40	45	50	55	60	65	70	80	90	100
M8	1.25			8	10	12	16	20	25	30	35	40	45	50	55	60	65	70	80	90	100
M10	1.50				10	12	16	20	25	30	35	40	45	50	55	60	65	70	80	90	100
M12	1.75						16	20	25	30	35	40	45	50	55	60	65	70	80	90	100
M14	2.00							20	25	30	35	40	45	50	55	60	65	70	80	90	100
M16	2.00							20	25	30	35	40	45	50	55	60	65	70	80	90	100
M20	2.50								25	30	35	40	45	50	55	60	65	70	80	90	100
M24	3.00								25	30	35	40	45	50	55	60	65	70	80	90	100

Note: Some lengths may be partially threaded. Check with the manufacturer.

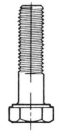

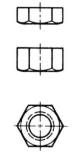

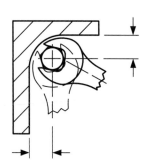

Table 6-51: Hex Nuts, Inch
(Up to 1 inch thread)
Part Complies With ASME B18.2.2 – 1987 (R1999)

Thread Size	Coarse Threads Per Inch	Decimal Size inch	Max Width Across Flats inch	Max Width Across Corners inch	Minimum Socket Drive Clearance Diameter inch	Open End Wrench Clearance Radius inch	Max Thickness inch	Max Thickness (Jam Nut) inch
1/4	20	0.2500	0.4380	0.505	0.713	0.470	0.226	0.163
5/16	18	0.3125	0.5000	0.577	0.727	0.520	0.273	0.195
3/8	16	0.3750	0.5620	0.650	0.808	0.590	0.337	0.227
7/16	14	0.4375	0.6880	0.794	0.998	0.770	0.385	0.260
1/2	13	0.5000	0.7500	0.866	1.140	0.770	0.448	0.323
9/16	12	0.5625	0.8750	1.010	1.280	0.970	0.496	0.324
5/8	11	0.6250	0.9380	1.083	1.340	0.970	0.559	0.387
3/4	10	0.7500	1.1250	1.299	1.570	1.140	0.665	0.446
7/8	9	0.8750	1.3120	1.516	1.850	1.390	0.776	0.510
1	8	1.0000	1.5000	1.732	2.055	1.470	0.887	0.575

Table 6-52: Hex Machine Screw Nuts, Inch
(Up to 1 inch thread)
Part Complies With ASME B18.6.3 – 1972 (R1991)

Thread Size	Coarse Threads Per Inch	Decimal Size inch	Max Width Across Flats inch	Max Width Across Corners inch	Minimum Socket Drive Clearance inch	Open End Wrench Clearance Radius inch	Max Thickness inch
#0	(80)	0.0600	0.1560	0.1800	0.540	0.220	0.050
#1	64	0.0730	0.1560	0.1800	0.540	0.220	0.050
#2	56	0.0860	0.1880	0.2170	0.540	0.250	0.066
#3	48	0.0990	0.1880	0.2170	0.540	0.250	0.066
#4	40	0.1120	0.2500	0.2890	0.540	0.280	0.098
#5	40	0.1250	0.3120	0.3610	0.540	0.380	0.114
#6	32	0.1380	0.3120	0.3610	0.540	0.380	0.114
#8	32	0.1640	0.3440	0.3970	0.577	0.420	0.130
#10	24	0.1900	0.3750	0.4330	0.627	0.420	0.130
#12	24	0.2160	0.4380	0.5050	0.713	0.470	0.161
1/4	20	0.2500	0.4380	0.5050	0.713	0.470	0.193
5/16	18	0.3125	0.5620	0.6500	0.808	0.590	0.225
3/8	16	0.3750	0.6250	0.7220	0.920	0.640	0.257

( ) = Fine pitch only available in this size

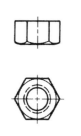

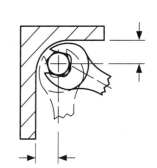

Table 6-53: Hex Nuts & Thin (Jam) Nuts, Metric
(Up to M24) Part Complies With ISO 4032:
1999 and ISO 4035: 1999

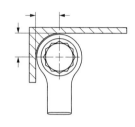

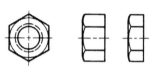

Thread Size	Coarse Pitch	Max Width Across Flats	Min Width Across Corners	Minimum Socket Drive Clearance Diameter	Minimum Box Wrench Clearance Radius	Max Thickness	Max Thickness (Jam Nut)
	mm	mm	mm	mm	mm	mm	mm
M1.6	0.35	3.2	3.41	9.0	4.56	1.3	1.0
M2	0.40	4.0	4.32	9.0	4.56	1.6	1.2
M3	0.50	5.5	6.01	11.0	6.66	2.4	1.8
M4	0.70	7.0	7.66	13.0	7.91	3.2	2.2
M5	0.80	8.0	8.79	15.0	8.26	4.7	2.7
M6	1.00	10.0	11.05	18.0	10.16	5.2	3.2
M8	1.25	13.0	14.38	24.0	12.31	6.8	4.0
M10	1.50	16.0	17.77	28.0	14.26	8.4	5.0
M12	1.75	18.0	20.03	33.0	15.41	10.8	6.0
M14	2.00	21.0	23.36	36.0	17.66	12.8	7.0
M16	2.00	24.0	26.75	40.0	19.81	14.8	8.0
M20	2.50	30.0	32.95	46.0	24.51	18.0	10.0
M24	3.00	36.0	39.55	58.0	28.81	21.5	12.0

Table 6-54: Hex Lock Nuts, Prevailing Torque, Nonmetallic Insert, Inch
(Up to 1 inch thread)

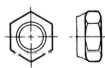

Thread Size	Coarse Threads Per Inch	Decimal Size	Nominal Width Across Flats	Max Width Across Corners	Minimum Socket Drive Clearance	Open End Wrench Clearance Radius	Max Thickness
		inch	inch	inch	inch	inch	inch
#2	56	0.0860	0.2500	0.289	0.540	0.280	0.1406
#3	48	0.0990	0.2500	0.289	0.540	0.280	0.1406
#4	40	0.1120	0.2500	0.289	0.540	0.280	0.1406
#5	40	0.1250	0.2500	0.289	0.540	0.280	0.1406
#6	32	0.1380	0.3125	0.361	0.540	0.380	0.1719
#8	32	0.1640	0.3438	0.397	0.577	0.420	0.2344
#10	24	0.1900	0.3750	0.433	0.627	0.420	0.2344
#12	24	0.2160	0.4375	0.505	0.713	0.470	0.3125
1/4	20	0.2500	0.4375	0.505	0.713	0.470	0.3125
5/16	18	0.3125	0.5000	0.577	0.727	0.520	0.3438
3/8	16	0.3750	0.5625	0.650	0.808	0.590	0.4531
7/16	14	0.4375	0.6250	0.722	0.920	0.640	0.4531
1/2	13	0.5000	0.7500	0.866	1.140	0.770	0.5938
9/16	12	0.5625	0.8750	1.010	1.280	0.970	0.6406
5/8	11	0.6250	0.9375	1.083	1.340	0.970	0.7500
3/4	10	0.7500	1.0625	1.227	1.510	1.090	0.8750
7/8	9	0.8750	1.2500	1.443	1.780	1.270	0.9844
1	8	1.0000	1.4375	1.660	1.985	1.470	1.0469

Table 6-55: Hex Lock Nuts, Prevailing Torque, Nonmetallic Insert, Metric
(Up to M24)
Part Complies With ISO 7040: 1997 and DIN 985

Thread Size	Coarse Pitch mm	Nominal Width Across Flats mm	Max Width Across Corners mm	Minimum Socket Drive Clearance Diameter mm	Minimum Box Wrench Clearance Radius mm	Max Thickness mm
M3	0.50	5.50	6.01	11.00	6.66	4.00
M4	0.70	7.00	7.66	13.00	7.91	5.00
M5	0.80	8.00	8.79	15.00	8.26	5.00
M6	1.00	10.00	11.05	18.00	10.16	6.00
M8	1.25	13.00	14.38	24.00	12.31	8.00
M10	1.50	17.00	19.63	28.00	15.41	10.00
M12	1.75	19.00	21.94	33.00	16.36	12.00
M14	2.00	22.00	25.40	36.00	18.56	14.00
M16	2.00	24.00	26.75	40.00	19.81	16.00
M20	2.50	30.00	32.95	46.00	24.51	20.00
M24	3.00	36.00	39.55	58.00	28.81	24.00

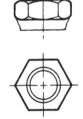

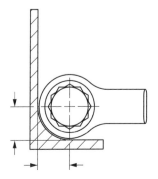

Table 6-56: *Plain Regular Flat Washers, Inch*
(Up to 1 inch screw size)
Part Complies With ASME B18.22.1 – 1965 (R2003) Type B

Screw Size	Decimal Size inch	Nominal Inside Diameter inch	Inside Diameter Tolerance inch	Nominal Outside Diameter inch	Outside Diameter Tolerance inch	Nominal Thickness inch	Thickness Tolerance inch
#0	0.0600	0.068		0.188			+0.003 / -0.003
#1	0.0730	0.084		0.219		0.025	
#2	0.0860	0.094	0 / -0.005	0.250	0 / -0.005	0.032	+0.004 / -0.004
#3	0.0990	0.109		0.312			
#4	0.1120	0.125		0.375			
#5	0.1250	0.141		0.406			+0.005 / -0.004
#6	0.1380	0.156	+0.008 / -0.005	0.438	+0.008 / -0.005	0.040	
#8	0.1640	0.188		0.500			
#10	0.1900	0.203		0.562			
#12	0.2160	0.234		0.625	+0.015 / -0.005		
1/4	0.2500	0.281		0.734			
5/16	0.3125	0.344		0.875	+0.015 / -0.007	0.063	+0.008 / -0.007
3/8	0.3750	0.406	+0.015 / -0.005	1.000			
7/16	0.4375	0.469		1.125			
1/2	0.5000	0.531		1.250			
9/16	0.5625	0.594		1.469			
5/8	0.6250	0.656		1.750	+0.030 / -0.007	0.100	+0.012 / -0.010
3/4	0.7500	0.812	+0.030 / -0.007	2.000			
7/8	0.8750	0.938		2.250		0.160	+0.014 / -0.014
1	1.0000	1.062		2.500			

Table 6-57: *Wide Washers, Inch*
(Up to 1 inch screw size)
Part Complies With ASME B18.22.1 – 1965 (R2003) Type B

Screw Size	Decimal Size inch	Nominal Inside Diameter inch	Inside Diameter Tolerance inch	Nominal Outside Diameter inch	Outside Diameter Tolerance inch	Nominal Thickness inch	Thickness Tolerance inch
#0	0.0600	0.068		0.250	+0 -0.005	0.025	+0.003 -0.003
#1	0.0730	0.084	+0 -0.005	0.281		0.032	+0.004 -0.004
#2	0.0860	0.094		0.344		0.032	
#3	0.0990	0.109		0.406		0.040	
#4	0.1120	0.125		0.438	+0.008 -0.005	0.040	+0.005 -0.004
#5	0.1250	0.141	+0.008 -0.005	0.500		0.040	
#6	0.1380	0.156		0.562		0.040	
#8	0.1640	0.188		0.625	+0.015 -0.005	0.063	+0.008 -0.007
#10	0.1900	0.203		0.734		0.063	
#12	0.2160	0.234		0.875	+0.015 -0.007	0.063	
1/4	0.2500	0.281	+0.015 -0.005	1.000		0.063	
5/16	0.3125	0.344		1.125		0.063	
3/8	0.3750	0.406		1.250		0.100	+0.012 -0.010
7/16	0.4375	0.469		1.469		0.100	
1/2	0.5000	0.531		1.750		0.100	
9/16	0.5625	0.594		2.000	+0.030 -0.007	0.100	
5/8	0.6250	0.656	+0.030 -0.007	2.250		0.160	+0.014 -0.014
3/4	0.7500	0.812		2.500		0.160	
7/8	0.8750	0.938		2.750		0.160	
1	1.0000	1.062		3.000		0.160	

Table 6-58: *Plain Regular Flat Washers, Metric*
(Up to M24 screw size)
Part Complies With ISO 7089: 2000

Screw Size	Nominal Inside Diameter mm	Inside Diameter Tolerance mm	Nominal Outside Diameter mm	Outside Diameter Tolerance mm	Nominal Thickness mm	Thickness Tolerance mm
M1.6	1.7	+0.140 / -0	4	+0 / -0.300	0.3	+0.050 / -0.050
M2	2.2		5		0.3	
M2.5	2.7		6		0.5	
M3	3.2	+0.180 / -0	7	+0 / -0.360	0.5	+0.100 / -0.100
M4	4.3		9		0.8	
M5	5.3		10		1.0	
M6	6.4	+0.220 / -0	12	+0 / -0.430	1.6	+0.200 / -0.200
M8	8.4		16		1.6	
M10	10.5		20		2.0	
M12	13	+0.270 / -0	24	+0 / -0.520	2.5	+0.300 / -0.300
M14	15		28		2.5	
M16	17		30		3.0	
M20	21	+0.330 / -0	37	+0 / -0.620	3.0	
M24	25		44		4.0	

Table 6-59: Wide Washers, Metric
(Up to M24 screw size)
Part Complies With ISO 7093: 2000

Screw Size	Nominal Inside Diameter mm	Inside Diameter Tolerance mm	Nominal Outside Diameter mm	Outside Diameter Tolerance mm	Nominal Thickness mm	Thickness Tolerance mm
M3	3.2	+0.180 / -0	9	+0 / -0.360	0.8	+0.100 / -0.100
M4	4.3		12	+0 / -0.430	1.0	
M5	5.3		15		1.0	
M6	6.4	+0.220 / -0	18	+0 / -0.520	1.6	+0.200 / -0.200
M8	8.4		24		2.0	
M10	10.5		30		2.5	
M12	13	+0.270 / -0	37	+0 / -0.620	3.0	0.300 / -0.300
M14	15		44		3.0	
M16	17		50		3.0	
M20	21	+0.330 / -0	60	+0 / -0.740	4.0	
M24	25	+0.520 / -0	72	+0 / -1.200	5.0	+0.600 / -0.600

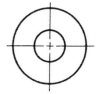

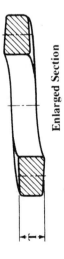

Enlarged Section

Table 6-60: High Collar Helical Spring Lock Washers, Inch
Part Complies With ASME B18.21.1 – 1999

Screw Size	Decimal Size inch	Inside Diameter inch		Maximum Outside Diameter inch	Nominal Section Thickness inch
		MIN	MAX		
#4	0.1120	0.114	0.120	0.173	0.022
#5	0.1250	0.127	0.133	0.202	0.030
#6	0.1380	0.141	0.148	0.216	0.030
#8	0.1640	0.167	0.174	0.267	0.047
#10	0.1900	0.193	0.200	0.294	0.047
1/4	0.2500	0.252	0.260	0.363	0.078
5/16	0.3125	0.314	0.322	0.457	0.093
3/8	0.3750	0.377	0.385	0.550	0.125
7/16	0.4375	0.440	0.450	0.644	0.140
1/2	0.5000	0.502	0.512	0.733	0.172
5/8	0.6250	0.628	0.641	0.917	0.203
3/4	0.7500	0.753	0.766	1.105	0.218
7/8	0.8750	0.878	0.894	1.291	0.234
1	1.0000	1.003	1.024	1.478	0.250

Table 6-61: High Collar Helical Spring Lock Washers, Metric
Part Complies With DIN 7980: 1987

Screw Size	Inside Diameter (mm)		Maximum Outside Diameter (mm)	Section Thickness (mm)	
	MIN	MAX		MIN	MAX
M3	3.1	3.4	5.6	0.9	1.1
M4	4.1	4.4	7.0	1.1	1.3
M5	5.1	5.4	8.8	1.5	1.7
M6	6.1	6.5	9.9	1.5	1.7
M8	8.1	8.5	12.7	1.9	2.1
M10	10.2	10.7	16.0	2.35	2.65
M12	12.2	12.7	18.0	2.35	2.65
M14	14.2	14.7	21.1	2.8	3.2
M16	16.2	17.0	24.4	3.3	3.7
M20	20.2	21.2	30.6	4.3	4.7
M24	25.5	25.5	35.9	4.8	5.2

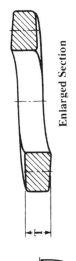

Enlarged Section

Table 6-62: Helical Coil Threaded Inserts, Inch
Part Complies With ASME B18.29.1 – 2010

| Thread Size | Coarse Threads Per Inch | Outside Diameter | | Nominal Length | | | | | | Installation Depth Below Surface |
		MIN inch	MAX inch	1 Dia. inch	1.5 Dia. inch	2 Dia. inch	2.5 Dia inch	3 Dia. inch		inch
#1	64	0.095	0.103	0.073	0.110	0.146	0.182	0.219		0.016
#2	56	0.110	0.119	0.086	0.129	0.172	0.215	0.258		0.018
#3	48	0.128	0.139	0.099	0.148	0.198	0.248	0.297		0.021
#4	40	0.144	0.159	0.112	0.168	0.224	0.280	0.336		0.025
#5	40	0.158	0.173	0.125	0.188	0.250	0.312	0.375		0.025
#6	32	0.178	0.193	0.138	0.207	0.276	0.345	0.414		0.031
#8	32	0.205	0.220	0.164	0.246	0.328	0.410	0.492		0.031
#10	24	0.244	0.259	0.190	0.285	0.380	0.475	0.570		0.042
#12	24	0.270	0.285	0.216	0.324	0.432	0.540	0.648		0.042
1/4	20	0.310	0.330	0.250	0.375	0.500	0.625	0.750		0.050
5/16	18	0.380	0.400	0.312	0.469	0.625	0.781	0.938		0.056
3/8	16	0.452	0.472	0.375	0.562	0.750	0.938	1.125		0.063
7/16	14	0.526	0.551	0.438	0.656	0.875	1.094	1.312		0.071
1/2	13	0.597	0.622	0.500	0.750	1.000	1.250	1.500		0.077
9/16	12	0.669	0.694	0.562	0.844	1.125	1.406	1.688		0.083
5/8	11	0.742	0.767	0.625	0.938	1.250	1.562	1.875		0.091
3/4	10	0.881	0.906	0.750	1.125	1.500	1.875	2.250		0.100
7/8	9	1.022	1.052	0.875	1.312	1.750	2.188	2.625		0.111
1	8	1.166	1.196	1.000	1.500	2.000	2.500	3.000		0.125

Table 6-63: Helical Coil Threaded Insert Hole Dimensions, Inch
Part Complies With ASME B18.29.1 - 2010

Thread Size	Drill Diameter		Min. Drill Depth (For Bottoming Tap)						Min. Tap Depth For Insert Length				
	Aluminum	Steel, Plastic	1 Dia.	1.5 Dia.	2 Dia.	2.5 Dia.	3 Dia.		1 Dia.	1.5 Dia.	2 Dia.	2.5 Dia.	3 Dia.
	inch	inch	inch	inch	inch	inch	inch		inch	inch	inch	inch	inch
#1	0.0785	0.0810	0.136	0.172	0.209	0.245	0.282		0.090	0.125	0.160	0.200	0.235
#2	0.0938	0.0960	0.157	0.200	0.243	0.286	0.329		0.100	0.150	0.190	0.230	0.280
#3	0.1065	0.1094	0.182	0.232	0.281	0.331	0.380		0.120	0.170	0.220	0.270	0.320
#4	0.1200	0.1200	0.212	0.268	0.324	0.380	0.436		0.140	0.190	0.250	0.310	0.360
#5	0.1339	0.1360	0.225	0.288	0.350	0.412	0.475		0.150	0.210	0.280	0.340	0.400
#6	0.1470	0.1495	0.263	0.332	0.401	0.470	0.539	Special	0.170	0.240	0.310	0.380	0.450
#8	0.1730	0.1770	0.289	0.371	0.453	0.535	0.617	Tap	0.200	0.280	0.360	0.440	0.520
#10	0.2031	0.2055	0.357	0.452	0.547	0.642	0.737	Required	0.230	0.330	0.420	0.520	0.610
#12	0.2280	0.2280	0.383	0.491	0.599	0.707	0.815		0.260	0.370	0.470	0.580	0.690
1/4	0.2660	0.2660	0.450	0.575	0.700	0.825	0.950		0.300	0.430	0.550	0.680	0.800
5/16	0.3320	0.3320	0.534	0.690	0.846	1.002	1.158		0.370	0.530	0.680	0.840	0.990
3/8	0.3970	0.3970	0.625	0.812	1.000	1.188	1.375		0.440	0.630	0.810	1.000	1.190
7/16	0.4531	0.4531	0.724	0.943	1.162	1.381	1.600		0.510	0.730	0.950	1.170	1.380
1/2	0.5156	0.5312	0.808	1.058	1.308	1.558	1.808		0.580	0.830	1.080	1.330	1.580
9/16	0.5781	0.5938	0.895	1.176	1.457	1.738	2.019		0.650	0.930	1.210	1.490	1.770
5/8	0.6562	0.6562	0.989	1.301	1.614	1.926	2.239		0.720	1.030	1.340	1.650	1.970
3/4	0.7812	0.7812	1.150	1.525	1.900	2.275	2.650		0.850	1.230	1.600	1.980	2.350
7/8	0.9062	0.9062	1.319	1.757	2.194	2.632	3.069		0.990	1.420	1.860	2.300	2.740
1	1.0312	1.0312	1.500	2.000	2.500	3.000	3.500		1.130	1.630	2.130	2.630	3.130

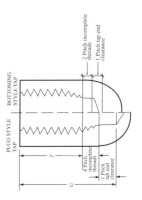

PLUG STYLE TAP
BOTTOMING STYLE TAP

2 Pitch incomplete threads
1 Pitch tap end clearance
4 Pitch incomplete threads
1 Pitch tap end clearance

Table 6-64: Helical Coil Threaded Inserts, Metric
Part Complies With ASME B18.29.2M – 2005

Thread Size	Coarse Pitch mm	Outside Diameter MIN mm	Outside Diameter MAX mm	Nominal Length 1 Dia. mm	Nominal Length 1.5 Dia. mm	Nominal Length 2 Dia. mm	Nominal Length 2.5 Dia mm	Nominal Length 3 Dia. mm	Installation Depth Below Surface mm
M2	0.40	2.50	2.70	2.0	3.0	4.0	5.0	6.0	0.40
M2.2	0.45	2.80	3.00	2.2	3.3	4.4	5.5	6.6	0.45
M2.5	0.45	3.20	3.70	2.5	3.8	5.0	6.3	7.5	0.45
M3	0.50	3.80	4.35	3.0	4.5	6.0	7.5	9.0	0.50
M3.5	0.60	4.40	4.95	3.5	5.3	7.0	8.8	10.5	0.60
M4	0.70	5.05	5.60	4.0	6.0	8.0	10.0	12.0	0.70
M5	0.80	6.25	6.80	5.0	7.5	10.0	12.5	15.0	0.80
M6	1.00	7.40	7.95	6.0	9.0	12.0	15.0	18.0	1.00
M7	1.00	8.65	9.20	7.0	10.5	14.0	17.5	21.0	1.00
M8	1.25	9.80	10.35	8.0	12.0	16.0	20.0	24.0	1.25
M10	1.50	11.95	12.50	10.0	15.0	20.0	25.0	30.0	1.50
M12	1.75	14.30	15.00	12.0	18.0	24.0	30.0	36.0	1.75
M14	2.00	16.65	17.35	14.0	21.0	28.0	35.0	42.0	2.00
M16	2.00	18.90	19.60	16.0	24.0	32.0	40.0	48.0	2.00
M18	2.50	21.30	22.00	18.0	27.0	36.0	45.0	54.0	2.50
M20	2.50	23.55	24.40	20.0	30.0	40.0	50.0	60.0	2.50
M22	2.50	25.90	26.90	22.0	33.0	44.0	55.0	66.0	2.50
M24	3.00	28.00	29.00	24.0	36.0	48.0	60.0	72.0	3.00

Table 6-65: Helical Coil Threaded Insert Hole Dimensions, Metric
Part Complies With ASME B18.29.2M – 2005

Thread Size	Drill Diameter		Min. Drill Depth (For Bottoming Tap)					Min. Tap Depth For Insert Length				
	Aluminum	Steel, Plastic	1 Dia.	1.5 Dia.	2 Dia.	2.5 Dia.	3 Dia.	1 Dia.	1.5 Dia.	2 Dia.	2.5 Dia.	3 Dia.
	mm	mm	mm	mm	mm	mm	mm	mm	mm	mm	mm	mm
M2	2.10	2.10	3.60	4.60	5.60	6.60	7.60	2.4	3.4	4.4	5.4	6.4
M2.2	2.30	2.35	4.00	5.10	6.20	7.30	8.40	2.7	3.8	4.9	6.0	7.1
M2.5	2.55	2.65	4.30	5.55	6.80	8.05	9.30	3.0	4.2	5.5	6.7	8.0
M3	3.15	3.20	5.00	6.50	8.00	9.50	11.00	3.5	5.0	6.5	8.0	9.5
M3.5	3.70	3.70	5.90	7.65	9.40	11.15	12.90	4.1	5.9	7.6	9.4	11.1
M4	4.20	4.25	6.80	8.80	10.80	12.80	14.80	4.7	6.7	8.7	10.7	12.7
M5	5.20	5.30	8.20	10.70	13.20	15.70	18.20	5.8	8.3	10.8	13.3	15.8
M6	6.25	6.30	10.00	13.00	16.00	19.00	22.00	7.0	10.0	13.0	16.0	19.0
M7	7.25	7.30	11.00	14.50	18.00	21.50	25.00	8.0	11.5	15.0	18.5	22.0
M8	8.30	8.40	13.00	17.00	21.00	25.00	29.00	9.3	13.3	17.3	21.3	25.3
M10	10.50	10.50	16.00	21.00	26.00	31.00	36.00	11.5	16.5	21.5	26.5	31.5
M12	12.50	12.50	19.00	25.00	31.00	37.00	43.00	13.8	19.8	25.8	31.8	37.8
M14	14.50	14.50	22.00	29.00	36.00	43.00	50.00	16.0	23.0	30.0	37.0	44.0
M16	16.50	16.50	24.00	32.00	40.00	48.00	56.00	18.0	26.0	34.0	42.0	50.0
M18	18.75	18.75	28.00	37.00	46.00	55.00	64.00	20.5	29.5	38.5	47.5	56.5
M20	20.75	20.75	30.00	40.00	50.00	60.00	70.00	22.5	32.5	42.5	52.5	62.5
M22	22.75	22.75	32.00	43.00	54.00	65.00	76.00	24.5	35.5	46.5	57.5	68.5
M24	24.75	24.75	36.00	48.00	60.00	72.00	84.00	27.0	39.0	51.0	63.0	75.0

Special Tap Required

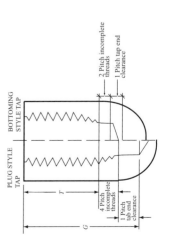

PLUG STYLE TAP
BOTTOMING STYLE TAP
2 Pitch incomplete threads
1 Pitch tap end clearance
4 Pitch incomplete threads
1 Pitch tab end clearance
T
G

7

WELDS
AND
WELDMENTS

Figures

7-1	Joint and Weld Types	360
7-2	Parts of a Weld	360
7-3	Basic Structure of an ANSI Weld Symbol	361
7-4	Basic Structure of an ISO Weld Symbol	362
7-5	Weldment Drawing Example	365
7-6	Gusset Best Practices	368

Tables

7-1	Weld Types and Their ANSI Symbols	362
7-2	Common ANSI Welding Callouts and Their Meaning	363
7-3	Common Materials for Weldments	366

One decision that must be made as part of the design process is whether or not a part or assembly should be made as a weldment (welded assembly). A weldment can replace a part that requires large amounts of material removal. Weldments are also often used to replace bolt-together assemblies in cases where the assembly need never be disassembled and where a weldment will be more cost effective than the assembly it replaces. In some cases, the cost of machining locating and fastening features into the parts of an assembly exceeds the cost of producing a weldment. However, weldments are labor intensive to produce, and that cost can sometimes exceed the cost of CNC machining of a single piece. When cost is a concern, the wise designer will quote it more than one way. It is important to note that a weldment will have lower fatigue strength than an equivalent single-piece component, so caution is warranted when an application involves dynamic loading.

It is important to note that the material contained in this section does not necessarily apply to structural weldments or pressure vessels, which are strictly governed by codes and standards. For typical machine design applications, commercial quality welding — rather than code quality welding — is expected. Consult the recommended resources for more information on code quality and commercial quality welds.

RECOMMENDED RESOURCES

- O. Blodgett, *Design of Weldments*, J.F. Lincoln Foundation, Cleveland, OH, 1963
- R. L. Norton, *Machine Design: An Integrated Approach*, 4th Ed., Prentice Hall, Upper Saddle River, NJ, 2011
- Oberg, Jones, Horton, Ryffel, *Machinery's Handbook*, 28th Ed., Industrial Press, New York, NY, 2008
- R. Parmely, *Standard Handbook of Fastening and Joining*, 3rd Ed., McGraw-Hill, New York, NY, 1997
- **American Welding Society** website: www.aws.org
- **ANSI/AWS 2.44:** "Standard Symbols for Welding, Brazing, Nondestructive Examination"
- **ISO 2553:** "Welded, Brazed and Soldered Joints-Symbolic Representation on Drawings"

WELD TYPES

There are five basic welded joint types: butt, tee, lap, corner, and edge. There are three basic weld types used to weld these joints: groove, fillet, and plug welds. These joints and welds are illustrated in Figure 7-1. The most common joint/weld combinations used in machine design are butt joints with groove welds and tee or corner joints with fillet welds. The parts and nomenclature of groove and fillet welds are shown in Figure 7-2.

A groove weld can fully or partially penetrate a joint. The weld size is equal to the throat dimension of the weld. For a fillet weld, the throat dimension is not equal to the weld size. Fillet weld size is given as the leg dimension, or both leg dimensions if they are not equal. For both weld types, the weld strength is limited by the throat dimension. For details on calculating weld strength, see the recommended resources.

The fusion zones shown in Figure 7-2 are volumes of material where the weld material commingles with the base material. Beyond the fusion zones lie the heat affected zones. A heat affected zone (HAZ) is a volume of base

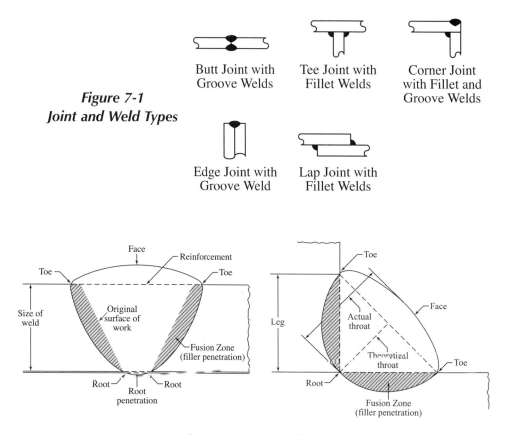

Figure 7-1
Joint and Weld Types

Butt Joint with Groove Welds

Tee Joint with Fillet Welds

Corner Joint with Fillet and Groove Welds

Edge Joint with Groove Weld

Lap Joint with Fillet Welds

Figure 7-2 Parts of a Weld

material around a weld that has had its microstructure altered by the heat of welding. Microstructural changes and tempering effects reduce the strength of the base metal in the HAZ. Because these zones will be weaker than the un-welded base material, they are vulnerable to cracking and other types of failure. Heat affected zones are particularly vulnerable to impact and fatigue failure. It is good practice to place welds at locations of minimal stresses and strains.

WELD SYMBOLS

American standard weld symbols are governed by ANSI/AWS 2.4, whereas the ISO standard for weld symbols is ISO 2553. Not all the information needs to be filled in on a weld symbol in a typical machine design application. It is common practice to specify only the weld type and size in non-critical applications. Figures 7-3 and 7-4 and Tables 7-1 and 7-2 illustrate ANSI and ISO weld symbols and typical callouts.

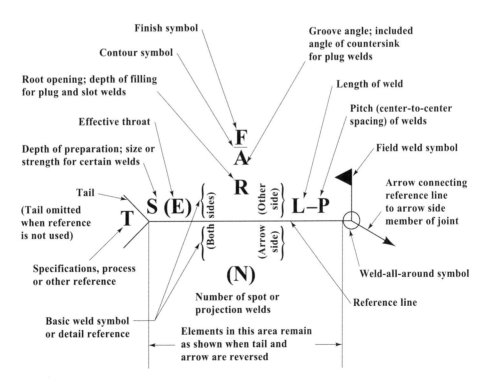

Figure 7-3 Basic Structure of an ANSI Weld Symbol

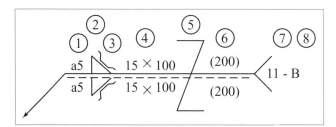

Information above reference line identifies weld on same side as symbolic representation.
Information below reference line identifies weld on opposite side to symbolic representation.

1. Dimension referring to cross section of weld
2. Weld Symbol
3. Supplementary symbol
4. Number of weld elements x length of weld element
5. Symbol for staggered intermittent weld
6. Distance between weld elements
7. Welding process reference
8. Welding class

Figure 7-4 Basic Structure of an ISO Weld Symbol

Table 7-1: Weld Types and Their ANSI Symbols

Groove Weld Symbols							
Square	Scarf[a]	V	Bevel	U	J	Flare V	Flare bevel

Other Weld Symbols							
Fillet	Plug or slot	Spot or projection	Seam	Back or backing	Surfacing	Flange	
						Edge	Corner

Weld all around	Field weld		Melt-thru	Backing or spacer material	Contour		
					Flush	Convex	Concave

[a]*This scarf symbol used for brazing only*

Table 7-2: Common ANSI Welding Callouts and Their Meaning

Desired Weld	Symbol	Symbol Meaning
		Symbol indicates two fillet welds, both with 1/2 inch leg dimensions.
		Symbol indicates a 1/2 inch fillet weld on *arrow side* of the joint and a 1/4 inch fillet weld on *far side* of the joint.
		Symbol indicates a V-groove weld with a groove angle of 65 degrees on the *arrow side* and 90 degrees on the *other side.*
		Symbol indicates plug welds of 1 inch diameter, a depth of filling of 1/2 inch, and a 60 degree angle of countersink spaced 6 inches apart on centers.
		Symbol indicates a 24 inch long fillet weld on the *arrow side* of the joint.
		Symbol indicates a series of intermittent fillet welds each 2 inches long and spaced 5 inches apart on centers directly opposite each other on both sides of the joint.
		Symbol indicates a series of intermittent fillet welds each 3 inches long and spaced 10 inches apart on centers. The centers of the welds on one side of the joint are displaced from those on the other.

continued on next page

Table 7-2: Common ANSI Welding Callouts and Their Meaning (Continued)

Desired Weld	Symbol	Symbol Meaning
		Symbol indicates a fillet weld around the perimeter of the member.
		Symbol indicates a fillet weld on the *other side* of the joint and a flare-bevel-groove weld and a fillet weld on the *arrow side* of the joint.
		Symbol indicates edge-flange weld on *arrow side* of joint and flare-V-groove weld on *other side* of joint.

WELDMENT DRAWINGS

Weldments are commonly drawn as either one or two drawings. When two drawings are used, one shows the part after welding, but before machining. The other drawing shows the weldment after machining. When a single drawing is used, it shows the weldment after machining, but provides all part structure and welding dimensions as well. Excess material that will be machined off is sometimes shown using phantom lines.

When drawing a weldment as a single drawing, it is important to think of it as two drawings with two sets of dimensions: as welded and as machined. It is helpful to first dimension the drawing as welded, using none of the machined surfaces for reference. Then dimension all the machined surfaces relative to one another. Finally, relate one "as welded" surface to one machined surface in each direction. See Figure 7-5 for illustration of this method. The figure shows a weldment that is machined on the top and bottom surface. In the figure, the 0.40 dimension links a machined surface to an unmachined weldment surface. The reference dimensions in parentheses specify the raw stock pieces used to create the weldment. The finished drawing should have everything the welder needs to cut and locate the pieces and weld them together without the benefit

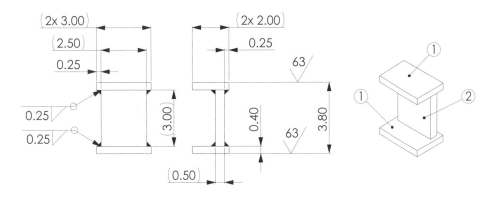

ITEM	DESCRIPTION	QTY
1	AISI A36 PLATE; 3" X 2" X 0.5" THK	2
2	AISI A36 PLATE; 3" X 2.5" X 0.5" THK	1

Figure 7-5 Weldment Drawing Example

of any of the machined surfaces. Machinists should have everything they need to reference the weldment to place their machined surfaces. Machining datums on weldments should normally be machined surfaces.

Weldment drawings normally contain a table of individual pieces used to construct the weldment. The table should include the dimensions, stock shape, and material for each piece. It is also common to dimension each individual piece on the drawing using reference dimensions wherever the dimension is the same as that given in the table. In cases where some stock is machined off, the reference dimension is not needed, and the stock list should account for that excess material or machining allowance. Figure 7-5 shows an example of this, where Item 1 calls for 0.5" thick stock yet the drawing shows the machined thickness to be 0.4 inches for both top and bottom plates. Section 8.1 of this book contains a list of some standard stock sizes.

MATERIALS AND TREATMENTS

Low carbon steel is the easiest of metals to weld. Low alloy steels and Aluminum are also commonly used in weldments. Table 7-3 is a table of commonly used materials for weldments. A proper steel weld can theoretically be as strong as the base material, but the fatigue and impact strength of a welded joint is less than expected from a single piece. This is due to residual stresses and the effects of heat on the base material in the HAZ as well as stress concentrations due to weld geometry. Aluminum welds tend to be weaker

Table 7-3: Common Materials for Weldments

Material	Description
ASTM A36	Steel, Structural, Hot Rolled
AISI 1018 HR	Steel, Hot Rolled, Low Carbon
AISI 1020 HR	Steel, Hot Rolled, Low Carbon
AISI 1026 HR	Steel, Hot Rolled, Low Carbon, Mechanical Tubing and Pipe
6061-T6	Aluminum, Structural, Solution Heat Treated, and Artificially Aged

than the base material. Careful analysis should be undertaken in cases of fatigue loading or high static loading. The recommended resources should be consulted for assistance in analyzing welded joints under static and dynamic conditions. Appropriate factors of safety should always be applied.

It is good practice to stress relieve weldments to improve strength and reduce distortion during later machining steps. Weldment drawings normally contain the note: "Stress relieve after welding" or "Stress relieve before machining." Weldments are normally coated after machining to prevent oxidation. Some good weldment coatings are black oxide, paint, or electroless nickel plate. When a coating with buildup is specified, be sure to indicate holes and surfaces to be masked during coating.

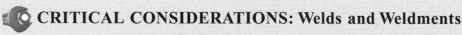

CRITICAL CONSIDERATIONS: Welds and Weldments

- Welding is a very large and specialized topic. Consult a welding expert for all critical applications.

- Weldments under high static loads should be analyzed thoroughly. Consult the recommended references for stress analysis assistance.

- The fatigue and impact strengths of weldments can be unpredictable and are normally less than that of a single piece component. When dynamic loading is expected, consider a machined part rather than a weldment. If a weldment is necessary under these conditions, consult the recommended references for fatigue calculation assistance.

- Over-welding can be as detrimental to weldment performance as under-welding. Size welds proportionally to the thinnest part being welded. A rule of thumb for fillet welds is to make the leg length of the weld equal to 3/4 of the thickness of the thinnest section in the weld. This results in a full-strength weld. Smaller welds can be used where full plate strength is not needed.

- Plan to machine all functional surfaces of a weldment and include enough material allowance to account for distortion that may have occurred during welding.

BEST PRACTICES: Welds and Weldments

- Design weldments such that welds are not subjected to bending loads.
- Grind butt welds flush with part surfaces to maximize fatigue strength.
- For best strength, taper mismatched parts so that their cross sections match at the joint. The AWS recommended angle of taper is 22° or less.
- To minimize distortion during welding, place welds symmetrically about the neutral axis of the part, and place welds on both sides of a joint. Weld together parts of similar thickness to minimize distortion.
- In the United States, it is common to use inch stock sizes even for metric weldments.
- Consider intermittent welds for cost and weight reduction, but with due consideration to the fact that intermittent welds produce stress concentrations that can lower fatigue strength.
- It is common practice for welders to drill a hole in capped tubing to release hot gases produced during welding. If the size and location of this hole matters, it is good practice to specify it on the drawing.
- To reduce cost or increase strength, consider machining your part or assembly from one piece rather than creating a weldment. Have weldments quoted as machined parts when possible to determine which is more cost effective.
- It is good practice to remove the internal corners of gussets to assist in assembly, and provide right angle outside corners to improve weldability and maximize the throat (and the strength) of the fillet weld. See Figure 7-6 for an illustration.

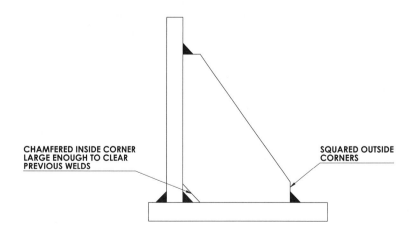

CHAMFERED INSIDE CORNER
LARGE ENOUGH TO CLEAR
PREVIOUS WELDS

SQUARED OUTSIDE
CORNERS

Figure 7-6: Gusset Best Practices

8

MATERIALS, SURFACES, AND TREATMENTS

Contents

8.1	MATERIALS	371
8.2	SURFACE FINISH	391
8.3	HEAT TREATMENT	398
8.4	SURFACE TREATMENT	407

Tables

8-1	Carbon and Alloy Steel Designations	372
8-2	Tool Steel Designations	372
8-3	Aluminum Designations	373
8-4	Common Material Choices for Light Industrial Machinery	377
8-5	Tool Steel Selector	383
8-6	Properties of Select Carbon and Alloy Steels	384
8-7	General Properties of Carbon and Alloy Steels	385
8-8	Properties of Select Tool Steels	385
8-9	Properties of Select Stainless Steels and Titanium Alloys	386
8-10	General Properties of Stainless Steels and Titanium Alloys	387
8-11	Properties of Select Aluminum Alloys	387
8-12	Properties of Select Non-Metals	388
8-13	Common Sheet Metal Thicknesses, Inch Series	389
8-14	Common Metal Plate Thicknesses, Inch Series	389
8-15	ASME Surface Texture Symbols & Construction	392
8-16	ASME Surface Finish Examples	392
8-17	ASME Lay Symbols	393
8-18	ISO Surface Texture Symbols	394
8-19	Common Surface Finish Values & Descriptions	395
8-20	Normal Maximum Surface Roughness of Common Machined Parts	396
8-21	Manufacturing Process Average Surface Finish	397
8-22	Hardness Scale Conversions	400
8-23	Typical Hardness Values	401
8-24	Heat Treatment Properties of Common Steels	404
8-25	Case Hardening Properties	405
8-26	Surface Treatment Selector	410
8-27	Surface Treatment Application Properties	411

Section

8.1

MATERIALS

A huge variety of materials are commercially available, and it's practically impossible to consider them all when working on a design. A typical machine design will draw material choices from the following: carbon and alloy steels, tool steels, stainless steels, aluminum alloys, and non-metals. Carbon and alloy steels are good for general use, whereas tool steels have elevated wear, hardness, hardenability, and/or toughness properties. Tool steels are normally used in a hardened state. Aluminum and its alloys have reduced weight and improved machinability compared to steel, but they often (but not always) have lower strength than steel and always have lower hardness. Non-metals can have unique properties like low coefficient of friction, low thermal conductivity, or innate lubricity. Material choice is critical to successful design.

RECOMMENDED RESOURCES

- Beer, Johnston, *Mechanics of Materials*, McGraw-Hill Book Company, New York, NY, 1981
- Juvinall, Marshek, *Fundamentals of Machine Component Design*, 2nd Ed., John Wiley & Sons, Inc., New York, NY, 1991
- R. L. Norton, *Machine Design: An Integrated Approach*, 4th Ed., Prentice Hall, Upper Saddle River, NJ, 2011
- Oberg, Jones, Horton, Ryffel, *Machinery's Handbook*, 28th Ed., Industrial Press, New York, NY, 2008
- **Matweb: Material Property Data**: www.matweb.com
- **eFunda: Materials**: www.efunda.com/materials

METALS NOMENCLATURE

Metals are subject to several numbering systems. The most widely used designations for carbon and alloy steels are the AISI/SAE codes, which are given in Table 8-1. Tool steels are identified using letter designations, the most common of which are shown in Table 8-2. Aluminum is designated by The Aluminum Association, and that numbering system is illustrated in Table 8-3.

When numbering carbon and alloy steels, the last two digits of the designation indicate the approximate mean carbon content of the material. Therefore, a mean carbon content of 0.30 – 0.37 will include steels such as 1330, 3135, and 4137.

Table 8-1: Carbon and Alloy Steel Designations

SAE Designation	Description
1XXX	Plain Carbon Steels
2XXX	Nickel Steels
3XXX	Nickel-Chromium Steels
4XXX	Molybdenum Steels
5XXX	Chromium Steels
6XXX	Chromium-Vanadium Steels
7XXX	Heat Resisting Casting Alloys (Tungsten Steels)
8XXX	Nickel-Chromium-Molybdenum Steels
9XXX	Silicon-Manganese Steels

Table 8-2: Tool Steel Designations

	Designation	Description
COLD WORK	A	Air Hardening
	D	Oil or Air Hardening (High Carbon, High Chromium)
	O	Oil Hardening
HIGH SPEED	M	High Speed (Molybdenum)
	T	High Speed (Tungsten)
HOT WORK	H	Hot Work (Chromium, Tungsten, Molybdenum)
SHOCK RESISTING	S	Shock Resisting
WATER HARDENING	W	Water Hardening
SPECIAL PURPOSE	F	Special Purpose (Carbon-Tungsten)
	L	Special Purpose (Low Alloy)
MOLD STEEL	P	Mold Making

MATERIAL PROPERTIES

Materials have many properties, some of which are commonly used during the design process to analyze parts and assemblies. Material data is statistical in nature, and published material data will vary. Consider whether the reported values correspond to an average value, a minimum, or a single test result. Understand that the tested specimens have a shape, size, and processing history that may differ from your application and may result in different properties. Please refer to the recommended resources for additional information and calculation methods. Always use appropriate factors of safety when performing material analysis.

Table 8-3: Aluminum Designations

	Alloy Series	Principal Alloying Element
WROUGHT ALLOYS	1xxx	99% Pure Aluminum
	2xxx	Copper
	3xxx	Manganese
	4xxx	Silicon
	5xxx	Magnesium
	6xxx	Magnesium & Silicon
	7xxx	Zinc
	8xxx	Other
CAST ALLOYS	1xx.x	99% Pure Aluminum
	2xx.x	Copper
	3xx.x	Silicon & Copper and/or Magnesium
	4xx.x	Silicon
	5xx.x	Magnesium
	6xx.x	Unused Series
	7xx.x	Zinc
	8xx.x	Tin
	9xx.x	Other
	Designation	**Meaning**
TEMPER DESIGNATION	-F	As Fabricated
	-0	Annealed
	-H	Strain Hardened
	-W	Solution Heat Treated
	-T	Thermally Treated

The following properties are of particular interest when selecting materials during the design process:

Carbon content is of concern with steels. Low carbon steels (less than 0.3%) are easy to work but have low strength and wear resistance. They can be case hardened in many cases to improve wear. Medium carbon steels (0.3 to 0.5%) have much higher strength and moderate hardness. High carbon steels (0.5% - 0.9%) have good wear resistance and high strength. Medium and high carbon steels can be further hardened through heat treating.

Machinability is the ease with which a material can be machined to a given surface finish. Machinability is often expressed as a percentage. The wise engineer will save cost by selecting materials with better machinability when possible. The steels with best machinability tend to have medium carbon content.

Mass density (ρ) is a measure of mass per unit area. It can be used to calculate the mass of a part using the formula:

$$Mass = density \times volume$$

Standard geometric techniques can be used to estimate the volume of a part. Many CAD packages will output volume, and even mass, given a density for the part.

Tensile yield strength is the stress at which a specimen of a given material begins to elongate significantly under tensile loading. The elastic limit of the material is exceeded slightly before the yield strength, so some deformation has already become permanent at this stress level. This stress value is often used in design analysis as the upper limit in evaluating whether a part will yield under tensile or bending loads. A factor of safety should always be used to account for uncertainty in the properties data, and to target a stress limit that is below the elastic limit. The tensile yield strength should not be used for evaluating shear loading. Recall that tensile stress is axial load per unit cross sectional area. For ductile metals, tensile yield strength is considered to be equal to compressive strength.

Compressive strength is of particular interest for plastics because, in machinery, plastics are often used in compression rather than in tension or bending. When compressive strength is exceeded, the material will begin to deform under the load. This value should not be used for shear, tension, or bending. It is also not a measure of hardness.

Ultimate tensile strength is the stress at which a tensile specimen begins to significantly neck down in preparation for failure. In practice, it is considered the maximum stress that a material can withstand without failure. For ductile materials, yielding will occur before failure. This value is used in evaluating not only direct failure due to loading, but also fatigue failure.

Shear yield strength is the shear stress that a specimen of material can withstand without significant deformation. For ductile metals, shear yield strength is often estimated to be 50% of tensile yield strength.

Ultimate shear strength is the shear stress that a specimen of material can withstand without failure. For ductile metals, shear ultimate strength is often estimated to be 75% of tensile ultimate strength.

Modulus of Elasticity (E) is an indicator of the stiffness of the material or its resistance to deformation under tensile loading. It defines the relationship between stress and strain for a given material using the relationship: E = stress/strain. This value is also used in beam bending equations to calculate beam deflection. Some of these equations are provided in Section 1.6.

Modulus of Rigidity (G) is an indicator of the stiffness of the material in shear. It is equal to the ratio of shearing stress to shear strain.

Ductility is a measure of the ability of a material to deform without fracture. Ductility matters when a part is to be subjected to shock loads, or when fracture must be avoided under unknown or significant loading conditions. A material that is not ductile is brittle. Often the factor of safety used with brittle materials is twice that used with ductile materials. Ductile materials are more often used in machinery than brittle materials.

Hardness is a measure of how resistant a material is to permanent deformation when a small round or conical shape is pressed forcibly into the surface. High hardness enhances a material's resistance to surface wear, scratches, or dings. Insufficient hardness can result in problems like bent cutting edges, indented die surfaces, or mushroomed punch faces. Hardness is linked to tensile strength, with harder materials having higher (but not linearly higher) strength. Hard materials tend to be less ductile and more prone to fracture. Carbides are very hard, but also brittle. One method of achieving a hard surface while maintaining a ductile core is to heat treat just the part surface. This method, called flame or induction hardening, is discussed in Section 8.3.

Corrosion resistance must be considered for every part. Oxidation of carbon and alloy steels is a major issue in machinery design. For steels, corrosion resistance is often achieved through the application of surface treatments. Coatings and surface treatments are discussed in Section 8.4. Stainless steels, titanium alloys, and aluminum alloys are resistant to most corrosion conditions. Galvanic corrosion is still an issue for aluminum, carbon steels, and alloy steels.

Coefficient of linear thermal expansion (α) is used to calculate the effects of a change in temperature on part dimensions. Thermal expansion is a function of temperature, material, and part geometry. When heat is a factor, calculate thermal expansion as follows:

$$\Delta L = \alpha L0(T - T0)$$

where L0 is the original length dimension and T–T0 is the change in temperature experienced by the part. T0 is typically room temperature (68°F or 20°C.) Specific material composition and temperature range affects thermal expansion properties, so be sure to check manufacturer's data for exact thermal expansion coefficients in critical applications.

Thermal conductivity is sometimes desirable and sometimes not. Metals are thermally conductive to some extent. Aluminum and copper are highly thermally conductive. Some plastics and ceramics are thermally insulative.

COMMON MATERIAL CHOICES

Choice of material depends on many things. Material properties are important, as are material availability and cost. There are countless materials to choose from. Table 8-4 shows some common choices of material for light industrial machinery, including fixtures and assembly machines. The properties of these and other material choices are detailed in Tables 8-6 through 8-12.

Non-Metals

Non-metals are quite common in machinery. They are often chosen to provide insulation, shock absorption, sacrificial wear, and a soft touch when contacting delicate parts. Many plastics are machinable, and are available in a variety of stock shapes. Some common machinable plastics used in machinery are listed with their properties in Table 8-12. Many plastics are sensitive to heat, but materials like PEEK can tolerate high temperatures. Many plastics can be considered to be soft, and therefore prone to scratches and scrapes. Acetals, like Delrin®[1], are a good choice for items like non-marring parts racks, and parts that are not highly stressed and don't see significant wear. PTFE is often chosen for its non-stick properties and chemical compatibility, though it is softer than most plastics.

In product design, where many copies will be manufactured, cast or molded plastic parts are often a good choice. With moldable plastics, a number of formulations are commercially available. Plastics that that are molded using heat to melt bulk material are called thermoplastics. Thermoplastics will re-melt when heat is again applied and the melting temperature is reached. Thermoset plastics can also be molded, and the plastic is created from component ingredients during the molding process. Thermoset plastics, once cured

---

[1] *Delrin® is a registered trademark of the DuPont Company.*

Table 8-4: Common Material Choices for Light Industrial Machinery

Material		Characteristic	Shafts	Weld-ments	Aluminum Machine Parts	Stainless Machine Parts	Assembly Tooling	Industrial Sheet Metal	Plastic Machine Parts
						Applications			
CARBON & ALLOY STEELS	ASTM A36	Structural, weldable		●					
	AISI/SAE 1018	Hardenable, weldable		●					
	AISI/SAE 1020	General use, weldable		●					
	AISI/SAE 1141	Deeper hardening	●						
	AISI/SAE 4140	General use, hardenable	●						
TOOL STEELS	A2	General use					●		
	A6	Wear resistant					●		
	M2	High speed steel					●		
	D2	Very wear resistant					●		
STAINLESS STEELS	303	Round shaft stock	●			●			
	304	General use		●		●	●	●	
	316	Extra corrosion resistant				●	●		
ALUMINUM ALLOYS	1100-H14	Sheet metal, weldable						●	
	2017-T4	Screw machine stock			●				
	6061-T6	High strength, weldable		●	●		●		
	440C	High strength, hardenable	●						
PLASTICS	UHMW	Wear resistant					●		●
	PEEK	High temperature					●		●
	Polycarbonate	Transparent							●
	Delrin®	General use							●

Delrin® is a registered trademark of the DuPont Company.

the first time, cannot be re-melted. When reheated, thermoset plastics may soften but will not flow or change shape significantly.

Plastics are not the only non-metals used in machine design. Ceramics and composites are becoming more and more common. Some ceramics are extremely hard, can tolerate extremely high temperatures, and are commonly used as thermal insulators. Many formulations are available. One commonly used ceramic used for very high temperature exposure is Machinable Bisque Alumina. Machinable, wear resistant, toughened Zirconia ceramics are used in high-strength applications with high temperatures.

Stainless Steels

Stainless steels are alloy steels that contain significant amounts of chromium. The three types of stainless steels are austenitic, ferritic, and martensitic. The austenitic grades are nonmagnetic as annealed, and can be hardened by cold working, but not by heat treatment. Austenitic stainless steels are more corrosion resistant than the other types. Ferritic grades are magnetic and nickel-free; like austenitic grades, they can be hardened by cold working, but not by heat treatment. Martensitic grades are also magnetic, but these grades can be hardened by heat treatment.

Stainless steels are often chosen for their corrosion resistance. Some grades are biologically compatible and appropriate for use on medical or food processing machinery. Stainless steel parts do not have to be surface treated to protect them from oxidation. These steels are not as hard or strong as carbon and alloy steels, and most are prone to scratches and dings. Stainless steels in the 300 (Austenitic) series are generally used, but many other grades are available. Further hardening through heat treatment is not an option with 300 series steels, but some grades like 440C and 17-7PH are heat treatable and can attain high hardness and strength. Properties of some common stainless steels are given in Tables 8-9 and 8-10.

The most commonly encountered grades of stainless steel in light industrial machinery are 301, 302, 303, 304, 316, 440C, and 17-7PH. 301 stainless is strong and ductile, making it suitable for structural applications. 303 is similar to 302, except it is much more readily machined. 304 is also similar to 302, except it has lower carbon content for improved weldability. 316 has superior corrosion resistance. 440C is heat treatable, and yields the highest hardness of any hardenable stainless steel. 17-7PH is a precipitation hardening grade of stainless that provides high strength and fatigue resistance.

Aluminum

Aluminum is lightweight, resists oxidation, and is very easy to machine. These reasons are the three most common for choosing it as a material. Prototype parts are often made in aluminum to keep cost and delivery time down. Aluminum is corrosion resistant and does not need to be coated to protect against oxidation, except in the presence of alkalis. Aluminum is generally considered to be a soft material. In many applications, aluminum parts are anodized to protect the surfaces against nicks and scratches, and prevent galvanic corrosion in the presence of dissimilar metals. Some heat treated aluminum alloys have high strength that can approach that of some steels. The properties of some commonly used aluminum alloys are listed in Table 8-11.

When using aluminum in assemblies, galvanic corrosion is a concern. Many steels, including those used for fasteners, can cause galvanic corrosion when put in contact with aluminum alloys in the presence of an electrolyte. Surface treatments such as coatings — and insulators such as tape, helical coil inserts, and other protective measures — are commonly used to prevent this from occurring.

Of the aluminum alloys, two commonly used for machine parts are 6061-T6 and 2017-T4. 6061-T6 is a strong, versatile alloy that has been heat treated and is weldable. 2017-T4 has also been heat treated, is good for general use, and is particularly suited for screw machine fabrication. 2000 series aluminums are less corrosion resistant than the other alloys.

Titanium

Titanium is not commonly used in machine design, but its special properties allow it to solve some types of special problems. Titanium is almost as light as aluminum, but stronger than some steels. It is a very stiff material, and some grades are biologically compatible. Titanium is capable of withstanding much higher temperatures than most other metals, and has a low thermal expansion coefficient.

The most common grades of titanium and their properties are given in Tables 8-9 and 8-10. Grade 2 titanium is an unalloyed titanium that has good strength, formability, and weldabililty. It generally has the lowest cost. Grade 5 titanium is an alloy that has excellent tensile strength but poor shear strength. It is heat treatable and is widely used. Grade 9 titanium is an alloy whose properties are a compromise between ease of machining and strength/weldability.

Carbon and Alloy Steels

Carbon and alloy steels are often chosen for their strength and durability. A broad range of properties are available through the use of alloys and/or heat treatment. Some common choices and their properties are listed in Tables 8-6 and 8-7. Carbon and alloy steels must be protected from corrosion, including oxidation from exposure to the atmosphere. Surface treatments that protect against corrosion are detailed in Section 8.4. Carbon steels have less than 1.65% manganese, less than 0.6% silicon, less than 0.6% copper, and no other significant alloying elements. They are typically used without additional heat treatment, but can be heat treated in a limited way to alter properties. Alloy steels have various alloying elements to enhance their material properties.

Very low carbon steels (SAE numbers up to 1015) are a good choice if ductility is a priority. Applications like cold forming or welding are readily accomplished with these lowest carbon steels. These steels are not easily machined. Low carbon steels with SAE numbers from 1016 and 1030 have increased machinability, strength, and hardness due to increased carbon content. This comes at the expense of ductility. These steels can be case hardened and readily welded. Medium carbon steels (SAE numbers 1030 through 1052) have higher strength and hardenability than low carbon steels. High carbon steels (SAE 1055 through 1095) have sufficient carbon content to hold a good edge. They are generally heat treated for use, and have good wear resistance when hardened.

Alloy steels are a good choice if higher strength, hardenability, or machinability is desired. Alloy steels in the SAE 1100 range have superior machinability. Many alloy steels are directly hardenable through heat treatment. Alloy steels with medium hardenability include SAE numbers 2330, 2340, 2345, 3130, 3135, 3140, 3141, 3145, 3150, 4053, 4063, 4137, 4140, 4145, 4640, 5135, 5147, and 5150.High hardenability grades include 4340 and 4150.

When choosing steel, one must decide between hot rolled steel or cold rolled (cold drawn). One factor will be the size and shape of the desired stock. Hot rolled shapes can be made in much larger sizes than cold rolled shapes. Many special stock shapes such as structural tubing, channels, and I-beams are hot rolled. Hot rolled rounds, plates, bars, flats, and strips are all readily available and an economical choice. Cold rolled rounds, bars, flats, and strips are readily available, but plates tend to be hot rolled only.

Another consideration is surface quality. Cold rolled steel has a high quality surface, and its stock shapes are generally held to close tolerances. Hot rolled shapes, on the other hand, are covered in an oxide layer which must be machined off if a high quality surface is needed. Hot rolled stock tolerances are looser than those for cold rolled.

The type of planned fabrication also matters when selecting between hot and cold rolled material. Cold rolled steel has residual stresses that can cause distortion during heavy machining, welding, or heat treating. Hot rolled steel is a good choice for these applications.

The material properties of hot and cold rolled steels are different. Steel that has been cold rolled is less ductile than hot rolled steel. Hot rolled steel tends to have lower strength and hardness than cold rolled steel, but can be heat treated to improve performance. If heat treating is planned, hot rolled material (or tool steel) is a good choice. If high strength is needed without heat treating, cold rolled is the way to go.

Tool Steels

Tool steels are fairly common in machinery, tools, and fixtures. They are used in applications where special properties like shock resistance, wear resistance, heat resistance, or superior hardenability are required. Tool steels are generally used in a hardened state achieved through heat treatment. The achievable hardness ranges for various tool steels are provided in Section 8.3. Tool steels are designated (as shown in Table 8-2) based on whether they are cold work, hot work, high speed, or shock resisting grades. These terms are defined as follows:

Cold work tool steels are suitable for ambient temperature applications. Some grades are highly wear resistant. These steels are often used for tools and machine parts. High-wear items such as slides and slideways are often made from cold work tool steels. When the heat of friction is expected to be significant, high-speed tool steels are recommended instead.

Hot work tool steels are suitable for high-temperature applications. They maintain their strength at very high temperatures. These steels are used in hot tools like casting dies.

High-speed tool steels are designated M (Molybdenum) or T (Tungsten). They are designed for use as high-speed cutting tools and are able to tolerate moderately elevated temperatures without losing strength.

Shock resisting tool steels have high toughness. Tougher steels have more shock and impact resistance. Tough materials are not brittle. These steels resist breakage, including chipping and cracking. Heat treatment can negatively affect shock resistance.

Wear resistance is resistance to erosion and abrasion. Wear resistance requires a high content of hard alloy carbides. Hardness will also improve wear resistance, but is not the primary factor. Parts that move against one another or parts encountering moving material should be wear resistant.

Surface treatments like nitriding or titanium nitride coating can be applied to tool steels, either without hardening or after hardening. These surface treatments can improve wear resistance or reduce friction. See Section 8.4 for more information on surface treatments.

Tool steels are very notch sensitive. Any sharp radii, notches, undercuts, or sudden changes in cross section can concentrate applied stresses and cause breakage at much lower than expected stress levels. Additional hardening processes can make notch sensitivity worse.

Table 8-5 can be helpful when comparing a few common tool steels and their properties. There are many more tool steels available. Some specific tool steel material properties are given in Table 8-8. Hardness ranges available through heat treatment are given in Section 8.3.

If tool steels are not wear resistant enough for an application, specialty steels and carbides are the next steps. Specialty steels are widely available from their manufacturers and have an array of properties. Surprising improvements can be had through special processing. Carbide parts are extremely wear resistant, but relatively brittle; therefore, care must be taken to support carbide parts as well as possible and keep stresses off of them. In highly abrasive environments, even carbides will wear. Inspection and replacement must be done in a timely manner for dimensionally critical applications. The two most commonly used carbides are tungsten carbide and titanium carbide. These carbides are highly wear resistant, hold an edge extremely well, and are commonly used on cutting tools.

MATERIALS DATA

Tables 8-6 through 8-12 list some commonly used materials, some of their properties, and some potential applications. Recall that for ductile metals, yield strength in shear is considered to be equal to 50% of their tensile yield

Table 8-5: Tool Steel Selector

TOOL STEEL DESIGNATION		WEAR RESISTANCE	SHOCK RESISTANCE (TOUGHNESS)	RESISTANCE TO DEFORMATION	RESISTANCE TO HEAT SOFTENING	SAFETY IN HARDENING	DEEPTH OF HARDENING	MACHINABILITY
COLD WORK AIR HARDENING	A2	•	•	•	●	●	●	•
	A6	○	•	•	○	●	●	○
COLD WORK OIL OR AIR HARDENING	D2	●	•	•	●	●	●	•
	D7	●	⊘	•	●	●	●	•
COLD WORK OIL HARDENING	O1	○	•	⊘	○	○	○	○
	O6	○	•	⊘	•	○	○	●
HOT WORK	H11	•	○	○	○	○	●	○
	H12	•	○	○	○	○	●	○
	H13	•	○	○	○	○	●	○
	H21	•	○	○	○	○	●	○
LOW ALLOY	L6	•	•	⊘	•	○	○	●
SHOCK RESISTING	S1	•	○	•	•	○	○	•
	S5	•	●	•	W ⊘ (a)	W ⊘ (a)	○	•
HIGH SPEED (MOLYBDENUM)	M1	○	⊘	○	○	•	●	●
	M2	○	⊘	○	○	•	●	○
	M3	●	⊘	○	○	•	●	•
	M4	●	⊘	○	○	•	●	⊘
HIGH SPEED (TUNGSTEN)	T1	○	⊘	○	○	○	●	●
	T5	○	⊘	●	○	•	●	○

(a) W indicates water quench

⊘ = Poor

• = Fair

○ = Good

● = Very Good

strength. Also, ultimate shear strength is considered to be equal to 75% of ultimate tensile strength. The equations are:

$$S_{y,shear} = 0.577 S_y$$

$$S_{u,shear} = 0.75 S_u$$

Some stock material thicknesses are provided in Tables 8-13 and 8-14. Other material stock sizes are commonly available, including rounds, flats, and bars. Metric stock is also available, though less commonly used. Check with your materials supplier for more information.

Table 8-6: Properties of Select Carbon and Alloy Steels

Designation	Characteristics and Use	Tensile Yield Strength		Tensile Ultimate Strength		Thermal Expansion Coefficient (32-572°F) $10^{-6}/°F$	(0-300°C) $10^{-6}/°C$	Brinell Hardness
		ksi	MPa	ksi	MPa			
ASTM A36	Structural, weldments	36.3	250	50	400	6.5	11.7	149
AISI/SAE 1018 HR	General Use	31.9	220	67	462	6.78	12.2	137
AISI/SAE 1018 CR		70	483	85	586	6.78	12.2	167
AISI/SAE 1020 HR	General Use	29.7	205	55	379	6.5	11.7	111
AISI/SAE 1020 CR		57	393	61	420	7.11	12.8	122
AISI/SAE 1026 HR	Structural Tubing	34.8	240	63.8	440	7.11	12.8	126
AISI/SAE 1045 HR	Machinery Parts	45	310	82	565	6.61	11.9	163
AISI/SAE 1045 CR		77	531	91	627	7.22	13	179
AISI/SAE 1141 HR	Shafts	52.2	360	94.3	650	6.78	12.2	187
AISI/SAE 1141 CR		87.7	605	105	725	6.78	12.2	212
AISI/SAE 1215 CD	Screw Machine Parts	60.2	415	87	600	6.78	12.2	187
A SI/SAE 4140 Annealed	High Strength	60.2	415	95	655	7.61	13.7	197
AISI/SAE 4140 Normalized		95	655	148	1020	7.61	13.7	302
AISI/SAE 5160 Annealed	Gears, Springs	39.9	275	105	724	7.39	13.3	197

Note: A wide range of material properties can be had with proper heat treatment of these materials.

Table 8-7: General Properties of Carbon and Alloy Steels

Designation	Density		Modulus of Elasticity (E)		Modulus of Rigidity (G)		Poisson's Ratio (ν)
	lb/in³	g/cc	ksi	GPa	ksi	GPa	
Carbon & Alloy Steels	0.284	7.87	30000	206.8	11700	80.8	0.28

Table 8-8: Properties of Select Tool Steels

Designation		Characteristics and Use	Density		Modulus of Elasticity (E)		Modulus of Rigidity (G)		Thermal Expansion Coefficient	
			lb/in³	g/cc	ksi	GPa	ksi	GPa	(32-572°F) 10^{-6}/°F	(0-300°C) 10^{-6}/°C
Air Hardening	A2	General Use, Moderate Wear	0.284	7.86	30000	207	11700	80.8	6.67	12
	A6	General Use	0.29	8.03					6.89	12.4
Oil or Air Hardening	D2	Wear, Heat Treatment	0.278	7.7					6.56	11.8
	D7	Wear Resistance, Cold Work	0.278	7.7					6.78	12.2
Oil Hardening	O1	General Use, Bushings	0.283	7.83					6	10.8
	O6	Free Machining, Anti-Friction	0.277	7.67					6.2	11.2
Hot Work	H11	Hot Work, Punches & Dies	0.282	7.8					6.9	12.4
	H12	Hot Work, Deep Hardening	0.278	7.7					6.5	11.7
	H13	Hot Work, Shock, Dies	0.282	7.8					6.39	11.5
	H21	Hot Work, Punches & Dies	0.296	8.19					6.1	11
Low Alloy	L6	General Use, Shock, Cutters	0.284	7.86					6.56	11.8
Shock Resisting	S1	Shock Resistance	0.285	7.89					7	12.6
	S5	Shock Resistance	0.28	7.75					7	12.6
High Speed (Molybdenum)	M1	Hot Work, High Speed Steel	0.285	7.89					5.6	10.1
	M2	Most Widely Used High Speed	0.295	8.16					5.56	10
	M3	High Wear, Cutting Tools	0.295	8.16					6.39	11.5
	M4	Wear, Punches & Cutters	0.288	7.97					5.39	9.7
High Speed (Tungsten)	T1	Hot Work, Wear, Knives	0.313	8.67					5.39	9.7
	T5	Wear, Hot Work	0.316	8.75					6.22	11.2

Note: Tool steels are generally used in a hardened state. Wide ranges of hardness, yield strength, and ultimate strength result from various heat treatments.

Table 8-9: Properties of Select Stainless Steels and Titanium Alloys

Designation	Characteristics and Use	Tensile Yield Strength		Tensile Ultimate Strength		Thermal Expansion Coefficient		Hardness
		ksi	MPa	ksi	MPa	(32-572°F) 10^{-6}/°F	(0-300°C) 10^{-6}/°C	
STAINLESS STEELS 303 CR	General Use	60.2	415	100	690	9.56	17.2	19 Rc
303 Annealed	General Use	34.8	240	89.9	620	10.1	18.2	160 Brinell
304 CR	General Use, Food Processing	160	1103	185	1276	9.56	17.2	40 Rc
304 Annealed	General Use, Food Processing	31.2	215	85	586	9.61	17.3	80 Brinell
316 Annealed	High Corrosion Resistance, Food Processing	34.8	240	90	621	8.89	16	85 Brinell
440C Annealed	Heat Treatable, High Strength	65	448	110	758	5.67	10.2	19 Rc
TITANIUM ALLOYS Ti-6Al-4V (Grade 5) Annealed	General Use, Heat Treatable, Poor Shear Strength	128	880	138	950	5.11	9.2	36 Rc
Titanium Grade 2	Lowest Cost, Easiest Machining, Weldable, Food Processing, Fluids Handling	39.9	275	49.9	344	5.11	9.2	138 Brinell
Ti-3Al-2.5V (Grade 9)	Compromise Between Strength & Ease of Machining & Welding	72.5	500	89.9	620	5.48	9.86	24 Rc

Note: A wide variety of material properties are available for heat treatable materials, but embrittlement will occur with hardening.

Table 8-10: General Properties of Stainless Steels and Titanium Alloys

Designation	Density (lb/in³)	Density (g/cc)	Modulus of Elasticity (E) (ksi)	Modulus of Elasticity (E) (GPa)	Modulus of Rigidity (G) (ksi)	Modulus of Rigidity (G) (GPa)	Poisson's Ratio (v)
Stainless Steels	0.289	8	28000	193	10700	74.1	0.28
Ti-6Al-4V (Grade 5) Annealed	0.16	4.43	16500	113.8	6380	44	0.34
Titanium Grade 2	0.163	4.51	15200	105	6530	45	0.37
Ti-3Al-2.5V (Grade 9)	0.162	4.48	14500	100	6380	44	0.3

(TITANIUM ALLOYS spans the three titanium rows above.)

Table 8-11: Properties of Select Aluminum Alloys

Designation	Characteristics and Use	Tensile Yield Strength (ksi)	Tensile Yield Strength (MPa)	Tensile Ultimate Strength (ksi)	Tensile Ultimate Strength (MPa)	Thermal Expansion Coefficient (32-572°F) 10⁻⁶/°F	Thermal Expansion Coefficient (0-300°C) 10⁻⁶/°C	Hardness
2017-T4	General Use, Screw Machine Stock	40	276	62	427	14.1	25.4	105 Brinell
2024-T4	High Strength, Structural	47	324	68	469	13.7	24.7	120 Brinell
6061-T6	General Use, High Strength, Weldable	40	276	45	310	14	25.2	95 Brinell
7075-T6	High Strength, Structural, Aircraft	73	503	83	572	14	25.2	150 Brinell
5052-H32	Sheet Metal, Weldable	28	193	33	228	14.3	25.7	60 Brinell

Designation	Characteristics and Use	Density (lb/in³)	Density (g/cc)	Modulus of Elasticity (E) (ksi)	Modulus of Elasticity (E) (MPa)	Modulus of Rigidity (G) (ksi)	Modulus of Rigidity (G) (GPa)	Poisson's Ratio (v)
2017-T4	General Use, Screw Machine Stock	0.101	2.79	10500	72.4	3920	27	0.33
2024-T4	High Strength, Structural	0.100	2.78	10600	73.1	4060	28	0.33
6061-T6	General Use, High Strength, Weldable	0.098	2.7	10000	68.9	3770	26	0.33
7075-T6	High Strength, Structural, Aircraft	0.102	2.81	10400	71.7	3900	26.9	0.33
1100-H14	Sheet Stock, Poor Machinability	0.097	2.68	10200	70.3	3760	25.9	0.33

Table 8-12: Properties of Select Non-Metals

Designation	Characteristics and Use	Density		Tensile Yield Strength		Compressive Strength		Max Operating Temperature	
		lb/in³	g/cc	ksi	MPa	ksi	MPa	°F	°C
Delrin®	General Use	0.051	1.41	9.5	65.5	15	103.5	180	82
Nylon	General Use, Wear Applications	0.042	1.15	8.7	60	10	69	200	93
UHMW	General Use, Wear Applications, Wet Environments	0.033	0.93	6.3	43.5	3.1	21.5	180	82
PEEK	High Temperature	0.047	1.31	14	96.5	17	117	480	250
PTFE	High Temperature, Non-stick, Soft	0.078	2.16	3.9	27	3.5	24	400	204
Pclycarbonate	Transparent	0.043	1.2	9.5	65.5	12	83	250	121

Note: Properties are typical for unfilled, virgin material. A large range of properties are available with the addition of fillers. Delrin® is a registered trademark of the DuPont Company. It is an Acetal with good mechanical properties.

Table 8-13: Common Sheet Metal Thicknesses, Inch Series

Gauge	Thickness		Gauge	Thickness	
	inch	mm (converted)		inch	mm (converted)
7	0.1793	4.5542	16	0.0598	1.5189
8	0.1644	4.1758	18	0.0478	1.2141
10	0.1345	3.4163	19	0.0418	1.0617
11	0.1196	3.0378	20	0.0359	0.9119
12	0.1046	2.6568	22	0.0299	0.7595
13	0.0897	2.2784	24	0.0239	0.6071
14	0.0747	1.8974			

Table 8-14: Common Metal Plate Thicknesses, Inch Series

Plate Thickness		Plate Thickness		Plate Thickness	
inch	mm (converted)	inch	mm (converted)	inch	mm (converted)
0.1875	4.7625	2.1250	53.9750	8.0000	203.2000
0.2500	6.3500	2.2500	57.1500	9.0000	228.6000
0.3125	7.9375	2.5000	63.5000	10.0000	254.0000
0.3750	9.5250	2.7500	69.8500	12.0000	304.8000
0.4375	11.1125	3.0000	76.2000	14.0000	355.6000
0.5000	12.7000	3.2500	82.5500	18.0000	457.2000
0.5625	14.2875	3.5000	88.9000		
0.6250	15.8750	3.7500	95.2500		
0.7500	19.0500	4.0000	101.6000		
0.8750	22.2250	4.2500	107.9500		
1.0000	25.4000	4.5000	114.3000		
1.1250	28.5750	4.7500	120.6500		
1.2500	31.7500	5.0000	127.0000		
1.3750	34.9250	5.5000	139.7000		
1.5000	38.1000	6.0000	152.4000		
1.6250	41.2750	6.5000	165.1000		
1.7500	44.4500	7.0000	177.8000		
2.0000	50.8000	7.5000	190.5000		

CRITICAL CONSIDERATIONS: Materials

- For applications where material strength is critical, consult several sources for strength data. Consider whether the reported values correspond to an average value, a minimum, or a single test result. Understand that the tested specimens have a shape, size, and processing history that may differ from your application and may result in different properties. Always use appropriate factors of safety when evaluating material performance.
- In applications where heat is applied, it is important to consider the coefficient of thermal expansion of each part in the assembly. Thermal expansion can change fits in an assembly and cause relative movement of parts. Selecting materials with matching thermal expansion coefficients is recommended in high temperature or thermal cycling applications.
- Be aware that material strength is reduced with elevated temperature, and materials can become brittle at low temperatures. If heat or cold is a factor, use material data specific to that temperature.
- When designing parts made of soft materials, especially plastics, be aware that creep may cause fits to loosen over time.
- Plastics can have any number of fillers and additives that will change their properties significantly. Always used the proper material data, which includes the filler type and percentage.
- Some plastics will swell when exposed to certain liquids.

BEST PRACTICES: Materials

- When using raw steel stock, keep in mind that cold rolled or cold drawn shapes have closer tolerances than their hot rolled counterparts. Cross-sectional tolerances of cold finished bars are usually held to within ±0.0025 inches (±0.06 mm.) Cold rolled steel is more costly than hot rolled, but does not have the mill scale and rough surface finish of hot rolled.
- For highly stressed steel parts, favor cold rolled material over hot rolled. Cold rolled or cold drawn steel has higher strength than hot rolled.
- For steel weldments and parts that need to be heat treated, specify hot rolled steel. Cold rolling results in residual stresses that can cause significant distortion during welding or heat treating.
- When designing parts, especially weldments, a common cost reduction method is to use raw stock sizes where tolerances allow. This reduces the need for machining operations. Manufacturer's catalogs and web sites will have a complete list of stock sizes to choose from. Some stock size data can be found later in this section.
- Use a thermoset plastic rather than a thermoplastic when material stiffness or high temperature performance is desired.

Section
8.2

SURFACE FINISH

All surfaces have a surface texture, which is commonly called surface finish. Surface texture is comprised of four components: roughness, waviness, lay, and flaws. Surface texture is primarily a result of the machining or production process used to create the surface. Surface finish can affect the appearance, function, wear, and failure of a part.

RECOMMENDED RESOURCES

- J. Bralla, *Handbook of Product Design for Manufacturing*, McGraw-Hill: New York, NY, 1986
- R. L. Norton, *Machine Design: An Integrated Approach*, 4th Ed., Prentice Hall, Upper Saddle River, NJ, 2011
- Oberg, Jones, Horton, Ryffel, *Machinery's Handbook*, 28th Ed., Industrial Press, New York, NY, 2008

SURFACE FINISH SYMBOLS

The specification of surface finish in the United States is governed by ASME B46.1. The ISO standard is ISO 1302. Tables 8-15 through 8-18 illustrate commonly used surface texture symbols. Note that the ASME standard has been updated recently, and that the following illustrations conform to the earlier standard in common use. Surface roughness values are measured in microinches and micrometers.

Table 8-15: ASME Surface Texture Symbols & Construction

Symbol	Meaning
√ Fig. 1a.	Basic Surface Texture Symbol. Surface may be produced by any method except when the bar or circle (Fig. 1b or 1d) is specified.
▽ Fig. 1b.	Material Removal By Machining Is Required. The horizontal bar indicates that material removal by machining is required to produce the surface and that material must be provided for that purpose.
3.5 ▽ Fig. 1c.	Material Removal Allowance. The number indicates the amount of stock to be removed by machining in millimeters (or inches). Tolerances may be added to the basic value shown or in general note.
⊘ Fig. 1d.	Material Removal Prohibited. The circle in the vee indicates that the surface must be produced by processes such as casting, forging, hot finishing, cold finishing, die casting, powder metallurgy or injection molding without subsequent removal of material.
√ Fig. 1e.	Surface Texture Symbol. To be used when any surface characteristics are specified above the horizontal line or the right of the symbol. Surface may be produced by any method except when the bar or circle (Fig. 1b and 1d) is specified.

Table 8-16: ASME Surface Finish Examples

1.6 √	Roughness average rating is placed at the left of the long leg. The specification of only one rating shall indicate the maximum value and any lesser value shall be acceptable. Specify in micrometers (microinch).	0.005-5 0.8 √	Maximum waviness height rating is the first rating place above the horizontal extension. Any lesser rating shall be acceptable. Specify in millimeters (inch). Maximum waviness spacing rating is the second rating placed above the horizontal extension and to the right of the waviness height rating. Any lesser rating shall be acceptable. Specify in millimeters (inch).	1.6 / 3.5 ▽	Material removal by machining is required to produce the surface. The basic amount of stock provided for material removal is specified at the left of the short leg of the symbol. Specify in millimeters (inch).
1.6 0.8 √	The specification of maximum and minimum roughness average values indicates permissible range of roughness. Specify in micrometers (microinch).			0.8 / 2.5	Roughness sampling length or cut-off rating is placed below the horizontal extension. When no value is shown, 0.80 mm (0.030 inch) applies. Specify in millimeters (inch).
1.6 / ⊘	Removal of material is prohibited.	0.8 /⊥	Lay designation is indicated by the lay symbol placed at the right of the long leg.	0.8 /⊥ 0.5	Where required maximum roughness spacing shall be placed at the right of the lay symbol. Any lesser rating shall be acceptable. Specify in millimeters (inch).

Table 8-17: ASME Lay Symbols

Lay Symbol	Meaning	Example Showing Direction of Tool Marks
=	Lay approximately parallel to the line representing the surface to which the symbol is applied.	
⊥	Lay approximately perpendicular to the line representing the surface to which the symbol is applied.	
X	Lay angular in both directions to line representing the surface to which the symbol is applied.	
M	Lay multidirectional	
C	Lay approximately circular relative to the center of the surface to which the symbol is applied.	
R	Lay approximately radial relative to the center of the surface to which the symbol is applied.	
P	Lay particulate, non-directional, or protuberant	

Table 8-18: ISO Surface Texture Symbols

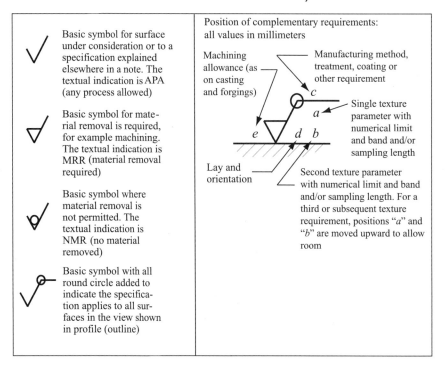

SURFACE FINISH AND TOLERANCE

It is important to understand the relationship between tolerance and surface finish. Because surface roughness has a direct impact on the measurement of deviation of surfaces, it is important to specify a surface finish that does not by itself take up the tolerance allowance. In general, if the surface roughness value specified takes up one eighth of the dimensional tolerance, the entire tolerance zone will be taken up by roughness. A rule of thumb is to specify surface roughness based on a 25:1 ratio between total tolerance and the roughness average for steel parts. For example:

- In inch units:

    Part Dimension: 0.25 ± 0.002 inches

    Total Part Tolerance: 0.002 x 2 = 0.004 inches

    Surface Finish Calculation: 0.004 / 25 = 0.000016 inches or 16 microinches

- In metric units:

    Part Dimension: 6 ± 0.01 mm

    Total Part Tolerance: 0.01 x 2 = 0.02 mm

    Surface Finish Calculations: 0.02 /25 = 0.0008 mm or 0.8 micrometers

SURFACE FINISH DATA

Tables 8-19 through 8-21 illustrate the most common surface finish specifications, some typical applications, and what roughness values machining processes are capable of producing.

Table 8-19: Common Surface Finish Values & Descriptions

Roughness		Description
Microinch	**Micrometer**	
1000	25.2	Extremely rough surface Typical of saw cutting or sand casting
500	12.5	Rough surface Typical of rough machining operations
250	6.3	Used for non-critical surfaces Typical for coarse grinding or milling
125	3.2	Commonly applied to non-sliding surfaces Often achieved in high speed machining
63	1.6	Commonly applied to parts with close tolerances Achievable with fine feeds and light cuts
32	0.8	Commonly used for sliding surfaces Typical for fine grinding operations
16	0.4	Finest commonly used finish Achievable with reaming and fine grinding
8	0.2	Uncommon, except for sealing surfaces Very fine machined finish Requires lapping or very fine honing
4	0.1	Uncommon, except for sealing surfaces Extremely smooth finish Achievable with lapping

Table 8-20: Normal Maximum Surface Roughness of Common Machined Parts

Machined Part	Surface Roughness	
	microinch	micrometer
Fluid Sealing Surfaces	5	0.13
Rolling Surfaces, Heavy Loads	8	0.2
Pistons	16	0.4
Gear Teeth, Normal Service, Small Pitch Diameter	16	0.4
Gear Teeth, Heavy Loads	16	0.4
Sliding Surfaces	32	0.8
Rotating or Pivoting Surfaces	32	0.8
Push Fit Surfaces	32	0.8
Datum Surfaces for Tolerances Under 0.001 inch (0.025 mm)	63	1.6
Rolling Surfaces, Normal Service	63	1.6
Press Fit Surfaces	63	1.6
Keys and Keyways	63	1.6
Gear Teeth, Normal Service	63	1.6
Mating Surfaces	125	3.2
Datum Surfaces for Tolerances Over 0.001 inch (0.025 mm)	125	3.2
Non-Mating Surfaces	250	6.3

Adapted from: Bralla, 1986[2]

[2]   *J. Bralla, Handbook of Product Design for Manufacturing, McGraw-Hill: New York, NY, 1986*

Table 8-21: Manufacturing Process Average Surface Finish

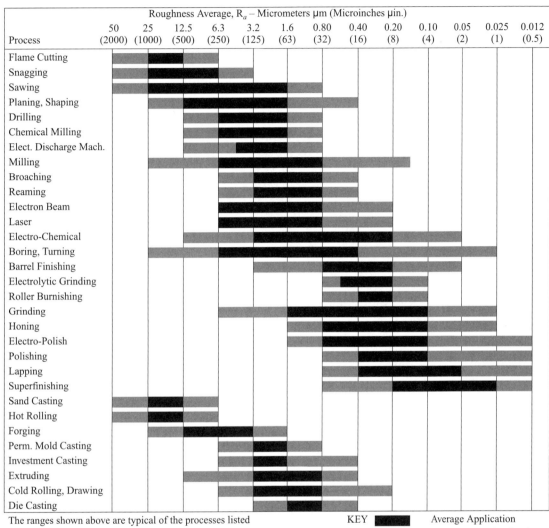

Process	Roughness Average, R_a – Micrometers μm (Microinches μin.)												
	50 (2000)	25 (1000)	12.5 (500)	6.3 (250)	3.2 (125)	1.6 (63)	0.80 (32)	0.40 (16)	0.20 (8)	0.10 (4)	0.05 (2)	0.025 (1)	0.012 (0.5)

The ranges shown above are typical of the processes listed
Higher or lower values may be obtained under special conditions

KEY ▮ Average Application
▮ Less Frequent Application

CRITICAL CONSIDERATIONS: Surface Finish

- Tighter tolerances necessitate finer surface finishes.
- Finer finishes tend to cost more and take longer to produce.
- Surface finish will affect the friction and stresses caused by that surface in a rolling or sliding joint. See the recommended resources for more information.
- Surface finish can affect stresses and deformation of mating surfaces.
- Surface finish can affect fatigue strength and crack propagation.

BEST PRACTICES: Surface Finish

- Include in the surface texture symbol the minimum information required to specify the required surface. It is common to specify only surface roughness. When required, lay, waviness, and machining allowance can be specified with the symbol as well.
- To simplify drawings, a default surface finish is usually listed in the drawing title block, with additional symbols attached to the part when improved finish over the default value is required.
- When a part is to be plated or coated, the drawing should indicate if the surface finish specifications apply before coating, after coating, or both.
- Specifying unmachined surfaces on a part can save cost when the part is designed and tolerance in order to be made from commonly available raw stock such as plate, sheet, round, or bar.
- Consider electropolishing (covered in Section 8.4) for stainless steel parts that require a fine finish all over.

HEAT TREATMENT

Section
8.3

Heat treatment is used to harden a metal or just the surface of a metal. Increased material strength, wear resistance, hardness, or shock resistance are achievable through heat treatment. Many ferrous steels are commonly heat treated, as are some aluminum alloys.

RECOMMENDED RESOURCES

- R. Leed, *Tool and Die Making Troubleshooter*, Society of Manufacturing Engineers, Dearborn, MI, 2003
- R. L. Norton, *Machine Design: An Integrated Approach*, 4th Ed., Prentice Hall, Upper Saddle River, NJ, 2011
- Oberg, Jones, Horton, Ryffel, *Machinery's Handbook*, 28th Ed., Industrial Press, New York, NY, 2008
- Shigley, Mischke, Brown, *Standard Handbook of Machine Design*, 3rd Ed., McGraw-Hill Inc., New York, NY, 2004

HARDNESS

Hardness is typically measured using one of the following scales: Brinell, Rockwell, or Vickers. In machine design, the most common way to specify the hardness of metal parts is using a Rockwell C value, usually indicated by the hardness value followed by "Rc" or "HRC." Table 8-22 is a conversion chart for several hardness scales and approximate tensile yield strengths. These strengths were calculated using the relationship:

$$Tensile\ Strength = BHN \times 490$$

These strength values are only approximate and not to be used for critical applications. It is important to note that the relationship between tensile strength and hardness is not the same for all materials, and can be any value between 450 and 1000. The relationship between hardness and tensile strength is complex. See the recommended resources for more information on hardness, its measurement, and its effect on tensile strength.

Hardness often matters to the machine designer. Harder materials will exhibit reduced wear and impact resistance. They will also have higher strength. However, harder materials tend to be less ductile (more brittle) and more prone to problems like cracking. Heat treatment is often called for when a part is to be hardened all the way through. When ductility matters, surface hardness can be achieved through some methods of heat treating, without hardening the core of the material. Hard surface treatments are another way to improve surface hardness while preserving the ductility of the part core.

Hardness ranges for each part should be carefully considered. Assemblies should be designed such that chosen parts will wear out or break

Table 8-22: Hardness Scale Conversions

Rockwell C Rc / HRC	Brinell BHN	Vickers HV / DPH	Approximate Tensile Strength Strength = BHN X 490	
			psi	Mpa
68		940		
66		865		
64	722	800	353780	2439.2
62	688	746	337120	2324.4
60	654	697	320460	2209.5
58	615	653	301350	2077.7
56	577	613	282730	1949.4
54	543	577	266070	1834.5
52	500	544	245000	1689.2
50	475	513	232750	1604.8
48	451	484	220990	1523.7
46	432	458	211680	1459.5
44	409	434	200410	1381.8
42	390	412	191100	1317.6
40	371	392	181790	1253.4
38	353	372	172970	1192.6
36	336	354	164640	1135.2
34	319	336	156310	1077.7
32	301	318	147490	1016.9
30	286	302	140140	966.2
28	271	286	132790	915.6
26	258	272	126420	871.6
24	247	260	121030	834.5
22	237	248	116130	800.7
20	226	238	110740	763.5

while the remaining parts are protected. This can be accomplished by applying high hardness (and therefore wear resistance) to valuable parts and applying lower hardness to easily replaced parts. If shear is a desired failure mode, through hardening will increase the brittleness of the part. Parts should have high hardness to resist wear, scratches, or act as bearing surfaces. Table 8-23 shows some typical hardness values found in machinery. Pay attention to the hardness of purchased components and the manufacturer's recommendations for the hardness of mating parts.

Table 8-23: Typical Hardness Values

Typical Hardness Range	
60+ Rc	ASME Hardened Ground Machine Dowel Pins - SURFACE
58+ Rc	Ball Bearing Components
58 - 62 Rc	Gear Contact Surfaces
55 - 62 Rc	Parts Under Heavy Wear or Acting as a Bearing Surface
54 - 58 Rc	Journal for an Aluminum Bronze Bearing
47 - 58 Rc	ASME Hardened Ground Machine Dowel Pins - CORE
38 - 58 Rc	Knife Blades
32 - 43 Rc	ASME Hex Socket Head Shoulder Screws
230 - 510 BHN	4140 Steel, Hot Rolled, Normalized, and Annealed
110 - 180 BHN	Cold Drawn Low Carbon Steel (As Drawn)
90 - 100 BHN	6061-T6, T651 Aluminum

Hardness of a material can be increased not only through heat treatment, but also through strain or work hardening. Some materials are sold in a "quarter hard," "half hard," "3/4 hard," or "full hard" state. These materials have been cold worked and are a good option when heat treatment is undesirable.

HEAT TREATMENT PROCESSES

The most common heat treating operations are annealing, quenching, tempering, case hardening, and flame or induction hardening. Annealing is a heat treatment process used to stress relieve, soften, and refine the grain structure of a material. Normalizing is similar to annealing, but performed at a higher temperature to yield a material that is stronger and more uniform in structure than that produced by annealing. Quenching is a process in which the cooling rate of a heated material is tightly controlled. The quenching process has a great effect on the final hardness of the material. Tempering is a process in which the quenched material is then reheated to relieve residual stresses and soften the material slightly. The tempering temperature also has a great effect on material hardness. Double tempering is sometimes used to further refine the grain structure of a material. Case, flame, and induction hardening are processes by which a surface layer of a part is hardened while leaving the core ductile. Case, flame, and induction hardening are usually followed by quenching and tempering.

There are three types of heat treatments commonly specified for machine parts: through hardening, case hardening, and flame or induction hardening. Through hardening exposes the entire part to heating and cooling processes to produce a specified hardness throughout the part. Case hardening exposes the entire part to heating and cooling processes, while exposing the surfaces to a liquid or gas atmosphere that induces a thin, hard case to form on the part surfaces. Flame or induction hardening expose only selected part surfaces to heating and cooling processes that cause only the outside layer of material to harden to some depth.

When specifying through hardening, it is common to specify simply the desired finished hardness of a part on a drawing without specifying the heat treatment process. There are many variables that must be controlled during heat treatment. The heating and cooling media (water, oil, and air are typical media) are factors, as are the temperature and duration of each stage in the heat treatment process. Different results can be had from different processes. Be sure to check with your heat treatment supplier when specifying process variables or when a specific result is desired.

Case hardening is used when the composition of the material precludes through hardening. Low carbon steels are case hardened, if hardened at all. Many alloy steels are case hardened through carburizing, where additional carbon is supplied to the part surfaces during heat treatment. Nitriding is a form of carburizing in which a very hard, thin case is produced in an ammonia environment with very little resultant distortion of the part. One advantage of case hardening is it leaves the core of the part ductile and resistant to fracture. It is necessary to specify the depth of the case layer when calling for case hardening. In applications where heavy impacts or high stresses are involved, it is possible for the substrate material to deform underneath the case layer.

Flame hardening or induction hardening is often used on gear teeth and cam surfaces. Flame hardening uses a flame, whereas induction hardening uses electrical energy to heat the material surfaces. Materials that can be through hardened can also be flame or induction hardened. When flame or induction hardening, it is important to specify not only the hardness and depth expected, but also to indicate which surfaces are to be heat treated. Avoid thin sections when flame or induction hardening, because these may end up through hardened and brittle without careful depth specification and control. Consult your chosen heat treatment vendor when dealing with thin sections or complex geometry. The image in Figure 8-1 shows a flame hardened rack tooth. The darker regions are hardened whereas the core remains soft.

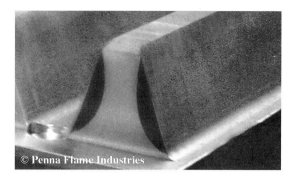

Figure 8-1: Flame Hardened Rack Tooth

Source: Penna Flame Industries (www.pennaflame.com)

HEAT TREATMENT AND DISTORTION

When parts are heat treated, cracks may develop or the part may warp. Some distortion due to heat treating is inevitable. Case hardening generally results in very low distortion and cracking. Through hardening or deep case hardening can result in significant distortion and cracking risk. Distortion and cracking can be minimized by avoiding sharp internal corners and sudden changes in part cross section. Holes and openings that are closely spaced should be avoided. Parts may need to be made into assemblies in some cases where sharp internal corners are needed. The individual parts in the assembly can then be made with radii and uniform sections. Material selection can also minimize distortion. Some tool steels, like D2, are particularly stable during heat treatment. Specifying "Stress Relieve Prior to Hardening" can reduce the chance of distortion. In extremely difficult cases, it is sometimes necessary to design a fixture to constrain the part during heat treatment to limit distortion. In the case of precision gages that must be stabilized against size and shape changes during their lifetime, a common stabilizing treatment is to subject the part to subzero temperatures either before or after tempering. Subzero treatment imparts long term stability, but can heighten the risk of cracking.

HARDNESS FROM HEAT TREATING

Table 8-24 lists materials commonly used for machine parts, typical hardness ranges achieved through heat treatment, and some relevant properties of the material as well as whether through or case hardening is indicated. Materials that can be through hardened can also be flame or induction hardened. Table 8-25 lists some typical case, flame, and induction hardening processes and their properties.

Table 8-24: Heat Treatment Properties of Common Steels

	Designation	Max Achievable Hardness Range Rc	Properties
TOOL STEELS	A2	62 - 63	Low distortion, better wear resistance than 01
	A6	60 - 62	Low distortion, moderate wear resistance and toughness, deep hardening
	D2	60 - 62	Low distortion, very high wear resistance, deep hardening
	D7	60 - 64	Highly wear resistant
	H11	53 - 57	High toughness, hot work
	H12	53 - 55	High toughness, hot work, dimensionally stable
	H13	50 - 54	High resistance to thermal shock
	L6	60 - 62	Improved toughness over 01 with some sacrifice of abrasion resistance
	M1	64 - 66	Hot work, high speed steel
	M2	64 - 66	High toughness, high speed steel
	M3	65 - 67	Good balance of wear and machinability
	M4	64 - 66	Superior wear resistance among high speed steels
	01	60 - 62	Versatile tool steel for general use
	06	62 - 66	Lubricity, machinability, good wear reisistance
	S1	54 - 57	Best wear resistance among shock resisting steels
	S5	60 - 62	Highest impact toughness
	T1	63 - 65	Versatile tool steel for general use. High red hardness, toughness, wear resistance
	T5	64 - 66	Highly wear resistant, hight speed steel, good hot hardness
CARBON & ALLOY STEELS	ASTM A36	20 - 25	Stress relieve or case harden for wear only, not practical to through harden
	AISI 1018 HR	25 - 30	Stress relieve or case harden for wear only, not practical to through harden
	AISI 1020	30 - 35	Stress relieve or case harden for wear only, not practical to through harden
	AISI 1026	35 - 40	Stress relieve or case harden for wear only, not practical to through harden
	AISI 1045	55 - 60	Case harden if part thickness is greater than 2" (50mm)
	AISI 1141	50 - 52	Tendency to corrode unless coated/plated due to sulphur inclusion
	AISI 1215	N/A	Stress relieve, not practical to through harden or case harden
	AISI 4140	50 - 55	Harden and then case harden for surface hardness up to Rc 55
	AISI 5160	58 - 60	High toughness blade steel
STAINLESS STEELS	303	N/A	Stress relieve only, will not through harden, case hardening layer very thin
	304	N/A	Stress relieve only, will not through harden, case hardening layer very thin
	316	N/A	Stress relieve only, will not through harden, case hardening layer very thin
	440C	58 - 60	Heat treatable. High corrosion and wear resistance

Table 8-25: Case Hardening Properties

Process	Typical Material	Case Hardness Rc	Typical Case Depth	
			inch	mm
Carburizing (Pack)	Low Carbon steels & alloys	50 - 63	0.005 - 0.06	0.125 - 1.5
Carburizing (Liquid)	Low Carbon steels & alloys	50 - 65	0.002 - 0.06	0.050 - 1.5
Carburizing (Gas)	Low Carbon steels & alloys	50 - 63	0.003 - 0.06	0.075 - 1.5
Nitriding	Alloy steels, stainless steels	50 - 70	0.0005 - 0.03	0.012 - 0.75
Cyaniding	Low Carbon steels	50 - 65	0.0001 - 0.005	0.0025 - 0.125
Salt	Ferrous metals including cast iron	50 - 70	0.0001 - 0.03	0.0025 - 0.75
Carbonitriding	Low Carbon steels & alloys, stainless steels	50 - 65	0.003 - 0.03	0.075 - 0.75
Flame Hardening	Steels with 0.35% - 0.5% Carbon	50 - 58*	0.125 - 0.75	3.175 - 19.05
Induction Hardening	Steels with 0.35% - 0.5% Carbon	50 - 58*	0.1 - 0.8	2.54 - 20

* depending on Carbon content

Hardness and hardenability are not the same thing. The achievable hardness of a material is related to the carbon content of the material. Hardenability of a material pertains to the depth and distribution of hardness achieved through the hardening process, and is related to alloy content.

For some grades of alloy steel, case hardening through carburizing is commonly used. SAE grades 2500, 3300, 4800, and 9300 are alloy steels that have high case hardenability. Grades 1300, 2300, 4000, 4100, 4600, 5100, 8600, and 8700 have medium case hardenability. In applications where high compressive stress or impacts are present, the core must be sufficiently hard to support the case layer. Case hardening alloy steels with medium core hardenability include SAE grades 1320, 2317, 2512, 2515, 3115, 3120, 4032, 4119, 4317, 4620, 4621, 4812, 4815, 5115, 5120, and 8620. Highly hardenable cores can be had with grades 2517, 3310, 3316, 4320, 4817, and 4820.

Through-hardening alloy steels with medium hardenability include SAE numbers 2330, 2340, 2345, 3130, 3135, 3140, 3141, 3145, 3150, 4053, 4063, 4137, 4140, 4145, 4640, 5135, 5147, and 5150. High hardenability grades include 4340 and 4150.

CRITICAL CONSIDERATIONS: Heat Treatment

- Know the hardness range achievable for a given metal in order to specify finished hardness properly. Specify a hardness range rather than a single value.
- Distortion due to heat treating is inevitable. Distortion can be minimized by avoiding sharp internal corners and sudden changes in part cross section. Material selection can also minimize distortion. Some tool steels, like D2, are particularly stable during heat treatment.
- Specifying "Stress Relieve Prior to Hardening" can reduce the chance of distortion.
- Coatings added after heat treatment can require special heat treatment process parameters, like high temperature tempering. Discuss coated parts and heat treatment with your coatings vendor.
- Heat treatment will affect the strength of the material. Consult the recommended resources for methods of estimating the strength of heat treated steels.

BEST PRACTICES: Heat Treatment

- Tool steels are normally through hardened, while mild carbon steels are sometimes case hardened. Aluminum commonly used in machinery is often purchased already heat treated.
- When wear is expected in an assembly, consider which element is cheaper/easier to replace when worn and specify that part's hardness such that it is softer than the mating parts. This will concentrate wear on the part that is easiest to replace and protect the other parts in the assembly.
- Consider specifying pre-hardened material for parts where distortion during heat treatment is particularly problematic. Pre-hardened materials are often "quarter hard," "half-hard," "¾ hard," or "full hard". The cost of machining pre-hardened materials up to about 50 Rc is not necessarily more than the cost of machining and then hardening.
- When only surface hardness is needed, consider coatings as an alternative to heat treatment. Coatings are discussed in Section 8.4 of this book.

Section
8.4

SURFACE TREATMENT

There are many reasons to apply surface treatments to parts: wear resistance, decreased friction, corrosion resistance, dimensional alteration, and cosmetic improvement. Many surface treatments are proprietary—and therefore not covered in this text—so be sure to check with your surface treatment supplier for critical or unusual applications. Corrosion-resistant surface treatments are often specified with instructions on minimum corrosion protection, such as a duration for passing the ASTM B-117 salt spray test.

RECOMMENDED RESOURCES

- Oberg, Jones, Horton, Ryffel, *Machinery's Handbook*, 28th Ed., Industrial Press, New York, NY, 2008

COMMON SURFACE TREATMENT TYPES

Black Oxide is an immersion coating for ferrous metals. It protects slightly against corrosion, imparts slight lubricity, and creates a uniform black appearance. It alters dimensions only slightly, with a buildup of 0.00008 inches (0.002 mm) which is typically ignored. Parts treated with black oxide can have subsequent surface treatments added for improved corrosion resistance or wear protection.

Zinc electroplating primarily protects against corrosion. It is most commonly available in clear and yellow, but other colors are occasionally used. A chromate coating is often added over the zinc to increase corrosion protection and is available in several colors. Zinc buildup is typically 0.0002 – 0.0005 inch (0.005 – 0.0127 mm).

Passivation is an immersion treatment used on stainless steel. It removes contaminating iron particles to improve corrosion resistance and produces a passive oxide film on the steel surface. It does not appreciably affect dimensions and is available in both clear and black. Passivation specification is detailed in ASTM A-967 and A-380.

Electropolishing is a surface conversion treatment that microscopically strips the part surfaces, potentially leaving a very fine finish. This treatment can be applied to ferrous and non-ferrous metals, but it is most commonly applied to stainless steel. This process can be expected to reduce surface roughness by half, so for example a 64 microinch surface will typically exit the electropolishing process with a surface roughness of 32 microinches. Typical material removal is 0.0001 – 0.0025 inch (0.0025 – 0.0635 mm.) Because this process removes material, it is not recommended for critical features. The specification for electropolishing should include material removed as well as both as-machined and post-treatment surface roughness.

Hard chrome plating can be used on ferrous and non-ferrous metals. It improves wear and corrosion resistance and leaves a bright reflective surface coating. Hard chrome plating is also used to restore worn parts to their proper dimensions. It can be ground after application. The coating hardness is 68 – 72 Rc and can be applied in thicknesses of 0.00002 – 0.05 inch (0.0005 – 1.27mm.) The standard thickness for Hard Chrome plating in non-salvage applications is 0.0002 – 0.0006" (0.00508 – 0.01524 mm) with high wear applications around 0.001 inch (0.03 mm.) It does affect final dimensions, and its thickness can be controlled to ±0.0005 inch (±0.013mm.) Indicate the target thickness range when specifying this surface treatment. Masking of critical features and threads is recommended before plating. Hard chrome plating specification is detailed in ASTM B-177.

Electroless Nickel plating can be applied to ferrous and non-ferrous parts. It provides corrosion and wear resistance. The coating hardness is 45 – 68 Rc and results in a bright black surface. This plating can be specified in thicknesses of 0.0001 – 0.005 inch (0.003 – 0.127 mm.) Typical thickness specification of electroless nickel plating for corrosion resistance is 0.0003″ – 0.0005″ (0.008 – 0.013 mm.) The plating thickness cannot be tightly controlled, so masking of critical features and threads is recommended before plating. Indicate the target thickness range when specifying this surface treatment. Electroless nickel plate specification is detailed in ASTM B-733.

Anodizing is applied to aluminum parts and comes in clear and a variety of colors. It provides corrosion resistance and moderate wear resistance, and can act as a galvanic barrier. This coating can be specified in thicknesses of 0.0001 – 0.0008 inch (0.003 – 0.02 mm) and has a hardness below 20 Rockwell C. Masking of critical features is recommended before anodizing due to buildup variation. Penetration depth and buildup thickness for this coating is generally a 50/50 split.

Hardcoat anodizing, or type III anodizing, is applied to aluminum parts and is similar to standard anodizing, except it results in a surface with superior wear and corrosion resistance. Coating hardness is around 60 – 70 Rc and is available in thicknesses of 0.001 – 0.003 inch (0.025 – 0.076 mm). Masking of threads and critical features is recommended due to coating buildup, and the finish is tan to dark gray. It is typical to specify both penetration depth and buildup thickness for this coating, which is a 50/50 split.

Hard lube anodize is hardcoat anodize with PTFE added for lubricity and non-stick properties.

Titanium nitride is a thin film coating that is extremely hard and protects against wear. It also protects moderately against corrosion and is non-stick. Coating hardness is around 80 Rc and the color is gold. This coating is suitable for precision features and threads without masking, with a typical coating thickness of 0.0001 – 0.0002 inch (3 – 5 µm.)

<u>Dry film lubricants</u> are a class of surface treatments that provide a permanent lubricious and wear resistant coating to metal parts. Some provide corrosion protection through the addition of rust inhibitors. Dry film lubricants are usually proprietary and characteristics vary by brand, so check with your surface treatment provider for information on performance and specification requirements.

<u>Paints</u> are generally applied to ferrous parts for corrosion protection and cosmetic improvement. Selection and specification of paint performance characteristics and colors will depend on the paint vendor selected. Paints have unpredictable buildup, so masking of threads and critical features is recommended before painting. When specifying color, check your paint manufacturer's available colors, or use a RAL color designation. RAL designations can be found online at www.ralcolor.com.

SELECTION OF SURFACE TREATMENTS

Tables 8-26 and 8-27 can be helpful in selecting among some common surface treatments for specific properties. Check with your chosen surface treatment vendor for further information on the application and specification of these and other treatments.

Table 8-26: Surface Treatment Selector

Surface Treatment	Ferrous Metal	Stainless Steel	Non-Ferrous Metal	Corrosion Protection	Wear Protection	Reduce Friction
Black Oxide	●	●	⊘	O	⊘	O
Zinc	●	●	O	●	⊘	⊘
Passivate	●	●	⊘	●	⊘	⊘
Electropolish	O	●	O	⊘	⊘	O
Hard Chrome Plate	●	O	●	●	●	O
Electroless Nickel Plate	●	O	●	●	●	O
Anodize	⊘	⊘	●	●	⊘	⊘
Hardcoat Anodize	⊘	⊘	●	●	●	O
Hard Lube Anodize	⊘	⊘	●	●	●	●
Titanium Nitride	●	●	●	O	●	O
Dry Film Lubricant	O	O	O	O	O	●
Paint	●	●	O	●	⊘	⊘

● = Suitable
O = Sometimes Suitable
⊘ = Unsuitable

Table 8-27: Surface Treatment Application Properties

Surface Treatment	High Temperature Process	Blind Holes Treated	Uneven Treatment on Corners	Masking Recommended	Gaps Due To Fixturing Possible
Black Oxide	⊘	●	⊘	⊘	○
Zinc	⊘	○	●	○	○
Passivate	⊘	○	⊘	⊘	⊘
Electropolish	⊘	○	○	⊘	○
Hard Chrome Plate	⊘	○	●	○	○
Electroless Nickel Plate	⊘	●	⊘	○	○
Anodize	⊘	○	⊘	○	○
Hardcoat Anodize	⊘	○	○	○	○
Hard Lube Anodize	⊘	○	○	○	○
Titanium Nitride	●	○	⊘	○	●
Dry Film Lubricant	○	○	○	○	○
Paint	⊘	○	○	●	●

● = Yes
○ = Sometimes, Depends on Dimensions and Process Steps
⊘ = No

CRITICAL CONSIDERATIONS: Surface Treatment

- Some surface treatment methods require high temperatures that may cause distortion or require specific heat treating process steps. Speak with your surface treatment provider.
- Most surface treatments will affect part dimensions to some extent. Specify that critical surfaces be masked before surface treatment if tolerances are incompatible with the surface treatment buildup variation. Specify that dimensions apply after coating in cases where the surface treatment buildup variation is compatible with feature tolerances. To save cost, mask as few features as possible and specify that dimensions apply before coating, as long as tolerances allow.
- Know the limitations of the treatment chosen. Many treatments have special requirements like fixturing surfaces, masking, or special treatment of blind holes. Work closely with the supplier as often as possible to evaluate parts for treatment.

BEST PRACTICES: Surface Treatment

- If possible, avoid specifying proprietary treatments or treatments by brand name. The use of brand names as part of the specification can limit the sources for producing that part.
- Specify surface treatment buildup (if any) and penetration (if any) on all parts. Use ranges rather than single values.

9

FORCE GENERATORS

Contents

9.1	SPRINGS	415
9.2	PNEUMATICS	466
9.3	ELECTRIC MOTORS	482

Tables

9-1	Common Spring Materials and Characteristics	419
9-2	Common Spring Wire Diameters	421
9-3	Compression and Extension Spring Fatigue Equations	424
9-4	Torsion Spring Fatigue Equations	425
9-5	Helical Coil Compression Spring Sizing Equations	431
9-6	Helical Coil Compression Spring Geometry Equations	432
9-7	Compression Spring Constant, Stress, and Frequency	432
9-8	Allowable Compression Spring Torsional Stress	433
9-9	Helical Coil Extension Spring Sizing Equations	443
9-10	Helical Coil Extension Spring Free Length, End Types	443
9-11	Extension Spring Constant, Stress, and Frequency	444
9-12	Allowable Extension Spring Stress	446
9-13	Helical Coil Torsion Spring Sizing Equations	451
9-14	Torsion Spring Constant, Force, Deflection, and Stress	452
9-15	Allowable Torsion Spring Bending Stress	452
9-16	Belleville Spring Washer Sizing Equations	460
9-17	Belleville Spring Force, Deflection, and Stress	461
9-18	Common Belleville Spring Materials	461
9-19	Belleville Spring Allowable Stress	461
9-20	General Pneumatic Symbols	469
9-21	Common Air Actuator Types	471
9-22	Cylinder Rod Buckling Equations	473
9-23	C_V and Related Equations	475
9-24	Directional Control Valve Functions	479
9-25	Common Valve Symbols	479
9-26	Common AC Motor Types and Characteristics	487
9-27	Common DC Motor Types and Characteristics	491
9-28	Electric Motor Sizing Equations	497

Section

9.1

SPRINGS

Springs are components that store energy when deflected and produce a restoring force. The restoring force produced by a spring is a function of the spring configuration and material; it may or may not be dependent on deflection amount. Springs of different types can exert a push, pull, or twisting force when deflected. The most commonly used spring types are helical compression, helical extension, helical torsion, and spring washers. Springs are normally selected from a catalog. However, custom designed springs are available from many manufacturers if standard catalog selections are not suitable for an application. Software to quickly design springs is commercially available, often from manufacturers.

RECOMMENDED RESOURCES

- R. L. Norton, *Machine Design: An Integrated Approach*, 4th Ed., Prentice Hall, Upper Saddle River, NJ, 2011
- Oberg, Jones, Horton, Ryffel, *Machinery's Handbook*, 28th Ed., Industrial Press, New York, NY, 2008
- Shigley, Mischke, Brown, *Standard Handbook of Machine Design*, 3rd Ed., McGraw-Hill, New York, NY, 2004
- *Design Handbook*, Associated Spring Barnes Group, Inc., 1981
- *The Spring Manufacturers Institute* website: www.smihq.org

SPRING TYPES

Helical compression springs and helical extension springs are the most common types used in machinery. Helical torsion springs and spring washers are also commonly used. The following are descriptions of these and other types of springs:

Helical coil compression springs (Figure 9-1a) are the most commonly used type of spring. These springs work to push things apart. Helical coil compression springs are constructed of a wire wrapped into a coil. The wire is usually, but not always, round in cross section. There are several types of helical compression springs, including: conical, barrel, hourglass, or cylindrical. Cylindrical springs are the most common of these, followed by conical springs, which have a reduced solid height and are great for tight spaces. Helical coil compression springs are detailed later in this section.

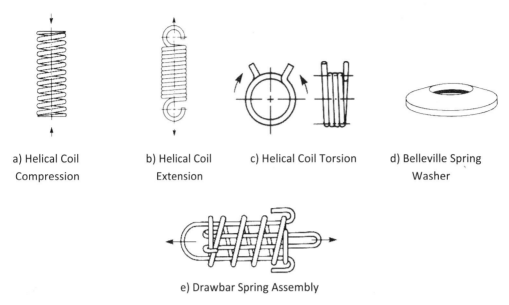

a) Helical Coil
Compression

b) Helical Coil
Extension

c) Helical Coil Torsion

d) Belleville Spring
Washer

e) Drawbar Spring Assembly

Figure 9-1: Compression, Extension, Torsion, and Belleville Springs

Source: Associated Spring Barnes Group Inc. (www.asbg.com)

Helical coil extension springs (Figure 9-1b) are the most commonly used type of spring to exert a pulling force. They are constructed of coiled wire, which usually has a round cross section. These springs are available with a variety of end attachment features, like hooks or loops. Helical coil extension springs are detailed later in this section.

Helical coil torsion springs (Figure 9-1c) are the most commonly used type of springs to exert a twisting force. They are constructed of wire coiled into a helix with both ends protruding at some angle. The wire is usually round in cross section. Helical coil torsion springs are detailed later in this section.

Spring washers come in many basic types, the most common of which are Belleville and wave. They act in compression. Belleville spring washers (Figure 9-1d) are conical in shape, whereas wave spring washers have a wavy shape. Most wave spring washers produce only light loading. Belleville spring washers are capable of producing high forces in a small axial space, and are detailed later in this section. Wave springs are often used to replace round wire compression springs when space is limited.

Drawbar spring assemblies (Figure 9-1e) resist being pulled apart. They are constructed of a helical coil compression spring and two components that transform the pulling motion into spring compression. Their main advantage over helical coil extension springs is their integral solid stop that prevents overloading of the spring. They also tend to have components that are stronger than extension spring hooks or loops.

Torsion bar springs consist of a single straight bar of some material. They usually have a round cross section. Torsion bars are usually supported between two radial bearings, and one end is fixed beyond its bearing. The other end of the torsion bar spring is subjected to an angular deflection, which produces a restoring torque. Torsion bar springs are sometimes used as counterbalancing springs for car trunk lids and hoods, and have been used in automobile suspensions.

Volute springs (Figure 9-2a) are a type of helical compression spring and are made from a conical coil of flat strip. They are particularly suited to impact loading and can produce a large amount of force in a small space.

Beam, or leaf, springs are constructed from one or more bars of spring material. They usually have a rectangular cross section and are capable of producing high forces in a small axial space. The properties of leaf springs can be customized by stacking individual leaf springs of different dimensions. Leaf springs have been used in many automotive suspensions.

Spiral springs, including hair springs, are made from a flat coil of strip with space between each coil. They are a type of torsion spring. These springs are generally limited to 3 or fewer revolutions of deflection. Their torque output is close to linear over their small range of deflections.

Constant force springs (Figure 9-2b) consist of a strip coiled into a spiral. The center of the coil is anchored whereas the end of the strip is pulled tangentially

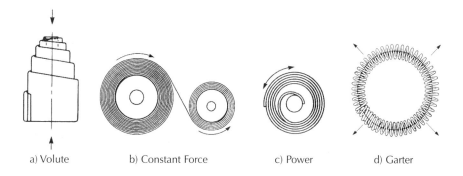

a) Volute   b) Constant Force   c) Power   d) Garter

Figure 9-2: *Volute, Constant Force, Power, and Garter Springs*

Source: Associated Spring Barnes Group Inc. (www.asbg.com)

to the coil. The force produced by this type of spring is virtually constant over a relatively long pull length.

Power, motor, or clock springs (Figure 9-2c) consist of a strip coiled into a spiral. The center of the coil is anchored whereas the end of the strip is pulled tangentially to the coil. These springs can store a large amount of energy.

Garter springs (Figure 9-2d) are essentially helical coil compression springs coiled into a ring shape. These springs exert a radial force that acts to restore the garter to its free shape. As a result, both inward and outward forces may be produced. These springs are commonly used in seals.

Gas springs are essentially a piston and rod arrangement with a volume of trapped gas inside. As the spring is compressed, the gas compression produces a restoring force. These springs have the advantage of long travel. Gas springs are often used in automobile hatches, overhead machine guarding, lids, and other long stroke applications. They look a lot like air cylinders without fittings.

SPRING MATERIALS

Any elastic material can be used for a spring. However, a limited set of materials make good springs. Some typical materials and their characteristics are shown in Table 9-1. A good spring material has a high yield point and low modulus of elasticity (E). A high ultimate strength is desired for springs undergoing fatigue loading. A graph of wire diameter vs. material and ultimate tensile strength is shown in Figure 9-3. Most springs are made from carbon and alloy steels. A few stainless steel alloys are used for springs, as are some copper alloys. For high temperature applications, and where material compatibility is required, stainless steel springs are recommended. Stainless steel helical coil springs exert only about 83.3% of the force exerted by music wire springs of the same geometry. Music wire is the most commonly used material for helical coil springs. Typical diameters of music wire and other common types of spring wire are shown in Table 9-2. Cold rolled steel strip is commonly used for other types of springs, such as constant force springs and Belleville washers.

Surface quality of the spring material can have an impact on properties, so it is essential to use spring wire quality grades that meet or exceed ASTM A227 and A229 standards to ensure a minimum level of pits, seams, and die marks on the wire surface. When elevated temperature is a factor, stress relaxation must be considered in critical applications.

Table 9-1: Common Spring Materials and Characteristics

Material Form	Common Name	Max. Service Temp.	Modulus of Rigidity (G)		Modulus of Elasticity (E)		Weight Density		Application Notes
			psi	MPa	psi	MPa	lbf/in³	N/mm³	
Wire	ASTM A228 Music Wire	250°F (120°C)	11.5×10^6	79.3×10^3	30×10^6	207×10^3	0.284	0.000077	General Use, Small Diameters
	ASTM A227 Cold Drawn or "Hard Drawn" Steel	250°F (120°C)	11.5×10^6	79.3×10^3	30×10^6	207×10^3	0.284	0.000077	General Use, Low Cost, Low Quality
	ASTM A229 Oil Tempered Steel	300°F (150°C)	11.5×10^6	79.3×10^3	30×10^6	207×10^3	0.284	0.000077	General Use, Large Diameters, Not For Shock Loading
	ASTM A401 (SAE 9254) Chrome Silicon Alloy Steel	475°F (245°C)	11.5×10^6	79.3×10^3	30×10^6	207×10^3	0.284	0.000077	Use For High Stress, Shock Loading, Hardenable
	ASTM A232 (SAE 6150) Chrome Vanadium Alloy Steel	425°F (220°C)	11.5×10^6	79.3×10^3	30×10^6	207×10^3	0.284	0.000077	Use For High Stress, Shock Loading, Fatigue Loading
	ASTM A313 (Type 302) Stainless Steel	500°F (260°C)	10×10^6	69×10^3	28×10^6	193×10^3	0.286	0.000078	General Use, High Temperatures
	17-7 PH Stainless Steel	600°F (315°C)	11×10^6	75.8×10^3	29.5×10^6	203×10^3	0.282	0.000077	General Use, High Temperatures
Strip	SAE 1074 Cold Rolled Steel	250°F (120°C)	11.5×10^6	79.3×10^3	30×10^6	207×10^3	0.284	0.000077	General Use, Formable and Hardenable
	ASTM A177 (Type 302) Stainless Steel	600°F (315°C)	10×10^6	69×10^3	28×10^6	193×10^3	0.286	0.000078	General Use, High Temperatures
	ASTM A177 (Type 301) Stainless Steel	600°F (315°C)	10×10^6	69×10^3	28×10^6	193×10^3	0.286	0.000078	High Temperatures
	17-7 PH Stainless Steel	700°F (370°C)	11×10^6	75.8×10^3	29.5×10^6	203×10^3	0.282	0.000077	Very High Temperatures, High Cost

Adapted From: Associated Spring Barnes Group Inc. (www.asbg.com)

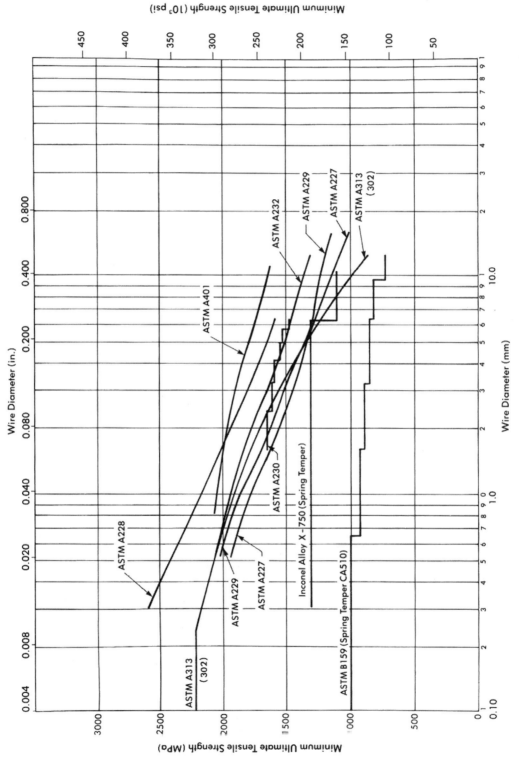

Figure 9-3: Ultimate Tensile Strengths for Spring Wire Materials

Source: Associated Spring Barnes Group Inc. (www.asbg.com)

Table 9-2: *Common Spring Wire Diameters*

Diameters up to 0.25 in (6.5 mm)

Music Wire (ASTM A228)		Cold (Hard) Drawn Wire (ASTM A227)		Oil Tempered Wire (ASTM A229)		Chrome Silicon Wire (ASTM A401)		Chrome Vanadium Wire (ASTM A232)		302 Stainless Steel (ASTM A313)	
Inch	mm	Inch	mm	Inch	mm	Inch	mm	Inch	mm	Inch	mm
0.004	0.1	-	-	-	-	-	-	-	-	0.004	0.1
0.005	0.12	-	-	-	-	-	-	-	-	0.005	0.12
0.006	0.16	-	-	-	-	-	-	-	-	0.006	0.16
0.008	0.2	-	-	-	-	-	-	-	-	0.008	0.2
0.01	0.25	-	-	-	-	-	-	-	-	0.01	0.25
0.012	0.3	-	-	-	-	-	-	-	-	0.012	0.3
0.014	0.35	-	-	-	-	-	-	-	-	0.014	0.35
0.016	0.4	-	-	-	-	-	-	-	-	0.016	0.4
0.018	0.45	-	-	-	-	-	-	-	-	0.018	0.45
0.02	0.5	-	-	0.02	0.5	-	-	-	-	0.02	0.5
0.022	0.55	-	-	0.022	0.55	-	-	-	-	0.022	0.55
0.024	0.6	-	-	0.024	0.6	-	-	-	-	0.024	0.6
0.026	0.65	-	-	0.026	0.65	-	-	-	-	0.026	0.65
0.028	0.7	0.028	0.7	0.028	0.7	-	-	-	-	0.028	0.7
0.03	0.8	0.03	0.8	0.03	0.8	-	-	-	-	0.03	0.8
0.035	0.9	0.035	0.9	0.035	0.9	-	-	0.035	0.9	0.035	0.9
0.038	1	0.038	1	0.038	1	-	-	0.038	1	0.038	1
0.042	1.1	0.042	1.1	0.042	1.1	-	-	0.042	1.1	0.042	1.1
0.045		0.045		0.045		-	-	0.045		0.045	
0.048	1.2	0.048	1.2	0.048	1.2	-	-	0.048	1.2	0.048	1.2
0.051		0.051		0.051		-	-	0.051		0.051	
0.055	1.4	0.055	1.4	0.055	1.4	-	-	0.055	1.4	0.055	1.4
0.059		0.059		0.059		-	-	0.059		0.059	
0.063	1.6	0.063	1.6	0.063	1.6	0.063	1.6	0.063	1.6	0.063	1.6
0.067		0.067		0.067		0.067		0.067		0.067	
0.072	1.8	0.072	1.8	0.072	1.8	0.072	1.8	0.072	1.8	0.072	1.8
0.076		0.076		0.076		0.076		0.076		0.076	
0.081	2	0.081	2	0.081	2	0.081	2	0.081	2	0.081	2
0.085	2.2	0.085	2.2	0.085	2.2	0.085	2.2	0.085	2.2	0.085	2.2
0.092		0.092		0.092		0.092		0.092		0.092	
0.098	2.5	0.098	2.5	0.098	2.5	0.098	2.5	0.098	2.5	0.098	2.5
0.105		0.105		0.105		0.105		0.105		0.105	
0.112	2.8	0.112	2.8	0.112	2.8	0.112	2.8	0.112	2.8	0.112	2.8
0.125	3	0.125	3	0.125	3	0.125	3	0.125	3	0.125	3
0.135	3.5	0.135	3.5	0.135	3.5	0.135	3.5	0.135	3.5	0.135	3.5
0.148		0.148		0.148		0.148		0.148		0.148	
0.162	4	0.162	4	0.162	4	0.162	4	0.162	4	0.162	4
0.177	4.5	0.177	4.5	0.177	4.5	0.177	4.5	0.177	4.5	0.177	4.5
0.192	5	0.192	5	0.192	5	0.192	5	0.192	5	0.192	5
0.207	5.5	0.207	5.5	0.207	5.5	0.207	5.5	0.207	5.5	0.207	5.5
0.225	6	0.225	6	0.225	6	0.225	6	0.225	6	0.225	6
0.25	6.5	0.25	6.5	0.25	6.5	0.25	6.5	0.25	6.5	0.25	6.5

HELICAL COIL SPRING TERMINOLOGY

Helical coil springs can be either left hand or right hand wound. The following terms are often used to describe the characteristics of helical coil springs and their behavior:

Spring constant (k,) or rate, is a value used to quantify the stiffness of the spring. Higher spring rates indicate a greater resistance to deflection and a greater restoring force. For compression and extension springs, the spring constant is linear between about 15% and 85% of the spring's total deflection range. In general, springs should be sized such that they operate within those limits. For most applications, use the smallest possible spring rate to provide the smallest force change over the range of movement. Stress relaxation occurs at elevated temperatures, resulting in reduced forces. Temperature also affects material properties. Catalog values for spring constants and loads are for normal ambient room temperature. Consult a manufacturer when sizing springs for use in elevated temperatures.

Wire diameter (d) is an essential characteristic of helical coil springs made with round wire. It has the greatest effect on the behavior, stress level, and performance of the spring. Typical wire diameters are listed in Table 9-2. Note that in Figure 9-3, it is clear that the ultimate tensile strength of spring wire generally increases with decreasing diameter.

Mean coil diameter (D_m) is the outer diameter of the spring minus the wire width (diameter, for round wire). It is also equal to the outer diameter plus the inner diameter, and the sum is divided by two. Mean coil diameter passes through the center of the wire that forms the coil.

Number of active coils (N_a) is the number of coils that are engaged in generating force under deflection. End coils that have been squared and/or ground will not be active coils.

Spring index (C) is the ratio of mean coil diameter to wire width (diameter, for round wire.) The spring index value is used in many spring calculations. A spring index greater than 4 is recommended for ease of manufacture. For compression springs, the spring index greater than 12 indicates that tangling is a risk. A value of 8 is a good target when designing a compression spring.

Maximum service deflection point, or operating point, is usually defined as the spring deflection at which the maximum service deflection load is measured. This value is often given in manufacturer catalogs for compression, extension, and torsion springs. For linear performance, a spring should not be deflected beyond the maximum service deflection point deflection value.

Spring surge is a term that describes the compression wave that travels the length of a helical coil spring when it is disturbed. Spring surge becomes violent and may cause spring failure if the frequency of deflection (cycle frequency) is near the spring's natural frequency. Formulas for natural frequency of helical coil compression springs and extension springs can be found in Tables 9-7 and 9-11. Compression, extension, and torsion springs are subject to this phenomenon. It is good practice to design or select springs such that the natural frequency is more than 15 times greater than the forcing frequency of the application. Natural frequency is essentially equal to the square root of stiffness divided by the square root of mass. To increase the natural frequency of a spring, reduce its mass or make changes to increase its spring constant.

HELICAL COIL SPRING FATIGUE

For springs expected to endure a very high number (millions) of cycles, the allowable working stresses should be limited by the fatigue strength. Fatigue can also be of concern when a spring is highly deflected. A cycling spring will operate between two force values, F_{min} and F_{max}. Both of these values should be of the same sign, since bidirectional loading should be avoided for most springs. Helical coil compression and extension springs undergo torsional shear stress when deflected, whereas helical coil torsion springs and spring washers undergo bending stress. Extension spring hooks often fail before the coils, and they are stressed in bending at one location (A), and torsion in another location (B) (See Figure 9-11). There are several common methods of finding the fatigue factor of safety. A modified Goodman diagram and related equations can be used to determine the allowable stress and safety factor. An excellent reference for this method is *Machine Design: An Integrated Approach* by R. L. Norton. This book is listed in the recommended resources at the beginning of this section.

Tables 9-3 and 9-4 contain calculations needed to find the fatigue factor of safety for different types of springs. For compression springs, calculate

Table 9-3: Compression and Extension Spring Fatigue Equations
Springs Made With Round Wire

D_m = Mean diameter (in or mm)	d = Wire diameter (in or mm)
F_{min} = Force at minimum operating deflection	F_{max} = Force at maximum operating deflection

ALTERNATING AND MEAN FORCE	$F_a = \dfrac{F_{max} - F_{min}}{2}$ $\qquad$ $F_m = \dfrac{F_{max} + F_{min}}{2}$
ALTERNATING AND MEAN COIL STRESS	$\tau_a = K_W \dfrac{8 F_a D_m}{\pi d^3}$ $\qquad$ $\tau_m = K_M \dfrac{8 F_m D_m}{\pi d^3}$
MINIMUM WORKING COIL STRESS Stress at minimum spring deflection	$\tau_{min} = K_M \dfrac{8 F_{min} D_m}{\pi d^3}$
ALTERNATING AND MEAN HOOK STRESS USE FOR EXTENSION SPRING HOOKS Full diameter hook or loop type ends	$\sigma_a = K_1 \dfrac{16 D_m F_a}{\pi d^3}$ $\qquad$ $\sigma_m = K_1 \dfrac{16 D_m F_m}{\pi d^3}$ $\tau_a = K_2 \dfrac{8 D_m F_a}{\pi d^3}$ $\qquad$ $\tau_m = K_2 \dfrac{8 D_m F_m}{\pi d^3}$
MINIMUM WORKING HOOK STRESS USE FOR EXTENSION SPRING HOOKS Stress at minimum spring deflection Full diameter hook or loop type ends	$\tau_{min} = K_2 \dfrac{8 D_m F_{min}}{\pi d^3}$ $\qquad$ $\sigma_{min} = K_1 \dfrac{16 D_m F_{min}}{\pi d^3}$
CORRECTION FACTORS C = Spring index (Table 9-5 or 9-9) R_1 = Mean radius of hook (Fig. 9-11) R_2 = Mean radius of bend in hook (Fig. 9-11)	$K_M = 1 + \dfrac{0.5}{C}$ $\qquad$ $K_W = \dfrac{4C-1}{4C-4} + \dfrac{0.615}{C}$ $K_1 = \dfrac{4 C_1^2 - C_1 - 1}{4 C_1 (C_1 - 1)}$ $\qquad$ $K_2 = \dfrac{4 C_2 - 1}{4 C_2 - 4}$ $C_1 = \dfrac{2 R_1}{d}$ $\qquad$ $C_2 = \dfrac{2 R_2}{d}$
TORSIONAL ENDURANCE STRENGTH Assumes repeated loading, room temperature, no corrosion, and 50% reliability. Use for spring wire diameters of 3/8" (10 mm) or less.	Non-Shot Peened Steel Springs $S_{ewt} \cong 45$ ksi (310 MPa) (for 50% reliability) Shot Peened Steel Springs $S_{ewt} \cong 67.5$ ksi (465 MPa)
FULLY REVERSED ENDURANCE LIMIT	$S_{es} = 0.5 \dfrac{S_{ewt} S_{us}}{S_{us} - 0.5 S_{ewt}}$
BENDING ENDURANCE STRENGTH	$S_e = \dfrac{S_{es}}{0.67}$
ULTIMATE TORSIONAL STRENGTH S_{ut} = See Fig. 9-3	$S_{us} = 0.67 S_{ut}$
TORSIONAL FATIGUE SAFETY FACTOR (INFINITE LIFE) USE FOR COILS USE FOR EXTENSION SPRING HOOKS τ_a = Alternating coil or hook stress τ_{min} = Minimum working coil or hook stress	$SF = \dfrac{S_{es}(S_{us} - \tau_{min})}{S_{es}(\tau_m - \tau_{min}) + S_{us} \tau_a}$ Determined using the modified Goodman diagram. It is assumed that the ratio of alternating to mean stress is constant over time.
BENDING FATIGUE SAFETY FACTOR (INFINITE LIFE) USE FOR EXTENSION SPRING HOOKS S_{ut} = See Fig. 9-3	$SF = \dfrac{S_e(S_{ut} - \sigma_{min})}{S_e(\sigma_m - \sigma_{min}) + S_{ut} \sigma_a}$ Determined using the modified Goodman diagram. It is assumed that the ratio of alternating to mean stress is constant over time.

Table 9-4: Torsion Spring Fatigue Equations
Springs Made With Round Wire

M_{max} = maximum working moment on spring M_{min} = minimum working moment on spring	σ_{OUT} = Bending stress at outside surface of coil
ALTERNATING AND MEAN COIL STRESS	$\sigma_{OUT,a} = 16K_O \dfrac{(M_{max} - M_{min})}{\pi d^3}$ $\sigma_{OUT,m} = 16K_O \dfrac{(M_{max} + M_{min})}{\pi d^3}$
MAXIMUM BENDING STRESS At outside of coil	$\sigma_{OUT,max} = K_O \dfrac{32M_{max}}{\pi d^3}$
MINIMUM WORKING STRESS Stress at minimum spring deflection	$\sigma_{OUT,min} = K_O \dfrac{32M_{min}}{\pi d^3}$
CORRECTION FACTORS C = Spring index (See Table 9-13)	$K_O = \dfrac{4C^2 + C - 1}{4C(C+1)}$
BENDING ENDURANCE STRENGTH (S_{ew}) Assumes repeated loading, room temperature, no corrosion, and 50% reliability.	Non-Shot Peened Steel Springs $\quad S_{ew} \cong 78$ ksi (537 MPa) Shot Peened Steel Springs $\quad S_{ew} \cong 117$ ksi (806 MPa)
ENDURANCE LIMIT (S_e) S_{ut} = See Fig. 9-3	$S_e = 0.5 \dfrac{S_{ew}S_{ut}}{S_{ut} - 0.5S_{ew}}$
BENDING FATIGUE SAFETY FACTOR (INFINITE LIFE) S_{ut} = See Fig. 9-3	$SF = \dfrac{S_e(S_{ut} - \sigma_{OUT,min})}{S_e(\sigma_{OUT,m} - \sigma_{OUT,min}) + S_{ut}\sigma_{OUT,a}}$ Determined using the modified Goodman diagram. It is assumed that the ratio of alternating to mean stress is constant over time.

the torsional factor of safety for the coils. For extension springs, the torsional factor of safety for the coils must be calculated in addition to the bending factor of safety for the hooks and the torsional factor of safety for the sharp bend that precedes each hook. For helical coil torsion springs, the bending factor of safety must be calculated for the coils. These factors of safety are examined at the points of maximum stress in the spring.

In general, fatigue life is negatively impacted by surface imperfections in the material. High cycle fatigue strength is improved by shot peening. Shot peening can increase the torsional fatigue strength by 20% or more. Testing by Zimmerli[1] shows that the torsional endurance limits for infinite life of spring steels subjected to repeated loading are unaffected by material up to a wire diameter of 3/8 inch (10 mm). Under these conditions, the torsional endurance limit depends only on whether or not the spring is shot peened.

[1] F.P. Zimmerli, "Human Failures in Spring Design" *The Mainspring,* Associated Spring Corp., Aug. - Sept. 1957

For compression springs, an alternative to fatigue safety factor calculation is to use Table 9-8 and Figure 9-3 in combination to determine the allowable stress for fatigue life, and then design the spring to stay within those stress levels. For extension springs, an alternative is to use Table 9-12 and Figure 9-3 in combination to determine design stress. For torsion springs, an alternative is to use Table 9-15 and Figure 9-3 in combination to determine design stress.

HELICAL COIL COMPRESSION SPRINGS

Helical coil compression springs act axially to push things apart, and are comprised of a coil of wire. Wire with round cross section is most commonly used. Rectangular wire is common for die springs and uses space more efficiently, particularly to reduce solid height. Free length of a compression spring is the length it assumes under zero loading. Solid height of a compression spring is the length it assumes when fully compressed. A compression spring cannot be compressed beyond solid height because its coils are all in contact with one another. The spring has essentially become a solid tube. The pitch of a compression spring is the distance between like points on adjacent coils. Most compression springs have constant pitch, but variable pitch springs are available. Formulas for calculating spring geometry can be found in Tables 9-5 and 9-6.

The range in which the spring rate of a compression spring is linear is between 15% and 85% of its total deflection range. Specifically, for linear performance, the gap between the solid height of the spring and its minimum operating length should be at least 15% of its total deflection range. If linear performance is not required but fatigue life is, keeping deflections within this range generally yields good results. Manufacturers usually recommend a maximum deflection for normal service rated compression springs that is within this range. Normal, or average service springs are rated for long life if the cycling frequency does not exceed 18,000 deflections per hour. Springs for light or static service can be compressed further, to within 15% of the working deflection range, though the spring rate will be nonlinear in the range near solid height. Compression to solid height is not recommended.

Compression springs can be used in parallel and series arrangements. When used in parallel (side by side), the spring rates of the individual springs are additive, as are the forces. When compression springs are arranged in series, their deflections are additive but the force at a given total deflection is the same as that of a single spring. A graph of these behaviors is given in Figure 9-4. Equations for series and parallel arrangements can be found in Table 9-7.

Helical compression springs can have the following ends: plain ends, plain ground ends, squared ends, and squared ground ends. Squared ends reduce tangling in storage. Squared ground ends are recommended in situations where squareness or buckling are concerns. The various end types are shown in Figure 9-5.

Compression springs must be captured radially and axially in an assembly. Normally, either a mandrel is passed through the center of the spring, or the spring is captured in a hole on at least one end. A common mandrel is a socket head shoulder screw, as shown in Figure 9-6b. Wide flat washers can sometimes be used in combination with shoulder screws or threaded rods and nuts to capture the spring. A center grooved spring pin, in combination with a wide washer, can be used to stop the end of a compression spring captured on a mandrel, as shown in Figure 9-6d. Commercially available parts, as shown in Figure 9-6a, can be used to capture a compression spring on a mandrel. When a spring is captured in a hole on one or both ends, the hole(s) should be chamfered. If the spring operates through an arc, the chamfer should reach the bottom of the hole(s). The angle of a full-depth chamfer should be selected such that the spring can move through the arc of operation without interference. With arcing movement, it is good practice to angle the seating surfaces for both ends of the spring to minimize the angle between them over the operating range of the device. This concept is illustrated in Figure 9-6c. Clearance must be present between a spring and either a mandrel or hole. This clearance can be calculated using the formula in Table 9-5.

Spring buckling is of concern when free length divided by mean coil diameter is greater than 4. As the ratio of deflection divided by free length goes up, buckling becomes more likely. End types and constraints also affect the tendency to buckle. A graph of spring critical buckling characteristics for two types of end conditions is shown in Figure 9-7. Buckling may be prevented by placing a compression spring over a rod or in a tube or deep hole. Due to friction, this will result in some loss in force exerted by the spring.

Helical coil compression spring coils undergo torsional stress when deflected. These springs must be operated within an allowable stress range. Table 9-7 contains common stress equations, and Table 9-8 shows allowable

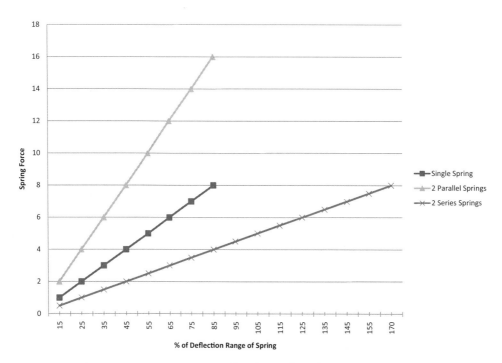

Figure 9-4: Force vs. Deflection of Spring Arrangements

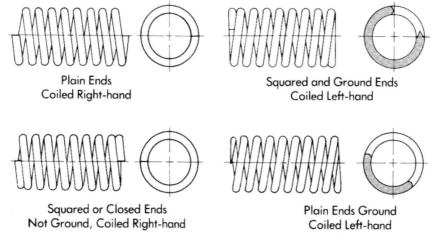

Plain Ends
Coiled Right-hand

Squared and Ground Ends
Coiled Left-hand

Squared or Closed Ends
Not Ground, Coiled Right-hand

Plain Ends Ground
Coiled Left-hand

Figure 9-5: Helical Coil Compression Spring End Types

Source: Associated Spring Barnes Group Inc. (www.asbg.com)

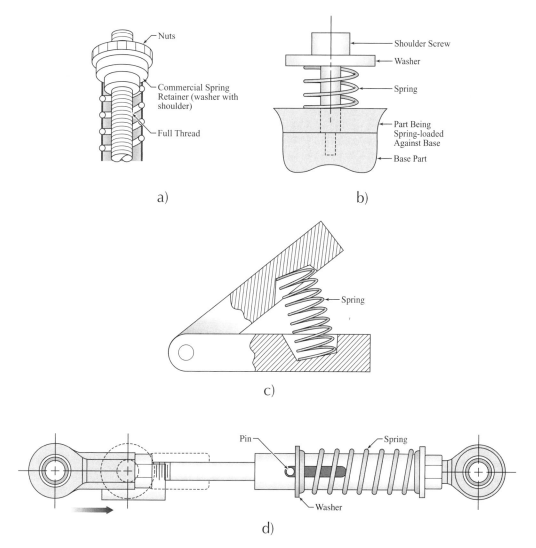

Figure 9-6: Spring Retaining Methods

stress levels for some common spring materials. The ultimate tensile strength of spring wire materials is a function of wire diameter, so Figure 9-3 must be used in combination with Table 9-8.

HELICAL COMPRESSION SPRINGS CATALOG SELECTION STEPS

Helical compression springs are usually specified in catalogs by coil diameter (either outer or inner), wire diameter, free length, solid height, and spring rate. Key spring characteristics also provided by the manufacturer

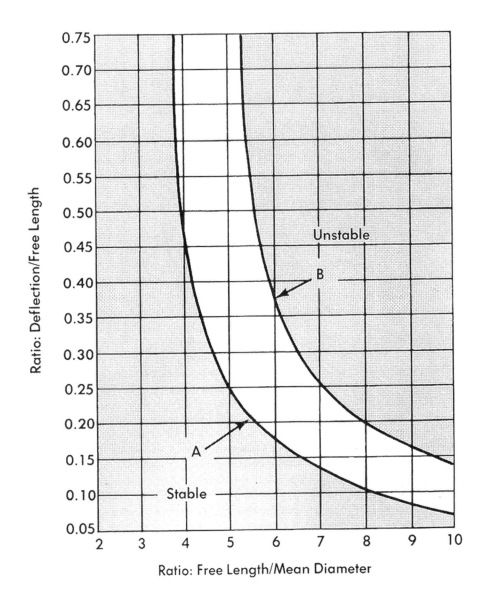

A: Springs are held flat against a plate and one end while the other end is free to tip.

B: Springs are held between parallel plates.

NOTE: To the right of each individual curve lies instability, and to the left of each curve lies stability.

Figure 9-7: Buckling Stability of Compression Springs
Source: Associated Spring Barnes Group Inc. (www.asbg.com)

Table 9-5: Helical Coil Compression Spring Sizing Equations

d = Wire diameter (in or mm) L_{max} = Maximum operating length L_{min} = Minimum operating length	k = Spring constant F_{max} = Force at maximum operating deflection
SPRING INDEX Tangling a concern when C > 12 Formability a concern when C < 4	For wire with round cross section: $$C = \frac{D_m}{d} \qquad \text{Starting Point: } C = 8$$
TESTS FOR BUCKLING See Fig. 9-7 for additional buckling criteria	Buckling is a concern if: $$\frac{L_f}{D_m} > 3.5$$
OPERATING DEFLECTION RANGE	$$x_w = L_{max} - L_{min} = x_{max} - x_{min}$$
MAXIMUM SOLID HEIGHT	$$L_{s,max} = L_{min} - 0.15 x_w$$ For linear performance: $$L_{s,max} = L_{min} - 0.215 x_w$$
FREE LENGTH (L_f) L_i = spring length at position i F_i = spring force at position i	$$L_f = L_i + x_i = L_i + \frac{F_i}{k}$$ $$L_f = L_s + 1.15 x_w$$ For linear performance: $$L_f = L_s + 1.43 x_w$$
TOTAL SPRING RANGE	$$x_s = L_f - L_s$$
MINIMUM WORKING LENGTH	$$L_{min} = L_s + 0.15 x_w$$ For linear performance: $$L_{min} = L_f - 0.85 x_s$$
MAXIMUM WORKING LENGTH	$$L_{max} = L_f - 0.05 x_s$$ For linear performance $$L_{max} = L_f - 0.15 x_s$$
TRIAL WIRE DIAMETER Based on uncorrected stress τ = Target stress at max force F$_{max}$ Be sure to check units for consistency	For wire with round cross section: $$d = \sqrt[3]{\frac{8 F_{max} D_m}{\pi \tau}}$$
MEAN SPRING DIAMETER OD = Outside diameter of spring coil ID = Inside diameter of spring, nominal	For wire with round cross section: $$D_m = OD - d \qquad D_m = \frac{OD + ID}{2}$$
NUMBER OF ACTIVE COILS G = Modulus of Rigidity (Table 9-1) (psi or MPa)	For wire with round cross section: $$N_a = \frac{d^4 G}{8 D_m^3 k}$$
CLEARANCE BETWEEN SPRING AND HOLE OR ROD D_c = diameter of spring hole or rod OD = spring outer diameter (nominal)	Minimum Diametral Clearance = 0.05D$_m$ when $D_c > 0.512$ inch (13mm) Minimum Diametral Clearance = 0.1D$_m$ when $D_c \le 0.512$ inch (13mm)
DIAMETER INCREASE DUE TO COMPRESSION OD_s = Outside diameter at solid height p = Spring pitch (See Table 9-6)	For wire with round cross section: $$OD_s = d + \sqrt{D_m^{\,2} + \frac{p^2 - d^2}{\pi^2}}$$

Table 9-6: Helical Coil Compression Spring Geometry Equations

Feature	End Type			
	Squared and Ground	Squared, Not Ground	Open & Ground	Open, Not Ground
NUMBER OF ACTIVE COILS	$N_a = \dfrac{L_f - 2d}{p}$	$N_a = \dfrac{L_f - 3d}{p}$	$N_a = \dfrac{L_f}{p} - 1$	$N_a = \dfrac{L_f - d}{p}$
TOTAL NUMBER OF COILS	$N_t = N_a + 2$	$N_t = N_a + 2$	$N_t = N_a + 1$	$N_t = N_a$
SOLID HEIGHT	$L_s = dN_t$	$L_s = d(N_t + 1)$	$L_s = dN_t$	$L_s = d(N_t + 1)$
PITCH d = Wire diameter	$p = \dfrac{L_f - 2d}{N_a}$	$p = \dfrac{L_f - 3d}{N_a}$	$p = \dfrac{L_f}{N_t}$	$p = \dfrac{L_f - d}{N_a}$
FREE LENGTH	$L_f = pN_a + 2d$	$L_f = pN_a + 3d$	$L_f = pN_t$	$L_f = pN_t + d$

Table 9-7: Compression Spring Constant, Stress, and Frequency

D_m = Mean diameter (Table 9-5) (in or mm)	d = Wire diameter (in or mm)
G = Modulus of Rigidity (Table 9-1) (psi or MPa)	N_a = Number of active coils (Table 9-5)

SPRINGS IN PARALLEL	$F = \sum F_i = x \sum k_i \qquad x = x_i \qquad k = \sum k_i$
SPRINGS IN SERIES	$F = \dfrac{1}{\sum \dfrac{1}{k_i}} \sum x_i \qquad x = \sum x_i \qquad \dfrac{1}{k} = \sum \dfrac{1}{k_i}$
SPRING CONSTANT	$k = \dfrac{F}{x} = \dfrac{\lvert F_2 - F_1 \rvert}{\lvert L_2 - L_1 \rvert}$ For round wire: $k = \dfrac{d^4 G}{8 D_m^3 N_a}$
SPRING DEFLECTION Deflection from free length	$x = \dfrac{F}{k}$ For round wire: $x = \dfrac{8 F D_m^3 N_a}{d^4 G} = \dfrac{\pi N_a \tau D_m^2}{Gd}$
SPRING FORCE Force at deflection x from free length	$F = kx$ For round wire: $F = \dfrac{Gd^4 x}{8 N_a D_m^3} = \dfrac{0.393 \tau d^3}{D_m}$
STRESS CORRECTION FACTORS C = Spring index (Table 9-5)	For Static Loading: $K = 1 + \dfrac{0.5}{C}$ For Dynamic Loading: (Wahl Factor) $K = \dfrac{4C - 1}{4C - 4} + \dfrac{0.615}{C}$
TORSIONAL SHEAR STRESS IN SPRING	$\tau = K \dfrac{8 F D_m}{\pi d^3} = K \dfrac{8 F C}{\pi d^2}$
NATURAL FREQUENCY f_n = natural frequency (Hz) g = 386.1 in/s^2 or 9810 mm/s^2 γ = weight density of material (Table 9-1)	Both spring ends against flat plates: $f_n = \dfrac{d}{\pi N_a D_m^2} \sqrt{\dfrac{Gg}{8\gamma}}$

Table 9-8: Allowable Compression Spring Torsional Stress

Application	Material	Maximum Allowable % of Ultimate Tensile Strength	
		Before Set Removed	After Set Removed
Static Loading	Music Wire (A228)	45%	65%
	Cold Drawn Wire (A227)	45%	
	Oil Tempered Wire (A229)	50%	
	Stainless Steels (Austenitic, 301, 302)	35%	
		Not Shot Peened	Shot Peened
Fatigue Loading (Life = 10^6 cycles)	Music Wire (A228)	33%	39%
	Oil Tempered Wire (A230)	40%	47%
	Chrome Vanadium Wire (A232)	40%	47%
	Stainless Steels (Austenitic, 301, 302)	33%	39%
Fatigue Loading (Life = 10^7 cycles)	Music Wire (A228)	30%	36%
	Oil Tempered Wire (A230)	38%	46%
	Chrome Vanadium Wire (A232)	38%	46%
	Stainless Steels (Austenitic, 301, 302)	30%	36%

Data Source: Associated Spring Barnes Group Inc. (www.asbg.com)

include load and deflection at the maximum service deflection point. The following procedure can be used to select a compression spring from a catalog.

1. Determine the application parameters. These include:
    *   Desired force at the most compressed position and any other known forces
    *   Operating deflection range or desired spring length at the extreme operating positions and any other operating positions
    *   Desired outer or inner diameter of the spring, if known
    *   End type
    *   Environmental factors, such as heat or material compatibility
2. Select a trial spring material based on performance parameters.
    Table 9-1 can be used to facilitate selection. Stainless steel springs are recommended for high temperatures or for corrosive environments. Stainless steel springs exert only about 83.3% of the force exerted by music wire springs of the same geometry.

3. If the minimum operating length and deflection range of the spring is known, calculate a maximum solid height for the spring. For linear performance, a gap equal to 15% of the spring's total deflection range should exist between the minimum operating height and the solid height. If linear performance is not needed, the gap (clash allowance) can be as low as 15% of the working deflection. The formulas for maximum solid height $L_{s,max}$ can be found in Table 9-5. If the spring must be compressed for installation, the solid height of the spring must be less than the installation length.

4. If the force at two positions is known, as well as the difference in length between the two positions, calculate the desired spring rate and the deflections at each load. Calculate the spring rate based on two loads at two positions using the spring constant formula in Table 9-7. Deflections from free length for the two positions can then be found using the formula: $F = kx$

5. Select a preliminary stock spring from a catalog that satisfies the desired load(s) at the desired deflection(s) and fits within the assembly geometry. First check the catalog to be sure what the given parameters mean.

   a. If the coil diameter is known, first look at springs that meet that requirement. Catalog springs are often organized by outside coil diameter.

   b. If the deflection range, forces at two positions, and spring constant are known, look for springs with nearly the required spring constant. It is expedient to look first at springs that exert greater than the maximum required force at their operating point or solid height. Spring deflection values for the given forces can be calculated by using the spring constant formula $F = kx$ (Table 9-7) and solving for x. That value of x will be equal to the deflection from free length of the spring at the given force. Now subtract these deflections from the selected spring free length to get the operating lengths of the spring at each force level.

   c. If the force at minimum operating length and deflection range are known:

      i. If the catalog lists spring force at maximum recommended deflection, look at catalog values for maximum service deflection point load and length to find a suitable spring.

- The force at the operating point listed in the catalog must be greater than or equal to the required maximum deflection force.
- The length at the operating point must be less than or equal to the desired minimum operating length.
- The solid height must be less than that calculated earlier.
- The free length must be longer than the minimum operating length plus the deflection range.
- For most applications, select a spring with the smallest possible spring rate that will fit within the application's geometry to provide the smallest force change over the range of movement.

ii. If the catalog lists spring force at solid height, look at catalog values for solid height load and length to find a suitable spring.
- The force at the solid height listed in the catalog must be greater than the required maximum deflection force.
- The solid height must be less than the desired minimum operating length.
- For linear performance, look for a spring where free length minus solid height is greater than 1.43 times the working length, according to the equation: $L_f - L_s > 1.43 x_w$
- If linear performance is not required, look for a spring where free length minus solid height is greater than 1.15 times the working length, according to the equation: $L_f - L_s > 1.15 x_w$
- For most applications, select a spring with the smallest possible spring rate to provide the smallest force change over the range of movement.

iii. If the catalog lists some other combination of parameters, solve for the required parameters using the design procedure detailed in this section. Select a spring with similar parameters.

6. <u>Analyze the spring's total deflection range</u>. Calculate the total range of the selected spring using the equation $x_s = L_f - L_s$ and then calculate the minimum and maximum working lengths using the formulas in Table 9-5. If the spring was selected from a catalog that provided a minimum length at maximum load, calculation of the minimum length can be skipped. The operating lengths for the application should be between the minimum and maximum working lengths of the spring. If operating

lengths are not yet known, set the minimum operating position length equal to or greater than the calculated minimum working length.

7. <u>Analyze the forces produced by the selected spring.</u> Calculate the deflection from free length at both extreme operating positions and any other critical positions. Use the force formula $F = kx$ (Table 9-7) and the value of spring rate to calculate the spring force at all critical operating positions. If the selected spring does not provide acceptable forces at all critical operating positions, a different spring must be selected and analyzed. To increase the amount of force a spring can exert, the spring requires more mass. This means increasing the wire size, coil size, free length, number of coils, or cross sectional geometry. Changing from round wire to rectangular cross section coils can increase the spring force capability without increasing the envelope size. Changing the spring material can also affect force capacity.

8. <u>Check the spring for buckling tendency.</u> The buckling tests in Table 9-5 and the graph in Figure 9-7 can be used to determine if the spring is likely to buckle in the application. If buckling is a concern, the application geometry and/or spring geometry will need to be changed. An alternative to changing the geometry is to provide a rod, deep hole, or tube to laterally constrain the spring.

9. <u>For applications with rapid cycling, check the natural frequency of the spring.</u> Natural frequency can be calculated using the formulas in Table 9-7. The natural frequency of the spring should be 15 or more times greater than the forcing frequency of the application. For a higher natural frequency, select a spring with reduced weight or a higher spring constant.

10. <u>Analyze the spring for fatigue life.</u> This analysis is necessary for severe service conditions with extremely rapid deflections. It is also needed for normal service conditions where the spring is compressed beyond the linear range or beyond the manufacturer's recommendation. Fatigue analysis is also necessary for applications with a high (millions of cycles) life expectancy. Calculate the torsional fatigue safety factor using the equations in Table 9-3.

11. <u>Calculate the required hole or mandrel diameter for the selected spring.</u> Use the diametral clearance formulas in Table 9-5.

HELICAL COMPRESSION SPRING DESIGN STEPS

The easiest way to design a spring is to use commercial software or seek the assistance of an applications engineer at a manufacturer. Spring design is an iterative process that is suited for spreadsheet automation at the very least. The following procedure is iterative and may be used to design a helical coil compression spring made from round wire. See the recommended resources for more information on spring design for static and dynamic loading.

1. Determine the application parameters. These include:
   * Desired forces at the most and least compressed operating positions
   * Desired operating deflection range or spring length at one position where force is known
   * Desired outer diameter of the spring
   * End type
   * Environmental factors, such as heat or material compatibility
2. Select a trial spring material type based on performance parameters. Find the shear modulus (G) for the chosen material. Table 9-1 can be used to facilitate selection. Stainless steel springs are recommended for high temperatures or for corrosive environments. Stainless steel springs exert only about 83.3% of the force exerted by music wire springs of the same geometry.
3. Select a trial wire diameter:
   a. Method 1: It is often expedient to find a likely wire diameter by looking in a catalog for springs that are nearly right for the application. Look for springs with listed loads that exceed what is needed. If more force is needed, the spring requires more mass. This means increasing the wire size, coil size, free length, number of coils, or cross-sectional geometry.
   b. Method 2: If outside diameter is known, set the spring index value at some acceptable value (8 is preferred). Using the spring index equation (Table 9-5), and using outer diameter for the value of D_m, calculate d. This calculation typically takes the form d=OD/8. Then calculate D_m = OD – d. Calculate spring index again. Reduce wire diameter or increase OD until the spring index is close to 8. Select an available wire diameter from stock sizes that is near what was calculated. Table 9-2 shows some available values of d.
   c. Method 3: Determine a target stress value at maximum operating force that is near but below the maximum allowable operating

stress for the spring material. Maximum allowable stress can be calculated using Table 9-8 and Figure 9-3 together. A wire diameter must be assumed when using Table 9-8. A trial value for d based on a target stress value can be calculated using the trial wire diameter equation in Table 9-5. If a target OD is unknown, a value of D_m must be assumed. A good place to start is to use the assumed value of d and solve the equation $D_m = 8d$. Once the trial diameter equation is solved, if the calculated value of d is very different from that assumed when selecting stress level, iterate by changing OD and/or d until they agree. If a trial value of d is calculated, an available wire size close to that value should then be selected, and then the spring index value (Table 9-5) checked to be sure it is between 4 and 12.

4. <u>Validate the material choice for the chosen wire diameter.</u> Calculate the torsional shear stress in the spring at maximum operating force using the equations in Table 9-7. Then compare the corrected maximum torsional stress against the maximum allowable torsional stress for the material and wire diameter. Table 9-8 and Figure 9-3 in combination can be used to determine maximum allowable stress. If allowable stress is exceeded, go back to either the wire diameter or material selection steps. Wire diameter may need to be increased. Be sure to re-check the spring index value and adjust D_m if the wire diameter is changed.

5. <u>Calculate the spring rate and solve for the number of active coils.</u> Spring rate may be calculated using the equation for spring constant as a function of two loads in Table 9-7. The spring rate may be changed by adjusting the expected load or length at one or more operating points. Use the calculated spring rate to calculate the number of active coils using the formula in Table 9-5. The active number of coils should be rounded to the nearest ¼ coil, which is a typical manufacturing accuracy for forming the spring. Then recalculate the spring rate based on the new number of active coils. Use the equation for spring rate found in Table 9-7.

6. <u>Calculate and verify the solid height of the spring.</u> Calculate solid height based on number of coils and end type using the formulas in Table 9-6. If a spring must be compressed for installation, the solid height of the spring must be small enough for installation. Solid height can be reduced by reducing wire diameter and/or increasing coil diameter. Solid height may also be reduced by switching from round wire to rectangular wire

7. <u>Calculate the free length of the spring.</u> Now that spring rate is known, use the maximum force value to calculate the deflection at that load using the formula: $F = kx$ Solve for x. If the length at any force is known, calculate the free length using the formula in Table 9-5. If a length at maximum deflection is not known, use this maximum deflection value of x, the calculated solid height, and the working deflection range in the free length formula in Table 9-5.

8. <u>Calculate the remaining unknown lengths, deflections, and forces at each operating position.</u> Operating length is free length minus deflection. Deflection can be calculated at each load using the formula $F = kx$ and solving for x. Force is calculated using the formula $F = kx$ where x is deflection from free length.

9. <u>Check the spring for buckling tendency.</u> The buckling test in Table 9-5 and the graph in Figure 9-7 can be used to determine if the spring is likely to buckle in the application. If buckling is a concern, the application geometry and/or spring geometry will need to be changed. A wider coil might be needed. An alternative to changing the geometry is to provide a rod, deep hole, or tube to laterally constrain the spring.

10. <u>For applications with rapid cycling, check the natural frequency of the spring.</u> Natural frequency can be calculated using the formula in Table 9-7. The natural frequency of the spring should be 15 or more times greater than the forcing frequency of the application. For a higher natural frequency, select a spring with reduced weight or a higher spring constant.

11. <u>Analyze the spring for fatigue life.</u> This is necessary for severe service conditions with extremely rapid deflections. It is also needed for normal service conditions where the spring is compressed beyond the linear range. It is also needed for applications with a high (millions of cycles) life expectancy. If the spring was designed using the fatigue allowable stress multipliers from Table 9-8, then the spring should be acceptable in fatigue. Otherwise, calculate the torsional fatigue safety factor using the equations in Table 9-3.

12. <u>Calculate the required hole or mandrel diameter for the spring.</u> Use the diametral clearance formulas in Table 9-5.

HELICAL COIL EXTENSION SPRINGS

Helical coil extension springs act axially to resist pulling forces and are comprised of a coil of wire, usually with a round cross section. Extension springs are usually configured with ends in the form of loops or hooks of wire to facilitate attachment to things. When a spring with ends is installed in an assembly, the hooks or loops are sometimes placed over pins (grooved pins with neck can be used, see Figure 9-8a) and are sometimes hooked onto retainers, which are commercially available as shown in Figure 9-8b. Extension springs are available without ends, allowing the spring body to be cut to any length. Their coils are threaded through a set of holes to form an attachment. Figure 9-8c shows a commercially available bracket with a hole for a screw and holes for the spring to thread onto. The mean diameter of the spring needs to be the same as the distance between the holes perpendicular to the spring's axis.

Some common end configurations are shown in Figure 9-9. The free length of an extension spring is the distance between the inside surfaces of the hooks or coils on each end when the spring is under zero load. When deflected, the spring length is understood to mean the distance between the inside surfaces of the coils. Formulas for calculating spring geometry can be found in Tables 9-9 and 9-10.

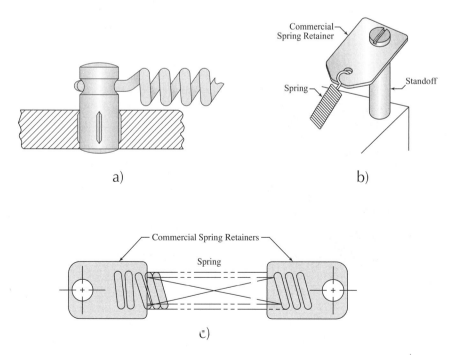

a)

b)

c)

Figure 9-8: Commercial Spring Retainers

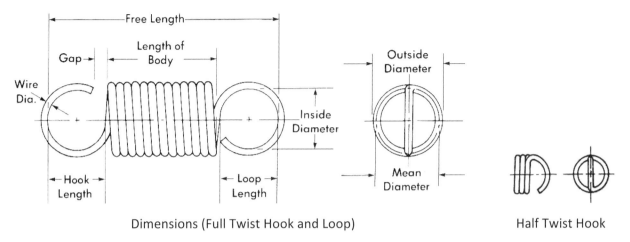

Dimensions (Full Twist Hook and Loop)   Half Twist Hook

Figure 9-9: Extension Spring Dimensions and Ends

Source: Associated Spring Barnes Group Inc. (www.asbg.com)

Initial tension of an extension spring is the pulling force that must be exceeded before the spring deflection becomes significant and linear with respect to applied load. The relationships between stress, force, and deflection can be found in Table 9-11. These springs must be operated within an allowable stress range. Helical coil extension spring coils undergo torsional stress when deflected. The hooks undergo maximum bending stress at location A in Figure 9-11. The hooks also undergo maximum torsional stress at location B in Figure 9-11. Both of these locations must be evaluated and may fail sooner than the coils do.

Unlike compression springs, extension springs are not inherently protected against overload. As a result, either stops must be built into the assembly to limit the spring to its maximum recommended length or the applied force must never exceed the recommended maximum load. Maximum extended length is based on stress level. A greater factor of safety is used for extension springs than for compression springs when defining the maximum allowable stress levels, as shown in Table 9-12. Most extension springs are not recommended for shock loading and are prone to spring surge.

Extension springs can be used in parallel and series arrangements. When used in parallel (side by side), the spring rates are additive, as are the forces. When extension springs are arranged in series, their deflections are additive, but the force at a given total deflection is the same as that of a single spring. A graph of these behaviors is given in Figure 9-10. Equations for series and parallel arrangements can be found in Table 9-11.

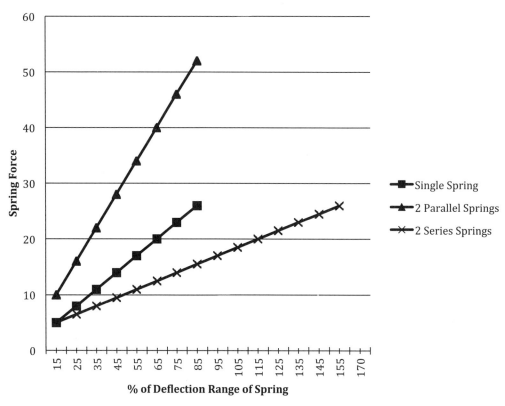

Figure 9-10: Force vs. Deflection of Spring Arrangements

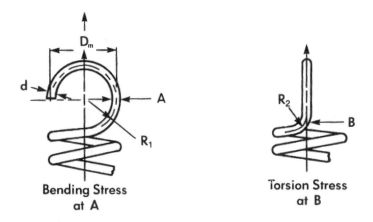

Figure 9-11: Extension Spring End Stress Locations

Source: Associated Spring Barnes Group Inc. (www.asbg.com)

Table 9-9: Helical Coil Extension Spring Sizing Equations

d = Wire diameter (in or mm)	
SPRING INDEX For round wire cross section springs Formability a concern when C < 4	$C = \dfrac{D_m}{d}$ Starting point: $C = 8$
MEAN SPRING DIAMETER OD = Outside diameter of spring coil ID = Inside diameter of spring, nominal	For wire with round cross section: $D_m = OD - d$ $D_m = \dfrac{OD + ID}{2}$
OPERATING DEFLECTION RANGE	$x_w = L_{max} - L_{min} = x_{max} - x_{min}$
TRIAL WIRE DIAMETER Based on uncorrected stress τ = Target stress at max force F_{max} Be sure to check units for consistency.	For wire with round cross section: $d = \sqrt[3]{\dfrac{8F_{max}D_m}{\pi\tau}}$
NUMBER OF ACTIVE COILS k = Spring constant (Table 9-11) G = Modulus of Rigidity (Table 9-1) (psi or MPa)	For wire with round cross section: $N_a = \dfrac{d^4 G}{8D_m^3 k}$

Table 9-10: Helical Coil Extension Spring Free Length, End Types

Feature	End Type	
	Full Twist Loops or Hooks	**Half Twist Hooks**
FREE LENGTH Length between attachment surfaces N_a = Number of active coils (Table 9-9) d = wire diameter D_m = Coil mean diameter (Table 9-9)	$L_f = 2(D_m - d) + (N_a + 1)d$	$L_f = (D_m - d) + (N_a + 1)d$

HELICAL EXTENSION SPRING CATALOG SELECTION STEPS

Helical coil extension springs are specified in catalogs by coil diameter (either outer or inner), wire diameter, initial tension, free length, and spring rate. Key spring characteristics also provided by the manufacturer include load and deflection at the maximum operating deflection point. The following procedure can be used to select an extension spring from a catalog:

1. Determine the application parameters. These include:
   - Desired force at the most extended position and any other known forces

Table 9-11: Extension Spring Constant, Stress, and Frequency

D_m = Mean diameter (Table 9-9) (in or mm)	d = Wire diameter (in or mm)
N_a = Number of active coils (Table 9-9)	G = Modulus of Rigidity (Table 9-1) (psi or MPa)

SPRINGS IN PARALLEL	$F = \sum F_i = x \sum k_i \qquad x = x_i \qquad k = \sum k_i$
SPRINGS IN SERIES	$F = \dfrac{1}{\sum \dfrac{1}{k_i}} \sum x_i \qquad x = \sum x_i \qquad \dfrac{1}{k} = \sum \dfrac{1}{k_i}$
SPRING CONSTANT	$k = \dfrac{\lvert F_2 - F_1 \rvert}{\lvert L_2 - L_1 \rvert} = \dfrac{F - F_{IT}}{x}$ For round wire: $k = \dfrac{d^4 G}{8 D_m^3 N_a}$
SPRING FORCE	$F = kx + F_{IT}$
SPRING DEFLECTION Deflection from free length Wire with round cross section	$x = \dfrac{8(F - F_{IT})D_m^3 N_a}{d^4 G} \qquad x = L - L_f$
INITIAL TENSION FORCE Wire with round cross section Check units for consistency.	$F_{IT} = \dfrac{\pi d^3 \tau_{IT}}{8 K D_m} \qquad K$ = Static loading factor
INITIAL TENSION STRESS Measured in psi When C=8: $\tau_{IT} \cong 14696.59\, psi$	$\tau_{IT} \cong \dfrac{\tau_{IT,1} + \tau_{IT,2}}{2}$ $\tau_{IT,1} \cong -4.231C^3 + 181.5C^2 - 3387C + 28640$ $\tau_{IT,2} \cong -2.987C^3 + 139.7C^2 - 3427C + 38404$
STRESS CORRECTION FACTORS C = Spring index (Table 9-9) R_1 = Mean radius of hook (Fig. 9-11) R_2 = Mean radius of bend in hook (Fig. 9-11)	For Static Loading: $K = 1 + \dfrac{0.5}{C}$ For Dynamic Loading: (Wahl Factor) $K = \dfrac{4C-1}{4C-4} + \dfrac{0.615}{C}$ For Extension Springs: $K_1 = \dfrac{4C_1^2 - C_1 - 1}{4C_1(C_1 - 1)} \qquad K_2 = \dfrac{4C_2 - 1}{4C_2 - 4}$ $C_1 = \dfrac{2R_1}{d} \qquad\qquad C_2 = \dfrac{2R_2}{d}$
TORSIONAL STRESS IN SPRING COIL Wire with round cross section	$\tau = K \dfrac{8 F D_m}{\pi d^3}$
BENDING STRESS IN HOOK, MAX Full diameter hook or loop type ends	$\sigma_h = \dfrac{16 D_m F_{max} K_1}{\pi d^3} + \dfrac{4 F_{max}}{\pi d^2}$
TORSIONAL STRESS IN HOOK BEND, MAX Full diameter hook or loop type ends	$\tau_b = \dfrac{8 D_m F_{max} K_2}{\pi d^3}$
NATURAL FREQUENCY f_n = natural frequency (Hz) g = 386.1 in/s^2 or 9810 mm/s^2 γ = weight density of material (Table 9-1)	Both spring ends constrained: $f_n = \dfrac{d}{\pi N_a D_m^2} \sqrt{\dfrac{Gg}{8\gamma}}$

- Operating deflection range or desired spring length at the extreme operating positions and any other critical positions
- Desired outer diameter of the spring, if known
- End type
- Environmental factors, such as heat or material compatibility

2. Select a trial spring material based on performance parameters.
   Table 9-1 can be used to facilitate selection. Stainless steel springs are recommended for high temperatures or for corrosive environments. Stainless steel springs exert only about 83.3% of the force exerted by music wire springs of the same geometry.

3. Select a preliminary stock spring from a catalog that satisfies the desired load(s) at the desired deflection(s) and fits within the assembly geometry. First check the catalog to be sure what the given parameters mean.
   a. If the coil diameter is known, first look at springs that meet that requirement. Catalog springs are often organized by outside diameter.
   b. If the deflection range and forces at two lengths are known:

      - Look first for springs with a lower initial tension than the minimum operating force.
      - Look for springs that exert greater than or equal to the maximum required force at maximum extension.
      - The length at maximum deflection listed in the catalog must be equal to or greater than the maximum operating length.
      - The difference between free length and most deflected length must be greater than the operating deflection range.
      - A solution may require changing geometric and load constraints for the application.

   c. If the maximum force and deflection range are known, but lengths are not specified:
      - Look for springs that exert greater than or equal to the maximum required force at maximum extension.
      - Look for springs where the difference between free length and maximum length is greater than the operating deflection range required.
      - Select a spring with a maximum operating length that fits the required application. Set the maximum operating length based on the spring that is selected.

Table 9-12: Allowable Extension Spring Stress

Springs with set not removed

Application	Material	Maximum Allowable % of Ultimate Tensile Strength		
		In Torsion		In Bending
		Body	End	End
Static Loading	Music Wire (A228)	45%	40%	75%
	Cold Drawn Wire (A227)			
	Oil Tempered Wire (A229)			
	Stainless Steels (Austenitic, 301, 302)	35%	30%	55%
Fatigue Loading (Life = 10^6 cycles)		**Not Shot Peened**		
	Music Wire (A228)	33%	30%	47%
	Stainless Steels (Austenitic, 301, 302)			
Fatigue Loading (Life = 10^7 cycles)		**Not Shot Peened**		
	Music Wire (A228)	30%	28%	45%
	Stainless Steels (Austenitic, 301, 302)			

Data Source: Associated Spring Barnes Group Inc. (www.asbg.com)

4. Analyze the forces produced by the selected spring. Calculate the deflection from free length at both extreme operating positions and any other critical positions. Use the force formula $F = kx + F_{IT}$ (Table 9-11) and the value of spring rate to calculate the spring force at all critical operating positions. If the selected spring does not provide acceptable forces at all critical operating positions, a different spring must be selected and analyzed. The force at minimum deflection must be greater than the initial tension force of the spring. To increase the amount of force a spring can exert, the spring requires more mass. This means increasing the wire size, coil size, free length, number of coils, or cross-sectional geometry. Changing spring material can also affect force capacity. If more force change over a deflection is needed, a higher spring constant should be sought.

5. For applications with rapid cycling, check the natural frequency of the spring. Natural frequency can be calculated using the formulas in Table 9-11. The natural frequency of the spring should be 15 or more times greater than the forcing frequency of the application. For a higher natural frequency, select a spring with reduced weight or a higher spring constant.

6. Analyze the spring for fatigue life. This is necessary for severe service conditions with extremely rapid deflections. It is also needed for normal service conditions where the spring is deflected beyond the linear range or beyond the manufacturer's recommendation. It is also needed for applications with a high (millions of cycles) life expectancy. The torsional factor of safety should be calculated for the coils and for the hooks. The bending factor of safety should be calculated for the hooks. Calculate the torsional and bending fatigue safety factors using the equations in Table 9-3.

HELICAL EXTENSION SPRING DESIGN STEPS

The easiest way to design a spring is to use commercial software or seek the assistance of an applications engineer at a manufacturer. Spring design is an iterative process that is suited for spreadsheet automation at the very least. The following procedure is iterative and may be used to design a helical coil extension spring made from round wire. See the recommended resources for more information on spring design using non-round wire.

1. Determine the application parameters. These include:
   • Desired forces at the most and least extended operating positions
   • Operating deflection range or spring length at one position where force is known
   • Desired outer diameter of the spring, if known
   • End type
   • Environmental factors, such as heat or material compatibility

2. Select a trial spring material type based on performance parameters. Find the shear modulus (G) for the chosen material. Table 9-1 can be used to facilitate selection. Stainless steel springs are recommended for high temperatures or for corrosive environments. Stainless steel springs exert only about 83.3% of the force exerted by music wire springs of the same geometry.

3. Select a trial wire diameter:
   a. Method 1: It is often expedient to find a likely wire diameter by looking in a catalog for springs that are nearly right for the application. Look for springs with listed loads that exceed what is needed. If more force is needed, the spring requires more mass. This means

increasing the wire size, coil size, free length, number of coils, or cross sectional geometry.

b.   Method 2: If outer diameter is known, set the spring index to some acceptable value (8 is a good place to start). Use the spring index equation (Table 9-9), and use outer diameter for the value of D_m, to calculate d. This calculation typically takes the form d=OD/8. Then calculate D_m = OD – d. Calculate spring index again. Reduce wire diameter or increase OD until the spring index is close to 8. Select an available wire diameter from stock sizes that is near what was calculated. Table 9-2 shows some available values of d.

c.   Method 3: Target a stress at maximum operating force that is near but below the maximum allowable operating stress for the spring material. Maximum allowable stress can be calculated using Table 9-12 and Figure 9-3 together. A wire diameter must be assumed when using Figure 9-3. A trial value for d based on a target stress value can be calculated using the trial wire diameter equation in Table 9-9. If a target OD is unknown, a value of D_m must be assumed. A good place to start is to use the assumed value of d and solve the equation D_m = 8d. Once the trial diameter equation is solved, if the calculated value of d is very different from that assumed when selecting stress level, iterate by changing OD and/ or d until they agree. If a trial value of d is calculated, an available wire size close to that value should then be selected, and then the spring index value checked to be sure it is greater than 4.

4. Validate the material choice for the chosen wire diameter: Calculate the maximum bending stress in the spring hook, the maximum torsional stress in the spring hook bend, and the maximum torsional stress in the spring coil at maximum operating force using the equations in Table 9-11. A trial value of C_2 = 5 can be used to estimate the bend in the hook if it is not known. Then compare these stresses against the maximum allowable stresses for the material and wire diameter. Table 9-12 and Figure 9-3 in combination can be used to determine maximum allowable stresses for all three cases. If any of the allowable stresses are exceeded, go back to either the wire diameter or material selection steps. Wire diameter may need to be increased.

5. Check the initial tension force of the spring. Calculate the initial tension force using the equation in Table 9-11. Use the correction factor for static loading. If the initial tension force exceeds the minimum

force required by the application, a different wire diameter and/or coil diameter must be selected. The simplest way to decrease initial tension is to decrease wire diameter.

6. Calculate the spring rate and solve for the number of active coils. If two loads and the distance between them are known, spring rate may be calculated using the equation for spring constant as a function of two loads in Table 9-11. If only one load is known, another must be assumed. The spring rate may be changed by adjusting the expected load or length at one or more operating points. Use the calculated spring rate to calculate the number of active coils using the formula in Table 9-9. The active number of coils should be rounded to the near-est 1/4 coil, which is a typical manufacturing accuracy for forming the spring. Recalculate the spring rate based on the number of active coils. Use the equation for spring constant as a function of active coils found in Table 9-11.

7. Calculate the free length of the spring using the formula in Table 9-10. Note that stresses were calculated assuming full loops or hooks.

8. Calculate the remaining unknown lengths, deflections, and forces at each operating position. Operating length is free length plus deflection. Deflection can be calculated at each load using the formula $F = kx + F_{IT}$ and solving for x. Force is calculated using the formula $F = kx + F_{IT}$ where x is deflection from free length.

9. For applications with rapid cycling, check the natural frequency of the spring. Natural frequency can be calculated using the formula in Table 9-11. The natural frequency of the spring should be 15 or more times greater than the forcing frequency of the application. For a higher natural frequency, select a spring with reduced weight or a higher spring constant.

10. Analyze the spring for fatigue life. This is necessary for severe service conditions with extremely rapid deflections. It is also needed for nor-mal service conditions where the spring is deflected beyond the linear range. It is also needed for applications with a high (millions of cycles) life expectancy. If the spring was designed using the fatigue allowable stress multipliers from Table 9-12, then the spring should be acceptable in fatigue. If not, the torsional factor of safety should be calculated for the coils and for the hooks. Also the bending factor of safety should be calculated for the hooks. Calculate the torsional and bending fatigue safety factors using the equations in Table 9-3.

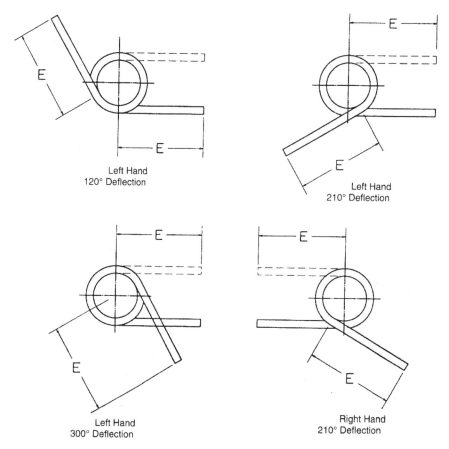

Left Hand
120° Deflection

Left Hand
210° Deflection

Left Hand
300° Deflection

Right Hand
210° Deflection

Figure 9-12: Helical Coil Torsion Springs

Source: Associated Spring Barnes Group Inc. (www.asbg.com)

HELICAL COIL TORSION SPRINGS

Helical coil torsion springs are made up of a helical coil of wire with its two ends extending tangentially from the coil. The ends can be positioned and shaped in a variety of ways. The most common ends are straight and positioned at some angle from each other. A few of the many standard configurations, including left and right hand winding, are shown in Figure 9-12. Torsion springs are normally made from round wire and are meant to be installed over a rod or mandrel. They are meant to be deflected in the direction that tends to wind the coils tighter. Deflection in the other direction is possible but not recommended. In calculating stresses, this text assumes deflection in the recommended direction only. See the recommended resources or consult an applications engineer if deflection in the other direction is desired.

Table 9-13: Helical Coil Torsion Spring Sizing Equations
Round wire

M = Torque at given deflection k = Spring constant (See Table 9-14)	d = Wire diameter (in or mm) $\theta_{max,rev}$ = Max. deflection from free (revolutions)
TRIAL WIRE DIAMETER Based on uncorrected stress σ_{max} = target stress at max torque M_{max} Check units for consistency.	$d = \sqrt[3]{\dfrac{32 M_{max}(1.15)}{\pi \sigma_{max}}}$
SPRING INDEX Formability a concern when C < 4	$C = \dfrac{D_m}{d}$ \quad Starting point: $C = 8$
MEAN DIAMETER OD = Outside diameter of coil (in or mm) ID = Inside diameter of coil (in or mm)	$D_m = OD - d$ \quad $D_m = \dfrac{OD + ID}{2}$
NUMBER OF ACTIVE COILS E = Modulus of elasticity (Table 9-1) (psi or MPa)	$N_a = \dfrac{E d^4}{10.8 D_m k}$ \quad $N_a = N_b + N_e$
EQUIVALENT NUMBER OF ACTIVE COILS L_1, L_2 = Leg lengths (in or mm)	For straight ends: $N_e = \dfrac{L_1 + L_2}{3\pi D_m}$
NUMBER OF BODY COILS	$N_b = N_a - N_e$
BODY LENGTH WHEN DEFLECTED	$L_b = d(N_b + 1 + \theta_{max,rev})$
MAXIMUM MANDREL DIAMETER	$d_{md} = 0.9(D_{m,min} - d) = 0.9\left(\dfrac{D_m N_b}{N_b + \theta_{max,rev}} - d\right)$

The spring rate of helical torsion springs is defined as a function of angular deflection in revolutions. When deflected, helical coil torsion springs undergo bending stresses. These springs must be operated within an allowable stress range. The maximum stress occurs at the inside surface of the coil. In fatigue, the stress at the outside surface is of concern. Torsion springs can be used in parallel. When arranged in parallel, the moment loads at a given deflection are additive, as are the spring rates. Formulas for calculating spring rate and stresses can be found in Table 9-14, and allowable stresses are given in Table 9-15.

HELICAL TORSION SPRING CATALOG SELECTION STEPS

Torsion springs are normally listed in catalogs with the following information: wire diameter, mandrel size, leg length, torque at maximum recommended deflection, deflection angle, end positions, and body length or minimum axial space requirement. The following steps are iterative and can be used to select a torsion spring from a catalog:

Table 9-14: Torsion Spring Constant, Force, Deflection, and Stress

M = Torque at given deflection	d = Wire diameter (in or mm)
E = Modulus of elasticity (See Table 9-1)	D_m = Mean coil diameter (Table 9-13) (in or mm)
N_a = Number of active coils (see Table 9-13)	

SPRINGS IN PARALLEL k_i = individual spring constant	$$M = \sum M_i = \theta_{rev} \sum k_i \qquad k = \sum k_i$$
SPRING CONSTANT	$$k = \frac{M}{\theta_{rev}} = \frac{\lvert M_2 - M_1 \rvert}{\lvert \theta_{2,rev} - \theta_{1,rev} \rvert}$$ For round wire: $$k = \frac{Ed^4}{10.8 D_m N_a}$$
SPRING TORQUE	$$M = k\theta_{rev}$$ From catalogs, torque at deflection θ_{rev} : $$M = \frac{M_{max} \theta_{rev}}{\theta_{max,rev}}$$
SPRING DEFLECTION Measured in revolutions Round wire	$$\theta_{rev} \cong \frac{10.8 M D_m N_a}{d^4 E}$$
STRESS CORRECTION FACTORS C = Spring index (See Table 9-13)	$$K_{IN} = \frac{4C^2 - C - 1}{4C(C-1)}$$
BENDING STRESS IN SPRING Round wire	$$\sigma_{IN} = K_{IN} \frac{32M}{\pi d^3}$$
YIELD STRENGTH S_{ut} = See Fig. 9-3	$S_y \leq S_{ut}$ See Table 9-15
STATIC SAFETY FACTOR AGAINST YIELDING	$$SF = \frac{S_y}{\sigma_{IN,max}}$$

Table 9-15: Allowable Torsion Spring Bending Stress

Application	Material	Maximum Allowable % of Ultimate Tensile Strength	
		Stress Relieved	**Not Stress Relieved**
Static Loading	Music Wire (A228)	80%	100%
	Cold Drawn Wire (A227)	80%	100%
	Oil Tempered Wire (A229)	85%	100%
	Stainless Steels (Austenitic, 301, 302)	60%	80%
		Not Shot Peened	**Shot Peened**
Fatigue Loading **(Life = 10^6 cycles)**	Music Wire (A228)	50%	60%
	Oil Tempered Wire (A230)	53%	62%
	Chrome Vanadium Wire (A232)	53%	62%
	Stainless Steels (Austenitic, 301, 302)	50%	60%

Data Source: Associated Spring Barnes Group Inc. (www.asbg.com)

1. Determine the application parameters. These include:
   - Desired torque at the most deflected operating position
   - Desired angle between legs at maximum deflection
   - Desired outer or inner diameter of the spring, or desired mandrel size
   - Environmental factors, such as heat or material compatibility

2. Select a trial spring material based on performance parameters. Table 9-1 can be used to facilitate selection. Stainless steel springs are recommended for high temperatures or for corrosive environments. Stainless steel springs exert less force than music wire springs of the same geometry.

3. Select a preliminary stock spring from a catalog that satisfies the desired maximum torque at the desired maximum deflection and fits within the assembly geometry. First check the catalog to be sure what the given parameters mean.
   - Torsion springs should not be deflected beyond the listed deflection value. Springs should be selected that have a maximum deflection range that is larger than required.
   - Depending on how a catalog is organized, it can be expedient to look first for springs with the desired free leg angle and a maximum deflection that is greater than what is needed.
   - Sometimes it might be best to look for springs that fit over the desired mandrel and then look at end configuration and maximum deflection.
   - Sometimes it may be faster to look for a torque at maximum deflection that exceeds what is needed and then look at end positions and deflection range.

4. Analyze the torques produced by the selected spring. Calculate the deflection from free position in revolutions for both extreme operating positions and any other critical positions. Using the maximum deflection range and maximum torque of the spring provided by the manufacturer, calculate the spring torque at all critical operating positions using the formula in Table 9-14. If the selected spring does not provide acceptable torques at all critical operating positions, a different spring must be selected and analyzed.

5. For applications with rapid cycling, determine the natural frequency of the spring. Consult the manufacturer for calculation guidance.

6. Analyze the spring for fatigue life. This is necessary for severe service conditions with extremely rapid deflections. It is also needed for

normal service conditions where the spring is deflected beyond the linear range or beyond the manufacturer's recommendation. It is certainly needed for applications with a high (millions of cycles) life expectancy. Calculate the fatigue safety factor using the equations in Table 9-4.

7. Integrate the spring into the application geometry. Be aware that catalog values of spring rate and torque are determined at some point along the leg lengths, usually at half of the extended length. Check the catalog to be sure where this is. The application geometry must apply force at this position in order for the spring to behave as expected.

HELICAL TORSION SPRING DESIGN STEPS

The easiest way to design a spring is to use commercial software or seek the assistance of an applications engineer at a manufacturer. Spring design is an iterative process that is suited for spreadsheet automation at the very least. The following procedure is iterative and may be used to design a helical coil torsion spring made from round wire.

1. Determine the application parameters. These include:
   - Desired torque at the extreme operating positions and any other critical positions
   - Desired deflection range (revolutions) between the extreme operating positions
   - Desired leg lengths
   - Desired leg positions at maximum or minimum deflection, if known
   - Desired outer or inner diameter of the spring, or desired mandrel size, if known
   - Environmental factors, such as heat or material compatibility
2. Select a trial spring material based on performance parameters. Table 9-1 can be used to facilitate selection. Stainless steel springs are recommended for high temperatures or for corrosive environments. Stainless steel springs exert less force than music wire springs of the same geometry.
3. Select a wire diameter:
   a. Method 1: It is often expedient to find a likely wire diameter by looking in a catalog for springs that are nearly right for the application. Look for springs with listed torques that exceed what is

needed. If more torque is needed, the spring requires more mass. This means increasing the wire size, coil size, free length, number of coils, or cross-sectional geometry.

b. Method 2: If D_m is known or can be approximated, set the spring index value at some acceptable value (8 is reasonable) and using the spring index equation (Table 9-13) calculate d. This equation takes the form $d = D_m/8$. If OD is known, use it instead of D_m in the spring index equation, calculate $D_m = OD - d$, and then calculate spring index again. Reduce wire diameter or increase D_m until the spring index is close to 8. Select an available wire diameter from stock sizes that is near what was calculated. Table 9-2 shows some available values of d.

c. Method 3: Target a stress at maximum operating torque that is near but below the maximum allowable operating stress for the spring material. Maximum allowable stress can be calculated using Table 9-15 and Figure 9-3 together. A wire diameter must be assumed when using Figure 9-3. A trial value for d based on a target stress value can be calculated using the trial wire diameter equation in Table 9-13. Once the trial diameter equation is solved, if the calculated value of d is very different from that assumed when selecting stress level, iterate by changing d until they agree. If a trial value of d is calculated, an available wire size close to that value should then be selected, and then the spring index value checked to be sure it is greater than 4.

4. Validate the material choice for the chosen wire diameter. Calculate the bending stress at the inside of the coil (Table 9-14) at maximum operating deflection. Then compare this maximum bending stress against the maximum allowable bending stress for the material. One method of comparison is to compute the static safety factor against yielding. Alternatively, Table 9-15 and Figure 9-3 can be used together to determine maximum allowable stress. If the allowable stress is exceeded, go back to either the wire diameter or material selection steps.

5. Calculate the spring rate and solve for the number of active coils. The torque and deflection must be specified or assumed for two operating positions. Spring rate may be calculated using the equation for spring constant as a function of two torques in Table 9-14. The spring rate may be changed by adjusting the expected load or angle at one or more operating points. Use the calculated

spring rate to calculate the number of active coils using the formula in Table 9-13. Calculate the number of body coils (Table 9-13), and round it to provide the end positions that are required for the application geometry. Now calculate the number of active turns based on the new body length and recalculate spring rate based on active turns (Table 9-14).

6. Calculate the deflections from free position at all operating positions. Use the calculated spring rate to find the deflections using the deflection formula in Table 9-14.

7. Calculate the axial length of the spring (body length) at maximum deflection. The formula for body length can be used from Table 9-13. This is the minimum axial space that must be provided for the spring.

8. Analyze the torques produced by the selected spring. Use the calculated spring rate to calculate the torques at all critical operating positions, such as the minimum deflection. The torque formula in Table 9-14 can be used. If the spring does not provide acceptable torques at all critical operating positions, a different spring material or geometry must be selected and analyzed.

9. Calculate and verify the mandrel size required for the spring. Calculate the maximum mandrel diameter using the formula in Table 9-13. Verify that this is acceptable for the application.

10. Calculate the spring inside diameter and outside diameter.
    $OD = D_m + d$ and $ID = D_m - d$

11. For applications with rapid cycling, determine the natural frequency of the spring. Consult with a manufacturer for calculation guidance.

12. Analyze the spring for fatigue life. This is necessary for severe service conditions with extremely rapid deflections. It is also needed for applications with a high (millions of cycles) life expectancy. If the spring was designed using the fatigue limits from Table 9-15, the spring should survive in fatigue to one million cycles. Otherwise, calculate the fatigue safety factor using the equations in Table 9-4.

BELLEVILLE SPRING WASHERS

Belleville spring washers are conical washers made of spring material. They are ideal for spring applications that require relatively high forces from low axial deflections. These springs are ideal for taking up "play" in assemblies

Figure 9-13:
Belleville Washer Geometry
Source: Associated Spring
Barnes Group Inc. (www.asbg.com)

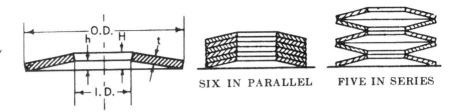

and compensating for small dimensional changes due to wear. They are suited for shock loading and can be used under fasteners to maintain proper joint tightness. Belleville spring washers have different spring rates depending on their height to thickness ratio. A graph of spring force as a function of percent deflection and ratio is shown in Figure 9-15.

Belleville washers may be stacked to produce different characteristics. The geometry of a Belleville washer is illustrated in Figure 9-13, with parallel and series stack configurations. A parallel stack of identical washers will have a deflection equal to that of a single washer, while the load at that deflection will be equal to the load of one washer multiplied by the number of washers in the stack. A series stack will have a load equal to the load of a single washer while the deflection at that load will be equal to the deflection of one washer multiplied by the number of washers in the stack. These relationships can be found in Table 9-17. Washers used in a series stack must each have an h/t ratio greater than 1.3. Series-parallel stacks are also possible. Figure 9-14 is a graph of the stacking behavior of Belleville springs.

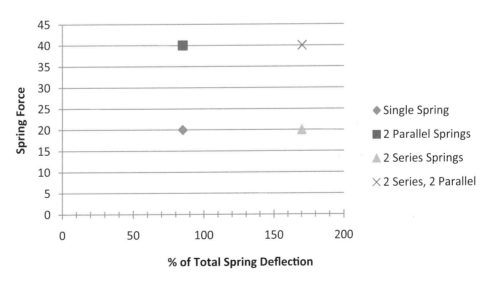

Figure 9-14: Force vs. Deflection of Spring Arrangements

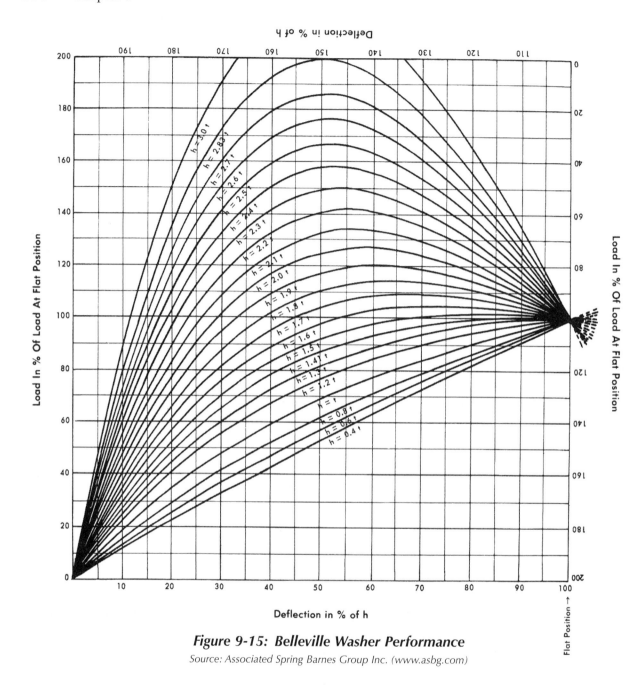

Figure 9-15: Belleville Washer Performance

Source: Associated Spring Barnes Group Inc. (www.asbg.com)

All series and series-parallel stacks have some danger of instability, so proper constraints must be designed into the assembly. Constraint by means of a hardened pin or hardened hole will produce friction that will reduce the load produced by the spring. Pins or holes with hardness values above 48 Rc are recommended.

Most Belleville washers have generous radii at their edges to reduce stress concentrations and improve fatigue life. These radii also act to reduce the moment arm of the spring and increase the spring rate. Belleville washers of large diameters are sometimes made with small flats at their top and bottom surfaces. These load bearing surfaces have a width approximately equal to the outside diameter of the spring divided by 150. The presence of these flats changes the moment arm in the spring and results in an increased spring force.

The target range of stable deflection for a Belleville washer is between 15% and 85% of its total deflection range, h. It is possible to use a Belleville washer in an application where it is deflected past flat, but special design considerations apply and fatigue is of concern. As a washer with an h/t ratio greater than 2.83 approaches flat, sudden movement occurs and the washer will snap past flat. The washer will require an external force to snap back to its original position.

Contact with a Belleville washer should be uniform around its entire circumference. When a washer is to be deflected within 15% of flat but is not allowed to snap past flat, a positive stop should be provided by placing the washer in a precisely made flat-bottomed hole or bore. This deflection range is often desirable because some h/t ratios produce a nearly constant force near the flat position.

When deflected, Belleville washers undergo stress. The compressive stress at the inside radius of the convex side of the washer will be the worst case. Table 9-17 contains formulas related to stress calculation, and Table 9-18 provides some common Belleville spring materials and their characteristics. Allowable stresses can be calculated using Table 9-19.

BELLEVILLE SPRING WASHER CATALOG SELECTION STEPS

Belleville springs are normally listed in catalogs with the following information: inside diameter, outside diameter, thickness, unloaded height, load at flat position, and recommended pin and/or hole size for proper fit. The following steps can be used to select a Belleville spring from a catalog. This procedure assumes that the washer will not be deflected past flat. The process is iterative.

1. <u>Determine the application parameters.</u> These include:
   - Desired force at the most compressed operating position
   - Desired force/deflection performance, if known (for example, constant force)

Table 9-16: Belleville Spring Washer Sizing Equations

x = Deflection from free at given force (in or mm) t = Material thickness of washer (in or mm)	h = Total deflection range of washer (in or mm)
DIAMETER RATIO D_{OUT} = Outside diameter (in or mm) D_{IN} = Inside diameter (in or mm)	$R_d = \dfrac{D_{OUT}}{D_{IN}}$ $R_d \cong 2$ for maximum energy storage
HEIGHT TO THICKNESS RATIO	$R_h = \dfrac{h}{t}$ $R_h = 0.4$ for approximately linear spring rate $R_h > 2.83$ for snap action $1.41 < R_h < 2.1$ for approximately constant force
TARGET MINIMUM DEFLECTION RANGE x_w = Total operating deflection range (in or mm)	$h_t = 1.43 x_w$
DEFLECTION RANGE OF WASHER	$h = R_h t$
OVERALL WASHER HEIGHT	$H = h + t$
HEIGHT AT OPERATING POSITION	$L = H - x$
MINIMUM OPERATING HEIGHT	$L_{\min} = H - 0.85h$
MAXIMUM OPERATING HEIGHT	$L_{\max} = H - 0.15h$
THICKNESS OF MATERIAL D_{OUT} = Outside diameter (in or mm) $F_{f\,lat}$ = Force at flat position	English: $t = \sqrt[4]{\dfrac{D_{OUT}^2 F_{f\,lat}}{R_h 19.2 \times 10^7}}$ Metric: $t = \dfrac{1}{10}\sqrt[4]{\dfrac{D_{OUT}^2 F_{f\,lat}}{R_h 132.4}}$

- Desired deflection range or spring height at the extreme operating positions
- Desired outer or inner diameter of the spring, if known
- Environmental factors, such as heat or material compatibility

2. Select a material based on performance parameters. Table 9-1 can be used to facilitate selection. Stainless steel springs are recommended for high temperatures or for corrosive environments. Stainless steel springs exert less force than spring steel springs of the same geometry.

3. If an operating range of deflection is known, calculate a target minimum deflection range (h_t) for the washer. The target range of deflection for a Belleville washer is between 15% and 85% of its total deflection range h. Deflections past this range are possible, but the stresses are considerably higher. The formula for target washer deflection range can be found in Table 9-16.

Table 9-17: Belleville Spring Force, Deflection, and Stress

SPRINGS IN PARALLEL	$F = \sum F_i \qquad x = x_i$
SPRINGS IN SERIES	$F = F_i \qquad x = \sum x_i$
COMPRESSIVE STRESS ν = Poisson's Ratio (Table 9-18) x = Deflection from free (in or mm) D_{OUT} = Outside diameter of washer (in or mm) h = Deflection height of washer (in or mm) t = Material thickness of washer (in or mm) E = Modulus of elasticity (Table 9-18) (psi or MPa)	$\sigma_c = -\dfrac{4Ex}{C_{c1}D_{OUT}^2\left(1-\nu^2\right)}\left[C_{c2}\left(h-\dfrac{x}{2}\right)+tC_{c3}\right]$
CONSTANTS USED IN COMPRESSIVE STRESS R_d = Diameter ratio (Table 9-16)	$C_{c1} = \dfrac{6}{\pi \ln R_d}\left[\dfrac{\left(R_d-1\right)^2}{R_d^2}\right]$ $C_{c2} = \dfrac{6}{\pi \ln R_d}\left(\dfrac{R_d-1}{\ln R_d}-1\right)$ $C_{c3} = \dfrac{6}{\pi \ln R_d}\left(\dfrac{R_d-1}{2}\right)$
SAFETY FACTOR FOR YIELDING	$SF = \dfrac{S_{uc}}{\sigma_c}$
ALLOWABLE COMPRESSIVE STRESS	S_{uc} = See Table 9-19 and Figure 9-3

Table 9-18: Common Belleville Spring Materials

Material	Ultimate Tensile Strength		Rockwell Hardness	Modulus of Elasticity (E)		Poisson's Ratio
	ksi	MPa	Rc	psi x 10^6	GPa	ν
Steel, Spring Temper	246	1700	50	30	207	0.3
302 Stainless Steel	189	1300	40	28	193	0.31
17-7 Stainless Steel	210	1450	44	29.5	203	0.34

Data Source: Associated Spring Barnes Group Inc. (www.asbg.com)

Table 9-19: Belleville Spring Allowable Stress

Application	Material	Maximum Allowable % of Ultimate Tensile Strength	
		Set Removed	Set Not Removed
Static Loading	Carbon & Alloy Steels	275%	120%
	Austenitic Stainless Steels	160%	95%

Data Source: Associated Spring Barnes Group Inc. (www.asbg.com)

4. If specific performance is required, select an h/t ratio based on performance requirements. This can be done using the manufacturer's graph or the graph in Figure 9-15. The target range of deflection for a Belleville washer is between 15% and 85% of its unloaded height, so look within that range on the x-axis.

5. Select a preliminary stock spring from a catalog that satisfies the desired maximum load at maximum deflection and fits within the assembly geometry. First check the catalog to be sure what the given parameters mean.
   - If a hole or mandrel size is known, it is expedient to locate springs with the desired diameter first.
   - If a height at maximum deflection is known, look for washers with height at maximum deflection that is less than what is desired.
   - Look for washers where the desired load divided by the load at flat position is greater than 15%.
   - If a target washer deflection height has been calculated, the value of H – t in the catalog should exceed it.
   - If specific force performance as a function of percent of deflection is required, look for springs with the h/t ratio selected earlier.

6. If the operating heights are not known, calculate the operating heights of the washer. Divide the desired load by the load at flat position percentage value for each position. Find the line for the chosen h/t ratio on the manufacturer's graph or the graph in Figure 9-15. Find the percentage of deflection for the calculated percent load for each operating position. Calculate the numerical value of deflection for each. Subtract the deflections from overall washer height H to find the operating heights.

7. Check the operating range of the washer. The typical target operating range for a Belleville washer is between 15% and 85% of its total deflection range h. Calculate L_{max} and L_{min} for the washer using the equations for maximum and minimum operating height in Table 9-16. If the operating heights are outside of this range, either the operating heights or the overall height of the washer will need to change.

8. Analyze the forces produced by the selected spring. Figure 9-15 is a graph of deflection vs. load as percentages of flat position values for various h/t values. The manufacturer will likely also supply a graph. Calculate the deflection as a percent of h for each operating position. Find the percent of flat position load on the graph for the deflection percent and

h/t value. Calculate the numerical load based on the percent of flat load for each position. If the selected spring does not provide acceptable forces at all critical operating positions, a different spring may be selected and analyzed. Changing the thickness ratio of the washer will change its load performance. Also consider stacking springs if a desired load and deflection combination cannot be found. Parallel stacks increase load whereas series stacks increase deflection. Series and parallel force and deflection relationships can be found in Table 9-17.

9. For applications with a high number (millions) of cycles, check the fatigue performance of the spring. Consult with the manufacturer.

BELLEVILLE SPRING WASHER DESIGN STEPS

The easiest way to design a spring is to use commercial software or seek the assistance of an applications engineer at a manufacturer. Spring design is an iterative process that is suited for spreadsheet automation at the very least. The following procedure is iterative and may be used to design a Belleville spring washer. It is assumed that this washer will not be deflected past flat. For dynamic applications with a high cycle count, consult with a manufacturer.

1. Determine the application parameters. These include:
   - Desired force at the extreme operating positions and any other critical positions
   - Desired deflection range or desired spring heights at the extreme operating positions
   - Desired outer or inner diameter of the spring
   - Environmental factors, such as heat or material compatibility

2. Select a material based on performance parameters. Table 9-1 can be used to facilitate selection. Stainless steel springs are recommended for high temperatures or for corrosive environments.

3. If inner diameter is known, calculate the outer diameter. Assume a diameter ratio of 2 for maximum energy storage. Solve the diameter ratio equation (Table 9-16) for outer diameter. This equation takes the form:

$$OD = 2(ID)$$

4. Select a height to thickness ratio based on performance requirements. This can be done using the graph in Figure 9-15. The target range of deflection for a Belleville washer is between 15% and 85% of its unloaded

height, so look in that range on the graph for most applications. As a starting point, it can be useful to target a minimum deflection of 50% h.

5. Calculate a flat position load. Based on the graph in Figure 9-15 and the target percent deflection (50% was assumed in the last step,) calculate the flat position load. Where F = desired load at the deflected position and %load is from the y-axis of the graph:

$$F_{flat} = \frac{F}{\%_{load}}$$

6. Calculate the material thickness. Use the flat position load calculated earlier and the formula for thickness in Table 9-16.

7. Calculate the washer deflection range and overall height. First calculate the total deflection range of the washer using the formula in Table 9-16. This takes the form:

$$h = t\left(R_h\right)$$

The h/t ratio R_h was determined earlier. Then use the overall height equation H = h + t to find overall height.

8. If the operating heights are not known, calculate the operating heights of the washer. Use the assumption from Step 4 for target minimum percent deflection to calculate the maximum height of the washer in operation using the formula (Table 9-16)

$$L = H - (\%_{height})h$$

So, if the assumption was 50%, then: $L = H - 0.5h$ where H = h + t. Also, calculate the minimum height using either the deflection range or the assumed percent of deflection.

9. Check the operating range of the washer. The target operating range for a Belleville washer is between 15% and 85% of its total deflection range. Calculate L_{max} and L_{min} for the washer using the equations for maximum and minimum operating height in Table 9-16. If the operating heights calculated in the last step are outside of this range, revisit the h/t selection step (Step 4) and assume a deflection target that is less than 50% but greater than 15%.

10. <u>Verify that the washer will not yield at maximum deflection.</u> First calculate the compressive stress (Table 9-17) at the inside radius of the convex side of the washer. This will be the point of maximum stress. The stress should be calculated for the maximum deflection. Then calculate the safety factor (Table 9-17) and check for acceptability.

CRITICAL CONSIDERATIONS: Springs

- Elevated temperatures will affect the performance of springs.
- Resonance (spring surge) is a concern for helical coil springs. Design or select springs such that the natural frequency is more than 15 times greater than the forcing frequency of the application.
- Design or select springs for a high number (millions) of cycles for infinite fatigue life.
- For linear performance, the operating deflection of a helical coil compression spring should be kept between 15% and 85% of its total deflection range.
- Analyze extension springs for failure at their hooks as well as in their coils. Failure often occurs at the hooks.
- The working deflection of a Belleville washer should be kept between 15% and 85% of its total deflection range.
- Extension springs must be protected against overload by limiting their travel with external stops

BEST PRACTICES: Springs

- Select stock springs from a catalog to minimize cost and maximize part availability.
- For most applications, use the smallest possible spring rate to provide the smallest force change over the range of movement.
- Favor compression springs over extension springs. Compression springs are not vulnerable to hook breakage.
- Use conical helical coil compression springs when solid height must be minimized.

continued on next page

> ### ◉ BEST PRACTICES: Springs (Continued)
>
> - Use Belleville spring washers for applications requiring high force in a small axial space.
> - When more force is needed from a helical coil compression spring, but space is at a premium, two or more compression springs can be nested inside of one another to increase force in a space-efficient manner. When this is done, the springs should be wound in opposite directions to prevent interlocking. These springs will act in parallel.
> - When a helical coil compression spring is captured in a hole on one or more ends, the hole(s) should be chamfered. If the spring operates through an arc, the chamfer should reach the bottom of the hole(s). The angle of a full-depth chamfer should be selected such that the spring can move without interference.
> - Take advantage of standard parts to reduce cost. Commercial components are available to make attachments for compression or extension springs. Washers, socket head cap screws, mandrels, spring retainers, spring nuts, etc. are commercially available for compression springs. Components for making connections with extension springs are also commercially available.

PNEUMATICS

Section
9.2

Pneumatic system components use compressed and flowing gas, usually air, to generate forces. Pneumatic systems generally are composed of actuators and valves, connected in circuits with tubing or pipe to a compressed air source. Actuators transform air pressure into force and/or motion. Pneumatic valves control the air to the actuators using automatic or manual switches. Pneumatic circuits often include air-tight threaded fittings. The special threads for these connections are detailed in Section 6.1.

> ### ◉ RECOMMENDED RESOURCES
>
> - **The International Fluid Power Society:** Technical Papers http://www.ifps.org/docs/education_training/technical_papers/
> - **Pneumatics Online Website:** http://www.pneumaticsonline.com
> - **ANSI Y32.10:** "Graphic Symbols for Fluid Power Diagrams"

PRESSURE AND REGULATION

Any nonvolatile gas can be used to power pneumatic systems. Air is typically used, especially for large systems. In permanent factory installations, air is often compressed at a central location and supplied to the pneumatic systems through pipes. Safety components such as pressure relief valves are installed at the compressor to prevent overpressure hazards.

Compressed air supplies for pneumatic systems normally have moisture removed to prevent fouling and corrosion of the system components. In some cases, a small amount of oil is added to the compressed air supply to lubricate components like air actuators. Manufacturers will specify whether their components are 'oil free' or not.

Compressed gas is often delivered at a much higher pressure than needed for a pneumatic circuit. This pressure can be brought down to a set level using a pressure regulator after the compressor. When supply pressure fluctuates, a lower pressure regulator installed between the source and a pneumatic circuit can effectively hide the high end fluctuation of the source pressure. An air reservoir installed close to a pneumatic circuit is often used to ensure a steady pressure supply and absorb pressure fluctuations due to supply or usage. A pressure regulator is installed at the output of the reservoir. The designer must ensure that the reservoir and/or regulators are sized properly to supply the affected circuits.

PNEUMATIC CIRCUITS

A pneumatic circuit is responsible for converting pressurized air energy into well controlled physical work. A circuit normally consists of a compressed gas supply, one or more control components like valves, one or more actuators, and all the connectors and tubing required to connect the components. Pneumatic circuits also normally include provisions for gas to be exhausted to the atmosphere, to a reclassifying filter, or to an exhaust manifold where it can be ducted elsewhere. Quick exhausting of gas is essential for fast circuit response. There are four main pneumatic circuit types:

- **Open center circuits** permit nearly unrestricted flow through the valve and back to the source when the valve is centered (de-energized).
- **Closed center circuits** block air flow through the valve when the valve is centered (de-energized). This type of circuit is recommended when you have multiple circuits on the same air tank.

- **Meter-in circuits** have a flow control valve before the actuator. With an open center circuit, meter-in control is required for actuator speed control. A regulated bypass circuit may be recommended to improve circuit efficiency.
- **Meter-out circuits** are required for applications with large negative loads, for example, when lowering a heavy weight. Flow control is placed after the actuator to restrict airflow coming out of the actuator. Because the actuator sees full tank pressure and flow, the mechanical system is very stable under loading.

PNEUMATIC SYMBOLS

Pneumatic symbols representing circuit components and functions are governed by ANSI Y32.10. The number of symbols is quite large and beyond the scope of this text. Consult the standard for a complete list of symbols. Some of the most general pneumatic symbols are shown in Table 9-20. Symbols for valves and actuators are presented later in this section.

AIR ACTUATORS

Two common types of pneumatic actuators are linear (air cylinders, Figures 9-16 and 9-17) and rotary (Figure 9-18). Linear air actuators can be either single or double acting, whereas rotary actuators tend to be double acting only. Single acting actuators with spring return use a spring to return the actuator to its free state when air pressure is removed. Single acting air cylinders use about half as much air as double acting cylinders, and can be operated by three-way valves. Double acting actuators require pressure on both the "extend" and "retract" strokes; when pressure is removed, the state of the actuator becomes indeterminate. Some common types of actuators and their standard symbols are shown in Table 9-21. Figure 9-19 shows single and double acting cylinders in cross section.

Air actuators are available with an assortment of options. One of the most commonly specified options is a magnetic piston that enables position sensing. For applications with side loading, a double rod cylinder or external bearing may be needed. When a cylinder is guiding a load that should not rotate, anti-rotation options such as a non-round rod or integral guide rods are commonly used. For heavy loads or high speeds, internal or external end cushions are usually offered to soften the impacts and allow the cylinder to

Table 9-20: General Pneumatic Symbols

DESCRIPTION	SYMBOL
Main Line Conductor, Physical Outline, and Shaft	Solid Line
Pilot Line for Control	Dashed Line
Exhaust or Drain Line	Dotted Line
Enclosure Outline	Center Line
Device, Connector, or Component	Circle
Rotary Device	Semi-Circle
Pressure Control Functions	Square
Accumulator or Receiver	Oval
Fluid Conditioner	Diamond
Indicates Flow Direction (A pneumatic direction symbol is not filled in whereas a hydraulic symbol is a solid triangle.)	Triangle
Indicates that the component can be adjusted	Arrow Through Symbol at 45°

Figure 9-16: Air Cylinder

Source: Clippard Instrument Laboratories(www.clippard.com)

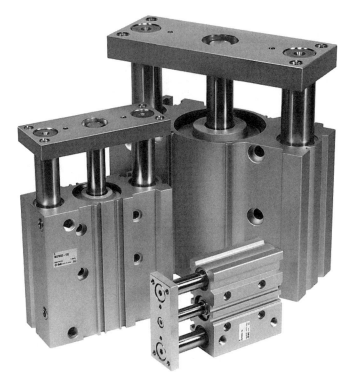

Figure 9-17: Guided (Non-Rotating) Air Cylinders

Source: SMC Corporation of America(www.smcusa.com)

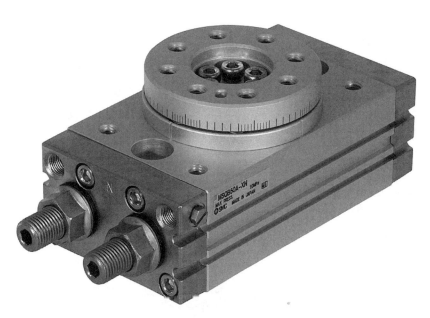

Figure 9-18: Rotary Actuator

Source: SMC Corporation of America(www.smcusa.com)

Table 9-21: Common Air Actuator Types

ACTUATOR TYPE	SYMBOL	FUNCTION
Single Acting Air Cylinder		The moving element (piston) sees pressure from one direction only. The images shown are for single rod cylinders.
Spring Return Single Acting Air Cylinder		The piston is returned to its start position by a spring when air pressure is relieved.
Double Acting Air Cylinder		Pressure is applied in either direction to move the piston to its stops.
Double Acting Rotary Actuator (Pneumatic Oscillator)		Pressure is applied in either direction to rotate the actuator between stops.

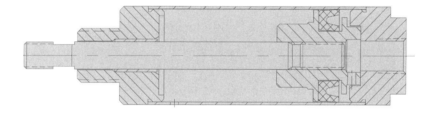

SINGLE-ACTING AIR CYLINDER.

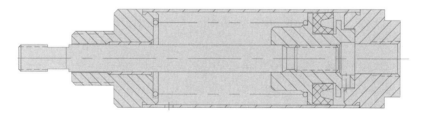

SPRING RETURN SINGLE-ACTING AIR CYLINDER.

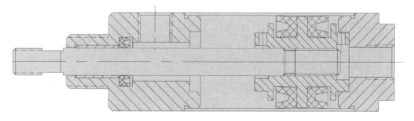

DOUBLE-ACTING AIR CYLINDER

Figure 9-19: Air Cylinders, Cross Section

Source: Clippard Instrument Laboratories(www.clippard.com)

withstand higher kinetic energy movements. Consult the manufacturer of your choice to find out what options and configurations are available.

SIZING AIR ACTUATORS

To determine the size of actuator needed, the force required of the actuator must be known, as well as the speed at which it is required to move.

The force generated by an air actuator to move a load should typically be 25% oversized to account for losses in the system and ensure smooth operation of the actuator. For high speed operation, choose an actuator capable of generating 50% more force than what is required. Note that for a spring return cylinder, the force of the retraction stroke will be equal to the spring force rather than a function of air pressure. For a linear actuator, the equation governing the relationship between air cylinder force and pressure is as follows:

F = PA
Where:
F = force generated by cylinder on attached load
P = air pressure supplied to the cylinder
A = area of the piston exposed to air pressure
Area of the piston during the extension stroke = π(radius of piston)^2
Area of the piston during the retraction stroke =
$$[π(\text{radius of piston})^2] - [π(\text{radius of piston rod})^2]$$

For a rotary actuator, this relationship is more difficult to calculate. Force values for both directions of rotation are normally provided by the manufacturer.

An air actuator takes time to move, and this is affected by both load and air supply. Actuator speed under load is difficult to calculate due to flow and pressure losses within the circuit. Most manufacturers can provide graphs of actuator speed vs. loading for their products. Consult these graphs when determining actuator size, but be aware that they are only approximations. Air supply can be characterized by C_V factor (coefficient of velocity factor). Valves and circuits with a higher C_V will be capable of moving an actuator faster because they can provide more air flow. Calculation of C_V is covered later in this section.

Kinetic energy is generated when a load is moved. When a moving load is stopped, the kinetic energy is absorbed by the mechanism that stops it. Air actuators can have either internal or external stops. External stops can generally withstand greater impacts than internal stops. Kinetic energy is of particular concern with rotary actuators. Manufacturers normally publish the kinetic energy limits of their actuators. Be sure to confirm that a selected cylinder or rotary actuator is capable of handling the kinetic energy of the application.

Rod failure and piston wear is of concern when linear actuators are used in compression or are subjected to side loading. Side loading can be a result of external forces, eccentrically mounted loads, or mounting and operating an actuator horizontally or at some angle. It is critical to align connections to a cylinder such that side loading due to misalignment or eccentric loading is minimized. Compliant connections are often used to eliminate overconstraint and resultant loading. When rod bending or buckling is a risk, the use of external bearings or double rod cylinders can help reduce the risk. For horizontal air cylinder installations, center trunnion mounting can help reduce the side loading on the cylinder. Another way to reduce premature failure due to side loading is to limit the extension stroke of an air cylinder so that it cannot fully extend.

The potential for buckling must be evaluated for any air cylinder in compression. To increase the buckling strength of an air cylinder, increase the piston rod diameter. Critical buckling load P_{cr} is calculated as shown in Table 9-22 using Euler's equation for an upright long column. Any load near or greater than this will cause buckling. More information on the variables E and I can be found in Sections 1.3 and 8.1 of this text.

Equivalent length is used to calculate critical buckling load rather than actual rod length. The value of L_e used in the critical buckling load equation is dependent on the end conditions of the column. The length L used to calculate L_e is assumed to be the distance at maximum extension between the exposed end of the piston rod and the nearest point of fixation. This means that L for a front-mounted cylinder is equal to the exposed length of rod at maximum

Table 9-22: Cylinder Rod Buckling Equations

CRITICAL BUCKLING LOAD E = Modulus of elasticity is a function of the rod material	$P_{cr} = \dfrac{\pi^2 EI}{L_e^2}$
AREA MOMENT OF INERTIA I = Area moment of inertia is a function of rod size and shape. r = Rod radius	$I = \dfrac{1}{4}\pi r^2$
EQUIVALENT LENGTH Cylinder is rigidly mounted, load is free (one end fixed, one end free): Cylinder and load are pin connected (both ends pinned): Cylinder is rigidly connected, load is pinned (one end fixed, one end pinned: Cylinder and load are rigidly connected (both ends fixed):	$L_e = 2L$ $L_e = L$ $L_e = 0.7L$ $L_e = 0.5L$

extension. L for a rear-mounted cylinder is equal to the length of the cylinder itself plus the exposed rod length at maximum extension. Note that a load is considered "free" on a cylinder rod if the load has no external supports or guides other than the cylinder itself. Table 9-22 provides some typical approximate values of L_e as functions of L for various configurations.

CALCULATING C_V

Many manufacturers have C_V calculation software available that can save time and effort. Be sure to understand any assumptions made by the software before trusting its results. C_V is itself unitless, but is typically used in the United States and calculated using Imperial (English) units. The SI system has its own flow coefficient, K_V which is unitless and calculated using metric units. These two values are not equivalent. C_V is defined as the flow rate in U.S. gallons per minute [gpm] of water at a temperature of 60°F with a pressure drop across the valve of 1 psi. K_V is defined as the flow rate in cubic meters per hour [m³/h] of water at a temperature of 16°C with a pressure drop across the valve of 1 bar. C_V is commonly used by manufacturers for both English and metric products. Table 9-23 contains equations used to calculate C_V.

The following step-by-step procedure can be used with the formulas in Table 9-23 to calculate air consumption of a cylinder as C_V factor:

1. Calculate the compression ratio of the compressed air supply.
2. Calculate the piston area exposed to air pressure on the extend stroke.
3. Calculate the volume of air per extend stroke.
4. Calculate required air flow during the extend stroke, which is volume per unit time.
5. Calculate the C_V needed to extend the cylinder.
6. Calculate the piston area exposed to air pressure on the retract stroke.
7. Calculate the volume of air per retract stroke.
8. Calculate required air flow during the retract stroke, which is volume per unit time.
9. Calculate the C_V needed to retract the cylinder.
10. Select the larger C_V and match your valve to be greater than or equal to that value.

The C_V of rotary actuators can also be calculated and used to size valves. The rotary swept volume is normally given in the manufacturer's data, and that volume can be used in the C_V calculations. Kinetic energy is of par-

Table 9-23: C_v and Related Equations

COMPRESSION RATIO P_1 = Pressure supplied to the actuator P_a = Ambient air pressure (normally 14.7psi)	$$C_f = \frac{(P_1 + P_a)}{P_a}$$
PISTON AREA r = Piston radius in feet r_{rod} = Rod radius in feet	Extend stroke: $$A_e = \pi r^2$$ Retract stroke: $$A_r = \pi(r^2 - r_{rod}^2)$$
VOLUME OF AIR PER STROKE	Extend stroke: $V_e = A_e$ x Stroke Length = ft^3 Retract stroke: $V_r = A_r$ x Stroke Length = ft^3
AIR FLOW DURING STROKE Q = ft^3/min (CFM) t_e = Time required for extend stroke (minutes) t_r = Time required for retract stroke (minutes)	Extend stroke: $$Q_e = \frac{V_e C_f}{t_e}$$ Retract stroke: $$Q_r = \frac{V_r C_f}{t_r}$$
C_v FACTOR T = Air temperature °R (°F + 460.67) G = 1 (specific gravity of air) P_a = Ambient air pressure (14.7psi) $P_2 = P_1 - \Delta P$ ΔP = Pressure drop across the cylinder in psi The value of ΔP can be assumed to be 10% of supply pressure for general use.	Extend stroke: $$C_v = \frac{Q_e}{22.48\sqrt{\dfrac{\Delta P(P_2 + P_a)}{TG}}}$$ Retract stroke: $$C_v = \frac{Q_r}{22.48\sqrt{\dfrac{\Delta P(P_2 + P_a)}{TG}}}$$

ticular concern with rotary actuators, so be sure to check with an applications engineer regarding kinetic energy in your given application. You will be asked to provide the actuation time, angle, and the moment of inertia of your applied load about the axis of rotation.

It is important to note also that if the C_v of the valve is known, the equations can be used to determine the time required to extend or retract the actuator. Some loss can be expected due to tubing runs and other factors; therefore, be aware that the time calculated will be best case and 25% longer might be more realistic.

Figure 9-20: Flow Control Needle Valves

Source: Clippard Instrument Laboratories(www.clippard.com)

PNEUMATIC VALVES

There are many different types of pneumatic valves. Some basic types include check, flow control (Figure 9-20), relief, and directional control valves. Directional control valves are used to stop, start, or change the direction of air flow through a circuit. Valves can be normally open or normally closed. Normally open valves allow air flow when the valve is not energized. Normally closed valves block air flow when not energized. Two different directional control valves and two valve manifold assemblies are shown in Figures 9-21 and 9-22.

Valves may be actuated manually or automatically. Some manual valves are shown in Figure 9-23. Automatic valves typically change state using energy from either internal or external pilot pressure. An external pilot air supply should be used to control a directional valve when the supply pressure to the valve is insufficient to shift the valve (when low pressure

Figure 9-21: Directional Control Valves for Manifold Mounting

Source: SMC Corporation of America (www.smcusa.com)

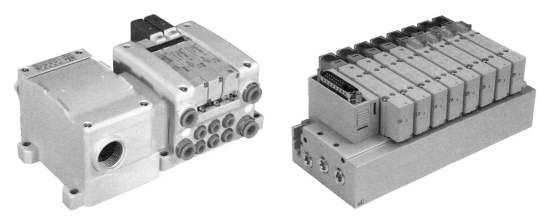

Figure 9-22: Directional Control Valve Manifold Assemblies
Source: SMC Corporation of America (www.smcusa.com)

Figure 9-23: Manual Toggle Valves
Source: Clippard Instrument Laboratories (www.clippard.com)

or vacuum is being controlled) or when the medium being controlled by the valve would damage the shifting mechanism. Automatic valves can be equipped with manual overrides which permit the operator to actuate the valve manually. Manual overrides are particularly useful when testing a pneumatic circuit.

Directional control valves are functionally described using the number of positions, ways, and ports. Most directional control valves have either two or three positions, symbolized by rectangles. Ways are the number of paths (including direction) the gas can take through the valve during all states of actuation. Three-way valves have one output and four-way valves have two outputs. Generally, three-way valves operate single acting cylinders and four-way valves operate double acting cylinders. Ports are the connections available to the power section of the valve. Some of the most commonly used directional control valves are listed here:

- **Two Port, Two Position (2/2,) Two-Way Directional Valve:** This type of valve has one inlet and one outlet port. It performs a basic on/off function.
- **Three Port, Two Position (3/2,) Two-Way Directional Valve:** This valve has one inlet, one outlet port, and an exhaust port. The valve switches flow between outlet and exhaust. The 3/2 normally closed valve is one of the most commonly used. This valve can be used to pressurize and exhaust a single actuator supply line.
- **Five Port, Two Position (5/2,) Four-Way Directional Valve:** This valve controls pressure and exhaust to two actuator supply lines. This is ideal for control of double acting cylinders. In any given position, the valve should be plumbed to supply pressure to one side of the cylinder and exhaust to the other. 5/2 valves are also available with four ports. Five ported valves have separate exhaust ports for each supply port. If exhaust silencers with built-in speed controls are used, the speed of the cylinder motion may be individually controlled in each direction.
- **Five Port, Three Position (5/3,) Four-Way Directional Valve:** This type of valve often has the center position normally closed when the valve is de-energized. This valve is excellent for control of double acting cylinders where a closed position is desired. The closed position makes it possible to 'jog' pneumatic actuators.

Directional control valve symbols show the valve's positions, methods of actuation, flow paths, and ports. The methods of left and right valve actuation are represented by symbols on both ends of the valve symbol. The flow path for each position is shown within its respective position box. Ports are marked only on the position box that corresponds to the de-energized state of the valve. Directional control valve symbols are illustrated in Table 9-24. A few of the most common ANSI valve symbols are shown in Table 9-25.

Table 9-24: Directional Control Valve Functions

DESCRIPTION	SYMBOL
Two Position Directional Valve, Base symbol	
Three Position Directional Valve, Base symbol	
Two Port, Two Position (2/2,) Two-Way Directional Valve, Normally Open	
Three Port, Two Position (3/2,) Two-Way Directional Valve, Normally Closed	
Five Port, Two Position (5/2,) Four-Way Directional Valve, Normally Open	
Five Port, Three Position (5/3,) Four-Way Directional Valve, Normally Closed	

Image Source: Clippard Instrument Laboratories (www.clippard.com)

Table 9-25: Common Valve Symbols

FUNCTION	SYMBOL
Manual On/Off Valve	
Check Valve	
Flow Control Valve, Adjustable, Bidirectional	
Air Pressure Regulator, Adjustable and Relieving	
Two Position, Three-Way Valve, Normally Open, With Manual Actuation and Spring Return	
Two Position, Three-Way Valve, Normally Open, With Single Solenoid and Spring Return	
Two Position, Three-Way Valve, Normally Open, With Double Solenoids and Manual Overrides	

SIZING VALVES

Valves are usually sized based on C_V factor, or coefficient of velocity factor. This factor is normally listed in manufacturer's catalogs as a specification for each valve. Using C_V to size valves is industry standard, but be aware that it is only an approximation. Factors like pipe or tubing runs and bends will cause some deviation of system performance from ideal. Valve switching response time can also be a factor to consider when selecting a valve.

To select the proper size valve for an actuator set to perform a specific task, first calculate the C_V of the actuator performing that activity. Then select a valve that has a corresponding or greater C_V.

CRITICAL CONSIDERATIONS: Pneumatics

- When designing pneumatic circuits, safety should be of paramount importance. The logic of all circuits should be checked for proper performance during startup, shutdown, pressure/vacuum loss, emergency stops, and other foreseeable events.

- Before selecting any pneumatic control element such as a valve, it is important to first know the safety category for the function being controlled by the pneumatic circuit. The manufacturer can assist in selecting components with the proper safety functions and ratings. Safety category is discussed in Chapter 12.

- When handling heavy loads using air actuators, consult manufacturer data on piston rod buckling loads when selecting air cylinders. Buckling is of particular concern when the cylinder is allowed to extend fully, or when the cylinder is pivot mounted.

- Be aware of any special air preparation requirements for components you select. Requirements such as filtration, drying, or lubricated air are not uncommon.

- All pneumatic actuators have a limit as to how much kinetic energy they can absorb using their internal or external stops. Consult the manufacturer for this information.

- It is important to note that any compressed gas other than air is an asphyxiation hazard. Take proper precautions, like air quality monitoring devices, when the volume of available gas has the potential to create an unsafe environment.

- Some gases, like CO_2, can be incompatible with materials used in the circuit. Check material compatibility of seals, tubing, and other components.

continued on next page

CRITICAL CONSIDERATIONS: Pneumatics (cont.)

- If a pneumatic component is expected to be exposed to elevated temperatures, check material compatibility and high temperature options for that component.
- Pneumatic actuators and the loads they carry can be extremely dangerous, and proper safety precautions such as guarding must be taken. See Chapter 2 for more information on machinery safety.

BEST PRACTICES: Pneumatics

- Pneumatic components are being continually updated and improved. Contact the vendor of your choice for a complete line of components and options.
- A regulated pressurized air tank is desirable as a compressed air source for most circuits because it can buffer pressure variations and supply multiple pneumatic circuits.
- Long runs of tubing and restrictive fittings should be avoided if possible because they contribute additional losses to the circuit supplying the actuator.
- Avoid reducing supply or exhaust tubing diameter, which will restrict flow and negatively impact performance.
- Loads should be in line with the piston rod in order to minimize piston wear. Consult manufacturer's specifications when an eccentric load must be accommodated. Most manufacturers provide performance charts as a function of load size and eccentricity.
- Speed control of air actuators is usually accomplished with restrictive flow controls. The flow either in or out of the actuator (meter in or meter out) can be controlled, depending on the application.
- When mounting cylinders, beware of overconstraint conditions that could cause premature wear of the cylinder. Clevis mounts and spherical bearings are often used to eliminate overconstraint.
- Select a cushioned stop for cylinders exposed to high amounts of kinetic energy. External stops may be required in some cases.
- To minimize the effect of side loading, limit the stroke of air cylinders so that they cannot fully extend.
- Compressed air (and other gases) cost money. It is good practice to continually work to find ways to limit the amount of air used.

ELECTRIC MOTORS

Electric motors are commonly used in industrial machinery and consumer products. Motors and gearboxes are often used in combination. Both AC and DC motors are available. The field of electric motors is constantly evolving, and this section serves as only a brief overview of motor types and selection. Contact your preferred manufacturer or distributor for complete information on the range of motors available, as well as assistance with selection and sizing.

RECOMMENDED RESOURCES

- R. Mott, *Machine Elements in Mechanical Design*, 5th Ed., Pearson/ Prentice Hall, Inc., Upper Saddle River, NJ, 2012
- Oberg, Jones, Horton, Ryffel, *Machinery's Handbook*, 28th Ed., Industrial Press, New York, NY, 2008
- **The Association of Electrical and Medical Imaging Equipment Manufacturers (NEMA)** website: www.nema.org
- **The International Electrotechnical Commission (IEC)** website: www.iec.ch

ELECTRICAL POWER

There are three types of power supplied to motors: Direct Current (DC), single phase Alternating Current (AC), and polyphase AC. DC power is unidirectional, whereas AC is alternating. DC power can be supplied in a variety of voltages. DC power for motors is normally generated by a converter or power supply, which gets its power from an AC source. Small motors can run on batteries.

AC power can be single phase or polyphase. Residential power is generally single phase, whereas industrial machine installations often use three-phase power. With single-phase power, a single waveform is present with a given frequency. Three-phase power has three waveforms of a given frequency, offset from one another by 120°. Common AC voltages in the USA are 120, 120/208, 240 and 480. In the United States, AC power is supplied at a frequency of 60 Hz. 50 Hz power is common in Europe. Both single-phase motors and three-phase motors are available. Motors using three phase power are generally more efficient and economical than those using single-phase power.

One dangerous power condition for AC motors is undervoltage. An undervoltage condition is one in which the supplied power is less than 90% of rated voltage for more than one minute. This can occur when more power is drawn from a line or transformer than it can deliver. Undervoltage is also known as a "brown-out." Because AC motor torque changes as a square of supplied voltage, a 10% reduction of voltage results in a 19% reduction of torque capability in the motor. More severe undervoltage conditions result in even more lost torque capability. If the torque capability of the motor drops below that required by the application load conditions, the motor will stall and begin to heat up rapidly. Overheating can quickly lead to permanent damage to the motor. Every motor running on an undervoltage circuit can be damaged simultaneously. To reduce or eliminate the risk of damage during an undervoltage incident, voltage monitoring devices and proper training (to turn off or throttle back loads on motors immediately if undervoltage occurs) are commonly employed.

MOTOR TERMINOLOGY

The following terms are used to describe both AC and DC motors, and may appear on motor nameplates:

Manufacturer's type is a general motor type descriptor and varies among manufacturers. Two commonly encountered types are "General Purpose" and "Special Purpose." General purpose motors are designed to handle mechanical loads and inertial loads. They generally have bearing arrangements that can handle radial and axial loading. Special purpose motors are designed for specific applications. Wash-down and hazardous location motors are special purpose motors. Motors for HVAC devices like fans and pumps are special purpose.

Rated horsepower (HP) is the power output at the motor shaft under full-rated load. Power is a function of motor speed and torque. International Electrotechnical Commission (IEC) designations measure rated power in kW.

Maximum ambient temperature (AMB) listed on the nameplate should never be exceeded while the motor is running.

Insulation class (INSUL CLASS) is a NEMA standard classification of the thermal tolerance of the motor. Insulation class H has higher thermal tolerance than class A. IEC generally uses the same designations as NEMA.

Rated full-load speed (RPM) is the speed in revolutions per minute that the motor will run at under full-load conditions when voltage and frequency are at rated values. On standard induction motors, the full-load speed is typically 96% or more of the synchronous speed. Sometimes the motor's maximum speed will be specified along with the rated full-load speed.

Enclosure type (ENCL) is sometimes listed on the motor nameplate. Enclosures protect the motor from the environment and some common types are detailed later in this section.

AC MOTORS

There are two main types of AC motor: synchronous and asynchronous. Synchronous motors are normally run as constant speed motors. They operate in synch with line frequency, and are capable of very precise constant speed operation. Speed regulation is possible with asynchronous motors through the use of variable frequency controls. The most common type of asynchronous motor is the induction motor, or "squirrel cage" motor. The characteristics of various types of AC motors are shown in Table 9-26.

The following information, in addition to those terms listed earlier, is normally given on the nameplates of single-phase and three-phase induction motors:

Frame size designation is of particular interest to mechanical designers because it relates to a set of standard dimensions for the motor and its mounting features. Frames are discussed later in this section. NEMA and IEC designations will differ.

Time rating or duty is the length of time the motor can safely operate at full-rated load. A NEMA time rating of CONT on the nameplate means that the motor can operate continuously at rated maximum rated load and ambient temperature. NEMA standard motors designed for intermittent use are given a time rating in minutes. IEC uses a classification system to indicate duty cycle. S1 is the IEC classification for continuous duty at constant load whereas S8 indicates continuous duty with variable load and/or speed. Intermittent duty motors under IEC standards are classified S2 through S5.

Cooling designation is an IEC standard classification of how a motor is cooled.

Rated frequency is specified in Hz (cycles per second) and is the frequency of the AC power for which the motor is designed.

Number of phases relates to the AC power source.

Rated load current, or full-load amps (FLA) is the expected amperage when the rated maximum horsepower is being produced by the motor. This value is used to size the electrical components associated with the motor.

Figure 9-24: General Purpose Servo Motor

Source: Rockwell Automation, Inc. (www.rockwellautomation.com)

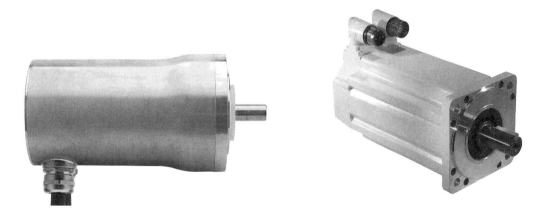

Stainless Steel Hygienic Environment Servo Motor   Food Grade Servo Motor

Figure 9-25: Special Purpose Servo Motors

Source: Rockwell Automation, Inc. (www.rockwellautomation.com)

Rated voltage is the voltage at which the motor is designed to operate and perform as expected. AC motors are available for a variety of voltages, the most common of which are 120 (motor voltage 115,) 240 (motor voltage 230), and 480 (motor voltage 460). Motors are designed to tolerate a voltage variation of ±10%, though performance may suffer. On some motors, two voltages and/or a range of voltages are provided. IEC standards allow voltage variation between 95% and 105% of rated voltage.

Code letter for locked rotor kVA defines the locked rotor amperage on a per-horsepower basis. Codes range from A to V. Different codes often require different associated electrical components, like starters.

Design code is provided for NEMA induction motors. The designs of induction motors are A, B, C, and D. The letter designation relates to the torque and current characteristics of the motor. These motors are characterized in Table 9-26. IEC design codes differ from NEMA design codes. The most common IEC design types are N (similar to NEMA B) and H (NEMA C) motors.

Nominal efficiency at full load is the ratio of power output to power input, expressed as a percentage. Values closer to 100% are preferred. Efficiency ratings are established through testing, and represent an average value. NEMA standards allow 20% variance from average. In general, larger motors are more efficient, and three-phase motors are more efficient than single phase motors.

Service factor is generally listed on NEMA motors only if it is higher than 1.0. IEC motors do not have a service factor rating. When service factor is multiplied by horsepower, the result is the allowable horsepower loading. A service factor greater than 1 allows the motor to be run at greater than rated horsepower. Motors should not be run continuously with this allowable horsepower loading, but intermittent operation is allowed.

Thermally protected (OVER TEMP PROT,) when stated on the nameplate, means that the motor is equipped with thermal protectors. The type of protection as specified by NEMA is normally given as a letter value.

Synchronous speed is the speed at which a motor will run with no load. When torque demand increases, the speed of most motors decreases due to internal slip.

Altitude listed on the nameplate specifies the maximum height above sea level at which the motor is rated to operate. At this altitude, the motor will remain

Table 9-26: Common AC Motor Types and Characteristics

MOTOR TYPE	CHARACTERISTICS	TYPICAL APPLICATIONS
Synchronous	• Runs at synchronous speed • Designed for continuous operation • Available in a full range of sizes • Stops if pull-out torque is exceeded	Compressors Pumps Blowers
Induction NEMA A	• Runs at essentially constant speed • Moderate starting torque • Suitable for brief, heavy overloads • High starting currents	Injection Molding Presses Die Presses
Induction NEMA B	• Runs at essentially constant speed, multi-speed is available but not adjustable • Moderate starting torque • Good speed control • High breakdown torque • Requires a high-current starting circuit	Centrifugal pumps Fans Blowers Lathes
Induction NEMA C	• Runs at essentially constant speed, multi-speed is available but not adjustable • High starting torque • Good speed control • High slip	Reciprocating compressors Conveyors with heavy loads
Induction NEMA D	• Runs at essentially constant speed • Very high starting torque • Poor speed control • "Soft" (high slip) response to shock loads	Cranes Elevators Punch Presses
AC Servo	• Full position and speed control • Available in a full range of sizes • Closed loop control system	Assembly machines Machine tools Robots
Wound-Rotor	• Can be tuned to an application • Speed and torque adjustment capability • Speed adjustable above 50% of synchronous speed	Conveyors Hoists
Universal	• Runs on either AC or DC power • Operates at high speeds • High power-to-size ratio • Poor speed regulation: power drops rapidly under increasing load	Hand-held devices Power tools Appliances
Split Phase **Single Phase**	• Designed for continuous operation • Moderate starting torque • Good speed regulation and efficiency • Requires a centrifugal switch	Machine tools Business machines Centrifugal pumps
Capacitor Start **Single Phase**	• Designed for continuous operation • High starting torque • Good speed regulation and efficiency • Requires a centrifugal switch	Conveyors with heavy loads Pumps for heavy fluids
Shaded Pole **Single Phase**	• Low starting torque • Low efficiency • Low cost • No start switch required	Small direct-drive fans Small gear motors

within its rated temperature rise. At higher altitudes, the motor is likely to run hotter and may need to be de-rated.

Rated torque is the torque output at the shaft that can be produced at rated speed. This is equal to full-load torque (See Figure 9-26).

Power Factor (PF) is the ratio of active power to apparent power at full rated load. A power factor closest to 1.0 or 100% is most desirable. The IEC code for power factor is "cos."

Bearings used in the motor are sometimes listed on the nameplate to facilitate maintenance.

Shaft type can be listed on a motor nameplate. The most common shaft types are round, keyed, with a flat, and threaded.

Mounting method describes the method of mounting the motor. Mounting methods are discussed later in this section.

Capacitor correction (MAX CORR KVAR) is sometimes listed on a nameplate to indicate the capacitor value in kilovars. Applying a value greater than listed may damage the motor.

The following terms are used to describe AC motor performance and related application characteristics:

Locked-rotor, or static torque is sometimes called breakaway torque. It is the minimum torque a motor will produce at rest. This is the starting torque for the motor (See Figure 9-26).

Locked rotor current is the steady-state current taken from the line with the motor's rotor locked and when rated voltage is applied at rated frequency.

Peak torque is the maximum instantaneous torque that a motor can produce.

Accelerating torque is the amount of torque needed to accelerate a load from rest to full-rated speed.

Full-load torque (See Figure 9-26) is the torque needed for a motor to produce its rated horsepower at its rated full-load speed. This is equal to rated torque.

Breakdown torque (See Figure 9-26) is the maximum torque a motor will produce without an abrupt change in speed when rated voltage is applied at rated frequency.

Pull-out torque of a synchronous AC motor is the maximum sustained torque it will produce at synchronous speed when rated voltage is applied at rated frequency and with normal excitation.

Pull-in torque of a synchronous AC motor is the maximum sustained torque under which the motor will pull its connected load into synchronism when rated voltage is applied at rated frequency and with normal excitation.

Pull-up torque of an AC motor is the minimum torque exerted by the motor during acceleration from rest up to the speed at which breakdown torque occurs. For motors which do not have a definite breakdown torque, the pull-up torque is the minimum torque developed up to rated speed.

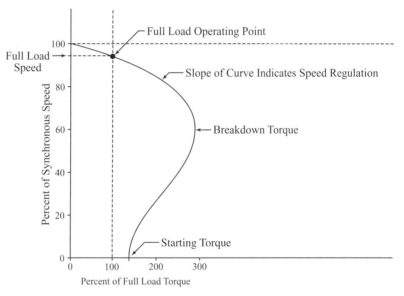

The true shape of this curve can vary significantly depending on the motor type.

Figure 9-26: General Form of Motor Performance Curve

DC MOTORS

DC motors are versatile and are often used in applications that require variable speed, reversals, or quick stops. Their speed control tends to be less accurate than that of AC motors. DC motors tend to have a high torque to inertia ratio and output a very smooth motion. The two basic types of DC motors are brush motors and brushless motors. Most DC motors have internal brushes, which wear over time. DC motors are capable of dynamic braking, which can minimize or eliminate the need for an external brake.

An appropriately specified DC power supply must be provided to power DC motors. Table 9-27 lists several of the most common types of DC motors used in industrial machinery.

The following information is usually provided on DC motor nameplates in addition to those terms listed earlier in this section:

Rated armature voltage in the United States is typically 250 VDC or 500 VDC. This is the voltage at which the motor will run at base speed. Varying the armature voltage will vary the motor speed.

Base speed is the speed at which the motor will run with rated armature voltage, rated amps, and rated flux.

Maximum speed is sometimes provided on the motor nameplate. If maximum speed is not listed, the motor should not be run over its base speed unless the manufacturer has validated the application.

Rated field voltage is the voltage of the field that turns the armature. This voltage is supplied by a different source than the armature voltage. In the United States, 150 VDC and 300 VDC are common. Permanent magnet motors do not require external field excitation.

Armature rated load current is the current that must be applied for the motor to operate at rated power. Varying the armature current will vary the motor torque.

Winding type expresses the type of field winding in the motor. Shunt winding is the most common type.

ELECTRIC MOTOR CONTROLS

Specifying and selecting the proper control components for electric motors should be done by a qualified motion control expert. The following are some basic control needs that must be addressed for motors:

A DC power supply must be provided for all DC motors and controls components.

Starting a motor usually requires a starter circuit and components. The configuration and components of a starter circuit will depend on the type of motor being started.

Table 9-27: Common DC Motor Types and Characteristics

MOTOR TYPE	CHARACTERISTICS	TYPICAL APPLICATIONS
Shunt-Wound	• Good speed control • Low starting torque • Capable of constant speed for a range of torques	Small fans Small blowers
Series-Wound	• Very high starting torque • "Soft" (high slip) response to shock loads • Speed varies with loading • Requires a safety device to prevent runaway speeds under low load conditions	Rollers Lathes
Compound-Wound	• High starting torque • Narrow adjustable speed range • "Soft" (high slip) response to shock loads	Hoists Cranes
Permanent Magnet	• Speed varies linearly with torque • Generally small • Often available with integral gear reducer	Small fans Small blowers Small actuators
Stepper	• Moves an angular increment in response to a single electrical pulse • Stepped position control • Open loop control system	Pick and place equipment Indexing applications
DC Servo	• Full position and speed control • Available in a full range of sizes • Closed loop control system	Industrial assembly machines Machine tools Robots
Brushless	• Very long life (no brush wear) • High precision • Maintenance free • High relative cost • No internal sparking	Hazardous environments Maintenance-free applications
Linear	• Many motor types available • Eliminates the need for a linear power transmission device	Instruments Positioning systems

Stopping a motor can be as simple as allowing it to coast down to rest. When a motor must be stopped suddenly or under control, a brake is often employed. Some motors are available with integral brakes. DC motors are capable of dynamic braking.

Overload protection can take several forms. Thermal overload can be prevented through the use of a temperature-sensitive switch located inside the motor enclosure. Electrical overload can be prevented with fuses, overcurrent detection devices, or solid state overload relays.

Speed control is possible for both AC asynchronous and DC motors. AC motor speed can be varied through the use of controls that produce variable frequency power. DC motor speed is controlled through the use of a rheostat or similar voltage control device.

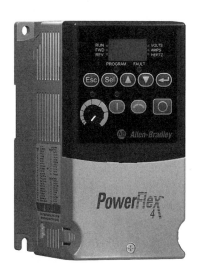

Figure 9-27: AC Motor Controllers

Source: Rockwell Automation, Inc. (www.rockwellautomation.com)

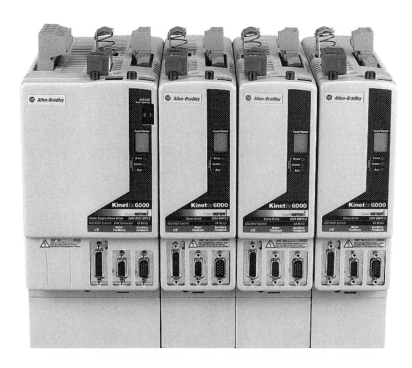

Figure 9-28: Servo Motor Controllers

Source: Rockwell Automation, Inc. (www.rockwellautomation.com)

Controls enclosures must be provided to protect controls components from contamination and accidental contact. NEMA standards are available governing controls enclosures.

ELECTRIC MOTOR FRAMES AND ENCLOSURES

The functional dimensions of motors are standardized by NEMA and the IEC. These two organizations specify frame sizes differently. The standardized dimensions control motor height, width, shaft diameter, keyway size, mounting features, and other key characteristics. Frame dimensions can be obtained either from the manufacturer or the appropriate standards.

Motors are available in a variety of standard frame types offering different mounting options. Custom designs are also available. The most common mounting options are as follows:

Foot mounted motors are the most common type found in industrial machinery. This mounting configuration provides mounting feet integral to the motor frame suitable for attachment to a flat surface. Bolt patterns and sizes are standardized. Foot mounted motors are available with or without vibration isolating features. This mounting method is especially appropriate for large motors.

Face mounted (C-Face) motors provide a mounting face around the motor output shaft. The face and bolt pattern are standardized. These motors are ideal for mounting directly to gearboxes or other equipment. For shaft alignment purposes, a precise pilot (circular pad) is provided on the mounting face. This pilot is made to fit inside a mating bore on the gearbox or other equipment.

Flange mounted (D-Flange) motors provide a mounting flange around the motor output shaft. The flange and bolt pattern are standardized. These motors are ideal for mounting directly to gearboxes or other equipment. For shaft alignment purposes, a precise pilot (circular pad) is provided on the mounting flange. This pilot is made to fit inside a mating bore on the gearbox or other equipment.

Extended through-bolt motors have bolts protruding from the front or rear. These are often used in fan and blower applications.

<u>Unmounted</u> motors have no mounting features and consist of the bare rotor and stator. These motors must be built in to an assembly with proper support, alignment, and thermal management.

Motor enclosures serve to protect and support the internal motor components while providing for cooling. NEMA and the IEC have different codes for the various enclosure types. The following are some common types of enclosures:

<u>Open frame</u> motors have an enclosure that is generally made of sheet metal and has openings to allow for ventilation. These motors are meant to have additional guarding added once integrated into a piece of machinery.

<u>Drip proof</u> (ODP) motors are open frame motors where the ventilation holes in the enclosure are limited to the underside of the motor only. These motors should have additional guarding added once integrated into a piece of equipment. A drip proof motor is shown in Figure 9-29.

<u>Totally enclosed non-ventilated</u> (TENV) motors have enclosures without ventilation holes. Fins usually protrude from the enclosure to provide cooling. Contaminants cannot enter the motor. This type of motor enclosure is very common in light industrial machinery. A TENV motor is shown in Figure 9-30.

<u>Totally enclosed fan cooled</u> (TEFC) motors are similar to TENV motors, with the addition of a fan. The fan is oriented such that it draws air over the cooling fins protruding from the enclosure.

<u>Totally enclosed explosion-proof</u> (TEFC-XP) motors are fan-cooled, totally enclosed motors. Explosion-proofing is accomplished through the use of special electrical connections. These motors are suited for hazardous environments.

ELECTRIC MOTOR SIZING

Motor size primarily depends on how much torque the application requires. Duty cycle and speed are also important factors. The formulas relating torque to power are provided in Table 9-28. It is also essential to calculate the inertia and resultant torque of any loads or mechanisms attached to the motor shaft. Section 11.4 contains formulas used to calculate the inertia of various mechanisms reflected back on a gearbox. These formulas can also be used if a motor is coupled directly to the mechanism instead of through a gearbox. All gearboxes have their own inertia; if a gearbox is used, this inertia must be applied to the motor when performing sizing calculations.

Root Mean Square (RMS) torque is the time-weighted average of the torque required during a complete machine cycle. It is calculated using the formula in Table 9-28. RMS torque is used to size motors for repeated intermittent motion. Motors designed for intermittent motion will have torque vs. speed graphs with two regions: continuous and intermittent. Use these graphs to determine whether a given motor will be able to run at the required speed and load continuously or not. The RMS torque for an intermittent motion application must fall within the motor's continuous region on the torque vs. speed graph. If RMS torque falls within the intermittent region, the motor will likely overheat and performance will suffer.

Motor selection and sizing is primarily based on two parameters: available torque and thermal characteristics under load. To select and size a motor properly, consult an applications engineer. Many manufacturers offer drive sizing software, often including gearbox selection as well as motor selection. One excellent example of such software, this one focusing on servo motors and gearboxes, can be found at www.ab.com/motion/software/analyzer.html. The general procedure for sizing a motor is as follows:

1. Determine the motor type based on performance objectives. An applications specialist should be able to assist with this. Some factors to consider are: available power source, speed range, allowable speed variation, allowable position variation, reversal capability, duty cycle, and environmental temperature. Tables 9-26 and 9-27 can be used to facilitate selection.

2. Calculate the maximum acceleration torque at the motor output shaft. Include non-inertial forces like friction. Exclude motor inertia at this time. The following calculations must be made to find the maximum torque at the motor:
   a. Calculate the maximum move speed and acceleration rate for your application. This must be done in terms of angular motion of the motor output shaft. Negative accelerations, if present, must be considered as well as positive accelerations. With servo motors, it may be necessary to calculate accelerations assuming some acceleration profile.
   b. Calculate the inertia of the load on the motor. This must take into account any gearbox inertia as well as anything else attached to

Figure 9-29: Drip Proof Motor

Source: Rockwell Automation, Inc. (www.rockwellautomation.com)

Figure 9-30: Totally Enclosed Non-Ventilated (TENV) Motor

Source: Rockwell Automation, Inc. (www.rockwellautomation.com)

the motor shaft. Table 9-28 contains gearbox inertia calculation formulas and Section 11.4 contains inertia and reflected torque calculation methods for various mechanical devices.

c. Calculate maximum torque due to the maximum acceleration of the load. The formula for acceleration torque can be found in Table 9-28.

3. Calculate the peak torque required of the motor. Peak torque is equal to the breakaway torque plus the acceleration torque of the motor.

4. Calculate the starting torque required of the motor. Starting torque is equal to acceleration torque plus friction torque.

5. Calculate the running torque required of the motor. If motion is continuous, calculate the torque required at operating speed. If motion is intermittent, calculate the RMS torque. Table 9-28 contains the formula

Table 9-28: Electric Motor Sizing Equations

POWER AS A FUNCTION OF TORQUE N = Shaft speed in RPM English: T = Torque in ft-lb Metric: T = Torque in N-m	English: $$hp = \frac{TN}{5252}$$ Metric: $$kW = \frac{TN}{9550}$$ $$hp = 0.746kW$$
AC MOTOR SYNCHRONOUS SPEED n_s = Synchronous speed (rev/min) f = Power supply frequency (Hz) p = Number of poles in the motor	$$n_s = \frac{120f}{p}$$
ACCELERATION TORQUE I = Mass moment of inertia of load at motor shaft (See Sections 11.4 and 1.3 for I of various devices and loads) α = Angular acceleration (radians/sec^2) Dependent on motion profile used. (See Section 1.3 for rotational acceleration equations)	$$T_A = I\alpha$$
BREAKAWAY TORQUE	T_B = Consult manufacturer
FRICTION TORQUE	T_F = Consult manufacturer
PEAK TORQUE T_B = Motor breakaway torque T_A = Acceleration torque	$$T_P = T_B + T_A$$
STARTING TORQUE T_A = Acceleration torque T_F = Friction torque	$$T_S = T_A + T_F$$
ROOT MEAN SQUARE (RMS) TORQUE Deceleration torques have negative sign T_i = Torque required during portion i of cycle t_i = Duration of portion i of cycle	$$T_{RMS} = \sqrt{\frac{\sum T_i^2 t_i}{\sum t_i}}$$
GEARBOX REFLECTED INERTIA AND TORQUE AT MOTOR $T_{L \to M}$ = Torque of load through gearbox reflected at motor T_L = Torque of load at gearbox output N_R = Gearbox ratio E_f = Gearbox efficiency $I_{\to M}$ = Mass moment of inertia seen at motor I_R = Mass moment of inertia of gearbox $I_{L \to M}$ = Mass moment if inertia of load reflected at motor I_L = Mass moment of inertia of load at gearbox output	$$T_{L \to M} = \frac{T_L}{N_R E_f}$$ $$I_{\to M} = I_R + I_{L \to M}$$ $$I_{L \to M} = \frac{I_L}{E_f}\left(\frac{1}{N_R}\right)^2$$
INERTIA RATIO Consult manufacturer for acceptable values. I_L = Mass moment of inertia of load at motor shaft I_M = Mass moment of inertia of motor rotor R_G = Gear ratio of reducer	Motor without gear reducer: $$IR = \frac{I_L}{I_M}$$ Motor with gear reducer: $$IR = \frac{I_L}{I_M R_G^2}$$

for RMS torque. If the motor is synchronous, the running torque must be expressed at synchronous speed.

6. <u>Make an initial motor selection.</u> Verify that the motor is capable of the full range of speeds required by the application. Graphs showing torque vs. speed are often provided for motors by the manufacturer to facilitate selection. The available torque must exceed your running torque or RMS torque calculations. The RMS torque for an intermittent motion application must fall within the motor's continuous region on the torque vs. speed graph. Rated peak and starting (static) torques must exceed the maximum peak and starting torques of the application by a factor of safety.

7. <u>Check the inertia ratio between the motor and its load.</u> Large inertia ratios will negatively affect starting and stopping times. Acceptable ratios will be dependent on motor model and manufacturer. If the inertia ratio is determined to be excessive, a gearbox should be inserted between the motor and load. Inertia ratios with and without a gear reducer can be calculated using Table 9-28. If a gear reducer is used, torques must be recalculated as reflected through the gearbox using the equations from Table 9-28.

8. <u>Recalculate the total torque, including the motor inertia, and verify that the motor is still capable.</u> Calculate the torque added by the motor's own inertia based on your motor selection. Motor inertia is normally provided in catalog information. If a gearbox was added, include that inertia as well.

9. <u>Calculate any overhanging moments on the motor shaft</u> due to the weight of components or any offset or radial loading. Verify that the motor can handle the calculated moment load. Common sources of overhanging moments are belt pulleys/sheaves/sprockets, gears, and chain sprockets. Some of the loads inducing these moments are detailed in Section 11.1.

10. <u>Select motor mounting and enclosure options.</u> These are discussed earlier in this section.

11. Because motor capability is very much influenced by heat generated under load and the motor's ability to dissipate it, a specialist should verify your motor selection.

CRITICAL CONSIDERATIONS: Electric Motors

- The field of commercial motors is constantly evolving. Consult with a manufacturer for a complete rundown of current offerings.
- Motor performance degrades with elevated temperature.
- AC motors run hot when lightly loaded or heavily loaded. Oversizing an AC motor beyond 20% is generally not beneficial.
- The RMS torque for an intermittent motion application must fall within the motor's continuous region on the torque vs. speed graph.
- Rotating shafts, couplings, attachments, etc., are extremely dangerous—proper safety precautions such as guarding must be taken. See Chapter 2 for more information on machinery safety.

BEST PRACTICES: Electric Motors

- Selection and sizing of motors is best accomplished with the help of an applications engineer at a manufacturer or distributor.
- Use standardized motors to reduce cost and improve availability of parts.
- Selection of electrical and controls components for use with motors is best accomplished with the help of an electrical or applications engineer.

10

BEARINGS

Contents

10.1	PLAIN BEARINGS	503
10.2	ROLLING ELEMENT BEARINGS	510
10.3	LINEAR BEARINGS	534

Tables

10-1	PV Formulas for Rotary Sleeve Bearings	508
10-2	Typical Plain Bearing Material Properties	508
10-3	Applications and Suitable Bearings	519
10-4	Selected Bearing Types and Characteristics	520
10-5	Select Radial Bearing Tolerance Classes	521
10-6	Shaft Fits for Radial Bearings (ABEC/RBEC 1)	522
10-7	Housing Bore Fits for Radial Bearings (ABEC/RBEC 1)	523
10-8	Bearing Seat Geometric Tolerances	524
10-9	Loads On Rolling Element Bearings	528
10-10	Life and Load Ratings for Radial and Angular Contact Bearings	529
10-11	Static Load Safety Factors for Bearings	530
10-12	Shock Service Factors for Bearings	530
10-13	PV Formulas for Plain Linear Bearings	537
10-14	Life and Load Ratings for Linear Bearings	538

Section **PLAIN BEARINGS**

10.1

Plain bearings provide sliding contact between two surfaces. The most common type of plain bearing is the sleeve bearing or bushing. Plain bearings are often chosen over rolling element bearings due to cost or space limitations. They are also more rigid and quieter in operation than rolling element bearings. The main disadvantages of plain bearings are their higher potential to wear (as compared to rolling element bearings) as well as their relative vulnerability to contaminants. This section serves as a general introduction to plain bearings, with a focus on sleeve bearings used in rotary motion applications with boundary lubrication conditions. Please consult the recommended resources for more information and calculation methods for other types of lubrication. Plain bearings used in linear motion applications are discussed in Section 10.3.

> **RECOMMENDED RESOURCES**
>
> - R. Mott, *Machine Elements in Mechanical Design*, 5th Ed., Pearson/ Prentice Hall, Inc., Upper Saddle River, NJ, 2012
> - R. L. Norton, *Machine Design: An Integrated Approach*, 4th Ed., Prentice Hall, Upper Saddle River, NJ, 2011
> - Oberg, Jones, Horton, Ryffel, *Machinery's Handbook*, 28th Ed., Industrial Press, New York, NY, 2008

LUBRICATION OF PLAIN BEARINGS

Plain bearings must be lubricated in order to have long life and low friction. There are four types of lubrication conditions under which plain bearings are run: hydrostatic lubrication (full film), hydrodynamic lubrication (full film), mixed film lubrication, and boundary lubrication (thin film). Full film lubrication occurs when the lubricant layer between surfaces is thick enough to prevent any surface contact. Boundary lubrication occurs when the lubricant layer is present but not thick enough to prevent contact between surfaces. A graph showing relative coefficients of friction for the different types of lubrication are shown in Figure 10-1. The horizontal axis is a function of lubricant viscosity (Z), journal speed (N), and bearing pressure (P).

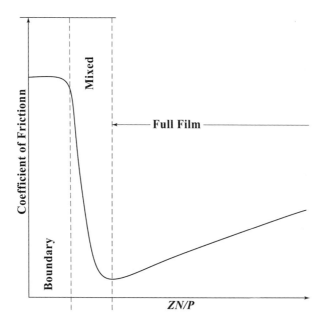

Figure 10-1: Plain Bearing Lubrication vs. Coefficient of Friction

Hydrostatic lubrication is full film lubrication, and occurs when high pressure lubricant is used to force the sliding surfaces apart. Plain bearings with hydrostatic lubrication can accommodate heavy loads at low speeds. Hydrostatic lubrication is normally used in planar or linear bearings rather than in sleeves. Design of an assembly using hydrostatic lubrication is extremely complex and must focus on lubricant feeding and containment. Hydrostatic lubrication is uncommon in light industry and is beyond the scope of this text.

Hydrodynamic lubrication is full film lubrication, and is most commonly employed with high-speed rotating shafts in plain sleeve bearings. Bearings with hydrodynamic lubrication are often called journal bearings. A wedge of lubricant is caught between the rotating shaft and bearing surface, providing sufficient pressure to carry the applied load. The shaft rides on a film of oil and does not contact the bearing except during periods of low speed or stasis. Typical coefficients of friction range from 0.002 to 0.010. Oil is typically used as the lubricant, and it must be supplied from a reservoir to maintain hydrodynamic lubrication. The lubricant also cools the bearing, and lubricant leakage and circulation enhances the cooling effect. Lubricant viscosity and temperature are important parameters in hydrodynamic lubrication performance, and temperature control is recommended. Hydrodynamically lubricated bearings go through periods of boundary lubrication during startup and shutdown periods. In light industrial machinery, relatively low speeds and/or intermittent movements mean that boundary lubrication of plain bearings is more common than hydrodynamic lubrication. Hydrodynamically lubricated journal bearings are beyond the scope of this text, but are detailed in the recommended resources.

Mixed film lubrication occurs when the lubricant film thickness is slightly greater than the surface roughness of the surfaces. This is not hydrodynamic lubrication because the surfaces are close enough together to create drag and some surface contact. The conditions for mixed film lubrication are between those of hydrodynamic and boundary lubrication.

Boundary lubrication occurs at low speeds and/or high pressures. Surfaces in contact are partially separated by a lubricant film and partially in direct rubbing contact. Bearings with boundary lubrication have higher coefficients of friction and shorter life expectancy than hydrostatically or hydrodynamically lubricated bearings. To reduce friction and reduce wear, lubricant is usually embedded in the bearing and/or added to the shaft. Values of friction coefficients vary depending on bearing material and construction, but typically range from 0.05 to 0.20. Sleeve bearings designed to run well without added lubricants are often called "oil free" bearings, and generally have lubricant embedded in them, or lubricity is a property of the material. Many plastics, for instance, are inherently lubricious and can provide low coefficients of friction without additional lubrication. Plain bearings with boundary lubrication are common in the machinery industry and will be examined further in this section.

Lubricant is often applied to boundary lubricated bearings to extend life and reduce drag. Check the manufacturer's recommendations for recommended or prohibited lubricants. Grease is often used instead of oil in boundary lubrication applications because it's convenient to use, stays in place, and allows for simpler design. Grease and other fixed lubricants do nothing to cool the bearing, so relatively slow speeds are a must with boundary lubrication.

When the intention is to add lubricant to a plain bearing, provide for lubricant proliferation through the assembly, even when designing for boundary lubrication. In a fixed shaft arrangement, the shaft is often cross drilled and has an axial or a radial groove to allow lubricant to enter the bearing. A grease fitting or oil supply is often placed at one end of the shaft. In a rotating or sliding shaft arrangement, the lubricant supply is usually routed through the bearing housing and relies on a hole or a hole and groove to allow lubricant to enter the bearing. A grease fitting or oil supply is then placed on the bearing housing. When planning for hydrodynamic or hydrostatic lubrication, a thorough analysis of the lubricant circulation must be undertaken.

BEARING MATERIALS FOR BOUNDARY LUBRICATION

Plain bearings designed to run with boundary lubrication are available in a variety of materials. There are many materials available, but the most common commercial bearing materials are tin, lead, bronze, copper, and plastic. Their general characteristics are as follows:

Bronzes and copper alloys: Bearings made of these materials are extremely common. They resist wear and temperature variation well. These bearings have a limited capacity to embed contaminants, so shaft scoring is a risk and requires shaft hardening. Alloys with higher lead content reduce shaft scoring and reduce friction during boundary lubrication, but will be more sensitive to elevated temperatures and require lower speeds. Copper alloy bearings often are available with dry lubricant embedded in the bearing surface. Sintered bronze bearings are porous and usually saturated with lubricant. Material properties vary, so consult the manufacturer's catalog for more information.

Plastics: In recent years plastic plain bearings have met or exceeded many characteristics of their metal counterparts. Plastic bearings can be formulated to meet a wide variety of requirements, including sanitary applications. When specifying a plastic bearing, it is important to consider that these bearings tend to run hotter than their thermally conductive metal counterparts. Elevated ambient temperatures can be of particular concern with plastic bearings. Confirm the temperature ratings of any plastic bearings with the manufacturer. It is important to note that some lubricants can attack plastic bearings or cause them to swell, so be sure to check material compatibility. Material properties vary greatly, so consult the manufacturer's catalog for more information.

BOUNDARY LUBRICATED SLEEVE BEARINGS

Sleeve bearings used with boundary lubrication are very common in modern machine design. They are ideal for bearing applications requiring small radial clearances, quiet running, low speeds, and oscillating or intermittent movements. They are often available in straight and single flanged types (Figure 10-2). The flange can be used as a thrust bearing, and is also subject to boundary lubrication conditions. Sleeve bearings are most commonly pressed into a housing bore, but some have features for retaining the bearing with screws. Follow the manufacturer's recommendations for fits and tolerances of

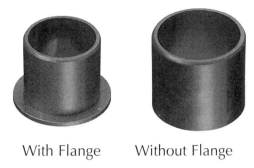

With Flange   Without Flange

Figure 10-2: Plain Sleeve Bearings
Source: igus Inc. (www.igus.com)

the bearing housing and shaft because proper fit is required to produce proper bore clearances, ensure even wear, and minimize drag.

Commercial catalogs often rate the load capacity of sleeve bearings intended for boundary and mixed film lubrication in terms of maximum allowable load, maximum allowable surface velocity, and PV factor. These values are used as maximum values to determine bearing size and materials. A safety factor of 2 is typical when examining PV factor. PV factor ratings will be highly variable depending on material, method of manufacture, and whether any embedded lubricants are present. Some manufacturers offer online software to select bearings and calculate life expectancy. One excellent example of such software can be found at www.igus.com.

Bearing pressure (P) is the radial force on the bearing divided by the projected area of contact. Projected area is equal to shaft diameter multiplied by length. Because PV is a function of bearing length, the use of a longer bearing is one method of getting the PV of the application lower than the PV rating for the bearing. In general, the length to bore ratio of plain sleeve bearings should never exceed 4:1.

PV factor is proportional to the heat of friction generated per unit of bearing area; it is used to measure the performance capability of a bearing. PV factor alone should not be used to size sleeve bearings. For low speed or high load applications, maximum allowable load will be the primary consideration. PV factor is calculated for rotating sleeve bearings according to the formulas in Table 10-1. Some common bearing material properties are listed in Table 10-2.

Selection of proper shaft material and hardness is critical for plain bearing and assembly performance. A smooth shaft surface makes the most of thin film lubrication and reduces friction. Sufficient hardness is essential to ensure that the shaft does not wear before the bearing does. A rule of thumb says that

Table 10-1: PV Formulas for Rotary Sleeve Bearings

PV FACTOR	PV factor = P x V
INCH UNITS P = Bearing pressure (psi) F = Load on the bearing (lbf) A = Bearing area (in^2) D = Bearing internal diameter (in) or shaft external diameter L = Bearing length (in) V = Bearing surface speed (ft/min) N = Revolutions per minute θ = Swept angle of oscillation (radians) f = Frequency of oscillation (oscillations per minute)	$P = \dfrac{F}{A}$ $A = DL$ Continuous rotation: $V = \dfrac{D\pi(N)}{12}$ Oscillating rotation: $V = \dfrac{\theta f D}{24}$
METRIC UNITS P = bearing pressure (MPa) F = load on the bearing (N) A = bearing area (m^2) D = bearing internal diameter (m) or shaft external diameter L = bearing length (m) V = bearing surface speed (m/min) N = revolutions per minute θ = Swept angle of oscillation (radians) f = Frequency of oscillation (oscillations per minute)	$P = \dfrac{F}{A(10^6)}$ $A = DL$ Continuous rotation: $V = D\pi(N)$ Oscillating rotation: $V = \dfrac{\theta f D}{2}$

Table 10-2: Typical Plain Bearing Material Properties

Bearing Material	Type	Temperature Range	Shaft Hardness	Dynamic P Max.	V Max.	PV Max
SAE 841 Bronze	Porous	10°F to 220°F	> 35 Rc	2,000 psi	1,200 ft/min	50,000 psi-ft/min
		-12°C to 104°C		14 MPa	6 m/s	1.8 MPa-m/s
SAE 660 Bronze	Non-porous	10°F to 450°F	> 35 Rc	4,000 psi	750 ft/min	75,000 psi-ft/min
		-12°C to 232°C		27 MPa	3.8 m/s	2.7 MPa-m/s
Nylon Plastic	Non-porous	10°F to 20°F	< 35 Rc	2,000 psi	600 ft/min	3,000 psi-ft/min
		-12°C to 93°C		14 MPa	3 m/s	0.11 MPa-m/s
PTFE Plastic	Non-porous	-350°F to 500°F	< 35 Rc	500 psi	50 ft/min	1,000 psi-ft/min
		-212°C to 260°C		3.5 Mpa	0.25 m/s	0.035 Mpa-m/s

the shaft should be 3 times harder than the bearing material. This difference in hardness ensures that any contaminant particles that enter the bearing will embed in the bearing material rather than score the shaft.

Hardened and ground steel shafts are generally recommended for use with bronze bearings. SAE 841 bronze bearings require steel shafts that have greater than 0.4% carbon content, such as SAE 1137 – 1141 steels. When corrosion protection is needed, hard chrome plating is a good choice. If stainless steel shafting must be used, choose a 400-series steel. When specifying stainless steel, a 416-series steel that is heat treated to maximum hardness is best. Shaft surface roughness should be 16 microinches (0.4 micrometers) or better.

Plastic bearings have different requirements depending on their makeup. Chrome plating may cause stick-slip behavior in some plastic bearings because the surface finish is too fine. Stainless steel in the 300 series may be used with some plastic bearings. For plastic bearings, always follow the manufacturer's recommendations for shaft material, hardness, and surface roughness.

SLEEVE BEARING SELECTION PROCEDURE FOR BOUNDARY LUBRICATION

When selecting a sleeve bearing, it can be beneficial to contact the manufacturers for application assistance. They often have specialists available to analyze the application and make recommendations to maximize bearing life and performance. The following procedure can be useful when selecting and sizing a sleeve bearing. This procedure assumes that the shaft is rotating in the bearing and there is boundary lubrication. This scenario is a very common one in light industrial machine design. Linear bearings are selected similarly, and are discussed in Section 10.3.

1. <u>Determine the design parameters.</u> These include the following:
   - Desired shaft diameter
   - Revolutions per minute of the shaft
2. <u>Select a bearing length.</u> A typical sleeve bearing should have a length of 1.5 to 2 times its inner diameter. An L/D as low as 1 is acceptable for some sintered bearings.
3. <u>Select a bearing material.</u> Compute the PV factor using the formulas in Table 10-1. Select a bearing in the desired size with a rated PV factor of at least twice (factor of safety) the computed value. Choosing a longer bearing will lower the PV factor for the application. Increasing the shaft size will also lower the PV factor.
4. <u>Check the bearing maximum pressure rating.</u> Verify that the chosen bearing maximum pressure rating is well above your calculated bearing pressure from Table 10-1. This includes checking the thrust load on the bearing flange if applicable.
5. <u>Check the bearing maximum velocity rating.</u> Verify that the chosen bearing maximum velocity rating is well above the calculated velocity of the shaft (journal) surface from Table 10-1.
6. <u>Choose a bearing outer diameter.</u> The bearing OD may already be set based on the results of previous steps. Once an OD is selected, verify that there is plenty of space for the OD of the chosen bearing in the bearing housing. Because the OD is generally pressed into the housing, avoid thin walls around the bearing.

7. <u>Dimension and tolerance your shaft and bearing housing</u> per the bearing manufacturer's recommendations. Specify the recommended shaft material and hardness. Add any required chamfers or radii, and specify any required surface finishes.

CRITICAL CONSIDERATIONS: Plain Bearings

- Plain bearing material selection is governed by the speed and loading conditions of the assembly.
- All plain bearings run under boundary lubrication conditions and will wear over time. Life can be maximized by staying well within the PV rating for the bearing.
- Shaft hardness and surface finish are critical for proper function and life expectancy.
- Heat is often the limiting factor with plastic bearings. This includes not only ambient temperature, but also the heat generated by friction during operation. Higher speeds generate more heat.
- Never store plain bearings with embedded lubricant in or on absorbent material, because the lubricant could wick out.

BEST PRACTICES: Plain Bearings

- Use a factor of safety of at least 2 when evaluating PV factors. This will ensure long bearing life.
- Add grease to plain bearings to prolong bearing life.
- Avoid ultra-soft bearing materials like PTFE unless absolutely needed. Soft materials are vulnerable to damage during installation and running, and tend to have shorter life spans than harder bearings.

ROLLING ELEMENT BEARINGS

Section
10.2

Rolling element bearings are extremely common in machinery. They create rolling contact between two parts to minimize the effects of friction and enable high relative speeds with very little wear. The most common types of rolling element bearings are ring shaped and provide rolling contact between a shaft and a housing bore. Another common rolling element bearing is a thrust bearing that takes an axial load while rotating about a shaft or pin. Both types use either balls or rollers to carry the applied loads. Design of bearings is beyond the scope of this text. This section will focus on the types, selection, and sizing of commercially available steel ball and roller bearings.

RECOMMENDED RESOURCES

- Juvinall, Marshek, *Fundamentals of Machine Component Design*, 2nd Ed., John Wiley & Sons, Inc., New York, NY, 1991
- R. Mott, *Machine Elements in Mechanical Design*, 5th Ed., Pearson/ Prentice Hall, Inc., Upper Saddle River, NJ, 2012
- R. L. Norton, *Machine Design: An Integrated Approach*, 4th Ed., Prentice Hall, Upper Saddle River, NJ, 2011
- Oberg, Jones, Horton, Ryffel, *Machinery's Handbook*, 28th Ed., Industrial Press, New York, NY, 2008
- *SKF General Catalogue*, SKF Group, June 2008
- *ISO 281:* "Rolling bearings — Dynamic load ratings and rating life"
- *American Bearing Manufacturers Association Website:* www.americanbearings.org

LUBRICATION, SEALS, AND SHIELDS

Rolling element bearings must be lubricated to ensure long life and low drag. Bearings are normally lubricated with grease for speeds below 500 rpm at normal ambient temperatures. High speeds and high temperatures both require a circulating oil lubrication system to keep the bearing cool. Manufacturers will normally specify the appropriate lubricant for each bearing.

Seals are available in many formats, but in general all are capable of fully sealing the bearing from contaminants and retaining lubricant in the bearing. "Lubricated for life" or "maintenance free" are phrases commonly used to describe rolling element bearings that have seals and are pre-lubricated at the manufacturer. If seals are not available for a bearing, provisions must be made in the assembly to direct and retain lubricant at the bearing. Some bearing manufacturers also sell a line of externally mounted seals for bearing arrangements.

Shields are intended to protect the bearing from large contaminants and accidental contact damage. They are generally metal and do not form a complete seal around the bearing. Bearings can have just seals, just shields, or both seals and shields. Not all bearings are available with seals or shields. Separable bearings are not sealed. In general, seals are available for deep groove ball bearings, angular contact ball bearings, self-aligning ball bearings, cylindrical roller bearings, needle roller bearings, and spherical roller bearings.

BEARING CHARACTERISTICS

Rolling element bearings are generally made up of two rings, called "races," separated by a set of rolling elements. Radial bearings are composed of an inner and outer race, whereas thrust bearings are composed of two (or more, in the case of double direction bearings) races. Rolling elements can be balls, cylindrical rollers, tapered rollers, or other specialized shapes.

A bearing cage is an internal component that captivates each rolling element in the bearing and prevents contact between individual rolling elements. Not all bearings have cages. Whether a bearing does or not will be clearly stated by the manufacturer. Cages help reduce internal friction in the bearing and ensure that the load bearing elements are evenly distributed for smooth running. "Full complement" bearings normally forego a cage so that more rolling elements can be fit into the bearing for greater load capacity. Cages are not always metal, so material and thermal compatibility must be verified if using caged bearings.

Some types of bearings, such as cylindrical roller bearings, are capable of axial displacement between the inner and outer races. When axial displacement is a function of the bearing, the maximum allowable displacement will be provided in the catalog information. Axial displacement can occur during normal running conditions as the shaft warms and elongates. This effect must be accommodated in the bearing arrangement through the use of either a bearing that allows axial displacement, or by designing the bearing arrangement and fits so that one bearing can easily move axially in its housing.

Most radial bearings are made with a cylindrical bore which is meant to be mounted on a cylindrical shaft. Some bearings are available with a tapered bore that is designed to be mounted on an adapter sleeve or tapered shaft. A tapered bore allows the bearing to be adjusted to remove internal clearance from the bearing. Cylindrical bores are much more commonly used and will be assumed for the purposes of this text unless otherwise specified. See the recommended resources for more information on the use of tapered bores.

Some bearings are separable, which means the inner and outer races can be separated axially from each other. Tapered roller bearings are a common example of separable bearings. Separability eases installation, but generally precludes integral sealing.

Temperature rating is not always stated in catalogs. In general, catalog values for life, speed, and loading assume an ambient temperature of 68°F (20°C). Most standard steel bearings have a maximum temperature of 250°F (121°C). The maximum temperature must include the elevated temperature of the bearing due to running conditions and friction.

Limiting speed is usually provided in catalog data for bearings. In all cases, the limiting speed should never be exceeded. The bearing can run up to limiting speed, but only for short periods of time. The reference or operating speed rating normally represents the speed at which the heat generated by the bearing is in equilibrium with the heat dissipated. It is assumed that operating above the rated speed for any length of time will cause the bearing to overheat. Reference or operating speed ratings are usually given for bearings in catalogs. Both the limiting and operating speed ratings assume a given set of operating and lubricant conditions. The standard ISO 15312: 2003 provides a set of reference conditions for the operating speed rating, which include an ambient temperature of 20°C (68°F), a bearing temperature of 70°C (158°F), and a constant load. These conditions also assume that the inner ring of the bearing is rotating. When the outer ring rotates, the speed ratings should be lowered. Note that when a bearing is operated at a very low speed or undergoes oscillating movements rather than continuous running, hydrodynamic lubrication does not occur in the bearing. In those cases, lubricants should be used that form a film on the bearings. Lubricants with EP additives form such a film.

Internal clearance is defined as the total distance that one race can be moved relative to the other in the stated direction. Bearings have radial and axial internal clearances. Normal clearance bearings are typically used and must be mounted with the recommended standard fits on the shaft and housing. If different mounting fits are used, the bearing must have a different internal clearance in order to work properly. Bearings with less internal clearance are available commercially for ultra-precision applications.

Basic dynamic load rating is the load for which each bearing is expected to meet its life expectancy. Static load rating is the maximum load the bearing can withstand without internal damage. These load ratings are discussed in more detail later in this section.

Starting torque is the minimum torque required to start a bearing rotating from rest. For small instrument bearings, starting torque and bearing friction is of concern. The addition of seals will increase starting torque as well as running friction. Values for starting torques and friction calculations can be found in manufacturer's data.

BEARING TYPES

The most common bearings in light industrial machine design are deep groove, angular contact, cylindrical roller, and spherical roller bearings. Deep groove ball bearings are by far the most commonly used type of rolling element bearing. The following are descriptions of some of the bearings found in machinery.

Deep groove ball bearings (Figure 10-3) are extremely versatile and are available fully lubricated and sealed. They can accommodate some axial loading in both directions and are capable of high speeds. They can be used as the locating bearing in bearing arrangements. Deep groove ball bearings typically can accommodate angular misalignment between 2 and 10 minutes of arc. Verify this with the manufacturer for every bearing. Double row deep groove ball bearings are used when the load capacity of single row deep groove ball bearings is insufficient. Their ability to handle axial loads and misalignment is about the same as that of single row bearings.

Single Row   Double Row

Figure 10-3: Deep Groove Ball Bearings

Source: SKF USA Inc. (www.skf.com)

Self-aligning ball bearings (Figure 10-4) were invented by SKF USA Inc. These bearings have a double row of balls that ride in a spherical raceway. This type of bearing is capable of high speeds and is particularly suited for applications with significant angular misalignment. Misalignment up to 3 degrees is allow-

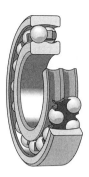

Figure 10-4: Self-Aligning Ball Bearing
Source: SKF USA Inc. (www.skf.com)

able with some bearings. Verify this with the manufacturer for every bearing. Self-aligning ball bearings are available with seals.

Angular contact ball bearings (Figure 10-5) are designed to handle combined loading. They are available in single and double row configurations. A single row angular contact ball bearing can handle axial loading in only one direction. These bearings are normally applied in a pair, arranged back to back, and adjusted against each other to provide a stiff arrangement that can handle axial loads in both directions, as well as radial loads and tilting moments. Angular contact ball bearings have a very limited ability to accommodate angular misalignment.

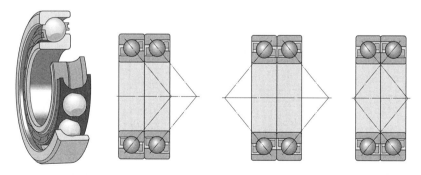

Figure 10-5: Angular Contact Ball Bearings
Source: SKF USA Inc. (www.skf.com)

Four-point contact ball bearings (Figure 10-6) are single-row, angular contact ball bearings that have special raceways designed to allow the bearing to support axial loads in both directions. These bearings are meant for applications with predominantly axial loads and can support only limited radial loading.

Figure 10-6: Four-Point Contact Ball Bearing

Source: SKF USA Inc. (www.skf.com)

Cylindrical roller bearings (Figure 10-7) are available in many different configurations. The most common configuration consists of a single row of cylindrical rollers, caged, with an inner race that is axially displaceable. These bearings are suitable for very heavy radial loads and can operate at high speeds. Cylindrical roller bearings are not suitable for axial loads and can accommodate angular misalignment up to only a few minutes of arc. These bearings are generally not used to locate a shaft axially, but special designs are available that allow the bearing to be used as a locating bearing.

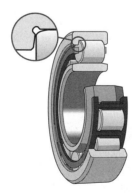

Figure 10-7: Cylindrical Roller Bearing

Source: SKF USA Inc. (www.skf.com)

Spherical roller bearings (Figure 10-8) are comprised of a double row of rollers within a spherical raceway. They are capable of accommodating significant angular misalignment up to about 3 degrees. Verify this with the manufacturer for every bearing. These bearings can handle heavy radial and axial loads acting in both directions. Sealed versions are available.

Tapered roller bearings (Figure 10-9) are available in a variety of configurations. These bearings have tapered rollers that are arranged at an angle to the

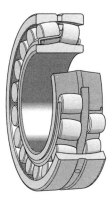

Figure 10-8: Spherical Roller Bearing
Source: SKF USA Inc. (www.skf.com)

Figure 10-9: Tapered Roller Bearing
Source: SKF USA Inc. (www.skf.com)

bore. Tapered roller bearings are particularly suitable for significant combined loading, but can accommodate axial loads in only one direction. Single-row tapered roller bearings are generally separable and are most often used in pairs arranged such that axial loads in both directions can be borne. Single-row tapered roller bearings can accommodate angular misalignment only up to a few minutes of arc. These bearings are usually preloaded at installation and require special care to install and run in.

Needle roller bearings are ideal for applications where radial space is limited. They have rollers that are very long and thin. These bearings have high radial load capacities, but cannot tolerate axial loads. These bearings are very intolerant of angular misalignment. Needle roller bearings are sometimes used without an inner race. This provides an even more compact bearing solution, but special care must be taken in the manufacture of the shaft that will become the inner race.

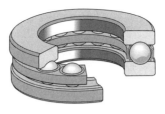
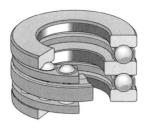

Figure 10-10: Ball Thrust Bearings
Source: SKF USA Inc. (www.skf.com)

Single Direction   Double Direction

Ball thrust bearings (Figure 10-10) are designed for axial loads only. Single-direction thrust bearings have a single row of balls and can handle axial loading in only one direction. Double-direction thrust bearings, as the name suggests, can handle axial loading in both directions and have two rows of balls. These bearings are generally separable.

Cylindrical roller thrust bearings have cylindrical rollers and are designed to handle axial loads acting in one direction only. These bearings are ideal for heavy axial loading and are fairly compact.

Spherical roller thrust bearings use sphered rollers arranged at an angle to the shaft axis to transmit loading. This enables the bearing to handle both axial and radial loading. These bearings are self-aligning and can handle angular mis-alignments up to 3 degrees. Verify this with the manufacturer for every bearing.

Needle roller thrust bearings are ideal for supporting heavy axial loads in a minimum of axial space. These bearings are similar to cylindrical roller thrust bearings in that they are intolerant of misalignment, cannot support radial loads, and are available in single-direction or double-direction versions. When space is at an absolute premium, the needles and cage can be used without upper and lower races. In this situation, the abutting parts must be carefully manufactured to serve as races.

Some common application conditions are listed in Table 10-3 with one or more suitable bearing types. Table 10-4 lists some of the commonly available bearing types and a few key characteristics. There are many more bearing types and options available. It is important to note that rolling element bearings are less appropriate than plain bearings for applications where the angle of rotation is small; for example, a shaft that undergoes a small oscillating motion. Rolling element bearings in such an application would lubricate and wear unevenly, causing premature failure.

Table 10-3: *Applications and Suitable Bearings*

Application Condition	Typical Bearing Types
High Misalignment	Self-Aligning Ball Bearings, Spherical Roller Bearings
Bearing Used To Locate Shaft Axially	Angular Contact Bearings, Deep Groove Ball Bearings (Light axial load only)
High Speeds (Radial Loads Only)	Deep Groove Ball Bearings, Self-Aligning Ball Bearings
High Speeds (Combined Loads)	Angular Contact Bearings
Light to Moderate Radial Loads	Deep Groove Ball Bearings
Heavy Radial Loads	Cylindrical Roller Bearings, Needle Roller Bearings
Combined Loads, Radial Dominant	Angular Contact Bearings, Tapered Roller Bearings
Combined Loads, Axial Dominant	Spherical Roller Thrust Bearings
Light to Moderate Axial Loads	Thrust Ball Bearings, Four-Point Contact Bearings
Heavy Axial Loads	Roller Thrust Bearings
Moment Loads	Paired Angular Contact Ball Bearings, Face-To-Face Tapered Roller Bearing Pair
Axial Displacement Expected	Cylindrical Roller Bearings, Needle Roller Bearings
Low Profile	Needle Roller Bearings
Maximum Stiffness	Cylindrical Roller Bearings, Tapered Roller Bearings, Needle Roller Bearings

Table 10-4: Selected Bearing Types and Characteristics

Key: ● Characteristic of Bearing • Lesser Characteristic of Bearing ○ Characteristic of Some Configurations		Characteristics of Bearing									
Bearing Type	**Typical Application**	Axial Load Capability	Radial Load Capability	Moment Load Capability	Accommodates Misalignment	Internal Axial Displacement	High Stiffness	Seals Available	Low Profile	Separable	Tapered Bore
Deep Groove Ball Bearing	General Use. Moderate Radial, Light Axial Loads	•	•		•			●			
Deep Groove Ball Bearing (Double Row)	Heavy Radial Loads	•	●								
Self-Aligning Ball Bearing	High Misalignment	•	•		●			●			●
Angular Contact Ball Bearings	Unidirectional Combined Loads	•	•		•						
Angular Contact Pair (Back To Back)	Bidirectional Combined Loads	•	•	•				○			
Four-Point Contact Ball Bearings	Light Axial Loads	•								●	
Cylindrical Roller Bearings	Heavy Radial Loads		●		•	●	●			●	
Full Complement Cylindrical Roller Bearings	Very Heavy Radial Loads		●		•	●	●			○	
Needle Roller Bearings	Low Profile		●			●	●	●	●	●	
Spherical Roller Bearings	High Misalignment, Heavy Combined Loads	●	●		●			●			●
Taper Roller Bearings	Unidirectional Combined Loads	•	●		•		●			●	
Taper Bearing Pair (Face to Face)	Bidirectional Combined Loads	•	●	●	•		●			●	
Thrust Ball Bearings	General Use, Thrust	•								●	
Cylindrical Roller Thrust Bearings	Heavy Loads, Thrust	●								●	
Needle Roller Thrust Bearings	Low Profile, Thrust	●							●	●	
Spherical Roller Thrust Bearings	High Misalignment, Thrust	●	•		●					●	

RADIAL BALL AND ROLLER BEARING DIMENSIONS AND TOLERANCES

Rolling element bearing dimensions and tolerances are heavily standardized. The Anti-Friction Bearing Manufacturers Association (AFBMA) and ISO have standards that are in agreement. Bearings from different manufacturers are generally interchangeable as a result. The Annular Bearing Engineers Committee (ABEC) defines the standard tolerances for bearings in the United States, and ISO standards are in agreement for metric bearings. Radial ball and roller bearings have five standard classes of tolerances, which are given in Table 10-5. Higher precision levels yield a longer life expectancy, but come

Table 10-5: Select Radial Bearing Tolerance Classes

Application	ANSI/ABMA Standard 20 Tolerance Class		ISO 492 Tolerance Class
	Ball Bearings	Roller Bearings	
Standard	ABEC 1	RBEC 1	Normal
Semi-Precision	ABEC 3	RBEC 3	Class 6
Precision	ABEC 5	RBEC 5	Class 5
High Precision	ABEC 7	RBEC 7	Class 4
Ultra Precision	ABEC 9	RBEC 9	Class 2

at a higher cost. Most machinery applications will use ABEC 1 class bearings. Applications that require extra smooth and accurate shaft running may require ABEC 5 or even ABEC 7 bearings. For tolerance values in each class, consult the recommended resources.

Machine designers will primarily be concerned with the dimensions they have control over: shaft diameter and housing bore diameter. In general, a light press fit on the rotating race is sought for bearing mounting. The rotating race is generally the inner race, but not always. It is essential to apply the appropriate fits to the shaft and housing because the bearing internal clearance in operation relies on the proper mounting fits. For bearings with an ABEC 1 (standard) tolerance class, Tables 10-6 and 10-7 are useful for determining the appropriate tolerances for the shaft and housing bore. For other types of bearings, such as needle roller bearings, consult with the manufacturer on tolerance information. It is typical for manufacturers to specify tolerances on bearing features and recommend tolerances for mating parts. These tolerances are typically very small (a few ten-thousandths of an inch) and are the responsibility of the machine designer.

Bearing seats should generally be ground and have a surface roughness of 16 to 32 microinches (0.4 to 0.8 micrometers). Maximum fillet dimensions for shoulders and steps are normally provided for each bearing in the manufacturer's catalog. Larger fillet radii reduce stress concentrations, so be sure to take advantage of the fillet allowance.

Proper geometric tolerancing of the shaft and housing bearing seats is essential for proper bearing fit and function. Both the diameters and the perpendicular abutments must be tightly controlled. Recommended geometric tolerances for metric bearing seats and abutments are shown in Table 10-8. IT (International Tolerance) grades are provided; they can be used to determine the numerical tolerances. The numerical values for tolerance grades are provided in Section 3.1 of this book. For cylindricity and total radial runout, the numerical tolerance represented by the tolerance grade is divided by 2 as shown.

Table 10-6: Shaft Fits for Radial Bearings (ABEC/RBEC 1)

Numerical values for tolerance designations are given in Section 3.1

Operating Conditions		Ball Bearings		Cylindrical Roller Bearings		Spherical Roller Bearings		ISO Tolerance Designation (a)
		mm	inch	mm	inch	mm	inch	
Inner Ring Stationary w/Respect to Load — All Loads	Inner ring has to be easily displaceable	All Diameters	All Diameters	All Diameters	All Diameters	All Diameters	All Diameters	g6
	Inner ring does not have to be easily displaceable	All Diameters	All Diameters	All Diameters	All Diameters	All Diameters	All Diameters	h6
RADIAL LOAD				**NOMINAL SHAFT DIAMETER**				
Inner Ring Rotating w/Respect to Load or Direction of Load Indeterminate — LIGHT: Loads up to 0.075C (C = dynamic load rating)		≤ 18	≤ 0.71	≤ 40	≤ 1.57	≤ 40	≤ 1.57	h5
		> 18	> 0.71					j6 (b)
				(40) - 140	(1.57) - 5.51	(40) - 100	(1.57) - 3.94	k6 (b)
				(140) - 320	(5.51) - 12.6	(100) - 320	(3.94) - 12.6	m6 (b)
NORMAL: Ball: 0.075C to 0.15C; Cylindrical Roller: 0.075C to 0.2C; Spherical Roller: 0.075C to 0.25C (C = dynamic load rating)		≤ 18	≤ 0.71					j5
		> 18	> 0.71	≤ 40	≤ 1.57	≤ 40	≤ 1.57	k5
				(40) - 100	(1.57) - 3.94	(40) - 65	(1.57) - 2.56	m5
				(100) - 140	(3.94) - 5.51	(65) - 100	(2.56) - 3.94	m6 (b)
				(140) - 320	(5.51) - 12.6	(100) - 140	(3.94) - 5.51	n6
				(320) - 500	(12.6) - 19.7	(140) - 280	(5.51) - 11.0	p6
HEAVY: Loads over 0.15C (C = dynamic load rating)		(18) - 100	(0.71) - 3.94					k5
		>100	> 3.94	≤ 40	≤ 1.57	≤ 40	≤ 1.57	m5
				(40) - 65	(1.57) - 2.56	(40) - 65	(1.57) - 2.56	m6 (b)
				(65) - 140	(2.56) - 5.51	(65) - 100	(2.56) - 3.94	n6 (b)
				(140) - 200	(5.51) - 7.87	(100) - 140	(3.94) - 5.51	p6 (b)
				(200) - 500	(7.87) - 19.7	(140) - 200	(5.51) - 7.87	r6 (b)
				> 500	> 19.7	> 200	> 7.87	r7 (b)
Pure Thrust Load		All Diameters	All Diameters	Consult Manufacturer				j6 (b)

a) Tolerances are shown for solid steel shafts. For hollow or nonferrous shafts, tighter fits may be needed.
b) When greater accuracy is required, use j5, k5, and m5 instead of j6, k6, and m6, respectively.

Table 10-7: Housing Bore Fits for Radial Bearings (ABEC/RBEC 1)
Numerical values for tolerance designations are given in Section 3.1

Design and Operating Conditions				ISO Tolerance Designation (a)
Rotational Conditions	**Loading**	**Outer Ring Axial Displacement Limitations**	**Other Conditions**	
Outer Ring Stationary w/Respect to Load	Light, normal, and heavy	Outer ring must be easily displaceable axially	Heat input through shaft	G7
			Housing split axially	H7 (b)
	Shock with temporary complete unloading			H6 (b)
Load Direction is Indeterminate	Light and normal	Transitional range (c)	Housing not split axially	J6 (b)
	Normal and heavy			K6 (b)
	Heavy shock			M6 (b)
Outer Ring Rotating w/Respect to Load	Light	Outer ring need not be axially displaceable	Split housing not recommended	K6 (b)
	Normal and heavy			M6 (b)
				N6 (b)
	Heavy		Thin wall housing not split	P6 (b)

a) For cast iron or steel housings. For nonferrous alloy housings tighter fits may be needed.
b) Where wider tolerances are permissible, use tolerances P7, N7, M7, K7, J7, and H7
c) The tolerance zones are such that the outer ring may be either tight or loose in the housing.

Table 10-8: Bearing Seat Geometric Tolerances

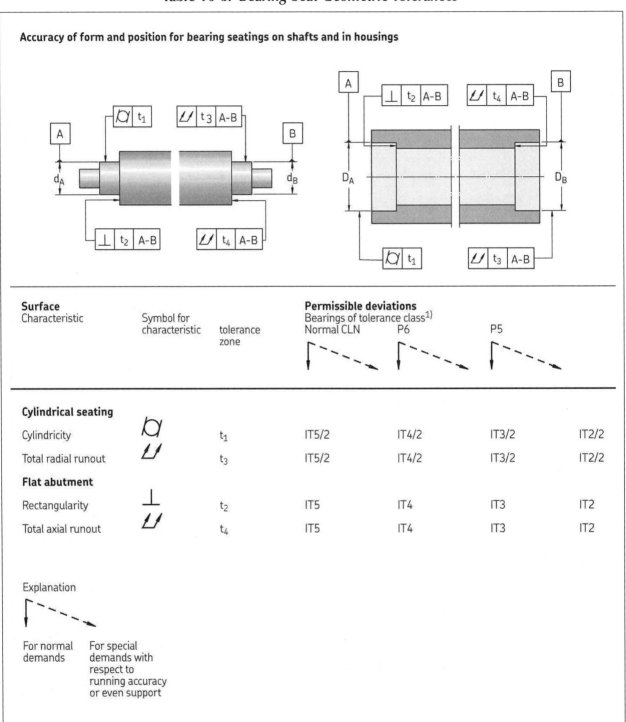

Accuracy of form and position for bearing seatings on shafts and in housings

Surface Characteristic	Symbol for characteristic	tolerance zone	Permissible deviations Bearings of tolerance class[1] Normal CLN	P6	P5	
Cylindrical seating						
Cylindricity	$\bigcirc$	t_1	IT5/2	IT4/2	IT3/2	IT2/2
Total radial runout	$\mathcal{L}\!\!\mathcal{L}$	t_3	IT5/2	IT4/2	IT3/2	IT2/2
Flat abutment						
Rectangularity	$\perp$	t_2	IT5	IT4	IT3	IT2
Total axial runout	$\mathcal{L}\!\!\mathcal{L}$	t_4	IT5	IT4	IT3	IT2

Explanation

For normal demands   For special demands with respect to running accuracy or even support

"For bearings of higher accuracy (tolerance class P4, etc.), please consult with the manufacturer.

Source: SKF USA Inc. (www.skf.com)

BEARING ARRANGEMENTS

Bearing arrangements can be broken into two basic types: preloaded and non-preloaded. Most bearing arrangements found in light industrial machinery will be of the non-preloaded variety. Preloaded bearing arrangements are often found in machine tools and other extremely high accuracy applications. Preloading is also called for when a bearing is run under extremely light loading conditions at high speeds. Preloading may be accomplished using springs or by adjusting two bearings against each other, using spacers, until zero operational clearance is achieved. Preloading is beyond the scope of this text, but is detailed in the recommended resources.

A non-preloaded bearing arrangement generally includes two bearings and associated hardware that support and locate a shaft, both radially and axially. Normally just one bearing, called the locating bearing, is tasked with axial location of the shaft in both directions. The locating bearing (or bearing combination) necessarily must be able to handle combined axial and radial loading. Suitable locating bearings include deep groove ball bearings when axial loading is light, angular contact bearings, spherical roller bearings, or a matched pair of tapered roller bearings. The second (non-locating) bearing usually constrains the shaft only radially. This type of arrangement requires that the locating bearing be axially and radially fixed on both the shaft and housing, while the other is allowed to float in the axial direction. Allowing one bearing to float axially prevents overloading of the bearings due to shaft expansion. If the non-locating bearing is of a type that allows axial displacement within the bearing, as do some types of cylindrical roller bearings, the bearing races may be axially constrained without fear of overloading.

Figure 10-11a shows a bearing arrangement consisting of a deep groove ball bearing and a cylindrical roller bearing capable of internal axial displacement. In this situation, both bearings may be fully captivated. Figure 10-11b shows an arrangement using two deep groove ball bearings. In this case, the non-locating bearing is free to move in the housing and allow axial displacement. Figure 10-11c illustrates an arrangement using two opposed angular contact bearings to locate the shaft axially, while a deep groove bearing is allowed to move axially in its housing. Figure 10-11d shows a pair of tapered roller bearings locating and supporting a shaft. In this type of arrangement, the bearing housings must be axially adjusted relative to one another to achieve proper preload on the bearings.

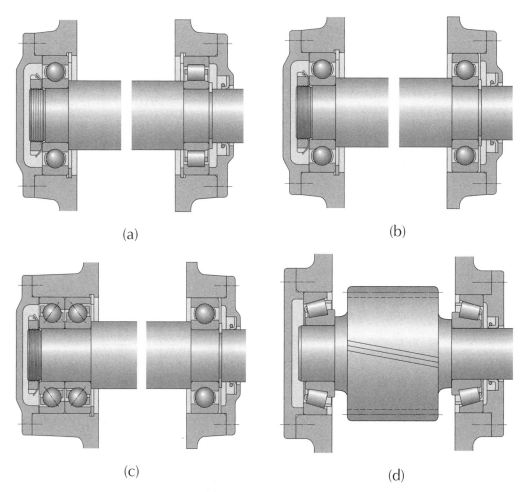

(a)   **(b)**

(c)   **(d)**

Figure 10-11: Typical Bearing Arrangements
Source: SKF USA Inc. (www.skf.com)

     Axial constraint of bearings cannot be accomplished merely through the bearing fit on the shaft and in the housing. When a bearing with a cylindrical bore must be located axially on a shaft, this is usually accomplished using a shoulder on the shaft. Another component, such as a nut, collar, or retaining ring, is used to captivate the bearing race against the shaft shoulder. This location method is illustrated in Figure 10-12a and b. The shoulder and other hardware must be properly sized to fully support the inner race without interfering with bearing rotation. The outer race of a bearing is normally captured between a shoulder in the housing bore and a clamping ring attached to the housing (Figure 10-12c). Bearings with tapered bores are less commonly used and require the use of adapter sleeves and spacers to mount and locate properly. Consult the recommended resources for information on installation of tapered bore bearings.

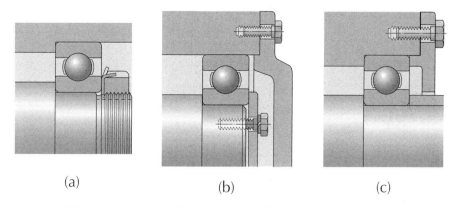

(a)   (b)   (c)

Figure 10-12: Typical Axial Bearing Location Methods
Source: SKF USA Inc. (www.skf.com)

LOADS ON BEARINGS

Axial and radial load carrying capacities are highly variable among different types of bearings. Many radial bearings have very limited axial load carrying capability. Some bearings can carry axial loads in only one direction. The size and direction of applied loads is of primary concern when selecting a bearing type.

When selecting and sizing a bearing, one must first quantify the axial and radial loads seen by the bearing during operation. Section 11.1 of this text includes some formulas for calculating loads on a shaft for common machine elements. Two types of loading on each bearing must be evaluated: static loading and dynamic loading. Static loading must take into account any maximum loads seen by the bearing, even if only for a short length of time. Dynamic loading considers normal operation loads. In both cases, an equivalent load must be calculated for the bearing. The equivalent static and dynamic loads are then compared with catalog static and dynamic load ratings to select the appropriate bearing.

Equivalent load is a function of both the loading on the bearing and the bearing configuration. Pure thrust bearings have an equivalent axial load equal to the applied axial load, and the radial load must be zero. Purely radial bearings have an equivalent radial load equal to the applied radial load and the axial load must be zero. Bearings designed for combined loading, like angular contact bearings, are configured such that applied axial loads induce radial loads, and applied radial loads induce axial loads. This means that even a purely axial or radial load condition will result in combined loading on the bearing. In many applications, a combined axial and radial applied load is present. Based on the bearing type and configuration, an equivalent load must

Table 10-9: Loads On Rolling Element Bearings

F_R = applied radial load		F_A = applied axial (thrust) load	
FLUCTUATING LOADS		Speed, load direction constant: $$F_R = \frac{F_{Rmin} + 2F_{Rmax}}{3}$$ $$F_A = \frac{F_{Amin} + 2F_{Amax}}{3}$$	Load direction rotating: $F_R = f_m(F_c + F_r)$ F_c = Constant direction load F_r = Rotating direction load $0.75 \leq f_m \leq 1$ (See manufacturer)
RADIAL BEARINGS AND SPHERICAL ROLLER BEARINGS	**STATIC EQUIVALENT LOAD** W = Consult manufacturer's catalog Z = Consult manufacturer's catalog		$F_{MAX} = WF_R + ZF_A$ If $F_{MAX} < F_R$ then $F_{MAX} = F_R$
	DYNAMIC EQUIVALENT LOAD $V = 1$ if inner bearing race rotates $V = 1.2$ if outer bearing race rotates e = Consult manufacturer's catalog X = Consult manufacturer's catalog Y = Consult manufacturer's catalog		Radial Load Only: $\quad F = VF_R$ Combined Radial And Thrust Loads: $\quad F = VXF_R + YF_A$ Test For Relatively Small Thrust: If $\dfrac{F_A}{VF_R} \leq e$ Then $X = 1, Y = 0$
ANGULAR CONTACT BEARINGS AND TAPERED ROLLER BEARINGS	**EQUIVALENT LOAD** Note: Radial loads on angular contact and tapered roller bearings induce axial loads. Consult with the manufacturer for axial load calculation methods.		Depends on bearing geometry. Consult Manufacturer
THRUST BEARINGS	**EQUIVALENT LOAD**		Axial (Thrust) Load Only: $\quad F = F_A$ Radial or Combined Loads: $\quad$ Consult manufacturer

be calculated from the applied loads. The formulas in Table 10-9, in combination with the manufacturer's catalog data, can be used to accomplish this.

Most bearings must be subjected to a minimum load in order to function properly (have the rolling elements roll inside the bearing rather than slide) and reach their expected lifespans. In most cases, the weight of the components is enough to meet this minimum load requirement. When dealing with very light components and loads, consult the manufacturer for calculation methods for minimum load requirements. A rule of thumb is to apply a minimum load of 0.01C to ball bearings and 0.02C to roller bearings. C is the dynamic load rating for the bearing, which is provided in catalogs and calculated in Table 10-10.

The radial loads used In Table 10-9 are assumed to be in a constant direction. When a load varies, its mean value must be determined and used as the load value in life and dynamic load calculations. When a load fluctuates while its direction is fairly constant, its mean value can be estimated using the equation in Table 10-9. If the direction of the load or part of the load rotates, as with an unbalance condition, the mean load can be calculated using the equation in Table 10-9.

BEARING LOAD RATINGS AND LIFE EXPECTANCY

Life expectancy is a statistical phenomenon. Basic dynamic load rating is typically defined as the load at which 90% of a population of identical bearings will complete 1 million revolutions without failure. This may vary among manufacturers, so it is essential to verify the assumptions behind any catalog information. Most manufacturers follow the standard ISO 281: 1990 which contains bearing life and dynamic load calculation methods. Life calculations based on the standard can be found in Table 10-10. Part of load calculation involves applying a service factor (Table 10-12) which accounts for shock loading in service.

Table 10-10: Life and Load Ratings for Radial and Angular Contact Bearings

L = Life expectancy in revolutions Ball Bearings: $k = 3$ F = Equivalent load on bearing Roller Bearings: $k = 3.33$	
BASIC STATIC LOAD C_0 = Required basic static load rating SF_0 = Factor of safety (See Table 10-11) F_{MAX} = Maximum equivalent load on bearing	$C_0 = (SF_0)F_{MAX}$
BEARING LOAD VS. LIFE	$\dfrac{L_2}{L_1} = \left(\dfrac{F_1}{F_2}\right)^k$
LIFE EXPECTANCY h = Lifetime hours of operation N = Speed in revolutions per minute For Variable Loads R_1 = Number of revolutions under Load 1 L_1 = Lifetime (revolutions) under Load 1 R_T = Total revolutions expected from bearing	Conversion from lifetime In hours : $\quad L = 60hN$ Variable loading: $\quad L = \dfrac{1}{\dfrac{R_1}{R_T L_1} + \dfrac{R_2}{R_T L_2} + \ldots}$
BASIC DYNAMIC LOAD $L_{rated} = 10^6$ revolutions typically $\quad$ <u>Always verify rated life value</u> K_S = Shock service factor (Table 10-12) These factors can be applied if the load rating is based on 90% reliability: $K_R = 0.62$ for 95% reliability $K_R = 0.33$ for 98% reliability $K_R = 0.21$ for 99% reliability	Dynamic load using rated (usually 90%) reliability: $\quad C = FK_S \sqrt[k]{\dfrac{L}{L_{rated}}}$ (If $L = L_{rated}$, $\; C = FK_S$) Dynamic load using adjusted (>90%) reliability: $\quad C = FK_S \sqrt[k]{\dfrac{L}{K_R L_{rated}}}$

Table 10-11: Static Load Safety Factors for Bearings

	Rotating Bearings						Non-Rotating Bearings	
	Quiet Running Importance							
	Low		Normal		High			
Application Conditions	Ball Bearings	Roller Bearings	Ball Bearings	Roller Bearings	Ball Bearings	Roller Bearings	Ball Bearings	Roller Bearings
Smooth, No Vibration	0.5	1	1	1.5	2	3	0.4	0.8
Normal	0.5	1	1	1.5	2	3.5	0.5	1
Significant Shock Loads	≥ 1.5	≥ 2.5	≥ 1.5	≥ 3	≥ 2	≥ 4	≥ 1	≥ 2

Table 10-12: Shock Service Factors for Bearings

Application Conditions	Typical Range of K_S	
	Ball Bearing	Roller Bearing
Smooth, Uniform Load	1	1
Gearing	1.0 - 1.3	1
Light Impact	1.2 - 1.5	1.0 - 1.1
Moderate Impact	1.5 - 2.0	1.1 - 1.5
Heavy Impact	2.0 - 3.0	1.5 - 2.0

Loads used in life calculations are assumed to be constant and radial to the rolling elements. When calculating the required dynamic load rating for an application, the equivalent load (Table 10-9) on the bearing must be used rather than the applied loading. Load ratings in catalogs are usually based on an assumed bearing material and hardness. Typical assumptions are chromium steel bearing material and a hardness of 58 Rc. If alternative materials are required, load ratings may be affected.

The effects of friction and lubricant viscosity affect the operating temperature and life of the bearing. Bearing failure in the field is usually due to contamination, misalignment, and lubrication failure. As a result, actual life expectancy will be shorter than the rated value. A modified life calculation is available in standard ISO 281:1990/Amd 2:2000 to account for contamination and lubrication conditions. This method is quite complicated and is not presented here. Drag caused by seals can also be important in critical applications. The manufacturer can assist in calculating lubrication or seal drag.

BEARING SELECTION PROCEDURE FOR A ROTATING SHAFT APPLICATION

Rotating shaft applications are the most common type of bearing selection situations. The best way to properly select and size a rolling element bearing is to speak with the manufacturer. Most bearing manufacturers have engineers on their staff ready to analyze your application thoroughly and make recommendations. When selecting a bearing for this type of application, the following procedure can be used:

1. Determine the design parameters. These include:
   - Expected lifetime of the bearing in hours or revolutions
   - Speed of the bearing
   - Bearing arrangement, defining the locating bearing(s)
   - Running load on the bearing, broken into axial and radial components
   - Expected shock load
   - Mounting and sealing requirements
   - Required precision

2. Select a bearing type based on performance or characteristics. Tables 10-3 and 10-4 can be useful when selecting a bearing type. One of the primary considerations will be whether radial, axial, or combined loading must be accommodated.

3. Select a bearing size from a catalog based on dynamic load rating. Calculate the equivalent load on the bearing using the equations in Table 10-9 and the manufacturer's catalog. Then calculate the required basic dynamic load needed for the application and required reliability level using the formulas in Table 10-10. Basic dynamic load rating (C) is normally given in catalogs. Select a bearing with a dynamic load rating that exceeds the calculated dynamic load for the application by a factor of safety. A factor of safety for dynamic load as high as 5 is not out of the question.

4. Check the basic static load rating for the chosen bearing. Using an appropriate factor of safety (Table 10-11) and the equations in Tables 10-9 and 10-10, calculate the basic static load for the application and compare it to the basic static load rating for the chosen bearing. Basic static load rating (C_0) is normally given in catalogs. Account for any peak loads including impact loads.

5. <u>Check the limiting and rated operating speeds</u> for the bearing. Speed in operation should never exceed the limiting speed of the bearing. Normal running speed should never exceed the rated operating speed.

6. <u>Select bearing options</u> such as seals, shields, lubricant, etc.

7. <u>Select bearing accessories</u> such as nuts, locking washers, spacers, wrenches, etc. The geometry for the shaft and housing should be finalized and all hardware accounted for on the bill of materials.

CRITICAL CONSIDERATIONS: Rolling Element Bearings

- An appropriate factor of safety must be used when evaluating static and dynamic loading of bearings.
- Equivalent load, not applied load, must be used when selecting bearings based on static and dynamic load ratings.
- Most bearings have some "play" or internal clearance. If this is unacceptable, consider preloaded bearings or stiffer bearings (like needle roller bearings).
- Be sure not to axially constrain any bearing that cannot tolerate axial loading. This could cause the bearing to become overloaded if expansion occurs in the assembly due to the heat of running.
- Lubrication is essential to bearing life. Bearings used in oscillating applications may have reduced life due to inadequate lubricant distribution.
- Over-lubrication causes bearings to fail. Observe the manufacturer's lubrication instructions. Calibrated grease guns, automatic lubrication systems, or ultrasonic monitoring can be used to prevent improper lubrication.
- Bearings should have a light press fit on the rotating race. If the fit is too loose, the bearing may rotate in/on its mount and cause wear and particulate contamination. If the fit is too tight, the bearing may be damaged, or internal clearance could be compromised. Tables 10-6 and 10-7 can be used to select proper fits on standard bearings.
- Tight geometric tolerancing is required on bearing support surfaces like bores, shafts, and seating surfaces. If the bearing is not fully and evenly supported, premature bearing wear and failure can occur. It is typical for manufacturers to specify these tolerances.
- Fine surface finishes are required on all bearing support surfaces. Rough surfaces can break down and cause fits to loosen.
- Care must always be used when installing bearings. Contamination and shock loading during installation must be avoided. Press bearings on with steady, even pressure around the entire race that is being fitted. Pressing forces should never be applied across the rolling elements in the bearing (i.e. never press on the inner race by applying force to the outer race, etc.).

> ### BEST PRACTICES: Rolling Element Bearings
>
> - Consult the manufacturer for best results when designing a bearing application.
> - Whenever possible, axial loading should be carried by thrust bearings. Use one thrust bearing per direction of axial load.
> - The use of lubricated, sealed bearings simplifies the assembly design.
> - Use standard precision class bearings (ABEC 1) for most applications.
> - Design provisions for bearing dismounting if possible. Slots for bearing removal tools, grease fittings and channels for pressure dismounting, and holes for jacking screws are all useful.
> - Use a reputable bearing distributor or manufacturer to avoid counterfeit components. Counterfeit bearings are a big business.
> - Avoid using rolling element bearings in locations where heavy vibration occurs while the bearing is not rotating. This condition can cause deformation and wear in the bearing.
> - Wave springs can be used to lightly bias a bearing in a housing or on a shaft with axial clearance.

LINEAR BEARINGS

Section
10.3

There are many types of linear bearings. Some examples are plain linear bushings, linear ball bushings, linear rails, and ball splines. Plain bearings are simple and cost effective, but are limited in speed and life by friction between the sliding elements. Ball or roller bearings replace sliding with rolling contact for higher speeds, load ratings, and life expectancies. Rollers tend to have higher load ratings than balls and are used in high load applications. For all linear bearings, life expectancy is measured in linear travel distance and is determined by component materials, hardness, lubrication, and load conditions.

> ### RECOMMENDED RESOURCES
>
> - Consult a manufacturer for more information on linear bearings.

LUBRICATION

Plain linear bearings are generally used in a state of boundary lubrication. Grease or other thick lubricants are applied to the shaft and must be re-applied periodically to reduce friction and extend bearing life. Contaminants have easy access to linear plain bearings, so proper precautions or soft bearing materials must be used. For more information on boundary lubrication and plain bearing materials, see Section 10.1.

Rolling element linear bearings usually incorporate internal lubrication pathways into the bearing housing. Grease fittings are usually present on the bearing housing for lubrication purposes. These bearings often have contact seals at the shaft to reduce lubricant loss and prevent contaminants from entering the housing. Linear bearings should be lubricated periodically since some lubricant loss is to be expected.

PLAIN LINEAR BEARINGS

Plain linear bearings are essentially the same as rotary plain bearings in design, materials, and static pressure ratings. See Section 10.1 for more information on these aspects of plain bearings. Linear plain bearings can take a variety of forms, including round sleeves, flat plates, and other shapes. In many cases, a linear plain bearing arrangement may be that of a slide and slideway; this arrangement allows movement along only one axis. It is not uncommon for both slide and slideway to be made out of hardened tool steel for long life and coated with dry film lubricants as well as thoroughly greased. Heavy loads may be carried with some configurations, and high precision achieved as long as wear is kept to a minimum. Heavy duty slide and slideways are found on machining centers, and light duty versions are commonly found on assembly machines and fixtures. Many versions are commercially available, but these are also easily designed in-house if needed. Some common slide and slideway configurations are shown in Figure 10-13, and some commercial plain bearing slide assemblies are shown in Figure 10-14.

Plain linear sleeve bearings are very common in industry. These bearings will allow rotation of the bearing on the shaft if torque is applied. It is important to prevent moment loading of the bearing because that will result in uneven wear and reduced life. Shafts for plain linear bearings are generally hardened and ground steel, and must be precisely aligned if used in parallel.

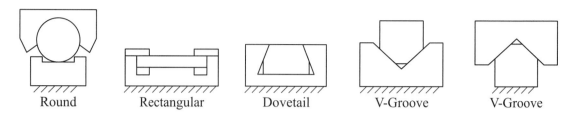

Round   Rectangular   Dovetail   V-Groove   V-Groove

Figure 10-13: Slide and Slideway Configurations

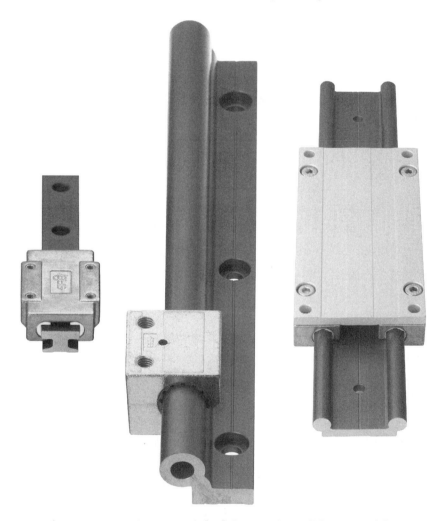

Figure 10-14: Commercial Plain Bearing Slide Assemblies

Source: igus Inc. (www.igus.com)

    Commercial linear plain bearings are normally selected based on PV factor, similarly to rotary plain bearings. However, linear plain bearings dissipate heat much more quickly than rotating bearings do. As a result, the PV factor rating for a linear plain bearing will normally be much higher than its rotary counterpart. A plain bearing that is meant for linear movement should be clearly labeled as such in the manufacturer's catalog and should have a linear PV factor.

Boundary lubrication is the norm with linear plain bearings. Selection and sizing of linear plain bearings is performed similarly to selection and sizing of plain sleeve bearings for rotating service. See Section 10.1 for that procedure, using the formulas in Table 10-13 for PV calculation.

Table 10-13: PV Formulas for Plain Linear Bearings

PV FACTOR	PV factor = P x V

INCH UNITS
    P = bearing pressure (psi)
    F = load on the bearing (lbf)
    A = bearing area (in^2)
    D = bearing internal diameter (in) or shaft external diameter
    L = bearing length (in)
	$P = \dfrac{F}{A}$
	$A = DL$
	
	V = bearing surface speed (ft/min)

METRIC UNITS
    P = bearing pressure (MPa)
    F = load on the bearing (N)
    A = bearing area (m^2)
    D = bearing internal diameter (m) or shaft external diameter
    L = bearing length (m)
	$P = \dfrac{F}{A(10^6)}$
	$A = DL$
	
	V = bearing surface speed (m/min)

ROLLING ELEMENT LINEAR BEARINGS

For rolling element linear bearings, rolling element bearing concepts and calculations discussed in Section 10.2 apply. When specifying linear bearings, the main criteria are bearing precision, internal clearance, static load rating, dynamic load rating, speed, and life expectancy. Equivalent load must be used when calculating dynamic load, and manufacturer's catalogs must be consulted in doing so. The life of linear bearings is often given in linear distance traveled. The ISO standard, which is commonly used by most manufacturers, is a dynamic loading life expectancy of 50 km. The formulas in Table 10-14 can be used to calculate dynamic load rating for linear bearings.

There are many types of rolling element linear bearings available commercially. The following are some common types and their descriptions.

Table 10-14: Life and Load Ratings for Linear Bearings

L = life expectancy in linear distance Ball Bearings: $k = 3$ F = equivalent load on bearing Roller Bearings: $k = 3.33$	

BASIC STATIC LOAD
    C_0 = required basic static load rating
    SF_0 = factor of safety (See Table 10-11)
    F_{MAX} = maximum equivalent load on bearing
	$C_0 = (SF_0)F_{MAX}$

BEARING LOAD VS. LIFE	$\dfrac{L_2}{L_1} = \left(\dfrac{F_1}{F_2}\right)^k$

LIFE EXPECTANCY
    h = lifetime hours of operation
    S = Stroke length
    n = number of strokes per minute
For Variable Loads
    R_1 = distance under Load 1
    L_1 = lifetime (distance) under Load 1
    R_T = total distance expected from bearing
	Conversion from lifetime In hours
	
	$L = 60h(2Sn)$
	
	Variable loading:
	
	$L = \dfrac{1}{\dfrac{R_1}{R_T L_1} + \dfrac{R_2}{R_T L_2} + \ldots}$

BASIC DYNAMIC LOAD
    L_{rated} = 50 km typically
          Always verify rated life value
    K_S = shock service factor (Table 10-12)
These factors can be applied if the load rating is
based on 90% reliability:
    $K_R = 0.62$ for 95% reliability
    $K_R = 0.33$ for 98% reliability
    $K_R = 0.21$ for 99% reliability
	Dynamic load using rated reliability:
	
	$C = FK_S \sqrt[k]{\dfrac{L}{L_{rated}}}$
	
	(If $L = L_{rated}$, $C = FK_S$)
	
	Dynamic load using adjusted reliability:
	
	$C = FK_S \sqrt[k]{\dfrac{L}{K_R L_{rated}}}$

<u>Ball bushings</u> are very commonly used and consist of a sleeve containing ball bearings that ride on a smooth shaft. They are generally meant for light loading conditions. Some designs allow rotation of the bearing on the shaft (Figure 10-15), while most are not intended to rotate. In general, torque loading must be avoided. Ball bushings are often used in pairs, running on parallel shafts. Ball bearings in the sleeve are usually re-circulated to provide continuous rolling contact with the shaft. Ball bushings are available in both closed and open configurations, with and without lubrication-preserving end caps. Some examples of these are shown in Figures 10-16 and 10-17. A closed sleeve is used with end supported shafts while an open sleeve is used with continuously supported shafts. Preloaded ball bushings are commercially available. Ball bushings have a limited capacity to carry moment loads (about an axis perpendicular to the bushing/shaft axis). Moment load ratings are often given for ball bushings, and using them in pairs is a common solution to eliminate moment loading. Longer bushings generally have greater moment load capacities.

Figure 10-15: Rotary Linear Ball Bushings

Source: Misumi USA Inc. (www.misumiusa.com)

With End Caps

Without End Caps

Figure 10-16: Ball Bushings

Source: Misumi USA Inc. (www.misumiusa.com)

Figure 10-17: Open Ball Bushings

Source: Misumi USA Inc. (www.misumiusa.com)

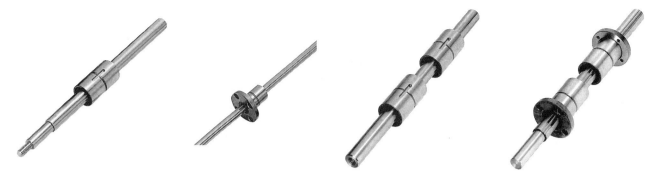

Figure 10-18: Ball Splines
Source: Misumi USA Inc. (www.misumiusa.com)

Ball splines (Figure 10-18) are recommended when the assembly must resist rotation of the bearing on the shaft under torsional loading. Engagement of the ball bearings within the sleeve with axial splines on the shaft prevents rotation. Ball splines are often rated in terms of allowable torque load, and are generally designed to transmit torque while allowing linear movement. Ball splines have a limited capacity to carry moment loads (about an axis perpendicular to the bushing/shaft axis). Moment load ratings are often given for ball splines, and using them in pairs is a common solution to eliminate moment loading. In general, longer spline nuts can carry greater moment loads.

Linear rails and guides (Figure 10-19) are a compact solution to the problem of preventing rotation while allowing linear movement. These assemblies normally consist of a rectangular rail and a bearing carriage that rides along the rail. In general, the carriage or guide is captive to the rail and can support loads in every

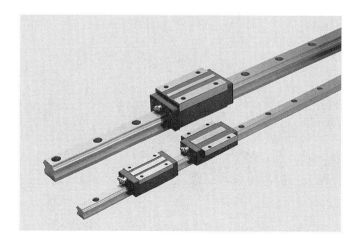

Figure 10-19: Linear Rail and Guide Assemblies
Source: SBC Linear Co., Ltd. (sbclinear.co.kr/eng/index.php) Courtesy of TPA (www.tpa-us.com)

direction. Consult the manufacturer for specific load carrying capabilities because these vary between models. These assemblies can generally carry moment loads as well, though this must be verified for each model. If large moment loads are present, the use of two guides on one rail, or two rails side by side should be considered. Both ball and roller versions are available commercially.

Crossed roller ways (Figures 10-20 and 10-21) are highly compact and rigid assemblies consisting of two angled rows of rollers between two linear races. They are capable of carrying very heavy loads. These assemblies can withstand significant moment loads and are often used as table bearings in machine tools. The rollers do not re-circulate, so movement is generally smoother than that of recirculating ball bushings. Due to their configuration, crossed roller ways must be twice as long as their expected stroke.

Extended Load Capacity

Needle Rollers

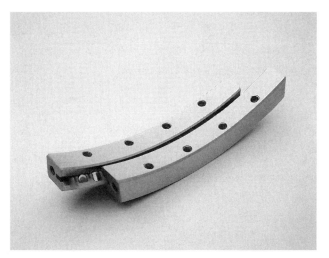
Curved Crossed Roller Way

Figure 10-20: Crossed Roller Ways

Source: PM Bearings (www.pmbearings.nl) Courtesy of TPA (www.tpa-us.com)

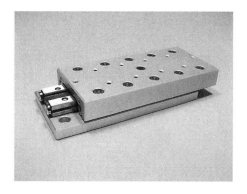 

Figure 10-21: Crossed Roller Slide Assemblies
Source: PM Bearings (www.pmbearings.nl) Courtesy of TPA (www.tpa-us.com)

Selection and sizing of linear rolling element bearings is performed in a similar way to the sizing and selection of rotating bearings. See Section 10.2 for that procedure, and use the formulas in Table 10-14 to calculate dynamic load ratings for linear bearings.

CRITICAL CONSIDERATIONS: Linear Bearings

- Use an appropriate factor of safety when performing bearing calculations.
- Proper alignment is critical to linear bearing function and life. When two linear bearings are used in parallel, overconstraint must be avoided.
- Lubrication is essential to long bearing life, and contamination of bearings must be avoided.

BEST PRACTICES: Linear Bearings

- Consult the manufacturer for best results when selecting and sizing linear bearings.
- Avoid applying moment loads to linear bearings.
- Prevent contamination of linear bearings as much as possible.

11

POWER TRANSMISSION DEVICES

Contents

11.1	SHAFTS	547
11.2	SHAFT COUPLINGS	563
11.3	GEARS	574
11.4	GEARBOXES	614
11.5	BELTS AND CHAINS	627
	11.5.1 BELTS	629
	11.5.2 CHAINS	655
11.6	LEAD, BALL, AND ROLLER SCREWS	664

Tables

11-1	Common Shaft Materials	549
11-2	Forces Exerted on Shafts	550
11-3	Shaft Deflection Calculations	553
11-4	Shaft Stress Calculations	555
11-5	Neuber's Constant for Notch Sensitivity	557
11-6	Shaft Critical Speed Calculations	559
11-7	Coupling Performance Comparison	573
11-8	Useful American-Standard ANSI/AMGA Standards	575
11-9	Useful Metric Gearing ISO Standards	575
11-10	Commonly Encountered Spur Gear Terms and Definitions	576
11-11	Equations of the Involute Tooth Profile	578
11-12	Minimum Number of Pinion Teeth to Avoid Interference Between 20° PA Full Depth Pinion and Gears of Various Sizes	581
11-13	Classification of Common Gear Types	583
11-14	Simple Gear Train Examples and Equations	587
11-15	Two Stage Compound Gear Train	588
11-16	Approximate Equivalencies of Gear Quality Standards	590
11-17	AGMA Recommended Quality Numbers, Applications, and Approximate ISO Equivalents	590
11-18	AGMA Recommended Quality Numbers for Pitch Line Speed with Approximate ISO Equivalents	591
11-19	Change in Center Distance to Backlash Ratio	593

11-20	AGMA Recommended Backlash Ranges	593
11-21	AGMA Recommended Backlash Settings for Metric Gears	593
11-22	Typical Backlash of Standard Commercial Gear Sets	593
11-23	AGMA Recommended Backlash Settings for Bevel and Hypoid Gears	595
11-24	Common Gear Materials, Features, and Applications	597
11-25	AGMA Recommended Lubricant Viscosities	598
11-26	Gear Lubrication Considerations	598
11-27	Formulas for Dimensions of Standard Spur Gears	600
11-28	Formulas for Dimensions of Standard Metric Spur Gears	601
11-29	Inch Spur Gear Sizing Formulas	602
11-30	Outline Factor (Y) for Use With Diametral Pitch	602
11-31	Example Safe Static Stress of Materials	603
11-32	Typical Service Factors (SF) for Calculating Design Horsepower	603
11-33	Advanced Gear Selection Problems and Possible Solutions	606
11-34	Common Equations Used with Helical Gears	606
11-35	Common Equations Used with Milled Bevel Gears	609
11-36	Formulas for Proportions of American Standard Fine Pitch Worms and Wormgears	611
11-37	Reflected Torque and Inertia on Gearboxes	617
11-38	Gearbox Service Factors	619
11-39	Gearbox Sizing Calculations	625
11-40	Belt and Chain Drive Formulas	628
11-41	Flat Belt Drive Load Formulas	632
11-42	Flat Belt Service Factors	633
11-43	NEMA Shaft Sizes	637
11-44	V-Belt Drive Load Formulas	637
11-45	V-Belt Hoop Tension Variable K_c	639
11-46	V-Belt Service Factors	639
11-47	NEMA Minimum Sheave Sizes	641
11-48	V-Belt Horsepower Rating Formulas	644
11-49	Narrow V-Belt Cross Section Parameters	645
11-50	Narrow V-Belt Speed Ratio Correction Factor	645
11-51	Narrow V-Belt Length Correction Factor	646
11-52	V-Belt Arc of Contact Correction Factor	646
11-53	Classical V-Belt Speed Ratio Correction Factor	647
11-54	Classical V-Belt Length Correction Factor	647
11-55	Synchronous Belt Equations	649
11-56	Synchronous Belt Service Factors	650
11-57	ANSI Standard Roller Chain Dimensions	656
11-58	ANSI Standard Plate Extensions and Extended Pins	657
11-59	Chain Equations	660
11-60	Chain Drive Service Factors	661

11-61 Recommended Roller Chain Sprocket Maximum Bore and Hub Dimensions 661
11-62 Screw Formulas 667
11-63 Common Lead Screw Coefficients of Friction 668
11-64 Screw Deflection and Buckling Formulas 670
11-65 Permissible and Critical Speed for Screws 671
11-66 End Fixity Factors for Critical Speed 671
11-67 Screw Selector 672
11-68 PV Values for Nuts 674
11-69 Ball Screw Formulas 676
11-70 Ball Screw Design Factor Interrelationships 677

Section ⊖ ⊕
11.1 ⊛ ⊙

SHAFTS

Shafts are typically long members with a round cross section. They are often stepped and have features for the attachment of parts like bearings and drive components. The steps are typically arranged to allow components to be assembled onto the shaft from both ends. There are two types of shafts commonly used in machinery: rotating (transmission) shafts and non-rotating shafts (axles). Both types must be carefully designed to carry the expected loads without shear failure or excessive deflection. Rotating shafts must also be designed to operate at the expected speeds without fatigue failure or excessive vibration. Design and analysis of shafts is an extremely complex subject with multiple methods of calculation. Commercial software is available to aid in shaft sizing and analysis. This section is a brief overview of some of the techniques and calculations. See the recommended resources for more information.

RECOMMENDED RESOURCES

- M. Lindeburg, *Mechanical Engineering Reference Manual for the PE Exam*, 11th Ed., Professional Publications, Inc., Belmont, CA, 2001
- R. Mott, *Machine Elements in Mechanical Design*, 5th Ed., Pearson/ Prentice Hall, Inc., Upper Saddle River, NJ, 2012
- R. L. Norton, *Machine Design: An Integrated Approach*, 4th Ed., Prentice Hall, Upper Saddle River, NJ, 2011
- Shigley, Mischke, Brown, *Standard Handbook of Machine Design*, 3rd Ed., McGraw-Hill Inc., New York, NY, 2004
- **Matweb: Material Property Data:** www.matweb.com
- **eFunda: Materials:** www.efunda.com/materials

METHODS OF ATTACHMENT

Components are attached to shafts using a variety of methods. Most rotating shafts are supported by two bearings. Rolling element bearings are typically mounted on shafts with a light press fit. Axial location of bearings is accomplished through the use of shaft shoulders, retaining rings, and bearing nuts. Other components are mounted on shafts using methods determined by their phasing and power transmission requirements. Location methods that accurately phase a component on a shaft are keys and splines. Both keys and splines transmit torque, but they need additional axial fixation. Location

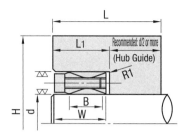

Figure 11-1
Hub to Shaft Clamping Assembly
Source: Misumi USA Inc.(www.misumiusa.com)

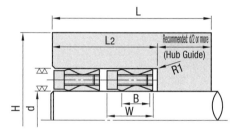

methods that transmit torque as well as lock components axially are set screws, clamps, pins, and press fits. Set screws are meant only for low torque transmission. Axial location methods that are not intended to transmit torque are nuts, retaining rings, and shoulders. Keys and retaining rings are detailed in Chapter 5 of this book.

Features added to shafts and hubs to provide location and power transmission all cause stress concentrations. Stress concentrations affect fatigue and stress calculations. Even clamps and press fits cause stress concentrations. To minimize stress concentrations on shafts, use generous radii. One component designed to minimize stress risers in bending near shaft shoulders is the Woodruff key. However, the Woodruff keyway still has significant stress concentrations along its sides. See the recommended resources for more information about shaft connections and stress risers.

MATERIALS AND TREATMENTS

Shafts are typically made from carbon and alloy steels, which all have comparable moduli of elasticity. Because of this similarity, material choice generally affects the shaft's ability to handle stress, but not the shaft's rigidity. Rigidity is controlled primarily through geometry. One way to preserve rigidity while minimizing mass is to use a hollow shaft. Hollow shafting is sometimes made from thick walled seamless tubing and then balanced. Round steel bars are often used for solid shafts. Shafts may be hardened or not, depending

Table 11-1: Common Shaft Materials

Material	Comments
Cold Rolled Steel Bars Up To 3.5 Inch Diameter	Can be left raw where machining is not required Residual stresses can cause deformation during machining of features
Hot Rolled Steel Bars Up To 6 Inch Diameter	Must be machined all over to remove carburized outer layer
SAE/ANSI 1018 – 1050	Use untreated when high strength is not required
SAE/ANSI 1018	General purpose commercial shafting
SAE/ANSI 11xx	Free machining steels
SAE/ANSI 1045	Commonly available as hardened and ground commercial shafting
SAE/ANSI 1340-50, 3140-50, 4140, 4340, 5140, 8650	Heat treat for added strength
SAE/ANSI 1020, 4320, 4820, 8620	Carburize for increased surface hardness
303 and 440C Stainless Steels	For special applications Commercially available as precision shafting in a large number of diameters

on material, strength requirements, and hardness requirements. High material strengths are achievable with hardening. When a shaft is heavily loaded, or when a smaller diameter is desired, hardening may be appropriate. The disadvantages of hardening are that it makes the shaft more brittle and also adds cost, so it should be avoided if not needed. Pre-hardened material can be used, and is recommended in cases where hardening after machining could cause unacceptable deformation. Many components, like plain bearings, require a minimum shaft hardness to protect the shaft from wear. Check all mating components for shaft hardness requirements.

Table 11-1 includes some typical material choices for shafts, along with general comments about each material. For those materials that may be heat treated, a wide range of strengths and hardnesses are achievable. More on material properties can be found in Section 8.1 and hardening information can be found in Section 8.3.

DEFLECTION, STRESS, AND FATIGUE

When the loads on an axle (non-rotating shaft) are not changing direction, the situation can be analyzed as a beam problem. Rotating shafts and axles with reversing loads must be analyzed for fatigue as well as stress and deflection. Shafts are subjected to loads delivered through attached components like bearings, pulleys, gears, and cams. Table 11-2 contains some equations used to calculate these loads. Be sure to understand the correct directions of these loads in each component. In cases where axial loads are induced on

Table 11-2: Forces Exerted on Shafts

TORQUE VS. POWER n = RPM of target component	Inch: $T = \dfrac{63025P}{n}$ T = Torque (in-lbf) P = Horsepower being transmitted	Metric: $T = \dfrac{9550P}{n}$ T = Torque (N-m) P = kW being transmitted
CHAIN DRIVE: FORCE EXERTED ON SHAFT F_{cx} = Bending force on shaft along line of centers F_{cy} = Bending force on shaft perp. to line of centers F_c = Tension in tight side of chain T = Torque transmitted by chain sprocket d_p = Sprocket pitch diameter	$F_{cx} = F_c \cos\theta \quad F_c = \dfrac{2T}{d_p}$ $F_{cy} = F_c \sin\theta$ θ = Angle of tight side of chain (deg.) from line of centers	
TIMING BELT: FORCE EXERTED ON SHAFT F_x = Bending force on shaft along the line of centers F_t = Tension in tight side of belt F_s = Tension in slack side of belt θ_1 = Angle of wrap (degrees) around smalle $^-$ sprocket	$F_x = \sqrt{F_t^2 + F_s^2 - 2F_t F_s \cos\theta_1}$ Where $\theta_1 = 180°$: $F_x = F_t + F_s$	
V-BELT DRIVE: FORCE EXERTED ON SHAFT F_x = Bending force on shaft along the line of centers F_t = Tension in tight sice of belt F_s = Tension in slack side of belt θ_1 = Angle of wrap (degrees) around drive sprocket d_p = Sheave pitch diameter T = Torque transmitted by belt sheave	$F_x = \sqrt{F_t^2 + F_s^2 - 2F_t F_s \cos\theta_1}$ Where $\theta_1 = 180°$: $F_x = F_t + F_s \cong \dfrac{3T}{d_p}$	

continued on next page

Table 11-2: Forces Exerted on Shafts (Continued)

continued on next page

FLAT BELT DRIVE: FORCE EXERTED ON SHAFT F_x = Bending force on shaft along the line of centers F_t = Tension in tight side of belt F_s = Tension in slack side of belt θ_1 = Angle of wrap (degrees) around drive sprocket d_p = Pulley pitch diameter T = Torque transmitted by belt pulley	$F_x = \sqrt{F_t^2 + F_s^2 - 2F_tF_s\cos\theta_1}$ Where $\theta_1 = 180°$: $F_x = F_t + F_s \cong \dfrac{4T}{d_p}$
SPUR GEAR: FORCE EXERTED ON SHAFT F_r = Radial force on shaft F_t = Tangential force T = Torque transmitted by gear	$F_r = F_t\tan\phi \qquad F_t = \dfrac{2T}{d_p}$ d_p = Gear pitch diameter ϕ = Gear pressure angle (deg.)
HELICAL GEAR: FORCE EXERTED ON SHAFT F_r = Radial force on shaft F_z = Axial force on shaft F_t = Tangential force T = Torque transmitted by gear	$F_r = F_t\tan\phi_t = F_t\dfrac{\tan\phi_n}{\cos\psi} \qquad F_t = \dfrac{2T}{d_p}$ $F_z = F_t\tan\psi$ $\tan\phi_n = \tan\phi_t\cos\psi$ d_p = Gear pitch diameter ϕ_n = Gear normal pressure angle (deg.) ψ = Gear helix angle (deg.) ϕ_t = Transverse pressure angle (deg.)

Table 11-2: Forces Exerted on Shafts (Continued)

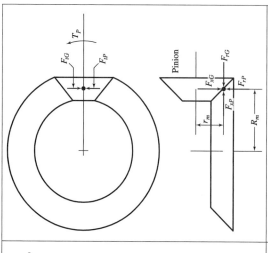

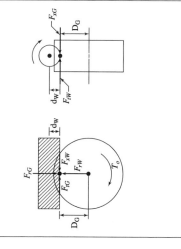

BEVEL GEAR: FORCE EXERTED ON SHAFT

F_{rG} = Radial force on gear shaft

F_{rP} = Radial force on pinion shaft

F_{xP} = Axial force on pinion at dist. r_m from shaft

F_{xG} = Axial force on gear at dist. R_m from shaft

F_{tP} = Tangential force on pinion at distance r_m

F_{tG} = Tangential force on gear at distance R_m

T = Torque transmitted by pinion or gear

d_p, D_p = Pitch diameters of pinion and gear

γ, Γ = Pitch cone angle for pinion and gear (deg.)

$$F_{rG} = F_{xP} = F_{tP}\tan\phi\sin\gamma = F_{tG}\tan\phi\sin\Gamma$$

$$F_{rP} = F_{xG} = F_{tP}\tan\phi\cos\gamma = F_{tG}\tan\phi\cos\Gamma$$

$$F_{tP} = F_{tG} \qquad F_{tP} = \frac{T_P}{r_m} \qquad F_{tG} = \frac{T_G}{R_m}$$

$$r_m = \left(\frac{d_p}{2} - \frac{F}{2}\sin\gamma\right)$$

$$R_m = \left(\frac{D_p}{2} - \frac{F}{2}\sin\Gamma\right)$$

F = Face width

ϕ = Pressure angle (degrees)

WORM GEAR: FORCE EXERTED ON SHAFT

F_{rG} = Radial force on gear shaft

F_{rW} = Radial force on worm shaft

F_{xG} = Axial force on gear at distance D_G

F_{xW} = Axial force on worm at distance d_w

F_{tG} = Tangential force on gear at distance D_G

F_{tW} = Tangential force on worm at dist. d_W

T_o = Output torque

μ = Coefficient of friction

ϕ_n = Normal pressure angle (deg.)

λ = Lead angle (deg.)

$$F_{rG} = F_{tG}\frac{\sin\phi_n}{\cos\phi_n\cos\lambda - \mu\sin\lambda}$$

$$F_{rG} = F_{rW} \qquad F_{tG} = F_{xW} \qquad F_{tG} = \frac{2T_o}{D_G}$$

$$F_{xG} = F_{tW}$$

$$F_{xG} = F_{tG}\frac{\cos\phi_n\sin\lambda + \mu\cos\lambda}{\cos\phi_n\cos\lambda - \mu\sin\lambda}$$

$$\mu = 0.124e^{\left(-0.074v_s^{0.645}\right)} \text{ when } v_s < 10 \text{ ft/min}$$

$$\mu = 0.012 + 0.103e^{\left(-0.11v_s^{0.45}\right)} \text{ when}$$
$v_s > 10$ ft/min

v_s = Pitch line speed of gear (ft/min)

$$v_s = \frac{v_{tG}}{\sin\lambda} = \frac{\pi D_G n_G}{12\sin\lambda}$$

D_G = Pitch diameter of worm gear (in)

n_G = Number of teeth on gear

a shaft by external loads or attachments like helical gears, it is good practice to take the axial loads to ground as close as possible to the source, using one thrust bearing per direction of load. Allow the remaining bearings to float axially to prevent undesirable thrust loads on the radial bearings.

Table 11-3: Shaft Deflection Calculations

y = Static deflection (in or m)	J = Polar second moment of area (in^4 or m^4)
L = Shaft length (in or m)	I = Mass moment of inertia (in-lb-sec^2 or kg-m^2)
θ_t = Angular deflection in radians	T = Applied torque (in-lbf or N-m)
g = Acceleration of Gravity (386.4 lbm-in/lbf-s^2 or 9.807 m/s^2)	
G = Material modulus of rigidity (Steels approx. 11.5x10^6 psi or 79 GPa)	
E = Modulus of elasticity (Steels approx. 30x10^6 psi or 207 GPa)	

STATIC BEAM TORSIONAL DEFLECTION

Uniform shaft: $\theta_t = \dfrac{TL}{GJ}$ 　　　　Stepped shaft: $\theta_t = \theta_{t1} + \theta_{t2} + ... = \dfrac{T}{G}\left(\dfrac{L_1}{J_1} + \dfrac{L_2}{J_2} + ...\right)$

POLAR SECOND MOMENT OF AREA

Uniform circular shaft: $J = \dfrac{\pi d^4}{32}$ 　　　　Stepped shaft: $J_{eff} = \dfrac{1}{\sum\limits_{i=1}^{n} \dfrac{L_i}{J_i}}$

SHAFT POWER　　　$P = T\omega$　Where ω is in radians per unit time

1hp = 550 ft-lbf/s = 33,000 ft-lbf/min = 745.7 W = 745.7 N-m/s

STATIC BEAM LATERAL DEFLECTION & FORCES

SIMPLY SUPPORTED, POINT LOAD	
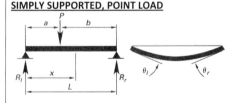	$y = \dfrac{Pa^2b^2}{3EIL}$ 　at x = a $y_{max} = \left(\dfrac{0.06415Pb}{EIL}\right)(L^2 - b^2)^{\frac{3}{2}}$ 　at $x = \sqrt{\dfrac{a(L+b)}{3}}$ $\theta_l = -\dfrac{Pab\left(1+\dfrac{b}{L}\right)}{6EI}$ 　　$\theta_r = \dfrac{Pab\left(1+\dfrac{a}{L}\right)}{6EI}$ Reactions: $R_l = \dfrac{Pb}{L}$ 　　　$R_r = \dfrac{Pa}{L}$ $M_{lx} = \dfrac{Pbx}{L}$ for x < a 　$M_{rx} = \dfrac{Pa(L-x)}{L}$ for x > a
SIMPLY SUPPORTED, DISTRIBUTED LOAD	
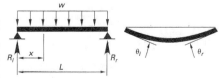 $W = wL$	$y_x = \left(-\dfrac{w}{24EI}\right)(L^3x - 2Lx^3 + x^4)$ $y_{max} = \dfrac{5wL^4}{384EI}$ 　at x = L/2 $\theta_l = \dfrac{-wL^3}{24EI}$ 　　$\theta_r = \dfrac{wL^3}{24EI}$ Reactions: $R_l = R_r = \dfrac{wL}{2}$ 　　　$M_l = M_r = \left(\dfrac{w}{2}\right)(x^2 - Lx)$

continued on next page

Table 11-3: Shaft Deflection Calculations (Continued)

CANTILEVERED, POINT LOAD	
	$$y_x = \left(-\frac{P}{6EI}\right)(2L^3 - 3L^2x + x^3)$$ $$y_{max} = -\frac{PL^3}{3EI} \text{ at } x = 0 \qquad \theta_l = \frac{PL^2}{2EI} \text{ End Slope}$$ Reactions: $$R_r = P \qquad\qquad M_x = -Px \qquad M_r = -PL$$
CANTILEVERED, DISTRIBUTED LOAD	
$W = wL$	$$y_x = \left(\frac{w}{24EI}\right)(3L^4 - 4L^3x + x^4)$$ $$y_{max} = -\frac{wL^4}{8EI} \text{ at } x = 0 \qquad \theta_l = \frac{wL^3}{6EI}$$ Reactions: $$R_r = W = wL \qquad M_x = \frac{-wx^2}{2} \qquad M_r = \frac{-wL^2}{2}$$
OVERHUNG LOAD	
	$$y_{tip} = (a+b)\left(\frac{Pa^2}{3EI}\right) \text{ at P}$$ $$y_{up} = (0.06415)\left(\frac{Pab^2}{EI}\right) \text{ at } x = 0.4226b$$ Reactions: $$R_l = \left(\frac{P}{b}\right)(b+a) \qquad R_r = \frac{-Pa}{b}$$ $$M_a = Px_a \qquad\qquad M_b = \left(\frac{Pa}{b}\right)(b - x_b)$$

Lateral shaft deflection is normally calculated using standard beam bending equations. The resultant deflection of several loads can be found by superposition. For stepped shafts, numerical integration is the most accurate method of evaluating deflection. Deflection due to transverse shear must also be evaluated for short shafts where length divided by diameter is less than 10. Lateral shaft deflection is important to quantify at each attached component or point of interest. Shaft deflection across a plain bearing must be small compared to the oil film thickness. Allowable shaft angular deflection at a rolling element bearing is on the order of 0.001 rad for cylindrical and tapered roller bearings, 0.004 rad for deep groove ball bearings, and 0.009 rad for spherical bearings. Self-aligning bearings allow significantly more angular deflection. Deflection of a shaft at gears with uncrowned teeth should be kept below a total separation of approximately 0.005" (0.13mm) or an angular difference below 0.0005 rad each. The values mentioned here should be considered general guidelines. Check the manufacturer specifications of all involved components for specific shaft deflection tolerances.

Torsional shaft deflection is a result of torque applied to the shaft. It is directly proportional to shaft length; therefore, long shafts are of special concern. Torsional deflection of stepped shafts with uniform material properties can be evaluated by summing the deflections of each section. Table 11-3 contains some equations commonly used to evaluate shaft deflections. More equations can be found in Section 1.3. Always use proper units and check units for consistency.

Rotating shafts are often loaded in both bending and torsion, which makes them complex multiaxial stress cases. Fluctuating torsion and fluctuating bending are a common load case in modern machine design. This requires calculation of the von Mises components of alternating and mean stresses, with appropriate stress concentration factors applied at points where the shaft is stepped or has stress-concentrating features. This can be expedited using FEA,

Table 11-4: Shaft Stress Calculations

MEAN AND ALTERNATING BENDING STRESSES of a solid circular shaft of diameter d as a result of fluctuating moment M.	$\sigma_a = k_f\left(\dfrac{32M_a}{\pi d^3}\right)$ $\sigma_m = k_{fm}\left(\dfrac{32M_m}{\pi d^3}\right)$
MEAN AXIAL STRESS In a solid circular shaft of diameter d as a result of axial force F	$\sigma_{m,axial} = k_{fm}\left(\dfrac{4F}{\pi d^2}\right)$
PERPENDICULAR SHEAR STRESS In a circular shaft of cross sectional area A V = Shear force	$\tau_{max} = k\dfrac{4V}{3A}$
MEAN AND ALTERNATING TORSIONAL SHEAR STRESSES Solid circular shaft of diameter d	$\tau_a = k_f\left(\dfrac{16T_a}{\pi d^3}\right)$ $\tau_m = k_{fm}\left(\dfrac{16T_m}{\pi d^3}\right)$
FATIGUE STRESS CONCENTRATION FACTORS Alternating and mean. k_f and k_{fm} must be calculated separately for shear and bending. q = Notch sensitivity factor $\sqrt{a}$ = Neuber's constant (Table 11-5) r = Notch radius (inches) k = Static stress concentration factor k_f = Fatigue stress concentration factor S_y = Yield strength (tensile)	$k_f = 1 + q(k-1)$ $q = \dfrac{1}{1+\dfrac{\sqrt{a}}{\sqrt{r}}}$ $k \cong 2$ for milled keyseat (worst case type) $k \cong 3$ for retaining ring groove $k \cong 2.5$ for sharp shoulder fillet with $\dfrac{r}{d} \cong 0.03$ $k \cong 1.5$ for large shoulder fillet with $\dfrac{r}{d} \cong 0.17$ $k_{fm} = k_f$ If $k_f\left|\sigma_{max}\right| < S_y$ $k_{fm} = \dfrac{S_y - k_f\sigma_a}{\left|\sigma_m\right|}$ If $k_f\left|\sigma_{max}\right| > S_y$ $k_{fm} = 0$ If $k_f\left|\sigma_{max} - \sigma_{min}\right| > 2S_y$

continued on next page

Table 11-4: Shaft Stress Calculations (Continued)

VON MISES COMPONENTS OF ALTERNATING AND MEAN STRESSES. Biaxial stress state	$\sigma_a' = \sqrt{\sigma_a^2 + 3\tau_a^2} \quad \sigma_m' = \sqrt{\left(\sigma_m + \sigma_{m,axial}\right)^2 + 3\tau_m^2}$
ESTIMATED ENDURANCE LIMIT With correction factors for ductile steel $S_{ut} \cong 68$ kpsi (469 MPa) for SAE 1020 CR $S_{ut} \cong 55$ kpsi (379 MPa) for SAE 1020 HR $S_{ut} \cong 91$ kpsi (627 MPa) for SAE 1045 CR $S_{ut} \cong 82$ kpsi (565 MPa) SAE 1045 HR $S_{ut} \cong 95$ kpsi (655 MPa) for SAE 4140 annealed $S_{ut} \cong 148$ kpsi (1020 MPa) for SAE 4140 normalized Heat treatment and manufacturing condition affect ultimate strength.	$S_e = C_{load} C_{size} C_{surf} C_{temp} C_{reliab} S_e'$ $S_e' \cong 0.5 S_{ut}$ when $S_{ut} < 200$ kpsi (1400 MPa) $S_e' \cong 100$ kpsi (700 MPa) when $S_{ut} \geq 200$ kpsi $C_{load} = 1$ for bending $C_{size} = 1$ for d ≤ 0.3 in (8 mm) $C_{size} = 0.869 d^{-0.097}$ for 0.3 in < d $\leq$ 10 in (d in inch) $C_{surf} = 1.34(S_{ut})^{-0.085}$ for ground surface (S_{ut} in kpsi) $C_{surf} = 2.7(S_{ut})^{-0.265}$ for machined or cold rolled surface (S_{ut} in kpsi) $C_{temp} = 1$ for T $\leq$ 840°F (450°C) $C_{reliab} = 0.868$ for 95% reliability $C_{reliab} = 0.620$ for 99.9999% reliability
FATIGUE FACTOR OF SAFETY (INFINITE LIFE) Determined using the modified Goodman diagram for ductile materials. It is assumed that the ratio of alternating to mean stress is constant over time. S_e = Estimated endurance limit S_{ut} = Ultimate tensile stress	$SF = \dfrac{S_e}{\sigma_a' + \left(\dfrac{S_e}{S_{ut}}\right)\sigma_m'}$

but can be calculated manually. These stresses are then normally entered into a modified Goodman diagram to examine the fatigue performance for infinite life and factor of safety. The modified Goodman line is valid for steels and aluminum alloys. The fatigue factor of safety can be calculated as well. Consult the recommended resources for more information on these methods and on stress concentration factors. Table 11-4 contains commonly used shaft stress calculations. Always check units for consistency and use appropriate factors of safety.

CRITICAL SPEEDS

Critical speeds and frequencies are those at which the shaft assembly will vibrate with ever-increasing amplitude. Critical frequencies are the natural frequencies of the system. Critical speeds are calculated from critical frequencies. Operating speed (frequency of rotation) is actually just one of many forcing frequencies applied to a shaft.

It is essential to design rotating shafts and their attachments such that the assembly's forcing frequencies are far away (either much higher or much

Table 11-5: Neuber's Constant for Notch Sensitivity

Material	S_{ut} (ksi)	Bending $\sqrt{a}\ \left(\sqrt{inch}\right)$	Torsion $\sqrt{a}\ \left(\sqrt{inch}\right)$
Steels	50	0.130	0.093
	55	0.118	0.089
	60	0.108	0.080
	70	0.093	0.070
	80	0.080	0.062
	90	0.070	0.055
	100	0.062	0.049
	110	0.055	0.044
	120	0.049	0.039
	130	0.044	0.033
	140	0.039	0.031
	160	0.031	0.024
	180	0.024	0.018
	200	0.018	0.013
	220	0.013	0.009
	240	0.009	0.004
Annealed Aluminums	10	0.500	
	15	0.341	
	20	0.264	
	25	0.217	
	30	0.180	
	35	0.152	
	40	0.126	
	45	0.111	

lower) from its critical frequencies. It is desirable to have the lowest critical frequency be much higher (by 3 to 10 times) than the forcing frequencies. In applications with very high masses the forcing frequencies are sometimes much higher than the critical frequencies. In cases where the operating speed is higher than critical speed, the shaft must be accelerated quickly through critical speed and never linger near it.

There are three types of vibration of concern in rotating shaft design: lateral vibration, shaft whirl, and torsional vibration. Shaft whirl is self-excited, whereas lateral vibration requires a forcing function. Unless another type of vibration is specified, critical speed is based on shaft whirl frequency. The critical frequencies for lateral vibration and shaft whirl are the same. If whirl is the sole concern, an operating speed up to half of the critical speed is generally acceptable. Lateral vibration can be excited by many sources, including vibration from nearby machinery. For this reason, the presence of lateral vibration can be assumed in most cases, and an operating speed of less than one third of the critical speed is more appropriate. The best way to increase the critical speed of a shaft is to increase its rigidity while minimizing the mass of the shaft and its attachments. Because most steels have a similar modulus of elasticity, rigidity is primarily influenced by geometry.

Lateral vibration and shaft whirl are usually calculated using Raleigh's method for lumped masses. More precise methods are available, but impractical to use. Raleigh's method overestimates the critical speed by about 5%. It assumes a horizontal shaft with sinusoidal vibration between two simple supports. This most closely represents a shaft between two self-aligning bearings. Raleigh's method ignores external forces and just uses the weights of the shaft and any attachments. The weight of the shaft is often ignored as well. Static deflection values calculated from standard beam equations are normally used, though integration methods yield more exact results. When more than two masses are present on a shaft, the Dunkerley approximation is recommended. This method is conservative and generally results in a critical speed that is lower than actual. The Dunkerley approximation allows the critical speed of a multi-mass system to be calculated using the critical speeds found when each mass is considered separately.

Torsional vibration is caused by variable torques applied to the shaft. Torsional vibration is of particular concern with long shafts. Any torsional forcing frequencies applied to the shaft must be well away from the critical torsional vibration frequency. The best way to increase the torsional critical frequency is to increase the rigidity of the shaft and minimize the mass of the shaft and its attachments.

The equations in Table 11-6 are commonly used to calculate approximate critical speeds. It is essential to use proper units and check units for consistency. Consult the recommended resources for more information on these equations and their proper use. Beam bending formulas and related calculations can be found in Table 11-3.

SHAFT DESIGN PROCEDURE

There are two commonly used procedures for shaft design. One procedure begins with stress and fatigue calculations to determine a suitable diameter, and then follows with deflection analysis. The other procedure begins by sizing the diameter for suitable deflections and then follows with stress and fatigue analysis. See Tables 11-2 – 11-6 for common formulas and values.

Table 11-6: Shaft Critical Speed Calculations

ω = Frequency (rad/sec)	g = Acceleration of Gravity (386.4 lbm*in/lbf*s² or 9.807 m/s²)
n = Critical speed (rpm)	y = Static deflection (in or m)
f = Frequency (Hz)	G = Modulus of rigidity (Steels approx. 11.5x10⁶ psi or 79 GPa)
w = Weight (lbf or N)	I = Mass moment of inertia (in-lb-sec² or kg-m²)
L = Shaft length (in or m)	E = Modulus of elasticity (Steels approx. 30x10⁶ psi or 207 GPa)

$$n_c = \frac{30\omega_n}{\pi} \qquad f_n = \frac{\omega_n}{2\pi}$$

RALEIGH'S METHOD

One Mass:
$$\omega_n = \sqrt{\frac{g}{y}}$$

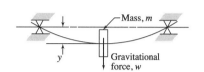

Multiple Masses:
$$\omega_n \cong \sqrt{g \frac{\sum\limits_{i=1}^{n} w_i y_i}{\sum\limits_{i=1}^{n} w_i y_i^2}}$$

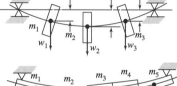

DUNKERLEY APPROXIMATION
$$\frac{1}{\omega_n^2} \cong \sum_{i=1}^{n} \frac{1}{\omega_{ii}^2}$$

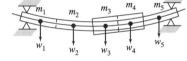

TORSIONAL FREQUENCY

Single disk on shaft: $\omega_n = \sqrt{\dfrac{JG}{I_m L}}$ 　　　 Two disks on shaft: $\omega_n = \sqrt{\dfrac{GJ}{L}\left(\dfrac{I_1+I_2}{I_1 I_2}\right)}$

POLAR SECOND MOMENT OF AREA

Smooth circular shaft: $J = \dfrac{\pi d^4}{32}$

Stepped circular shaft: $J_{eff} = \dfrac{1}{\sum\limits_{i=1}^{n} \dfrac{L_i}{J_i}}$

MASS MOMENT OF INERTIA

Solid circular disk: $I = \dfrac{mr^2}{2}$

1. <u>Begin with preliminary values for the following</u>: shaft length, bearing position and type (self-aligning or not), speed, attachments, applied force profiles, and applied torque profiles.

2. <u>Determine the alternating, peak, and mean values of shear, bending moment, and torque</u> at critical points along the shaft. Critical points can include bearings, attached loads, and stress risers. This is made simpler by constructing free body, shear, bending moment, and torque diagrams. Consult the recommended resources for guidance if needed. Self-aligning bearings act like simple supports and theoretically

eliminate moments at the bearings. Wide roller bearings and double rows of bearings tend to behave like a fixed support.

3. Select a preliminary material and obtain its properties. Ultimate strength, modulus of elasticity, and modulus of rigidity will be needed. Density may also be needed. Material properties will depend on the material condition including rolling and heat treatment. Decide whether the shaft will be machined all over or not.

4. Determine the shaft diameter based on either deflection or stress/ fatigue analysis, which are both detailed under separate headings, and which follow this general procedure.

Author's note: I have found it expedient to begin with a trial shaft diameter d and solve for deflection or stress rather than trying to solve for d. When a shaft is relatively short, I often size the support bearings for the expected loads and service conditions first. Next, I select an inner race diameter of my chosen bearing as a starting point for d, which I then adjust as needed based on the results of deflection and fatigue analysis.

5. Verify and adjust the shaft diameter based on the remaining analysis (deflection or fatigue).

6. If needed, round the shaft diameter at the bearings and other components up to the required sizes. Finalize the detail design of the shaft. Manufacturers of bearings and other components normally provide dimensions and tolerances for the mating shaft.

7. Calculate the critical frequencies of the shaft and verify that operating and forcing frequencies are within acceptable ranges.

Deflection Analysis

1. Identify the locations where lateral deflection must be held within known limits. Critical considerations include angular deflection at bearings and lateral deflection at gears.

2. Begin by assuming a shaft with uniform diameter (not stepped) and solve the deflection equations. Table 11-3 contains some of the more frequently needed formulas. Include deflection due to shear in short shafts where L/d is less than 10. The deflection equations can be solved for d, or a trial diameter can be selected and the deflections checked. Use an appropriate factor of safety when sizing a shaft for deflection. The diameter that satisfies the deflection conditions can be considered the minimum diameter of a uniform shaft used in this application. Evaluation of stepped shafts will require integration or FEA for accurate results.

3. Calculate the torsional deflection of the shaft. Deflections of stepped shaft segments can be summed. Verify that the deflection is acceptable, and iterate to find a new minimum diameter if it is not. Torsional deflection can be calculated using the formula in Table 11-3.

Stress And Fatigue Analysis

1. Select critical locations for potential failure based on the loads and geometry. Critical locations can include load points and stress risers like shoulders, grooves, and keyways.
2. Calculate stress concentration factors at the points of interest.
3. If a diameter has not been selected and axial stress is zero, solve for diameter (*d*) using the following equation, which applies for ductile materials. SF is the factor of safety. This equation was developed for the case of variable torsion and variable bending moment. It simplifies if alternating components of stress are not present. For fully reversed bending, the mean component of moment becomes zero. Table 11-4 has formulas and values for the fatigue stress concentration factors k.

$$d = \left\{ \frac{32(SF)}{\pi} \left[\frac{\sqrt{\left(k_f M_a\right)^2 + \frac{3}{4}(k_{f,shear} T_a)^2}}{S_e} + \frac{\sqrt{\left(k_{fm} M_m\right)^2 + \frac{3}{4}(k_{fm,shear} T_m)^2}}{S_{ut}} \right] \right\}^{\frac{1}{3}}$$

4. If a diameter has been selected rather than calculated, then calculate the mean and alternating components of bending stress, axial stress, and shear stress. The appropriate equations can be found in Table 11-4. These must be calculated for each point of concern along the shaft, and should be examined at the outside surface of the shaft where stresses will be greatest. Shear stresses are of special concern for shafts with *L/d* less than 10.
5. For shafts experiencing bending and torsion, calculate the von Mises components of stress—alternating and mean. For variable loading, both alternating and mean components will be used. For the less common case of constant loading these equations simplify. The appropriate equations can be found in Table 11-4.
6. Find the fatigue factor of safety using a modified Goodman diagram and related equations. Table 11-4 has the factor of safety equation.

CRITICAL CONSIDERATIONS: Shafts

- Analyze all shaft designs for deflection, fatigue, and critical frequencies using appropriate safety factors. Consult the recommended resources for a detailed treatment of shaft analysis.
- Rotating shafts, couplings, hubs, attachments, etc., are extremely dangerous, and proper safety precautions such as guarding must be taken. See Chapter 2 for more information on machinery safety.

BEST PRACTICES: Shafts

- Design the shortest possible shaft. Place bearings close to applied loads. This reduces deflections and increases critical speeds.
- Avoid cantilevered shafts if possible. Minimize any overhang.
- If stress risers are needed (as with keyways, etc.), place them well away from highly stressed regions of the shaft.
- Use generous radii near highly stressed areas to reduce stress concentrations.
- Use fine surface finishes, especially around stress risers and in highly stressed regions of the shafts.
- Take axial loads to ground as soon as possible using only one thrust bearing per direction. Allow the remaining bearings to float axially to prevent undesirable thrust loads on the bearings.
- An operating speed of less than one third of the critical speed is recommended because the presence of lateral vibration can usually be assumed.
- A good way to increase the critical speed of a shaft is to increase its rigidity. Because most steels have a similar modulus of elasticity, rigidity is influenced by geometry.
- Design and dimension shafts and related components so that their center of gravity is as close to the center of rotation as possible. This will result in higher critical speeds. Controlling shaft total runout is one way to influence center of gravity. In shaft assemblies, things like protruding set screws and unbalanced clamp types will affect center of gravity of the assembly.
- When weight is a concern, consider hollow shafting. Hollow shafting has a higher stiffness to weight ratio (specific stiffness) than solid shafting, which results in a higher critical speed.

Section

11.2

SHAFT COUPLINGS

Mechanical shaft couplings are designed to connect two shafts together and minimize relative movement while transmitting torque. They can be attached to shafts in a variety of ways. Some couplings are rigid, whereas others are compliant and allow some misalignment between the shafts.

> ### 🔍 RECOMMENDED RESOURCES
>
> - R. Mott, *Machine Elements in Mechanical Design*, 5th Ed., Pearson/ Prentice Hall, Inc., Upper Saddle River, NJ, 2012
> - R. L. Norton, *Machine Design: An Integrated Approach*, 4th Ed., Prentice Hall, Upper Saddle River, NJ, 2011
> - Oberg, Jones, Horton, Ryffel, *Machinery's Handbook*, 28th Ed., Industrial Press, New York, NY, 2008
> - Shigley, Mischke, Brown, *Standard Handbook of Machine Design*, 3rd Ed., McGraw-Hill Inc., New York, NY, 2004

ATTACHMENT TO SHAFTS

There are several ways in which couplings are typically attached to shafts. For low torque applications, set screws are often used to secure the coupling hubs to the shafts. For high torque applications, a combination of keys and set screws or keys and clamping hubs are often used. The set screws or clamping hubs are essential to provide zero backlash operation of the coupling, where keys alone will not. It is important to note that keyways create significant stress risers in both shafts and hubs. Interference fit hubs are used in very high torque applications and are the strongest method of connection. Set screws are prone to loosening and backing out if not secured with a thread locking compound, jam nut, or second (jamming) set screw. It is also important to note that many set screw types will mar the surface they are tightened against, and that a flat is often machined on the shaft to accommodate the set screw and any burrs it may create. For more information on set screws and keys, see Chapter 6.

Keys are not necessarily used with interference fit or clamping couplings. It is important to note that any keyless coupling has the advantage of allowing shaft phase adjustment. A keyed coupling requires a fixed relationship between both keyed shafts. Other less common methods of attachment include splines and pressure bushings.

SHAFT MISALIGNMENT

Shafts can be misaligned in four basic ways, illustrated in Figure 11-2. Axial, angular, and parallel misalignment may be static or dynamic; they are a function of the geometry and tolerances of the assembly. Torsional misalignment is usually a dynamic effect caused by phase errors between the shaft and whatever it is connected to. Phase errors are most often caused by load fluctuations during operation.

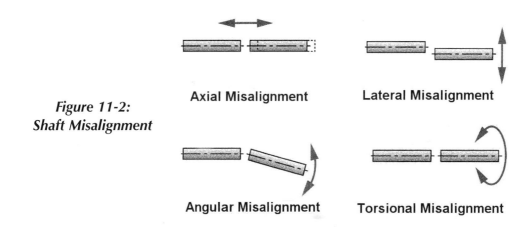

Figure 11-2:
Shaft Misalignment

Axial Misalignment   **Lateral Misalignment**

Angular Misalignment   **Torsional Misalignment**

COUPLING TYPES

There are three basic types of couplings commonly used in machine design: rigid couplings, universal joints, and flexible couplings. Rigid couplings are used when shafts can be perfectly aligned. Universal joints are used when the shafts to be joined are angularly misaligned more than 3 degrees. Flexible couplings can accommodate one or more types of misalignment, which makes them particularly useful in a variety of applications.

Rigid couplings have the advantage of extreme torsional rigidity, but they should only be used when there is no misalignment between the shafts. If the shafts are not perfectly aligned, the use of a rigid coupling will exert extreme stresses on the bearings and shafts, causing component failure over time. These couplings should not be used in applications where high speed or environmental conditions may cause thermal expansion of either shaft because the coupling cannot accommodate axial end play or misalignment once installed. Rigid couplings can take many forms. Some examples are rigid clamping sleeves (Figure 11-3), sleeves with set screws, keyed sleeves, pairs of mating flanges, interference fit sleeves, or shaft clamping assemblies. There are

Clamping Sleeve   Sleeve With Set Screws

Figure 11-3: Rigid Couplings

Source: Misumi USA Inc. (www.misumiusa.com)

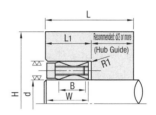

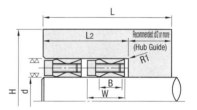

Figure 11-4: Hub to Shaft Clamping Assembly

Source: Misumi USA Inc. (www.misumiusa.com)

many types of clamping assemblies available. Double-cone and single cone clamping couplings are easily assembled compared to traditional interference fit couplings. They tend to be bulky, but result in an interference fit that can transmit high torque. Figure 11-4 shows a commercially available clamping coupling for attaching a hub.

Universal joints are also known as Cardan joints. They have the unique ability to transmit torque when the two shafts are extremely angularly misaligned. The shaft angular misalignment should be limited to 25°, but a universal joint is capable of functioning up to about 45° at very low speeds. The greater the operating angle, the slower the joint should be rotated. For instance, one automotive u-joint can operate at 11.5° at 1500 RPM, but at 3.25° it can operate at 3000 RPM without excessive vibration and reduced life. Every universal joint will generate vibration as it runs. How serious the vibration is depends on the working angle of the joint and the speed of rotation. Vibration leads to lifetime reduction of the joint. As a general rule of thumb, each doubling of the operating angle, speed, or load decreases the lifetime of the joint by half.

U-joints can be either a "block and pin" type or they can be outfitted with needle bearings to reduce friction and wear. Lubrication is essential for both types. Low angles of operation can interfere with lubricant distribution and cause premature wear of the joint. As a result, moderate operating angles are recommended for joints carrying heavy loads or being run at high speeds. Another consideration with U-joints is shaft balancing. An unbalanced or under-supported shaft can cause vibration that can negatively affect the joint.

One drawback of the universal joint is that its output angular velocity is not equal to its input shaft velocity at all degrees of rotation. A discussion of this effect and possible solutions can be found in the recommended resources. Universal joints should not be used when output angular velocity values are critical. Two universal joints back to back form a double U-joint, which has very high misalignment (up to 70°) capability and better velocity transmission fidelity. Constant velocity (CV) joints consist of a ball and socket arrangement and are an alternative to double U-joints. CV joints and U-joints are both

Figure 11-5: Single and Double Universal Joints
Source: Misumi USA Inc. (www.misumiusa.com)

commonly found in automobiles. They are normally covered by a flexible covering, called a "boot," that prevents debris from entering the joint.

Flexible couplings are the most commonly used couplings in modern machinery design. They have the ability to accommodate small angular and lateral misalignments between shafts. Some also have the ability to accommodate axial shaft movement or expansion. In general, commercial flexible couplings can accommodate angular shaft misalignments up to 3 degrees. Another function of some flexible couplings is to reduce noise and vibration. Flexible couplings take many forms. The most common of these are bellows, slit or beam, flexible disc, and Oldham.

FLEXIBLE COUPLING TYPES

Bellows couplings are comprised of two collars with a flexible metal bellows between them. They can provide zero backlash when secured with set screws or clamping hubs. The bellows has the ability to accommodate axial shaft movement as well as small angular and lateral misalignments. These couplings are generally the most torsionally rigid of any flexible coupling, yet allow misalignment without exerting high lateral forces on the shafts. Mass balanced bellows couplings can be purchased suitable for very high speed applications.

Clamping   Set Screws

Figure 11-6: Bellows Couplings
Source: Misumi USA Inc. (www.misumiusa.com)

Slit couplings are also called helical or beam couplings. They are usually made out of a single cylindrical piece, with spiral slits arranged to provide some flexibility in the coupling. These can accommodate angular misalignment, as well as some axial shaft movement. They are also capable of accommodating lateral shaft misalignment, but not as well as other types of couplings unless the coupling has multiple slits. The extra slit or slits create multiple beams in the coupling that, in combination, can accommodate much greater lateral misalignment than a single beam coupling. Slit couplings have zero backlash when secured to the shafts with set screws or clamping hubs. Slit or beam couplings are a good general purpose coupling that is economical and versatile. The main drawback of slit couplings are their relatively low torsional stiffness that will result in some output shaft position inaccuracy, especially during startup and stop cycles. Multiple beam styles can have significantly higher torsional stiffness than single beam couplings. Beam couplings in general are best suited for lower torque applications.

Upper: Double Beam
Lower: Single Beam

Figure 11-7: Slit or Beam Couplings

Source: Misumi USA Inc. (www.misumiusa.com)

Flexible disc or metallic disc couplings are comprised of two shaft hubs with one or more thin metallic or composite discs and supporting structures in the middle. They are often used in high speed applications due to their all-metal construction and good balance of mass. These are extremely torsionally stiff and provide zero backlash when secured to the shafts with set screws or

Double Disc   Single Disc

Figure 11-8: Disc Couplings

Source: Misumi USA Inc. (www.misumiusa.com)

clamping hubs. Flexible disc couplings can accommodate angular misalignment and axial shaft movement. Couplings with multiple discs are better at accommodating lateral misalignment than single disc types. Disc couplings are often recommended for use with servo motors. When accommodating misalignment, these couplings tend to exert fairly low lateral loads on the shafts, making them a good choice for sensitive assemblies. These couplings can be damaged easily, so care must be taken during installation and use.

Oldham couplings, sometimes called jaw couplings, use an intermediate member between the two shaft collars. The intermediate member is usually plastic or rubber, but is sometimes metal. These couplings are very easy to install and disassemble. Depending on the material used for the intermediate member, they have the ability to provide noise, shock, and vibration dampening as well as electrical isolation. These couplings can tolerate a small amount of angular misalignment and shaft end play, but are excellent at accommodating lateral shaft misalignment. Oldham couplings generally allow more backlash as they wear than many types of flexible couplings, but have the ability to accommodate greater misalignment. To maintain zero backlash, the intermediate member will need to be replaced regularly. These couplings, when fitted with a relatively soft intermediate member, tend not to apply large loads

Upper: With Keyways and Set Screws
Lower: With Set Screws, No Keyways

Figure 11-9: Oldham Couplings

Source: Misumi USA Inc. (www.misumiusa.com)

to the shafts as shaft misalignment increases. This makes them a good choice for highly misaligned or sensitive bearing arrangements. Oldham couplings can be temperature sensitive, depending on the properties of the intermediate material in the coupling. These couplings tend to be speed limited, and cannot support a shaft alone due to their construction. Both shafts must be fully supported for use with an Oldham coupling. These couplings have the advantage of disallowing torque transmission if the intermediate member breaks. This can provide some overload protection for the power train.

Other types of couplings include chain couplings, which use a double strand chain wrapped around a pair of sprockets; one sprocket at the end of each shaft to be joined. Gear couplings use a similar concept, where a gear is placed at the end of each shaft and an internally-toothed housing is placed around both. Flexible couplings with an elastomeric member, arranged to look like a tire, are good for shock absorption and misalignment. These and many more are commercially available. Proprietary couplings abound as well.

COUPLING SELECTION

Several factors must be considered when choosing a coupling: backlash, shaft loading, amount and type of misalignment, transmitted torque, inertia, speed, and equality of input and output shaft position and velocity.

Backlash is defined as the clearance between mating components. Backlash, as it applies to couplings, is the rotational movement between the input and output shafts allowed by the coupling. When a drive moves in one direction only, backlash only comes into play during slowdowns or stops. When a drive reverses direction, backlash will be seen not only during slowdowns or stops, but also at each reversal. The effects of backlash during a reversal are position (phase) errors, velocity errors, and noise. Severe backlash at the shaft can lead to set screw loosening, keyway impact damage, and key shear. Backlash can be a function of the coupling type as well as the method of shaft attachment. Select couplings and attachment methods with zero backlash for reversing applications, critical positioning applications, or those with variable loads or frequent stops and starts.

Shaft loading is the result of misalignment between the shafts. A rigid coupling can generate extremely high forces when misaligned shafts are forced into alignment. A flexible coupling has an effective spring constant based on its construction and generates forces proportional to the amount of misalignment.

When shaft loading will cause problems with system components like light duty bearings or slender shafts, select a coupling that is easily misaligned with minimal force, such as an Oldham coupling or a relatively soft flexible coupling of another type.

The amount and type of misalignment between the shafts is a critical consideration when choosing a coupling. Most catalogs will provide the allowable declination angle (angular misalignment) and an allowable shaft eccentricity (lateral misalignment) for each coupling. Choose a coupling that is rated for your expected misalignment. It is important to consider tolerance stacks when estimating misalignment for a given design. End play is another misalignment type to consider. When one shaft can move or expand axially with respect to the other, it is important to select a coupling that can accommodate that movement. Most catalogs will provide allowable end play values for those couplings that can accommodate it.

Transmitted torque is essential to know when sizing a coupling. Always include a factor of safety when estimating torque requirements. Most catalogs will provide an allowable torque value for each coupling. It is important to note that startup and stopping torque values may be many times greater than running torque values. A good rule of thumb is to select a coupling with a torque capacity 5 – 10 times the operating torque of the driving motor. In applications where stopping or reversing direction is expected, it is good practice to choose a coupling with at least 5 times greater torque capacity than the expected operating torque.

Moment of inertia of the coupling will increase the acceleration load on the drive. This is of special concern with low power drive units (like pneumatic rotary actuators) and/or high accelerations. Always calculate drive parameters to include the coupling. The diameter of the coupling and its mass distribution will directly influence inertia. To minimize added load on the drive, select the appropriate coupling with lowest moment of inertia. A bellows coupling has a low moment of inertia due to the thinness (and therefore low mass) of the bellows. A slit or beam coupling will tend to have a lower moment of inertia than a disc coupling due to its smaller diameter. An Oldham coupling with a plastic or rubber insert will also tend to have a lower moment of inertia than a disc coupling.

Speed can be a limiting factor for some couplings. If a speed rating is given in a catalog for a coupling, be sure to stay well within that limit during the lifetime of the coupling.

Equality of input and output shaft position and velocity is a consideration for precision applications. In these cases, the torsional spring constant of the coupling must be considered. This value is often provided in catalogs for couplings that are not torsionally rigid. This value can be used to estimate shaft phase misalignment as a function of torque and this spring constant. In critical applications, it is easier to select a torsionally rigid coupling than it is to estimate this misalignment.

Couplings are normally chosen from a catalog rather than designed from scratch. It can be helpful to speak with a manufacturer's representative about the application. In general, the following steps are followed when selecting a commercial coupling:

1. Select a coupling type based on performance characteristics.
   Table 11-7 can be helpful in selecting a coupling type. In modern machinery, zero backlash flexible couplings are the most commonly used. Rigid couplings are generally the least costly and most precise, but have the disadvantage of having no tolerance for misalignment. Flexible disc couplings are commonly used for servo motor applications. Oldham couplings are often seen on applications with high lateral misalignment or light duty bearings. Consult manufacturer data for complete information on any commercial coupling.

2. Select a manufacturer or brand. Special features are available, depending on the make and model of coupling selected. For example, many proprietary couplings have features to allow easy assembly and disassembly.

3. Choose the method of shaft attachment based on application requirements. The basic types are keyed, clamping, and set screw. Proprietary attachment methods are commercially available. Keyless clamping hubs are a very common method of shaft attachment, especially for use with small and medium sized servo motors.

4. Size the coupling based on torque, speed, and shaft size requirements. Know your peak and running torque requirements, and use a factor of safety. If the design dictates that the coupling should fail at some value (if it is a sacrificial element protecting other machinery) check with the manufacturer. An Oldham coupling may be a good choice in that situation because it results in no power transmission if the intermediate member breaks.

Table 11-7: Coupling Performance Comparison

Coupling Type	Zero Backlash	High Torsional Rigidity	High Torsional Strength	Angular Misalignment	Lateral Misalignment	Shaft End Play	Low Lateral Loads	No Maintenance
Rigid	●	●	●	⊘	⊘	⊘	⊘	●
Bellows	●	●	O	●	O	O	O	●
Slit/Beam	●	O	O	●	O	●	O	●
Disc - Single	●	●	●	O	⊘	O	O	⊘
Disc - Double	●	●	●	O	O	O	O	⊘
Oldham	O	O	O	●	●	●	O	O

● = Good Choice
O = May Be Acceptable
⊘ = Poor Choice

CRITICAL CONSIDERATIONS: Shaft Couplings

- Always include a factor of safety when sizing couplings. Factors of safety between 5 and 10 are commonly used to account for starting and stopping torque.
- Some couplings have speed ratings. A factor of safety should be applied.
- For applications where position or velocity are critical, take the torsional spring constant of the coupling into consideration.
- Use a zero backlash coupling and attachment method when the coupling is subjected to variable loads, variable speeds, or reversals.
- Remember to include the inertia of the coupling when calculating the load seen by the drive.
- Rotating shafts, couplings, hubs, attachments, etc., are extremely dangerous, and proper safety precautions such as guarding must be taken. See Chapter 2 for more information on machinery safety.

BEST PRACTICES: Shaft Couplings

- Choose the type of coupling first, and then look at sizes and capacities. Coupling type will be determined based on shaft misalignment expectations.
- Consult manufacturers early in the design process for expert assistance in selecting and sizing a coupling.

GEARS
Written by Gregory Aviza

Section 11.3

Gears are complex and highly standardized elements used to transmit power between different axes. They also have the ability to change the rotational speed, torque, and direction of shafts. The axes may both be rotational, or rotational and linear (if used in combination with a rack). Standards for the design and manufacture of American Standard gears are defined by the American Gear Manufacturer's Association (AGMA) and referenced in this text. For metric gears, AGMA has adopted many ISO standards which are also referenced. Other standards such as British Standards (BS), German (DIN), and Japanese (B) are not included in this section. The amount of material that is available on gear design and standards is staggering. This text provides a brief overview and highlights some of the more relevant information that is useful to the practicing machine designer. The design of custom gear sizes, manufacture, and inspection methods of gears are topics beyond the scope of this text and are covered thoroughly in most design of machine elements text books.

RECOMMENDED RESOURCES

- G. Michalec, *Precision Gearing: Theory and Practice*, John Wiley & Sons, New York, NY, 1966
- R. Mott, *Machine Elements in Mechanical Design*, 4th Ed., Pearson Education, Inc., Upper Saddle River, NJ, 2004
- R. L. Norton, *Machine Design: An Integrated Approach*, 4th Ed., Prentice Hall, Upper Saddle River, NJ, 2011
- Oberg, Jones, Horton, Ryffel, *Machinery's Handbook*, 28th Ed., Industrial Press, New York, NY, 2008
- **American Gear Manufacturers Association (AGMA):** www.agma.org
- **International Organization of Standardization (ISO):** www.iso.org

For further reading, Tables 11-8 and 11-9 list some of the more relevant standards issued for American Standard gears by AGMA, and metric gears by ISO.

Table 11-8: Useful American-Standard ANSI/AMGA Standards

Standard	Description
AGMA 2015/915-1-A02	Accuracy Classification System — Tangential Measurement Tolerance Tables for Cylindrical Gears
AGMA 911-A94	Guidelines for Aerospace Gearing
AGMA 912-A04	Mechanisms of Gear Tooth Failure
AGMA 913-A98	Method for Specifying the Geometry of Spur and Helical Gears
AGMA 917-B79	Design Manual for Parallel Shaft Fine-Pitch Gearing
AGMA 923-B05	Metallurgical Specifications for Steel Gearing
AGMA 933-B03	Basic Gear Geometry
ANSI/AGMA 1102-G05	Gear Nomenclature, Definitions of Terms with Symbols
ANSI/AGMA 2004-C08	Gear Materials, Heat Treatment and Processing Manual
ANSI/AGMA 2005-D03	Design Manual for Bevel Gears
ANSI/AGMA 9005-E02	Industrial Gear Lubrication

Table 11-9: Useful Metric Gearing ISO Standards

Standard	Description
ISO 53:1998	Cylindrical gears for general and heavy engineering — Basic rack
ISO 54:1996	Cylindrical gears for general and heavy engineering — Modules and diametral pitches
ISO 677:1976	Straight bevel gears for general and heavy engineering — Basic rack
ISO 678:1976	Straight bevel gears for general and heavy engineering — Modules and diametral pitches
ISO 701:1998	International gear notation — Symbols for geometrical data
ISO 1122-1:1998	Vocabulary of gear terms — Part 1: Definitions related to geometry
ISO 1328-1:1995	Cylindrical gears — ISO system of accuracy — Part 1: Definitions and allowable values of deviations relevant to corresponding flanks of gear teeth
ISO 1328-2:1997	Cylindrical gears — ISO system of accuracy — Part 2: Definitions and allowable values of deviations relevant to radial composite deviations and run-out information
ISO 1340:1976	Cylindrical gears — Information to be given to the manufacturer by the purchaser in order to obtain the gear required
ISO 1341:1976	Straight bevel gears — Information to be given to the manufacturer by the purchaser in order to obtain the gear required
ISO 9085:2002	Calculation of load capacity of spur and helical gears — Application for industrial gears
ISO/TR 18792:2008	Lubrication of industrial gear drives

TERMS AND DEFINITIONS

There are many terms used for gear definition and specification. Many of these terms sound similar, but can have different meanings. The terms most commonly encountered by the machine designer using American Standard gears are described in Table 11-10 (Adapted from AGMA 112.05, 115.01, and 116.01) and illustrated in Figure 11-10. A complete list of over 75 terms is available in *Machinery's Handbook* 28th Ed., pages 2030-2034.

Table 11-10: Commonly Encountered Spur Gear Terms and Definitions

Term	Variable	Definition
Addendum	a	The height that a tooth projects above the pitch circle
Backlash	B	The space or clearance between mating teeth *Backlash is described either as the distance along the pitch circle arc or as the shortest normal distance between teeth. Backlash is covered more extensively later in this section.*
Center distance	C	The distance between the axes of rotation of mating gears
Circular pitch	p	The distance along the circumference of the pitch circle between corresponding points on adjacent teeth
Clearance	c	The radial distance between the top of a tooth and the bottom of a mating tooth *Clearance is important, and sometimes increased in processing equipment, to provide an escape for debris and lubrication that gets caught in the mesh.*
Dedendum	b	The depth that a tooth projects below the pitch circle
Diametral pitch	P	The ratio of the number of teeth to the diameter of a specified pitch circle *For the inch system, this is the number of teeth in the gear for each inch of pitch diameter. Most American gearing is categorized by its diametral pitch. In layman's terms, it is the tooth size. Diametral pitch is sometimes represented by the variable DP.*
Gear ratio	m_g	Often simply called ratio, it is the relationship of the number of teeth on the driven gear divided by the number of teeth on the drive gear
Module	m	In the metric system: The ratio of the pitch diameter in millimeters to the number of teeth
Pitch Diameter	D	The diameter of the pitch circle, which is the circle through the pitch point that has its center at the center of the gear axis
Outside diameter	D_o	The outer (maximum) diameter of the gear
Pressure angle (PA)	ϕ	The angle at a pitch point between the line of pressure, which is normal to the tooth surface, and the plane tangent to the pitch surface *Although there are three main pressure angle standards (14.5°, 20°, and 25°), the 20°PA has become the most used standard due to the increased load capacity and reduced undercutting issues over the 14.5° standard, and the reduced backlash change due to center distance changes and quieter operation over the 25° standard. The relative shapes of gear teeth with three different pressure angles are shown in Figure 11-11.*
Number of teeth	N	The number of teeth on the gear *Subscripts are used to denote different gears*

The standards for metric gears are based on the millimeter and a slightly different design standard than American Standard gears. This means that even if dimensional conversions lined up, American Standard and metric gears are not interchangeable. Metric gears use a system known as the module system and are governed by ISO standards. The basis of the metric module system is the pitch diameter in millimeters, divided by the number of teeth. Therefore, a gear with a pitch diameter of 60 mm, and 20 teeth, has a module of 3, meaning that there are 3 mm of pitch diameter for every tooth. Fortunately, American metric gears have adopted these ISO standards.

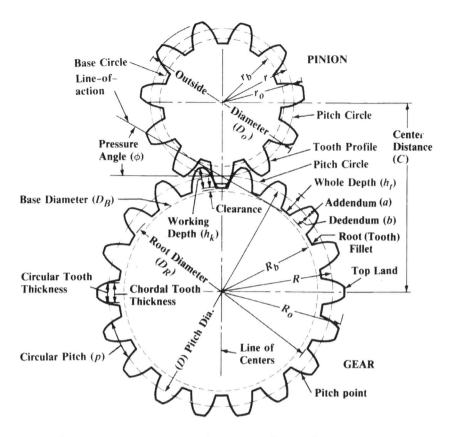

Figure 11-10: Gear Tooth Nomenclature for Spur Gears

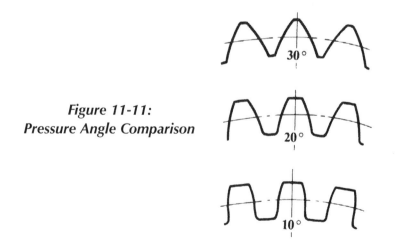

Figure 11-11:
Pressure Angle Comparison

The fundamental law of gearing states that the angular velocity ratio of all gears of a meshed gear system must remain constant (due to conjugate action). In addition, the common normal at the point of contact must pass through the pitch point. Plainly speaking, this means that power or motion is transmitted in a smooth and continuous manner. Although there are many curves that could be developed to meet this criterion, the involute tooth profile has become standard for most modern machine applications (with the notable exception of clock mechanisms). There are three major benefits to the involute profile:

- Conjugate action is not affected by changes in gear center distances. This ensures constant velocity ratio between gears even if center-to-center position tolerance is compromised.
- The tooth form for a rack has straight sides, making rack and cutter tool manufacture lower cost than complex curves, while at the same time imparting a higher level of accuracy onto the cut gear.
- A rack cutter can make a gear with any tooth count, as opposed to other methods that require separate cutters for different gear tooth count ranges.

The classic illustration of the involute curve is the arc that is obtained if a taut string is unwrapped from a base circle, as shown in Figure 11-12. Table 11-11 provides the equations that govern the involute profile for those who need to create an exact CAD model of the tooth face. Additional research and work will be required to add the appropriate fillet radii and tooth lands to create a complete model. For most common gears, the base curve is a circle; however, other curve shapes have been employed to alter the gear ratio while in operation.

As the diameter of a gear approaches infinity, the pitch circle becomes a straight line and we are left with a rack. Rack and pinion drives, like the one

Table 11-11: Equations of the Involute Tooth Profile

	α = Rolling tangent angle (radians)
	r = radius of the point on the tooth
	r_b = Base circle radius
	$\text{inv}\,\alpha = \theta - \alpha$
	$\theta - \alpha = \tan\alpha - \alpha$
	$\alpha = \cos^{-1}\left(\dfrac{r_b}{r}\right)$
	$x = r\cos(\theta - \alpha)$
	$y = r\sin(\theta - \alpha)$

Image Source: SDP/SI (www.sdp-si.com)

shown in Figure 11-13, are often used to turn rotary motion into linear motion and vice-versa. For simplicity, the rack is often used to describe the tooth profile dimensions. Rack profiles for American Standard and metric module gears are shown in Figures 11-14 and 11-15 respectively.

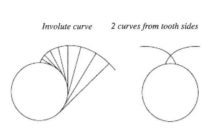

Figure 11-12: Involute Curve Generation Through String Unwinding

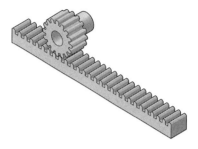

Figure 11-13: Rack and Pinion

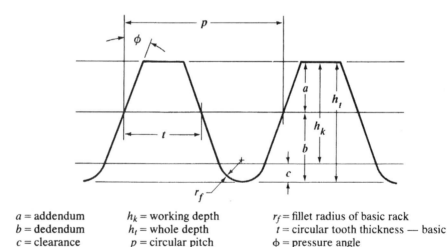

a = addendum	h_k = working depth	r_f = fillet radius of basic rack
b = dedendum	h_t = whole depth	t = circular tooth thickness — basic
c = clearance	p = circular pitch	ϕ = pressure angle

Figure 11-14: Basic Rack Profile for 20° and 25° Full-Depth American Standard Gears

There are two standard pitch systems used for American Standard gears, diametral pitch and circular pitch, with the diametral pitch system dominating most gearing produced in the United States. Although the diametral pitch system is most common, the machine designer must be aware of the existence of other standards to prevent errors in specification or application. There are special occasions, such as with cast gearing, when the tooth size of the gear is specified by circular pitch.

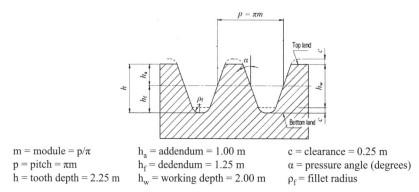

m = module = p/π h_a = addendum = 1.00 m   c = clearance = 0.25 m
p = pitch = πm h_f = dedendum = 1.25 m   α = pressure angle (degrees)
h = tooth depth = 2.25 m h_w = working depth = 2.00 m $ρ_f$ = fillet radius

Figure 11-15: Basic Rack Profile for Metric Module Gears

For American Standard gearing, there are three common tooth forms: full depth, American stub, and fellows stub. Both stub gear formats have teeth that are shorter, fit in a tighter package, and carry higher loads, but at the cost of a reduced contact ratio. Full-depth gears will be the focus of this text.

The term *coarse pitch* is used for diametral pitches up to, but not including, 20, where *fine pitch* is used to describe diametral pitches 20 and over. This break point is often used in backlash recommendations. The relative sizes of three different coarse pitches are shown in Figure 11-16.

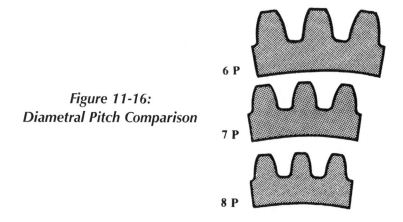

Figure 11-16:
Diametral Pitch Comparison

6 P

7 P

8 P

<u>Interference</u> and the need for <u>undercutting</u> must be considered when designs employ pinions with small tooth counts. Interference occurs between the tooth tip of the pinion and the lower portion of the mating gear tooth. This is most pronounced when the pinion is small and the mating gear is large. To accommodate this interference, material is removed (undercut) from the root of the tooth, as shown in Figure 11-17. The disadvantage to this practice is that tooth strength is reduced.

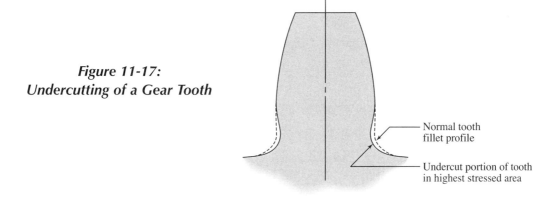

Figure 11-17:
Undercutting of a Gear Tooth

Normal tooth fillet profile

Undercut portion of tooth in highest stressed area

Interference between a rack and a pinion begins for 14½°, 20°, and 25° pressure angle gears when pinion tooth counts fall below 32, 18, and 12 respectively. The amount of material that needs to be removed by undercutting increases as tooth count decreases, resulting in weaker and weaker teeth. Gear manufacturers do sell undercut gears and set practical guidelines for the smallest gears to use; however, it is best practice to avoid them if possible. Table 11-12 provides minimum pinion teeth counts to avoid interference and the need for undercutting for various gear pairings of 20° PA gears.

If a design with zero or minimal undercutting cannot be achieved, the designer can increase the gear set pressure angle or switch to stub tooth gears (to maintain the use of standard gears). Another option is to use a custom cut set of gears called "profile shifted" gears. Profile shifted gears move the pitch circle away from the pinion and closer to the gear. The total tooth depth remains the same as standard gears. However, the addendum of the pinion increases while decreasing the gear addendum an equal amount. When considering profile shifted gears, consult a reputable gear manufacturer. Additional information on profile-shifted gears is available in the recommended resources.

Table 11-12: Minimum Number of Pinion Teeth to Avoid Interference Between
20° PA Full Depth Pinion and Gears of Various Sizes

Pinion Teeth	Maximum Teeth on Mating Gear
18	Rack
17	1309
16	101
15	45
14	26
13	16

Internal gear sets can also suffer from undercutting when the tooth count of the gears is close. It is good practice to avoid this condition by making sure that the tooth count difference is at least 12 for 20° PA internal gear sets.

GEAR TYPES

Table 11-13 lists the characteristics and suitability of the major commercially available gear types. The machine designer is most likely to use spur gears, due in part to their simplicity and relatively low cost. Once spur gears are understood, it is a simple jump to understanding helical gears, which are essentially spur gears with angularly cut teeth. Straight cut bevel gears are another family of gears that are useful to the machine designer since they allow for intersecting axes. They are often employed in gearboxes. If the ratio of the bevel gear set is 1:1, they are often referred to as miter gears. Worm gears, which are often employed in high ratio gearboxes and hoists, are also likely to be encountered.

GEAR TRAINS

When two or more gears are connected together, the set is called a gear train. There are three major classifications of gear trains: simple, compound, and epicyclic (planetary). The purpose of most gear trains used in power transmission applications is to alter the velocity ratio (m_V) between the input shaft and the output shaft. Altering this ratio also alters the torque in an inversely proportional manner (i.e., lowering the speed increases the torque). Basic examples with the corresponding velocity ratio equations are provided in Table 11-14 for each train type. These basic building blocks can be used to create more complex gear trains as required for different applications.

In a simple gear train, sometimes referred to as a <u>single stage gear train</u>, each gear in mesh resides on its own shaft. Here, the only gears that impact the velocity ratio are the first gear (input) and the last gear (output). However, the number of gears in mesh influences the rotation direction of the output gear. An even number of gears from input to output results in output gear rotation opposite of the input gear; an odd number results in output gear rotation in the same direction as the input gear. Table 11-14 shows some simple gear trains and the corresponding velocity ratio equations.

Simple gear trains with the same size gears are often used by machine designers as a method of synchronizing the position of tooling, such as grippers, rather than for traditional power transmission. In those cases, the velocity ratio is 1.

Table 11-13: Classification of Common Gear Types

Image Source: SDP/SI (www.sdp-si.com)

Type	Image	Description	Applications
EXTERNAL SPUR		• The simplest of the gear types with straight cut teeth • Adjacent parallel shafts in mesh will rotate in opposite directions • Suitable for high speeds and loads • No axial thrust present • Highest efficiency of all the gear sets • Noise increases with speed • High levels of precision and accuracy are obtainable due to the simplicity and accessibility of the tooth during manufacture • Teeth are in rolling contact	All types of gear trains and a wide range of velocity ratios
INTERNAL SPUR		• Similar to external gears except one gear has its teeth on the inside • Both shafts of mating internal gear sets rotate in the same direction • Suitable for high speeds and loads • No axial thrust present • More costly to produce than external spur gears	• Internal drives with high speeds and loads • Commonly used in planetary gear sets

continued on next page

Table 11-13: Classification of Common Gear Types (Continued)

Type	Image	Description	Applications
HELICAL		<ul><li>Similar to spur gears except the teeth are cut on an angle to the shaft axis</li><li>Adjacent parallel shafts in mesh will rotate in opposite directions</li><li>Suitable for very high speeds and loads</li><li>Slightly less efficient that spur gears</li><li>Greater load transmission capability, and run quieter than spur gears due to a greater number of teeth in mesh at any time.</li><li>Axial thrust is present and must be accounted for in shaft mounting design</li><li>Care must be taken in choosing because there are right and left hand variants (externally mated helix gears must be opposite hand; however, internally mated helix gears must be of the same hand)</li><li>Double helical gearing (called Herringbone) is available to negate the axial thrust, at the expense of higher cost</li></ul>	Most applications where spur gears are used and quieter operation or higher load capacity required
CROSSED-HELICAL		<ul><li>Similar to normal helical gears except the shafting does not have to be parallel</li><li>Less efficient than normal helical gears due to a point contact and high sliding action</li><li>45° standard helix angle allow for shafts at 90 degrees, other angles by custom design</li><li>Not suitable for precision applications</li><li>Axial thrust is present and must be accounted for in design</li></ul>	<ul><li>Relatively low ratios, speeds, and load applications</li><li>Non-precision applications</li></ul>

continued on next page

Table 11-13: Classification of Common Gear Types (Continued)

Type	Image	Description	Applications
RACK		• Racks can be considered gears with an infinite radius • Available in straight tooth variations for spur gears and angled tooth for helical gearing • Often the rack is used to define the gear profile dimensions	• Used in combination with a pinion to turn rotary motion into linear motion or vise-versa • Steering mechanisms in automobiles
STRAIGHT BEVEL		• The simplest of the bevel gears where teeth are straight cut on a conical blank • Intersecting shafts • Generally used only at speeds below 1000 ft/min (5 m/s), or for small gears operating below 1000 RPM • Most common application used with shafting at 90 degrees to each other, but other angles possible with custom gear design • Axial thrust is present and must be accounted for in design	Economical right angle drives
SPIRAL BEVEL		• Similar to straight bevel gears except the tooth is cut in a curved path across the face (similar to the relationship between spur and helical gears) • Intersecting shafts • More teeth in mesh resulting in quieter operation and higher load capacity over straight bevel gears • Care must be taken when selecting as there are right and left hand variants • The spiral of the pinion is used to identify the pair, with the other gear being the opposite hand • The hand and angle of the pinion angle affects the direction of axial thrust and should be selected so that the thrust is pushing the mesh apart to avoid jamming • Axial thrust is present and must be accounted for in design	Higher load capacity right angle gearboxes

continued on next page

Table 11-13: Classification of Common Gear Types (Continued)

Type	Image	Description	Applications
ZEROL BEVEL		• A special type of spiral bevel gear where the spiral angle is zero • Intersecting shafts • More teeth in mesh resulting in quieter operation and higher load capacity over straight bevel gears • Care must be taken when selecting as there are right and left hand variants • The spiral of the pinion is used to identify the pair, with the other gear being the opposite hand • The hand and angle of the pinion angle affects the direction of axial thrust and should be selected so that the thrust is pushing the mesh apart to avoid jamming • Axial thrust is present and must be accounted for in design	Higher load capacity right angle gearboxes
HYPOID BEVEL		• Non-intersecting and non-parallel shafts • Used for higher gear reductions than typically feasible with regular bevel gears • Cross between a spiral bevel and a worm gear • Axial thrust is present and must be accounted for in design	Commonly used in automotive applications such as differential drives
WORM GEARS		• Orthogonal shafting • High ratios possible in single stages • Lower efficiency in part due to heat generation caused by sliding contact in the mesh • Non-back-drivable and multi-start gears possible • Axial thrust is present and must be accounted for in design • Worm gear sets must be of the same pressure angle, lead, and hand to operate together • Current trends in sustainability are pushing designers to helical gearing (unless non-back-drivability is required for safety)	Used primarily for power transmission

Table 11-14: Simple Gear Train Examples and Equations

Image Source: SDP/SI (www.sdp-si.com)

Types and Characteristics	Illustration	Input vs. Output Shaft Velocity Ratio
SINGLE STAGE Input and output gears rotate in opposite directions.	Gear 2 (N_2) Gear 1 (N_1)	$m_v = \left(\dfrac{-N_1}{N_2} \right)$
SINGLE STAGE SPUR GEAR WITH AN IDLER GEAR Input and output gears rotate in the same direction.	Gear 3 (N_3) Gear 2 (N_2) Gear 1 (N_1)	$m_v = \left(\dfrac{-N_1}{N_2} \right)\left(\dfrac{-N_2}{N_3} \right)$

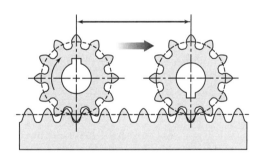

Figure 11-18: Rack and Pinion

Another special application of the single stage gear train is a rack and pinion arrangement. Here the designer often needs to calculate the amount of rack movement for a given pinion rotation, or vice-versa. The relationship is simple to see in Figure 11-18 where one complete revolution of the pinion equates to a linear rack travel distance equal to the circumference of the pitch circle diameter. The rack movement or pinion rotation can then be determined as a function of this distance, based upon the input to either the pinion or the rack.

Compound gear trains are those in which at least one shaft carries more than one gear. Gears that are on the same shaft rotate in the same direction and speed. These gear trains are capable of higher velocity ratios than single stage gear trains. Table 11-15 shows a compound gear train with one pair of gears on a common shaft, and the resultant velocity ratio.

For multiple stage gear trains the gear ratio can be expressed as:

$$m_V = \pm \frac{\text{(product of number of teeth on the driver gears)}}{\text{(product of number of teeth on the driven gears)}}$$

Table 11-15: Two-Stage Compound Gear Train

Illustration	Input vs. Output Shaft Velocity Ratio
Gear 4 (N_4) Gear 3 (N_3) Gear 2 (N_2) Gear 1 (N_1) *Image Source: SDP/SI (www.sdp-si.com)*	$m_v = \left(\dfrac{-N_1}{N_2}\right)\left(\dfrac{-N_3}{N_4}\right)$

The direction of rotation of the final gear in relation to the input gear is dependent upon the total number of meshes, and whether the gear pairs are internal or external. Care must be taken for each gear pair to note the sign of each pair so that the final rotation direction can be reached.

The final type of common gear train is the epicyclic (or planetary) gear train. These types of gear trains are able to take two inputs and create one output from them. They are popular in gearboxes, where one of the inputs (usually the ring) is held stationary. Gearboxes are discussed in Section 11.4 of this book. For most machine designers, this will be the extent of their exposure to these gear trains. For this reason, the full explanation of these gear trains is left to the recommended resources. A typical planetary gear set and major components are shown in Figure 11-19.

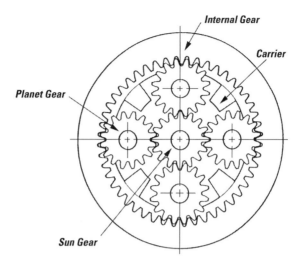

Figure 11-19: Planetary Gear Train

Source: SDP/SI (www.sdp-si.com)

SHAFT ATTACHMENT METHODS

The shaft attachment methods for gears are the same as for other power transmission elements such as sprockets and pulleys. However, for gears to function properly with their designed backlash, there must be a sufficient degree of shaft support, minimal shaft deflection (0.001″ — 0.025 mm — or less), and minimal run-out of the system while under load. For high-speed systems, dynamic balancing of assemblies must also be considered.

The most commonly utilized method of gear attachment to shafts is with keys and setscrews. For greater power transmission capability, splines and other custom lobed shaft methods can be used. If packaging constraints are tight the gear teeth can be cut directly onto a shaft or hub, making a one-piece component.

GEAR QUALITY RATINGS

Gear quality is classified with a numerical rating based on the deviation from perfect form. Higher quality gears allow for higher precision of transmitted motion due to tighter tolerances during manufacturing. Higher quality gears are also able to transmit higher loads due to the more even load distribution of the teeth in the mesh. Coarse pitch gears are often checked analytically, whereas fine pitch gears are often tested functionally on test apparatus. AGMA has created classifications (outlined in AGMA Gear Handbook 390.03) where a higher number indicates greater quality. For metric gears,

German DIN and ISO standards are often used. The former AGMA 2008 standard has been replaced with a newer standard, AGMA 2015. For reference, Table 11-16 provides an approximation of equivalencies across the current standards (AGMA 2015 and ISO 1328) and some of the past AGMA standards (AGMA 236.04 and 2008). Quality ratings listed at the top of the table are the coarser grades and become finer as you progress down the chart. It is impor-

Table 11-16: Approximate Equivalencies of Gear Quality Standards

Past	Transitioning Out	Current	Current
AGMA 236.04	AGMA 2008	AGMA 2015	ISO 1328
Commercial 1	Q5	*	12
Commercial 1 or 2	Q6	A11	11
Commercial 2	Q7	A10	10
Commercial 3	Q8	A9	9
Commercial 3	Q9	A8	8
Precision 1	Q10	A7	7
Precision 1	Q11	A6	6
Precision 2	Q12	A5	5
Precision 3	Q13	A4	4
Precision 3	Q14	A3	3
-	Q15	A2	2

* no equivalent value

Table 11-17: AGMA Recommended Quality Numbers, Applications, and Approximate ISO Equivalents

Application	Quality Number AGMA 2008	Quality Number AGMA 2015	Quality Number ISO 1328
Cement mixer drum drive	Q3 - Q5	*	* - 12
Cement kiln	Q5 - Q6	* - A11	12 - 11
Steel mill drive	Q5 - Q6	* - A11	12 - 11
Grain harvester	Q5 - Q7	* - A10	12 - 10
Cranes	Q5 - Q7	* - A10	12 - 10
Punch press	Q5 - Q7	* - A10	12 - 10
Mining conveyor	Q5 - Q7	* - A10	12 - 10
Paper box making machine	Q6 - Q8	A11 – A9	11 - 9
Gas meter mechanism	Q7 - Q9	A10 – A8	10 - 8
Small power drill	Q7 - Q9	A10 – A8	10 - 8
Clothes washing machine	Q8 - Q10	A9 – A7	9 - 7
Printing Press	Q9 - Q11	A8 – A6	8 - 6
Automotive transmission	Q10 - Q11	A7 – A6	7 - 6
Radar antenna drive	Q10 - Q12	A7 – A5	7 - 5
Marine propulsion drive	Q10 - Q12	A7 – A5	7 - 5
Aircraft engine drive	Q10 - Q13	A7 – A4	7 - 4
Gyroscope	Q12 - Q14	A5 – A3	5 - 3

* no equivalent value

Table 11-18: AGMA Recommended Quality Numbers for Pitch Line Speed with Approximate ISO Equivalents

Pitch Line Speed (fpm)	Pitch Line Speed (m/s)	Quality Number AGMA 2008	Quality Number AGMA 2015	Quality Number ISO 1328
0–800	0–4	Q6–Q8	A11–A9	11–9
800–2000	4–11	Q8–Q10	A9–A7	9–7
2000–4000	11–22	Q10–Q12	A7–A5	7–5
Over 4000	Over 22	Q12–Q14	A5–A3	5–3

tant to note that the AGMA quality standards are changing to align in number with the ISO quality standards, and that this is a reversal in the relationship between numerical ranking and quality. AGMA 2008 uses the "Q" prefix where AGMA 2015 uses the "A" prefix. The designer is advised to be aware of this difference and communicate clearly the standard used when specifying quality ratings during this transition period, especially when using printed catalogs, which may not reflect the latest standards.

As with most instances of tolerance and quality, a higher rating comes with higher cost. Table 11-17 lists typical quality requirements for various applications. Table 11-18 provides quality number recommendations as a function of pitch line speed.

BACKLASH

Backlash is the clearance between the tooth width and the tooth space measured on gears in a non-loaded condition. It is a function of tooth thickness and profile tolerance, center distance, gear run-out, lead, and operating temperature. Due to run-out tolerances, the backlash will cycle between a minimum value when the tolerances stack up in one direction and a maximum when they stack up in the other direction. For practical application of tooling design, the minute differences are academic. However, for applications deemed critical, the AGMA and DIN standards should be consulted.

For spur gears it is common to measure backlash along the operating pitch circle as shown in Figure 11-20. For bevel gears, the backlash is measured along normal planes, as shown in Figure 11-21. Backlash is commonly measured in the field by using feeler gages in the mesh, or by locking one shaft and measuring the angular displacement allowed on the other shaft. This second method requires precise setups and back-calculations for proper readings.

Gear sets are designed to have some amount of backlash to allow for tolerances in shaft location, shaft deflection, gear manufacture errors, as well as thermal expansion during operation. This helps to provide smoother and quieter operation with lower power requirements because both sides of the tooth are not grinding into contact at the same time. Backlash also allows for lubrication to enter the mesh and prolong gear life. However, a large amount of backlash is not desired either because rotation reversal will cause position inaccuracies, louder operation, and shock loading on the tooth.

There are two main methods to increase backlash: increasing the gear center distance or reducing the gear tooth thicknesses. For most applications, it is acceptable to tolerance shaft center distance such that it can increase by half of the average backlash. Tolerances that allow a decrease of center distance should carefully checked or avoided because they can eliminate the backlash, or even create an interference condition. The relationships between center distance and backlash for the 14½° and 20° PA systems are given in Table 11-19.

When reducing tooth thicknesses for a desired backlash, it is customary to reduce the tooth thickness of each gear in the mesh by half of the desired backlash if the gears are close in size. For higher ratio gear sets, the full tooth reduction is often taken completely from the gear, allowing the pinion to retain its full size teeth for strength.

Table 11-20 lists the AGMA recommended backlash settings based on diametral pitch and center distance for American Standard spur, helical, and herringbone gearing. Table 11-21 provides similar information for metric gears. Deviations from these guidelines are acceptable based on functional requirements and the discretion of the designer.

Gear manufacturers cannot realistically keep every possible tooth thinning option in stock. As a result, gear manufactures make gears that, when mounted on nominal center distances, exhibit backlash that is in rough agreement with the AGMA recommendations as shown in Table 11-22. Achieving a different backlash is possible by changing tooth thickness or center distance, as described previously.

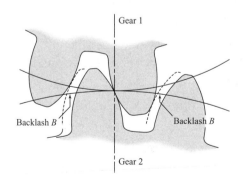

Figure 11-20:
Spur Gear Pitch Circle Backlash

Figure 11-21:
Bevel Gear Normal Backlash

Table 11-19: Change in Center Distance to Backlash Ratio

For 14½° PA	Change in Center Distance = 1.933 x Change in Backlash
For 20° PA	Change in Center Distance = 1.374 x Change in Backlash

Table 11-20: AGMA Recommended Backlash Ranges
Coarse Pitch Spur, Helical, and Herringbone Gearing

Center Distance (inches)	Normal	Diametral	Pitches		
	0.5–1.99	2–3.49	3.5–5.99	6–9.99	10–19.99
	Backlash		Normal	Plane,	inches
Up to 5	–	–	–	–	0.005–0.015
Over 5 to 10	–	–	–	0.010–0.020	0.010–0.020
Over 10 to 20	–	–	0.020–0.030	0.015–0.025	0.010–0.020
Over 20 to 30	–	0.030–0.040	0.025–0.030	0.020–0.030	–
Over 30 to 40	0.040–0.060	0.035–0.045	0.030–0.040	0.025–0.035	–
Over 40 to 50	0.050–0.070	0.040–0.055	0.035–0.050	0.030–0.040	–
Over 50 to 80	0.060–0.080	0.045–0.065	0.040–0.060	–	–
Over 80 to 100	0.070–0.095	0.050–0.080	–	–	–
Over 100 to 120	0.080–0.110	–	–	–	–

Table 11-21: AGMA Recommended Backlash Settings for Metric Gears

Center Distance (mm)	Module						
	1.5	2	3	5	8	12	18
	Backlash		Normal	Plane	mm		
50	0.13	0.14	0.18	–	–	–	–
100	0.16	0.17	0.20	0.26	0.35	–	–
200	–	0.22	0.25	0.31	0.40	0.52	–
400	–	–	0.35	0.41	0.50	0.62	0.80
800	–	–	–	–	0.70	0.80	1.00

Table 11-22: Typical Backlash of Standard Commercial Gear Sets

Diametral Pitch	Backlash (Inches)	Diametral Pitch	Backlash (Inches)
3	0.013	8–9	0.005
4	0.010	10–13	0.004
5	0.008	14–32	0.003
6	0.007	33–64	0.0025
7	0.006		

Metric gear manufacturers often stock gears that give backlash in a similar range when using the formula:

$$a = \frac{m(Z1 + Z2)}{2}$$

Where: a = Center distance with a H7-H7 tolerance in mm

m = Gear module

Z1 = Number of pinion teeth

Z2 = Number of gear teeth

For applications that require reduced or zero backlash, such as with instrumentation or precision control, the designer has a couple of options:

- Use a special set of gears where the teeth are cut on a taper and axial adjustment of the gears closes the backlash spacing (Figure 11-22a).
- Use a special pair of backlash compensating gears. Here the gear is split in half—one half is fixed to the shaft and the other half is connected through a spring pack to the fixed gear. The effect of the spring pack is that it essentially creates a gear that has teeth that expand to fill the backlash gap (Figure 11-22b).

Backlash is controlled on bevel gears by altering the mounting distance of the pinion or gear as shown in Figure 11-23. Often this is done with shim packs under the pinion.

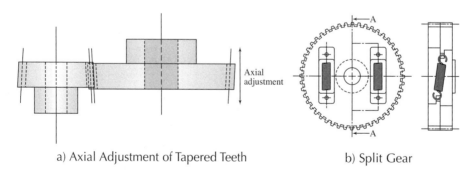

a) Axial Adjustment of Tapered Teeth   b) Split Gear

Figure11-22: Anti-Backlash Gears

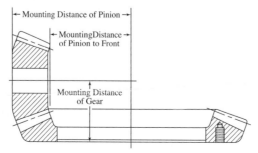

Figure 11-23: Bevel Gear Mounting Distance

Table 11-23: AGMA Recommended Backlash Settings for Bevel and Hypoid Gears

Measured at Tightest Point of Mesh

Diametral Pitch	Normal Backlash (Inch)		Diametral Pitch	Normal Backlash (Inch)	
	Quality Numbers Q7 to Q13	Quality Numbers Q3 to Q6		Quality Numbers Q7 to Q13	Quality Numbers Q3 to Q6
1.00 to 1.25	0.020–0.030	0.045–0.065	5.00 to 6.00	0.005–0.007	0.006–0.013
1.25 to 1.50	0.018–0.026	0.035–0.055	6.00 to 8.00	0.004–0.006	0.005–0.010
1.50 to 1.75	0.016–0.022	0.025–0.045	8.00 to 10.00	0.003–0.005	0.004–0.008
1.75 to 2.00	0.014–0.018	0.020–0.040	10.00 to 16.00	0.002–0.004	0.003–0.005
2.00 to 2.50	0.012–0.016	0.020–0.030	16.00 to 20.00	0.001–0.003	0.002–0.004
2.50 to 3.00	0.010–0.013	0.015–0.025	20 to 50	0.000–0.002	0.000–0.002
3.00 to 3.50	0.008–0.011	0.012–0.022	50 to 80	0.000–0.001	0.000–0.001
3.50 to 4.00	0.007–0.009	0.010–0.020	80 +	0.00–0.0007	0.00–0.0007
4.00 to 5.00	0.006–0.008	0.008–0.016	–	–	–

Table 11-23 lists the AGMA recommended backlash settings for bevel and hypoid gears with backlash measured at the tightest point of the mesh.

MATERIALS AND TREATMENTS

There are many materials available for gear production. Material selection reduces to a balance of cost and functional requirements. The lowest cost and most commonly available gears are made from cast iron and steel. When higher strength, corrosion resistance, or other special properties are needed, various alloys and exotic materials are employed. Bronze is recommended for worm and crossed helical gearing due to the increased sliding contact in these sets. Gear pairs are sometimes specified as different materials or with different treatment to compensate for higher stresses in the pinion (such as making the pinion steel and the gear bronze) or to limit wear to one lower cost or easily replaceable component in the pair (such as a bronze pinion with a steel gear). This choice is dependent on the design, service life, and maintenance requirements. Table 11-24 summarizes typical gear materials used by machine designers. Plastic and aluminum gears are not included due to their lack of presence in industrial environments.

After material selection, material and surface treatments must be considered. Material treatments, such as hardening, cause molecular conversion. Surface treatments are applied after material treatments, and involve the application of coatings for increased wear or increased lubricity. Sections 8.3 and 8.4 of this book provide information on many common material and surface treatments.

LUBRICATION AND WEAR

For most gear train applications, lubrication is a must. Lubrication provides four key benefits:

- Reduces wear in the mesh by providing a separating film and reducing friction
- Provides cooling to the gear train by transferring heat out of the mesh
- Helps to flush dirt and debris out of the mesh (especially effective when used in combination with filtering or simple magnetic drain plugs)
- Protects gear teeth from corrosion

The two most common methods of providing lubrication to the gear pitch line are submerged bath and splash. These methods are for gear trains that are enclosed in housings. For low-speed gears, or gears that are exposed, a moly-disulphide grease is often used if the application allows. For high power and high-speed applications it is often necessary to pump lubrication directly into the pitch line of the gears.

Older lubrication recommendations centered on mineral oils; however, lubrication is a field that has undergone recent advancement. The introduction of synthetics as well as specialized additives has allowed for reduced maintenance, while increasing the efficiency of the gear trains. ANSI/AGMA 9005-E02 "Industrial Gear Lubrication" provides recommendations for gearing and defines the following types of lubrication:

Rust and oxidation inhibited gear oils (R&O) are petroleum based with special chemical additives to resist oxidation.

Compounded gear lubricants (Comp) are petroleum oil based with 3%–10% blend of fatty oils.

Extreme pressure lubricants (EP) have special additives that resist scuffing and wear of the gear tooth faces.

Synthetic gear lubricants (S) are chemical formulas used mostly in the most severe applications.

AGMA has defined 14 viscosities of oils for gear lubrication, rated 0 to 14, where a higher number corresponds to greater viscosity. The two greatest factors that govern lubrication selection are the ambient temperature and the pitch line velocity of the lowest running speed gear pair in the train. Table 11-25 provides the AGMA recommended lubricant grades for spur, helical, herringbone, and bevel gear drives at different temperature and speed combinations.

Table 11-24: Common Gear Materials, Features, and Applications

Material	Features	Applications
Low-Carbon Steel	• Can be case-carburized • Good machinability	Low to moderate power transmission
Medium-Carbon Steel	• Can be induction or flame hardened	Moderate to high power transmission
High-Carbon Steel	• Fair machinability • Greater hardenability options • Steels with carbon content above 0.7% are typically not used due to brittleness issues	High power transmission
Stainless Steel 300 Series	• Best corrosion resistance • Non-magnetic • Non-hardenable • Fair machinability • Oxide layer damaged by sliding contact applications	Low power applications where superior corrosion resistance is required or sanitary requirements exist such as food processing
Stainless Steel 400 Series	• Moderate corrosion resistance • Magnetic • Heat treatable • Fair machinability • Oxide layer damaged by sliding contact applications	Low to medium power applications where superior corrosion resistance is required
Brass Alloys	• Low cost • Excellent machinability • Non-corrosive	Light duty applications
Bronze Alloys	• Low cost • Excellent machinability • Low Friction • Good compatibility with steel mating gears	• Mating gears to steel pinion gears • Use as sacrificial gear
Cast Iron	• Low cost • High internal damping • Very good machinability • Ability to cast gears in segments	• Moderate power transmission • Very low shock applications • Large gears
Nodular (ductile) Iron	• Relatively low cost • Better impact and fatigue strength than cast iron • Good machinability • Ability to cast gears in segments	• Moderate power transmission • Relatively low shock applications • Large gears
Chrome-Molybdenum Alloy Steel	• Can be nitrided and flame hardened • Excellent machinability	Special applications
Nickel-Chrome-Molybdenum Alloy Steel	• Can be nitrided and induction hardened • Poor machinability • Low coefficient of expansion	Special applications where thermal expansion must be minimized

Table 11-25: AGMA Recommended Lubricant Viscosities
Spur, Helical, Herringbone, and Bevel Gear Drives

Pitch Line Velocity		Ambient Temperature			
		-40°F to 14°F	-14°F to 50°F	-50°F to 95°F	-95°F to 130°F
ft/min	m/s	-40°C to -10°C	-10°C to 10°C	10°C to 35°C	-35°C to 55°F
< 1000	< 5	3S	4	6	8
1000 to 3000	5 to 15	3S	3	5	7
3000 to 5000	15 to 25	2S	2	4	6
> 5000	> 25	0S	0	2	3

Although lubrication is beneficial, there can be some negative consequences. Gearbox efficiency can be reduced through fluid churning (windage) or losses (drag) with highly viscous fluids, or if a gearbox is over-filled. However, the positive benefits from lubrication outweigh the negatives, and lubrication remains a must. Table 11-26 provides some high-level considerations and guidelines when selecting lubrication. When in doubt, a competent lubrication supplier should be consulted.

Even with the best lubrication, wear is a factor that the designer must take into account. The amount of accepted wear in a gear system is dependent upon

Table 11-26: Gear Lubrication Considerations

Considerations	Guidelines
Lubrication Method	• Enclosed baths are preferred to reduce contamination and provide adequate lubrication. • Spray or splash lubrication of enclosed or open-gear trains is a second choice. • For high-power and high-speed applications, it is often necessary to pump lubrication directly into the pitch line of the gears.
High Loads	• Higher loads require synthetic oils or Extreme Pressure (EP) additives. Viscosity is also often increased. • High load can also generate excess heat. • It is not uncommon for external oil chillers to be used.
High Speeds	• Higher-speed gear trains operate better with lower-viscosity oils. • High speed can also generate excess heat and oil foaming. • It is not uncommon for external oil chillers to be used.
Compatibility	• Some lubricants may attack gear, seal, or bearing materials. • Lubrication must be selected that works for the entire system.
Operating Temperature	• Lubrication must be specified to work across the range of operating temperatures that the gear train will see, including start-up. • Typically higher temperatures demand a more viscous lubrication. • Multi-viscosity blends are rarely used in gear trains. • Gear train operating temperatures should not exceed 150° F (65.5 °C) during normal operation, although brief periods of operation up to 200°F (93.3 °C) are acceptable.

the application. As teeth wear, there is a tooth-thinning effect that will increase backlash, reduce bending strength, and ultimately lead to tooth failure. Due to this wear, gears of equal material and treatments should be replaced in pairs. However, to avoid the need to replace both gears in a mesh, one gear can be designed to be sacrificial by downgrading the material or treatment of that gear.

Even with lubrication, mild wear of the oxide layer is present. This has the beneficial effect of increasing surface contact areas and equalizing load pressure across the tooth mesh. In addition the wear has a tendency to smooth or polish the tooth surface, allowing for a more uniform lubrication film thickness.

Gear wear is unavoidable, but can be minimized by following these recommendations:

- Gear teeth should be of high quality and smooth.
- The initial run-in of gears should be performed at half load if possible. Often in critical applications, lubrication is filtered or changed after run-in.
- Run at the highest speeds possible for the application. Highly-loaded and slow-running gears (pitch line speed less than 30 m/min or 100 ft/min) are boundary lubricated and suffer from excessive wear. Here, liberal amounts of the highest viscosity lubricant permissible should be used.
- For very slow speed gears (pitch line speed less than 3 m/min or 10 ft/min), lubrications with sulpher-phosphorus additives should be avoided.

For more detailed descriptions and examples of the different wear failure modes, please consult a machinery troubleshooting guide or contact your local gear manufacturer.

SPUR GEARS

Spur gears are the type most commonly encountered by the machine designer. Tables 11-27 and 11-28 list the most common formulas and equations used in specifying and categorizing American Standard and metric gears. A spur gear sizing and selection example appears later in this section.

SPUR GEAR SELECTION AND SIZING

Gear selection from catalogs involves a short series of steps and calculations using factors selected from tables and graphs. Table 11-29 contains formulas

Table 11-27: Formulas for Dimensions of Standard Spur Gears

N_G = Number of teeth in gear	N_P = Number of teeth in pinion	ϕ = Pressure angle
D_P = Pitch diameter of pinion	D_G = Pitch diameter of gear	
a = Addendum	b = Dedendum	
a_G = Addendum of gear	a_P = Addendum of pinion	

GEAR RATIO	$m_G = \dfrac{N_G}{N_P}$	BASE CIRCLE DIAMETER	$D_B = D\cos\phi$
CIRCULAR PITCH	$p = \dfrac{\pi D}{N} = \dfrac{\pi}{P}$	NUMBER OF TEETH	$N = \dfrac{\pi D}{p} = PD$
PITCH DIAMETER	$D = \dfrac{Np}{\pi} = \dfrac{N}{P}$	DIAMETRAL PITCH	$P = \dfrac{\pi}{p} = \dfrac{N}{D}$ $P = \dfrac{N_P(m_G+1)}{2C}$
OUTSIDE DIAMETER	$D_O = D + 2a$ Full Depth Teeth: $D_O = \dfrac{N+2}{P}$ $D_O = \dfrac{p(N+2)}{\pi}$ American Standard Stub Teeth: $D_O = \dfrac{N+1.6}{P}$ $D_O = \dfrac{p(N+1.6)}{\pi}$	CENTER DISTANCE	$C = \dfrac{N_P(m_G+1)}{2P}$ $C = \dfrac{D_P + D_G}{2}$ $C = \dfrac{N_G + N_P}{2P}$ $C = \dfrac{p(N_P + N_G)}{2\pi}$
ROOT DIAMETER	$D_R = D - 2b$	WHOLE DEPTH	$h_t = a + b$
		WORKING DEPTH	$h_k = a_G + a_P$

that are commonly used to size American Standard (inch) spur gears. This is a simplification over the procedure of designing a gear from scratch, as described in the machine design texts included in the recommended resources. The gear manufacturers have done these calculations in the background, and provide tables with ratings of what power and torque their gears are able to withstand.

The primary method that manufacturers use to calculate tooth strength is the Lewis formula with Barth revision. A full description of this equation and its history is available in the recommended resources. Plainly speaking, this method models the tooth as a cantilever beam and determines how much load it can take without breaking based on the bending stress. However, this is not the only factor that determines the power that a gear set can transmit.

Table 11-28: Formulas for Dimensions of Standard Metric Spur Gears
All linear dimensions in mm

N_1 = Number of teeth in gear 1		N_2 = Number of teeth in gear 2	
ϕ = Pressure angle		P = Diametral pitch	
D_P = Pitch diameter of pinion		D_G = Pitch diameter of gear	

MODULE	$m = \dfrac{25.4}{P}$	**BASE PITCH**	$p_B = m\pi\cos\phi$
CIRCULAR PITCH	$p = m\pi = \dfrac{\pi D}{N} = \dfrac{\pi}{P}$	**BASE CIRCLE DIAMETER**	$D_B = D\cos\phi$
PITCH DIAMETER	$D = mN$	**NUMBER OF TEETH**	$N = \dfrac{D}{m}$
OUTSIDE DIAMETER	$D_O = D + 2m$ $D_O = m(N+2)$	**CENTER DISTANCE**	$C = \dfrac{m(N_1 + N_2)}{2}$
ROOT DIAMETER	$D_R = D - 2.5m$	**TOOTH THICKNESS** At standard pitch diameter	$T_{std} = \dfrac{\pi m}{2}$
ADDENDUM	$a = m$	**DEDENDUM**	$b = 1.25m$
BACKLASH (ANGULAR) Arc-minutes	$B_a = 6880\dfrac{B}{D}$	**MINIMUM NUMBER OF TEETH TO AVOID UNDERCUTTING**	$N_c = \dfrac{2}{\sin^2\phi}$
BACKLASH (LINEAR)	Along Pitch Circle: $B = 2(\Delta C)\tan\phi$ $B = \Delta T$	Along Line of Action: $B_{LA} = B\cos\phi$	
CONTACT RATIO	$m_p = \dfrac{\sqrt{R_{O,1}^2 - R_{B,1}^2} + \sqrt{R_{O,2}^2 - R_{B,2}^2} - C\sin\phi}{m\pi\cos\phi}$		

Recall that material, quality rating, surface finish, and material treatment all influence load capacity. Also, gear sets with long service life requirements often require special materials or surface treatments (not included in this selection procedure) to protect against failure through surface pitting and subsequent wear. Unfortunately, many gear catalogs only publish cursory data on some of these additional factors because they have already worked some of the details into the calculations for you. Therefore, sizing calculations should be performed using the recommended values from each supplier for their own gears. For applications that are close to the limits of size and power or require a long service life, the gear manufacturer should be consulted for available options.

Table 11-29: Inch Spur Gear Sizing Formulas

Y = Outline Factor, example given in Table 11-30 S = Safe Static Stress (psi) from Table 11-31	
DESIGN HORSEPOWER HP_d = Design Horsepower (ft-lb$_f$/min) HP_a = Actual Horsepower (ft-lb$_f$/min) SF = Service Factor (See Table 11-32)	$HP_d = (HP_a)(SF)$
PITCH DIAMETER D_1 = Pitch Diameter of Pinion (inches) D_2 = Pitch Diameter of Gear (inches) C = Center Distance (inches) m_G = Gear ratio = input/output	$D_1 = \dfrac{(2C)}{m_G + 1}$ $D_2 = D_1 m_G$
PITCH LINE VELOCITY V = Pitch Line Velocity (ft/min) D_i = Pitch Diameter of Pinion or Gear (inches) RPM_i = Revolutions per minute of Pinion or Gear Note: This must be the RPM for selected pinion or gear used for pitch diameter.	$V = 0.262 D_i (RPM_i)$
DIAMETRAL PITCH P = Diametral Pitch	$P = \sqrt{\dfrac{0.75 \pi S V}{27.5(HP_d)(1200 + V)}}$
NUMBER OF TEETH N = Number of teeth D = Pitch Diameter	$N = PD$
FACE WIDTH F = Face Width (inches)	$F = \dfrac{33000 P(HP_d)}{VSY \left(\dfrac{600}{600 + V} \right)}$
LEWIS FORMULA WITH BARTH REVISION **LOAD AT PITCH LINE, HORSEPOWER** L = Load at pitch line (lb$_f$) HP = horsepower	$L = \dfrac{SYF}{P} \left(\dfrac{600}{600 + V} \right)$ $HP = \dfrac{LV}{33000}$

Table 11-30: Outline Factor (Y) for Use With Diametral Pitch

Number of Teeth	14-1/2° PA Involute	20° PA Involute	Number of Teeth	14-1/2° PA Involute	20° PA Involute
10	0.176	0.201	26	0.308	0.344
11	0.192	0.226	28	0.314	0.352
12	0.210	0.245	30	0.318	0.358
13	0.223	0.264	35	0.327	0.373
14	0.235	0.276	40	0.336	0.389
15	0.245	0.289	45	0.340	0.399
16	0.255	0.295	50	0.346	0.408
17	0.264	0.302	60	0.355	0.421
18	0.270	0.308	70	0.360	0.429
19	0.277	0.314	80	0.363	0.436
20	0.283	0.320	90	0.366	0.442
21	0.289	0.326	100	0.368	0.446
22	0.292	0.330	150	0.375	0.458
23	0.296	0.333	200	0.378	0.463
24	0.302	0.337	Rack	0.390	0.484
25	0.305	0.340	-	-	-

Table 11-31: Example Safe Static Stress of Materials

Material	S psi
Steel – 0.4 Carbon	25000
Steel – 0.2 Carbon	20000
Steel – 0.4 Carbon Heat Treated	35000
Cast Iron	12000
Bronze	10000
Non-Metallic	6000

Table 11-32: Typical Service Factors (SF) for Calculating Design Horsepower

LOAD CONDITIONS	OPERATION CONDITIONS		
	Intermittent or <8 hours/day	8 – 10 hours/day	>10 hours/day or Continuous
Uniform, No Shock	0.8	1	1.25
Light Shock	1	1.25	1.5
Medium Shock	1.25	1.5	1.8
Heavy Shock	1.5	1.8	2

To select a set of inch stock spur gears for an application, the following steps can be used:

1) Determine the design parameters:
   - Exact center distance (inches)
   - Required ratio
   - Required speeds (revolutions per minute)
   - Actual horsepower to transmit (ft-lbf/min)
   - Shaft sizes (inches)

2) Calculate the design horsepower by applying a service factor to the actual horsepower. Typical service factors are shown in Table 11-32. Design horsepower is actual horsepower multiplied by the appropriate service factor. The design horsepower will be used in all subsequent calculations.

3) Determine the pitch diameter of each gear in the pair by using the formulas in Table 11-29. You can verify that you performed this calculation correctly by checking that your calculated pitch diameters satisfy the following equation:

$$C = \frac{(D_1 + D_2)}{2}$$

4) Refer to the available gears in the catalog to select a pinion that meets the following criteria:
   (a) Meets or exceeds design horsepower at operating speed
   (b) Has required pitch diameter

(c) Has a bore allowance that can accommodate the required shaft size

Available gears are typically grouped by pressure angle, then accenting diametral pitch. For standard gears there is usually only one face width available for each diametral pitch.

5) Now that you have selected a pinion, you are locked into a tooth count for the gear to maintain the desired ratio. Calculate the required tooth count for the ratio and find that gear in the catalog (because it has to be the same diametral pitch, it will likely be on the same catalog page). Repeat step 4 for this gear.

If you are able to select a set of gears that meets your initial criteria you should consider yourself fortunate. However, more than likely, you will be required to either make some design changes, or specify a set of custom gears. Assuming that you are not able to change the ratio, speed, or horsepower requirements, consider altering the center distance to open up range of available gear sets, such as those with a lower diametral pitch (larger tooth), or larger tooth count gears of the original pitch.

If you are not able to select a set of gears from stock gearing, you can specify a custom set of gears. Gear manufacturers are set up to respond rapidly to some minor modifications to the existing offerings. Here we are describing minor modifications such as requesting a wider face width gear, a different material from a small list of choices, or a different tooth count, all while maintaining the standard diametral pitches. Essentially you are designing a gear that is not a commonly stocked item, but still follows all of the existing standards and preferred values for things such as pitch diameter. To design a minimally custom set of gears, the following steps should be followed with the aid of the manufacturer's catalog of values for their gears. Steps 1–3 are identical to those above. Once complete, the manufacturer should be contacted to verify the calculations and ability to provide the required gear.

To select a set of non-stock inch spur gears for an application the following steps can be followed:

1) Collect the following design factors:
   • Exact center distance (inches)
   • Required ratio
   • Required speeds (revolutions per minute)
   • Actual horsepower to transmit (ft-lbf/min)
   • Bore size (inches)
2) Calculate the design horsepower by applying a service factor to the actual horsepower. Refer back to Table 11-32 for typical service

factors. Design horsepower is actual horsepower multiplied by the appropriate service factor. The design horsepower will be used in all subsequent calculations.

3) Determine the pitch diameters of each gear in the pair by using the formulas in Table 11-29.

4) Determine the pitch line velocity using the equation in Table 11-29.

5) Calculate the required diametral pitch. Use the formula in Table 11-29 to calculate the approximate pitch, and round up to the nearest whole number for pitches greater than 3. For pitches lower than 3, standard available pitches are 1, 1¼, 1½, 1¾, 2, and 2½.

6) Calculate the number of teeth on each gear using the equation in Table 11-29.

7) Calculate the required gear face width for the pinion, and round up to the nearest half inch width. The formula for face width can be found in Table 11-29.

8) Verify the horsepower rating of the selected pinion using the Lewis Formula with Barth Revision (Table 11-29) for the tooth load L. This HP value does not include any required service factors (SF) and should be compared to the design horsepower requirement calculated in Step 2.

If the required HP rating has not been achieved, there are a few more design parameters that can be altered to increase the power rating. The following may be tried:

- Increase the face width
- Increase the tooth diametral pitch
- Harden the pinion
- Select a higher tolerance class gear
- Switch to helical gears
- Select different materials and treatment

If, after a few design iterations, a suitable set of minimally modified gears cannot be found that meet the application requirements, the designer may be forced into selecting a different style of gears, creating a gear train, or designing truly custom gears that will be made just for this application. Here development with the gear supplier is a must. Some potential advanced problems and solutions for custom designed gears are listed in Table 11-33. Greater detail of these methods is provided in the recommended resources, and should be discussed with the gear supplier.

Table 11-33: Advanced Gear Selection Problems And Possible Solutions

Problem	Possible Solution	Notes
Inability to meet required ratio or fixed center distance.	Design of custom diametral pitch gears.	This will necessitate the creation of custom cutters and should be avoided if possible.
Tooth strength too low	Use profile shifted gears	Here the pitch diameter is shifted away from one gear and to the other, creating teeth on the smaller gear that are wider and stronger. This is often used with pinions that suffer from undercutting.

HELICAL GEARS

Helical gears are similar to spur gears with the exception that the tooth is cut on an angle, called the helix angle. This increases the tooth contact ratio and allows for the transmission of greater loads. Two mating helical gears must have an equal helix angle. Externally mated helix gears must be opposite hand; however, internally mated helix gears must be of the same hand.

One side effect of the angled tooth is that thrust loads are generated, as shown in Figure 11-24 for various gear configurations. Note that the thrust vector reverses with reversing rotation. Table 11-34 lists the most common equations used for helical gear geometry. Here, because the teeth are cut on an angle, the term "normal" is often used to modify the gear terminology to specify the plane normal to the tooth surface at a point of contact. Helical gear strength and power transmission calculations are not covered in this text and can be found in the recommended resources.

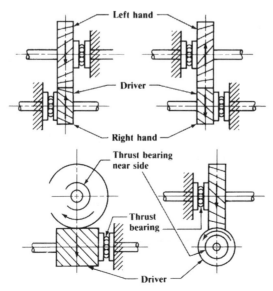

Figure 11-24: Resultant Thrust Directions for Helical Gear Pairs

BEVEL GEARS

Straight cut bevel gears are the simplest form encountered by the machine designer and are often used in right angle gear boxes. When bevel gears are of the same size, they are referred to as miter gears. Straight cut bevel gears are often used in applications that are low speed, open, and intermittent use, or for machine adjustment controls where liberties can be taken with the

Table 11-34: Common Equations Used with Helical Gears
Rules and Formulas for Helical Gear Calculations

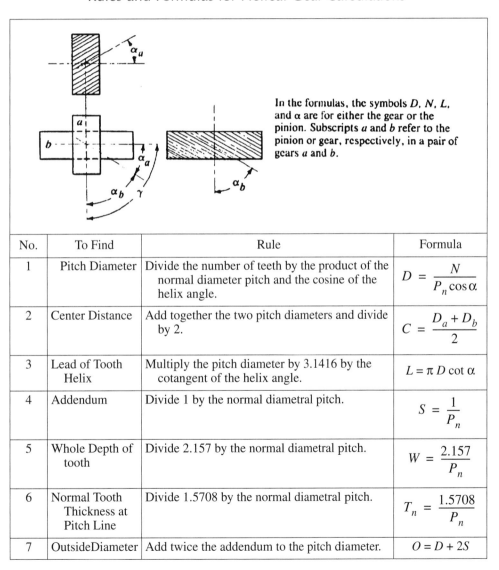

In the formulas, the symbols D, N, L, and α are for either the gear or the pinion. Subscripts a and b refer to the pinion or gear, respectively, in a pair of gears a and b.

No.	To Find	Rule	Formula
1	Pitch Diameter	Divide the number of teeth by the product of the normal diameter pitch and the cosine of the helix angle.	$D = \dfrac{N}{P_n \cos \alpha}$
2	Center Distance	Add together the two pitch diameters and divide by 2.	$C = \dfrac{D_a + D_b}{2}$
3	Lead of Tooth Helix	Multiply the pitch diameter by 3.1416 by the cotangent of the helix angle.	$L = \pi D \cot \alpha$
4	Addendum	Divide 1 by the normal diametral pitch.	$S = \dfrac{1}{P_n}$
5	Whole Depth of tooth	Divide 2.157 by the normal diametral pitch.	$W = \dfrac{2.157}{P_n}$
6	Normal Tooth Thickness at Pitch Line	Divide 1.5708 by the normal diametral pitch.	$T_n = \dfrac{1.5708}{P_n}$
7	OutsideDiameter	Add twice the addendum to the pitch diameter.	$O = D + 2S$

requirements of precision mounting. Figure 11-25 and Table 11-35 show the common nomenclature and equations used for the geometry of straight cut milled bevel gears. Bevel gear strength and power transmission calculations are not covered in this text and can be found in the recommended resources.

The most important responsibility of the designer when applying commercially available bevel gears is to ensure that the mounting is accurate enough to ensure that the pitch apex of both gears are coincident. This requires careful setting of the "pitch apex to back" dimensions for both gears. If the application requires high load capability with quiet operation, then hypoid, spiral, or Zerol gears should be selected. More information on the application of these specialized gears can be found in *Machinery's Handbook*, or with a competent gear manufacturer.

WORM GEARS

Worms and worm gears, like helical gears, also have left and right hand variants and produce thrust forces on their shafts. Figure 11-26 shows the resultant thrust forces for various configurations. Note that a reversal of

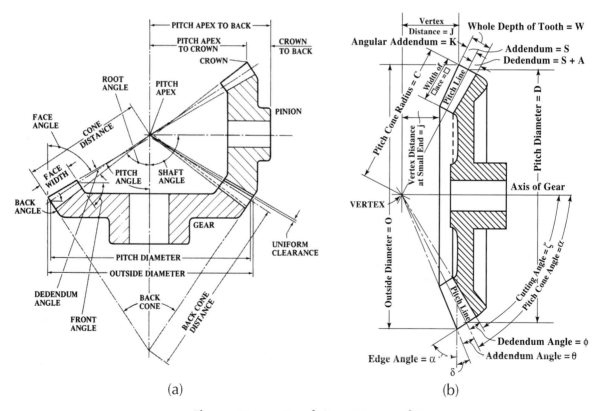

(a)   (b)

Figure 11-25: Bevel Gear Nomenclature

rotation will also reverse thrust forces. Worm gearing has high levels of sliding action and consequently very low efficiency due to increased friction and heat. Additional methods of heat dissipation (such as oil coolers) are sometimes required. The reduced efficiency must also be considered when specifying the power requirements of the overall drive system.

Table 11-35: Common Equations Used with Milled Bevel Gears

To Find		Rule	Formula
Pitch Cone Angle Of Pinion		Divide the sine of the shaft angle by the sum of the cosine of the shaft angle and the quotient obtained by dividing the number of teeth in the gear by the number of teeth in the pinion; this gives the tangent. *Note*: For shaft angles greater than 90° the cosine is negative.	$\tan \alpha_P = \dfrac{\sin \Sigma}{\dfrac{N_G}{N_P} + \cos \Sigma}$ For 90° shaft angle: $\tan \alpha_P = \dfrac{N_P}{N_G}$
Pitch Cone Angle Of Gear		Subtract the pitch cone angle of the pinion from the shaft angle.	$\alpha_G = \Sigma - \alpha_P$
Pitch Diameter		Divide the number of teeth by the diametral pitch.	$D = \dfrac{N}{P}$
These dimensions are the same for both gear and pinion.	Addendum	Divide 1 by the diametral pitch.	$S = \dfrac{1}{P}$
	Dedendum	Divide 1.157 by the diametral pitch.	$S + A = \dfrac{1.157}{P}$
	Whole Depth of Tooth	Divide 2.157 by the diametral pitch.	$W = \dfrac{2.157}{P}$
	Thickness of Tooth at Pitch Line	Divide 1.571 by the diametral pitch.	$T = \dfrac{1.571}{P}$
	Pitch Cone Radius	Divide the pitch diameter by twice the sine of the pitch cone angle.	$C = \dfrac{D}{2 \sin \alpha}$
	Addendum of Small End of Tooth	Subtract the width of face from the pitch cone radius, divide the remainder by the pitch cone radius, and multiply by the addendum.	$s = S \dfrac{C - F}{C}$
	Thickness of Tooth at Pitch Line at Small End	Subtract the width of face from the pitch cone radius, divide the remainder by the pitch cone radius, and multiply by the thickness of the tooth at pitch line.	$t = T \dfrac{C - F}{C}$
	Addendum Angle	Divide the addendum by the pitch cone radius to get the tangent.	$\tan \theta = \dfrac{S}{C}$
	Dedendum Angle	Divide the dedendum by the pitch cone radius to get the tangent.	$\tan \phi = \dfrac{S + A}{C}$
	Face Width (Maximum)	Divide the pitch cone radius by 3 or divide 8 by the diametral pitch, whichever gives the smaller value.	$F = \dfrac{C}{3}$ or $F = \dfrac{8}{P}$
	Circular Pitch	Divide 3.1416 by the diametral pitch.	$\rho = \dfrac{\pi}{P}$
	Face Angle	Add the addendum angle to the pitch cone angle.	$\gamma = \alpha + \theta$

continued on next page

Table 11-35: Common Equations Used with Milled Bevel Gears (continued)

Compound Rest Angle for Turning Blank	Subtract both the pitch cone angle and the addendum angle from 90°.	$\delta = 90° - \alpha - \theta$
Cutting Angle	Subtract the dedendum angle from the pitch cone angle.	$\xi = \alpha - \phi$
Angular Addendum	Multiply the addendum by the cosine of the pitch cone angle.	$K = S\cos\alpha$
Outside Diameter	Add twice the angular addendum to the pitch diameter.	$O = D + 2K$
Vertex of Apex Distance	Multiply one-half the outside diameter by the cotangent of the face angle.	$J = \dfrac{O}{2}\cot\gamma$
Vertex Distance at Small End of Tooth	Subtract the width of face from the pitch cone radius; divide the remainder by the pitch cone radius, and multiply by the apex distance.	$j = J\dfrac{C - F}{C}$
Number of Teeth for Which to Select Cutter	Divide the number of teeth by the cosine of the pitch cone angle.	$N' = \dfrac{N}{\cos\alpha}$

Fifteen different lead angles have been standardized: 0.5, 1, 1.5, 2, 3, 4, 5, 7, 9, 11, 14, 17, 21, 25, and 30 degrees. Pressure angle for cutters has been standardized at 20°, although in use pressure angles are slightly higher due to the relationship between gear diameters. The exact tooth profile of the worm gear is highly dependent upon the manufacturing method and requires

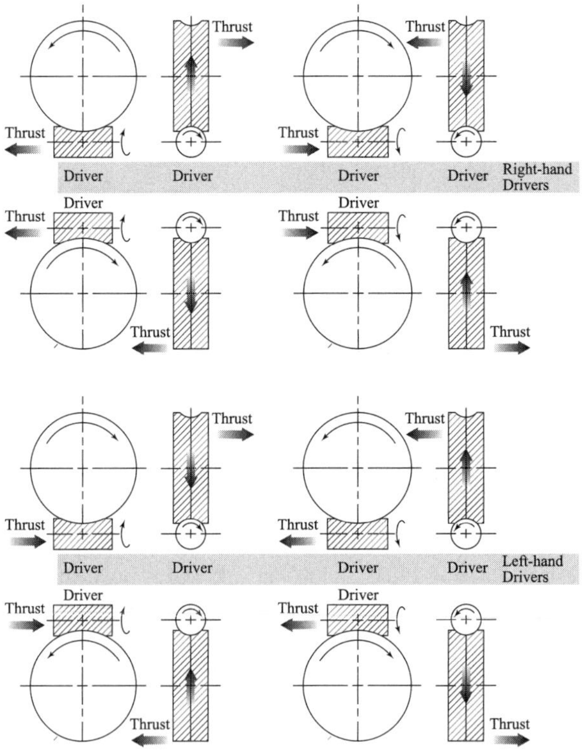

Figure 11-26:
Thrust Components of Various
Worm Gear Configurations

the specification of the mating worm, and the diameter of cutter used. Using standard gear sets from catalogs is recommended to avoid the potential for error in designing and specifying mating pairs.

Table 11-36 lists the most common equations used for worm gears that will help the machine designer in packaging the set. Worm and worm gear strength and power transmission calculations are not covered in this text and can be found in the recommended resources.

Table 11-36: Formulas for Proportions of American Standard Fine Pitch Worms and Wormgears
ANSI B6.9-1977

LETTER SYMBOLS

P = Circular pitch of wormgear

P = axial pitch of the worm, P_x, in the central plane

P_x = Axial pitch of worm

P_n = Normal circular pitch of worm and wormgear = P_x cos $\lambda = P$ cos ψ

λ = Lead angle of worm

ψ = Helix angle of wormgear

n = Number of threads in worm

N = Number of teeth in wormgear

$N = nm_G$

m_G = Ratio of gearing = $N \div n$

Item	Formula	Item	Formula
WORM DIMENSIONS		WORMGEAR DIMENSIONS	
Lead	$l = nP_x$	Pitch Diameter	$D = NP \div \pi = N\Pi_\xi \div \pi$
Pitch Diameter	$d = l \div (\pi \tan\lambda)$	Outside Diameter	$D_o = 2C - d + 2a$
Outside Diameter	$d_o = d + 2a$	Face Width	$F_{Gmin} = 1.125 \times \sqrt{(d_o + 2c)^2 - (d_o - 4a)^2}$
Safe Minimum Length of Threaded Portion of Worm	$F_W = \sqrt{D_o{}^2 - D^2}$		
DIMENSIONS FOR BOTH WORM AND WORMGEAR			
Addendum	$a = 0.3183P_n$	Tooth thickness	$t_n = 0.5P_n$
Whole Depth	$h_t = 0.7003P_n + 0.002$	Approximate normal pressure angle	$\phi_n = 20$ degrees
Working Depth	$h_k = 0.6366P_n$		
Clearance	$c = h_t - h_k$	Center distance	$C = 0.5 (d + D)$

CRITICAL CONSIDERATIONS: Gears

- Gears with different pitch, mode, tooth form, and pressure angle are not interchangeable. Each gear in the mesh must share the same characteristics.
- Externally mated helix gears must be of the opposite hand; however, internally mated helix gears must be of the same hand.
- Worm gear sets must be of the same pressure angle, lead, and hand to operate together.
- Helical, bevel, and worm gears all produce thrust components when in operation. Shaft support and mounting design must take this into account.
- Lubrication is a must for metal gears. Some plastic gears are considered "self-lubricating," but are not typically used for industrial machinery applications.
- Rotating gears, shafts, couplings, hubs, attachments, etc., are extremely dangerous, and proper safety precautions such as guarding must be taken. See Chapter 2 for more information on machinery safety.

BEST PRACTICES: Gears

- Consider if the application truly requires gears. Other methods of power transmission, such as belts and chains, have lower tolerance and set-up requirements, resulting in lower total cost.
- Use standard commercially available gears when possible instead of custom designed gears to reduce cost.
- Straight cut gear teeth should be used when the application allows because they are the most economical gears to produce and do not have additional thrust components to be designed around.
- Gear tooth interference between a rack and pinion begins for $14\frac{1}{2}°$, $20°$, and $25°$ pressure angle gears when pinion tooth counts fall below 32, 18, and 12 respectively. Increase pressure angle or use profile shifted gears to avoid tooth undercutting.
- Worm gearing is notoriously inefficient, even when using synthetic lubricants.

continued on next page

BEST PRACTICES: Gears (Continued)

- Total operating costs during service life should be considered when selecting gear sets.
- Larger gear teeth (lower diametral pitch) generally operate with lower stress due to tooth size. However, larger teeth gears are often louder in operation than their smaller tooth counterparts.
- Increasing the pinion size will decrease the transmitted load, which also lowers the stresses.
- Try to select a pinion size that is at least twice the diameter of the shaft.
- Be mindful of the trade-off of gear quality level and cost, but also consider maintenance costs. The cost savings from lower quality gears may be quickly lost if a machine needs to be removed from service for gear changing. Higher quality gears generally have better surface finishes that promote proper lubrication. They also have a more precise mesh that allows for a more accurate transfer of tooth loads, minimizing the overloading and deflection of teeth.
- Consider making the pinion two points harder on the Rockwell C scale than the gear. This will help to equalize the wear on the set because the pinion teeth will see more action over their life, proportional to the gear ratio.
- Shaft center distance tolerances should create a 0/+ condition so that backlash is not eliminated by tolerance.
- Gear noise can be reduced by one or more of the following methods: higher precision, better surface finish on teeth, lower loads, smaller teeth, use of proper lubricant, and/or reduce backlash (but not to the point of interference).
- Replace gears in sets, unless one gear has been designed to be sacrificial, either through material or treatment. Gear sets generally wear in together.
- Specify standard materials and treatments over custom and proprietary offerings to avoid being locked into a single source supplier. Using single source suppliers can increase cost and impact delivery times.
- $14\frac{1}{2}$ degree pressure angle gears are considered obsolete for new designs, but may be encountered on older equipment.

GEARBOXES

Gearboxes are enclosed gear arrangements, usually with an input shaft, output shaft, and provisions for lubrication of the gears and bearings. Gearboxes are common in machinery and readily available commercially. They are normally configured as speed reducers and torque multipliers. Special designs and functions are available. Gearboxes are usually purchased rather than designed, and come in a variety of shaft configurations, reduction ratios, and gear types. This section will describe the characteristics and selection criteria of commercial gearboxes. Design of gearboxes is beyond the scope of this text.

RECOMMENDED RESOURCES

- Speak with your chosen gearbox supplier for more information.

GEARBOX CHARACTERISTICS

Gearboxes generally have an input shaft and an output shaft. The input shaft usually operates at high speeds and relatively low torques compared to the output shaft. This configuration is known as a gear reducer. The input shaft is typically coupled to a motor either directly or through a coupling of some type. The output shaft is used to drive a mechanism.

There are two major types of commercially available gearbox: in-line and right angle. In-line gearboxes have either parallel offset, or collinear input and output shafts. Right angle gearboxes have the input and output shafts perpendicular, either in the same plane or offset. In-line gearboxes typically employ spur gears, planetary gears, cycloidal mechanisms, or harmonic wave generators. Right angle gearboxes typically feature worm gearing or bevel gearing.

Efficiency of a gearbox is output power divided by input power. Power loss in a gearbox is primarily due to friction, which generates heat. Efficiency is a function of gear type, number of engagements, and load torque. Gearbox efficiency is usually stated in catalogs as a single value. It is important to realize that efficiency actually varies with a number of factors, including speed and load. As a general rule, lightly loaded gearboxes have lower efficiencies than heavily loaded gearboxes. As a gearbox is more heavily loaded, it tends

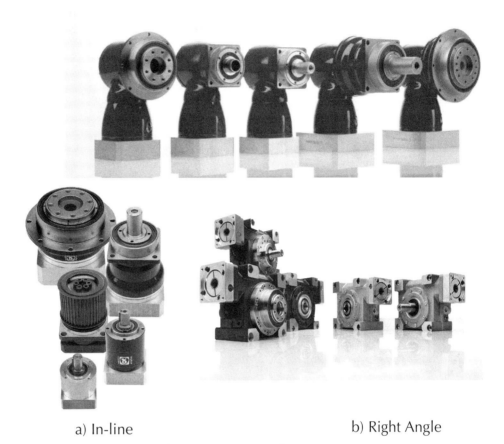

a) In-line   b) Right Angle

Figure 11-27: Gearboxes

Source: WITTENSTEIN (www.wittenstein-us.com)

to approach its stated efficiency. When gearboxes are operated outside of their recommended speed range, gear lubrication is sub-optimal and results in additional efficiency loss.

Reduction ratio is input speed divided by output speed. Ratio is normally stated as input:output. A 5:1 gearbox has an input speed five times greater than its output speed. Its numerical gear ratio is 5. This reduction ratio also applies to inversely to torque, though some torque is lost through inefficiency.

Gearboxes can have one or more reduction stages. Each stage represents one gear mesh or one reduction ratio. A general rule of thumb is that each reduction stage should not have greater than a 10:1 ratio. Reduction stages multiply. For example, a 4:1 stage driving a 5:1 stage will yield an overall ratio of 20:1. Therefore, a high reduction ratio can be achieved by combining two or more stages with lower ratios. In general, multi-stage gearboxes are capable of much higher reduction ratios than single stage gearboxes. Stages can also be used to change the direction of rotation of the output shaft. Multiple stages add to the gearbox size, complexity, and often cost.

LOADS ON GEARBOXES

The forces and torque reflected on a gearbox must be thoroughly evaluated. Section 11.1 contains some calculations of reflected forces on shafts caused by common machine components.

Lateral forces that will act to shear or bend the gearbox output shaft can be calculated based on the application geometry. Gearboxes usually have a radial (lateral) or overhanging load rating. Common sources of radial loads are pulleys, sheaves, sprockets, and cranks. These components should be mounted as close as possible to the gearbox to minimize overhanging moment. When a radial load is present, ideally it is minimized through the use of external shaft support bearings. Lateral forces can also result from misalignment between the gearbox and the driven mechanism. Proper alignment is essential, and the use of a flexible coupling is recommended to minimize these forces.

Thrust loads on gearbox shafts typically result from either orientation effects (vertical shaft) or from helical or bevel gears attached to the output shaft. The weight of components on a vertical shaft must be calculated, as well as the thrust components of gear forces. Most gearboxes will have a thrust load rating.

Output torque of a gearbox can be calculated based on input torque using the formula found in Table 11-39. Torque required by the application at the gearbox output shaft can be difficult to calculate. The torque must be sufficient to perform the required work. Enough torque must be provided by the gearbox to also overcome the effects of inertia, friction, and gravity to drive the attached machinery. This can be a complex task. The mass moment of inertia of the driven load or machine must be calculated at the gearbox output shaft. This mass moment of inertia will create torque loads on the gearbox output shaft during acceleration. The torque required to drive the attached load against gravity and friction must also be calculated at the gearbox output shaft. Some basic equations used for these calculations are provided in Table 11-37. Be sure to check all units for consistency.

Table 11-37: Reflected Torque and Inertia on Gearboxes

Driven Mechanism	Torque	Mass Moment of Inertia	Motion	Additional Considerations
DIRECT DRIVE	$T_{\to GB} = T_L$	$I_{\to GB} = I_L$ See Section 1.3 for I_L calculation methods For a circular Cross Section: $I = \dfrac{\pi d^4}{32}$	$\theta_{GB} = \theta_L$ $\omega_{GB} = \omega_L$	If applicable: Overhanging load Coupling inertia Bearing drag Braking requirements
GEAR DRIVE	$T_{\to GB} = \dfrac{T_L}{N_r E_f}$	$I_{\to GB} = I_{GM} + I_{GL\to GB} + I_{L\to GB}$ $I_{GL\to GB} = \dfrac{I_{GL}}{E_f}\left(\dfrac{1}{N_r}\right)^2$ $I_{L\to GB} = \dfrac{I_L}{E_f}\left(\dfrac{1}{N_r}\right)^2$	$\theta_{GB} = N_r \theta_L$ $\omega_{GB} = N_r \omega_L$ $N_r = \dfrac{N_{tL}}{N_{tM}}$	Lubricant drag If applicable: Coupling inertia Braking requirements
LEAD SCREW	$T_{\to GB} = T_P + \left(\dfrac{F_P + F_g + F_{fr}}{2\pi P_s E_f}\right)$ T_P = Torque due to preload $F_g = (W_L + W_T)\sin\gamma$ $F_{fr} = \mu(W_L + W_T)\cos\gamma$	$I_{\to GB} = I_C + I_S + I_{L\to GB}$ $I_{L\to GB} = \dfrac{W_L + W_T}{gE_f}\left(\dfrac{1}{2\pi P_s}\right)^2$ Where: P_s = pitch of lead screw	$\theta_{GB} = P_S X_L$ $\omega_{GB} = P_S V_L$	Bearing friction Lubricant drag If applicable: Braking requirements
SYNCHRONOUS BELT OR CHAIN DRIVE	$T_{\to GB} = \dfrac{T_L}{N_r E_f}$	$I_{\to GB} = I_{PM} + I_{PL\to GB} + I_{B\to GB} + I_{L\to GB}$ $I_{PL\to GB} = \dfrac{I_{PL}}{E_f}\left(\dfrac{1}{N_r}\right)^2$ $I_{B\to GB} = \dfrac{W_B}{gE_f}\left(\dfrac{d_{pM}}{2}\right)^2$ $I_{L\to GB} = \dfrac{I_L}{E_f}\left(\dfrac{1}{N_r}\right)^2$	$\theta_{GB} = N_r \theta_L$ $\omega_{GB} = N_r \omega_L$ $N_r = \dfrac{d_{pL}}{d_{pM}} = \dfrac{N_{tL}}{N_{tM}}$	Belt or chain inertia Bearing drag If applicable: Overhanging load Coupling inertia Braking requirements

continued on next page

Table 11-37: Reflected Torque and Inertia on Gearboxes (Continued)

Driven Mechanism	Torque	Mass Moment of Inertia	Motion	Additional Considerations
CONVEYOR	$T_{\rightarrow GB} = \dfrac{d_{p1}}{2}\left(\dfrac{F_P + F_g + F_{fr}}{E_f}\right)$ $F_g = (W_L + W_B)\sin\gamma$ $F_{fr} = \mu(W_L + W_B)\cos\gamma$	$I_{\rightarrow GB} = I_{P1} + \dfrac{I_{P2}}{E_f}\left(\dfrac{d_{p1}}{d_{p2}}\right)^2 + \dfrac{I_{P3}}{E_f}\left(\dfrac{d_{p1}}{d_{p3}}\right)^2 + I_{L\rightarrow GB}$ $I_{L\rightarrow GB} = \dfrac{W_L + W_B}{gE_f}\left(\dfrac{d_{p1}}{2}\right)^2$	$\theta_{GB} = \dfrac{X_L}{C_{P1}}$ $\omega_{GB} = \dfrac{V_L}{C_{P1}}$ $C_{P1} = \pi d_{p1} = \dfrac{N_t}{P_G}$	Belt inertia Bearing drag If applicable: Overhanging load Braking requirements Coupling inertia
RACK AND PINION	$T_{\rightarrow GB} = \dfrac{d_{pG}}{2}\left(\dfrac{F_P + F_g + F_{fr}}{E_f}\right)$ $F_g = (W_L + W_T)\sin\gamma$ $F_{fr} = \mu(W_L + W_T)\cos\gamma$	$I_{\rightarrow GB} = I_G + I_{L\rightarrow GB}$ $I_{L\rightarrow GB} = \dfrac{W_L + W_T}{gE_f}\left(\dfrac{d_{pG}}{2}\right)^2$	$\theta_{GB} = \dfrac{X_L}{C_G}$ $\omega_{GB} = \dfrac{V_L}{C_G}$ $C_G = \pi d_{pG} = \dfrac{N_t}{P_G}$	Bearing drag Lubricant drag If applicable: Overhanging load Braking requirements Coupling inertia

VARIABLES

$E_f =$ Efficiency of device

$N_r =$ Numerical ratio of device

$\mu =$ Coefficient of friction

$\mu \cong 0.25$ for steel on steel, dry

$\mu \cong 0.15$ for steel on steel, greased

$\mu \cong 0.45$ for steel on aluminum

$d_p =$ Pitch diameter

$W =$ Weight of component

$V =$ Velocity

$P =$ Pitch of gear, screw or sprocket

$F_P =$ Push or pull load

$\gamma =$ Load angle from horizontal (deg.)

$g =$ Accel. of gravity (9.8 m/s² or 386 in/s²)

$C =$ Circumference of pulley, gear, or sprocket

$N_t =$ Number of teeth

$\theta =$ Angular position

$\omega =$ Angular velocity

$X =$ Distance traveled by load

Once the output torque required from the gearbox is known, a service factor must be applied to account for service conditions. Some typical service factors that account for run time and shock loads are listed in Table 11-38. Multiply the required torque by the service factor to determine the design torque required of the gearbox. Some manufacturers employ more complex and conservative service factors based on temperature, starting frequency, duty cycle, load type, and others. Consult with a manufacturer's representative for gearbox sizing assistance.

GEARBOX SELECTION

Gearboxes are normally selected from a catalog of commercial units. The geometry and conditions of the application will dictate the gearbox configuration, type, and reduction ratio. More stringent gearbox requirements such as high accuracy and rigidity will tend to increase the cost of the gearbox. The following criteria must be considered when selecting and sizing a gearbox:

<u>Configuration</u> of a gearbox is dictated by the application geometry. In-line and right angle gearboxes are commonly available. In-line gearboxes are much more compact than right angle gearboxes, and tend to have lower inertia. If one of the shafts must be vertical, special lubrication provisions are likely needed. Consult the manufacturer whenever faced with a vertical shaft application.

<u>Direction of rotation</u> relates to whether or not the input and output shafts rotate in the same direction. This will depend on the number of stages in the gearbox as well as the type of gear mesh.

Table 11-38: Gearbox Service Factors

Shock of Driven Machine	Service Factors											
	Electric Motor				Piston Engine or Hydraulic				1-Cyl Piston Engine			
	Running Time (hours/day)				Running Time (hours/day)				Running Time (hours/day)			
	0.5	3	8	24	0.5	3	8	24	0.5	3	8	24
I	0.5	0.8	1	1.25	0.8	1	1.25	1.5	1	1.25	1.5	1.75
II	0.8	1	1.25	1.5	1	1.25	1.5	1.75	1.25	1.5	1.75	2
III	1.25	1.5	1.75	2	1.5	1.75	2	2.25	1.75	2	2.25	2.5

I = Relati vely shock free (electric generators, conveyor screws, light elevators, sti rrers)
II = Moderate shock (heavy elvators, piston pumps, cable winches)
III = Heavy shock (punch presses, shears, mills)

Ratio is often a primary selection factor when choosing a gearbox. Ratio relates to both speed and torque. Very high ratios are achievable with multi-stage gearboxes. Ratio calculations can be found in Table 11-39.

Output torque rating is often a primary selection factor when choosing a gearbox. Gearbox torque ratings are normally based on fatigue and bearing performance characteristics. If the application runs continuously, this corresponds to the running torque. If the application follows a motion profile, some manufacturers use the root mean cube (RMC) output torque for comparison against output torque rating. Be aware that the RMC method is designed for bearing life calculations and may not apply to all gearbox components. Sometimes gearboxes are rated separately for intermittent service. Always apply the appropriate service factor when sizing a gearbox for torque. The formula for RMC torque can be found in Table 11-39.

Peak torque is the maximum torque that will ever be expected in service. Peak torque could be encountered during emergency stops or maximum acceleration. Always apply the appropriate service factor when sizing a gearbox for torque. If a peak operating torque is part of the designed standard operation cycle of the gearbox, as it may be if the motion follows a profile, a good rule of thumb is to keep that peak operating torque below the *rated output torque* of the gearbox.

Output speed and input speed are sometimes limiting factors for a gearbox in high speed applications. High speed operation can require advanced cooling techniques to handle the heat generated from friction. When used in intermittent repeating applications, the mean output speed must not exceed the rated output speed of the gearbox. Mean output speed can be calculated using the formula in Table 11-39. Some commercial gearboxes have cooling built in, which increases their speed capacity.

Gear type can influence things like transmission error, efficiency, backlash, noise, and reduction ratio. In-line gearboxes typically employ spur gears, planetary gears, cycloidal mechanisms or harmonic wave generators. Spur gear reducers tend to be the most cost effective for in-line configurations. Planetary gearboxes generally produce the highest torque in the smallest package. Cycloidal and harmonic gearboxes can offer higher ratios than most others while maintaining a fairly compact size. Right angle gearboxes typically feature worm gearing or bevel gearing. Worm gears are the most cost effective solution for right angle configurations. However, worm gear drives usually have a minimum ratio of 5:1 and lose efficiency quickly as the ratio increases. Bevel gear reducers have high efficiency, but are limited to about a 6:1 reduction ratio. Spiral bevel, hypoid, and helical bevel gearboxes offer low noise and precise transmission.

Efficiency is primarily a function of gear type, number of engagements, and load torque. It is sometimes speed dependent as well. Because efficiency is a factor in calculating input torque required to produce an output torque, high efficiency will require less drive power to achieve the same output.

Backlash must be minimized if a gearbox requires high positional accuracy and will see reversing or variable torque on either the input or output shaft. Backlash is the result of clearance in the gear mesh and can lead to premature wear, noise, and transmission error. If torque is unidirectional, backlash is only of concern during starts and stops.

Moment of inertia of a gearbox will be reflected back to the drive, which is usually a motor. The additional moment of inertia will increase the motor torque required during acceleration and deceleration. Multi-stage gearboxes tend to have a higher moment of inertia than single stage gearboxes. Sometimes it is desirable to select a gearbox based on moment of inertia in order to match the motor and its load. The equation for inertial matching can be found in Table 11-39.

Transmission error is the difference between input and output shaft position. Transmission error during normal operation is a function of gear type and assembly quality. Transmission error during changes in torque is a function of backlash and the torsional rigidity of the gearbox assembly.

Torsional rigidity is the resistance of the gearbox to torsional deflection during acceleration and deceleration. Transmission error can result from inadequate torsional rigidity.

Mounting method must be specified for both the input and output sides of the gearbox. Most gearboxes offer features such as standard adapter plates to allow the motor frame to be mounted directly onto the gearbox frame. The motor shaft is connected to the gearbox input shaft using some coupling

Figure 11-28: Gearbox Output Options
Source: WITTENSTEIN (www.wittenstein-us.com)

method, which is usually built in to the gearbox. The gearbox frame must also be securely mounted to the ground or machinery frame. Most gearboxes have precision surfaces and mounting holes to allow mounting and alignment.

Radial load rating or overhanging moment rating is the amount of radial (lateral or bending) force or moment that can be applied to the output shaft. This is a critical factor to consider when attaching overhung loads like pulleys and gears directly to a gearbox output shaft. If excessive overhanging or radial load is present, it will need to be reduced to acceptable levels through the use of external bearings to support the output shaft.

Output shaft type is often selectable. Smooth, keyed, and flanged shafts are common in industry, but many options are available.

Lubrication type and method are normally stated in catalogs. Some gearboxes are lubricated for life and sealed. Others require oil circulation. Gearbox orientation will affect lubrication effectiveness, so be sure to check with the manufacturer if orientation requirements are not clear.

Service life is often related to the life of the bearings in the gearbox as well as its load conditions. Gearboxes have a finite life, and regular maintenance is required to achieve the rated service life.

Noise is generated by gears in mesh. The operating noise of a gearbox is often given in decibels and is a function of the gear type, operating speed, and gearbox construction. Noise indicates vibration in the system, which can be a factor in sensitive equipment.

Protection class describes the unit's resistance to the ingress of solids and liquids. Protection classes are prefixed IP and have two digits. The first digit describes solids ingress protection and the second digit describes liquids ingress protection. A protection class of IP65 offers full protection against dust and low pressure liquids. This is a typical class for light industrial machinery. IP64 offers full dust protection and liquid splash protection. IP66 offers full dust protection and high pressure liquids protection.

Maximum ambient temperature is often stated in catalogs. Temperature affects the ability of the gearbox to cool itself and maintain proper lubrication. Elevated temperatures can negatively affect the load rating of the gearbox.

     Gearboxes are often sized simultaneously with motors to find an optimal combination. Some manufacturers have excellent software packages that will allow the input of motion profiles and recommend motor/gearbox combinations for a given application. Some software will take CAD assemblies as input and perform all the calculations needed to select several motor/gearbox combinations. One excellent example of this type of software is Motion Analyzer, available on the Allen Bradley (Rockwell Automation, Inc.) web site at www.ab.com/motion/software/analyzer.html

     Speaking with an applications engineer at a manufacturer or motion control distributor is often the most efficient and safest way to select and size a gearbox. The following is a basic procedure covering the fundamental selection and sizing steps:

1. Define the application parameters. These include gear ratio (Table 11-39,) required torques and speeds, geometric requirements, drive motor size and type, and any special requirements like wash-down capability.
2. Calculate the required maximum (peak) torque for the application. The formulas in Table 11-37 are helpful in calculating the torque required by various mechanical devices. Maximum torque will be required under maximum acceleration or deceleration. Table 11-39 contains the equation for acceleration torque. Manufacturers will provide

peak torque ratings for all gearboxes. The gearbox should not be run continuously with peak torque.

3. <u>Calculate the required running torque for the application.</u> If the application is continuous, the torque output rating of a gearbox will need to be greater than the running torque of the application by some safety factor. If the application is intermittent and repeating, the RMC torque (Table 11-39) is normally calculated and expected to be less than the gearbox torque rating by some factor of safety. It is best to consult the manufacturer or use appropriate software to calculate the RMC torque. Be aware that not all manufacturers will use RMC torque for gearbox sizing.

4. <u>Calculate any lateral (radial) forces or overhanging moments applied to the gearbox output shaft.</u> The manufacturer will provide maximum acceptable values for these loads.

5. <u>Determine the output speed required of the application.</u> When motion is continuous, the running speed must not exceed the rated output speed of the gearbox. When motion is intermittent and repeating, the mean output speed must not exceed the rated output speed of the gearbox. A factor of safety is recommended. Mean output speed can be calculated using the formula in Table 11-39.

6. <u>Select a gearbox based on geometric and performance parameters.</u> Performance parameters and features will be manufacturer and model specific. The basic geometric types are inline and right angle. Torque and ratio combinations will be a primary consideration. Overhanging moments and lateral loads will also be of particular concern. Output shaft configuration may also drive type selection. Service factors that may be manufacturer-specific are usually applied to the output torque requirements. Some general service factors can be found in Table 11-38. Speak with an applications engineer for more information on service factors, gear types, and other options.

7. <u>Finalize gearbox and motor selection together.</u> Inertia matching can be examined if applicable. Motor mounting options must be selected for the gearbox based on motor dimensions and are often dependent on the motor manufacturer.

Table 11-39: Gearbox Sizing Calculations

OUTPUT TORQUE N_r = Numerical gear ratio E_f = Gearbox efficiency	$$T_{out} = T_{in} N_r E_f$$
RATIO ω_{in} = Input angular velocity ω_{out} = Output angular velocity T_{out} = Output torque T_{in} = Input torque	$$N_r = \frac{\omega_{in}}{\omega_{out}} = \frac{T_{out}}{T_{in}}$$
POWER AS A FUNCTION OF TORQUE N = Shaft speed in RPM English units: T = Torque in ft-lb Metric units: T = Torque in N-m	$hp = 0.746 kW$ English units: $$hp = \frac{TN}{5252}$$ Metric units: $$kW = \frac{TN}{9550}$$
ACCELERATION TORQUE I = Mass moment of inertia of load at output shaft α = Angular acceleration (radians/sec^2) Dependent on motion profile used	$T_A = I\alpha$ (See Table 11-37 and Section 1.3 for I of various devices and loads) (See Section 1.3 for rotational acceleration equations)
RMC TORQUE n_i = Speed during cycle portion i t_i = Duration of cycle portion i T_i = Torque required during cycle portion i	$$T_{RMC} = \sqrt[3]{\frac{\sum n_i t_i T_i^3}{\sum n_i t_i}}$$
MEAN OUTPUT SPEED n_i = Speed during cycle portion i t_i = Duration of cycle portion i	$$n_m = \frac{\sum n_i t_i}{\sum t_i}$$
INERTIA MATCHING N_r = Numerical gear ratio	$$I_{motor} \cong I_{gearbox} + \frac{I_{load}}{N_r^2}$$

CRITICAL CONSIDERATIONS: Gearboxes

- Apply the proper service factors when selecting a gearbox for an application.
- Rated maximum speeds, torques, radial loads, and thrust loads generally cannot safely be applied simultaneously to a gearbox.
- For high speed applications, cooling is essential. Consider external cooling or an oil circulation system.
- Proper installation and maintenance are essential to gearbox life.
- Gearbox life is finite. Plan for service and replacement accordingly. Avoid shocks during installation and operation that could reduce life expectancy or damage the gearbox bearings.
- Gearboxes must not be operated outside their rated temperature range.
- Rotating shafts, couplings, hubs, attachments, etc., are extremely dangerous, and proper safety precautions such as guarding must be taken. See Chapter 2 for more information on machinery safety.

BEST PRACTICES: Gearboxes

- Talk to a manufacturer or motion control specialist. Proper selection and sizing is complex, and ideally involves also selecting and sizing a prime mover. Solutions will be manufacturer-specific.
- Torque increases with shaft acceleration. To reduce the torque required (and therefore gearbox cost), reduce the acceleration and deceleration of the application.
- Speed reduction and torque multiplication can be accomplished cost effectively using belt and chain drives. Consider sharing the overall reduction requirements between belt or chain drive and gearbox. Lower reduction and torque ratios can be met by a more cost effective gearbox.
- The use of a high efficiency gearbox can reduce the motor size required and save cost.
- Using a high ratio reducer can reduce the motor size required and save cost.
- Lateral (radial) forces resulting from a pulley, sheave, sprocket, or gear can be reduced by increasing the diameter of the pulley, sheave, sprocket, or gear. Be sure to account for the added weight of the larger diameter component.

Section

11.5

BELTS AND CHAINS

Belts and chains are used to transmit power from one shaft to another, or to move items along their length. Belts are often chosen over chains because they have higher top speeds, run quieter, have lower inertia, and require no lubrication. Some synchronous belts have high load carrying capacities that rival that of chains. Belts generally stretch less than chains, though pre-stretched chains are available. Chains have the advantage of having very high load carrying capacities, are repairable, and can be run in dirty conditions. Belts and chains are both more economical than gears.

RECOMMENDED RESOURCES

- Collins, Busby, Stabb, *Mechanical Design of Machine Elements and Machines,* 2nd Ed., John Wiley & Sons, Inc., Hoboken, NJ, 2010
- R. Mott, *Machine Elements in Mechanical Design,* 5th Ed., Pearson/ Prentice Hall, Inc., Upper Saddle River, NJ, 2012
- Oberg, Jones, Horton, Ryffel, *Machinery's Handbook,* 28th Ed., Industrial Press, New York, NY, 2008
- Shigley, Mischke, Brown, *Standard Handbook of Machine Design,* 3rd Ed., McGraw-Hill Inc., New York, NY, 2004
- Shigley, Mischke, Budynas, *Mechanical Engineering Design,* 7th Ed., McGraw-Hill Inc., New York, NY, 2004
- **"The Complete Guide to Chain"** Website: www.chain-guide.com
- **American Chain Association** Website: www.americanchainassn.com

DRIVE CALCULATIONS

Belt and chain drives consist of two or more wheels (pulleys, sheaves, or sprockets) set some distance apart. These wheels are often different sizes, resulting in a reduction or increase in speed of the driven shaft. The ratio between shaft speeds is known as the speed ratio. For most belt drives, a speed ratio of 6:1 or less is recommended. A ratio of 8:1 is possible, but at some risk. Double reduction drives can be employed to yield greater ratios. A basic diagram of a two-shaft belt or chain drive is shown in Figure11- 29. For a flat belt drive, the wheels on which the belt rides are called pulleys. In V-belt drives, the wheels are called sheaves. For timing belt and chain drives, the wheels are called sprockets. In most applications, there are two shafts and the drive

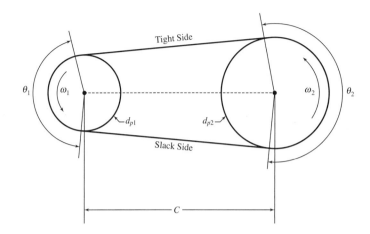

Figure 11-29: Basic Speed Reducing Belt or Chain Drive Geometry

wheel is smaller and faster than the driven wheel. This is referred to as a speed reducing drive arrangement.

Pitch diameter is defined as the diameter at which the tension carrying part of the belt or chain rides. Pitch diameter is not equal to the outside diameter of a pulley, sheave, or sprocket and is normally given in manufacturer data or calculated. The formulas in Table 11-40 can be used to determine and analyze belt and chain drive geometry. Always check all units for consistency. Unsupported span can be useful in evaluating the risk of whip or vibration.

Table 11-40: Belt and Chain Drive Formulas

d_{p1} = Pitch diameter of smaller pulley, sheave, or sprocket	ω_1 = Angular velocity of smaller wheel
d_{p2} = Pitch diameter of larger pulley, sheave, or sprocket	ω_2 = Angular velocity of larger wheel
θ_1 = Wrap angle of belt around driving wheel (radians)	C = Center distance between shafts
θ_2 = Wrap angle of belt around driven wheel (radians)	

SPEED AND DIAMETER RELATIONSHIPS To calculate speeds or pitch diameters in a compound drive (more than two shafts), see the recommended resources.	$d_{p1} = \dfrac{\omega_2 d_{p2}}{\omega_1} \qquad d_{p2} = \dfrac{\omega_1 d_{p1}}{\omega_2}$ $\omega_1 = \dfrac{\omega_2 d_{p2}}{d_{p1}} \qquad \omega_2 = \dfrac{\omega_1 d_{p1}}{d_{p2}}$
DRIVE SPEED RATIO	$SR = \dfrac{d_{p2}}{d_{p1}} = \dfrac{\omega_1}{\omega_2}$
LINEAR BELT OR CHAIN SPEED At pitch line n = RPM or RPS of shaft	$v = \dfrac{d_{p1}\omega_1}{2} = \dfrac{d_{p2}\omega_2}{2} \qquad v = \pi d_p n$

continued on next page

Table 11-40: Belt and Chain Drive Formulas (continued)

ANGLES OF WRAP On each pulley, sheave, or sprocket	$$\theta_1 = \pi - 2\sin^{-1}\left[\frac{d_{p2} - d_{p1}}{2C}\right]$$ $$\theta_2 = \pi + 2\sin^{-1}\left[\frac{d_{p2} - d_{p1}}{2C}\right]$$
BELT OR CHAIN LENGTH Based on drive geometry	$$L \cong 2C + \frac{\pi(d_{p2} + d_{p1})}{2} + \frac{(d_{p2} - d_{p1})^2}{4C}$$
DRIVE CENTER DISTANCE Based on belt length	$$C \cong \frac{B + \sqrt{B^2 - 32(d_{p2} - d_{p1})^2}}{16}$$ $$B = 4L - 6.28(d_{p2} + d_{p1})$$
UNSUPPORTED BELT OR CHAIN SPAN	$$S = \sqrt{C^2 - \left(\frac{d_{p2} - d_{p1}}{2}\right)^2}$$

Angles of wrap are very important for both friction and positive belt drives. A minimum wrap angle of 120° around the smaller pulley is recommended for friction drives to minimize slip. On positive drives, a minimum wrap angle of 90° is recommended to prevent tooth jumping and ensure that enough teeth are in engagement to carry the load. In drives with only two pulleys/sheaves/sprockets, wrap angles are generally not an issue.

11.5.1: BELTS

Three of the most commonly used types of belts are flat belts, V-belts, and timing (synchronous) belts. Flat belt drives and V-belt drives are considered "friction drives" whereas synchronous belt drives are considered "positive drives." Friction drives are subject to slip and creep, which can result in speed differences between the input and output sides of up to 2%. The power transmission capability of friction drives is limited by the friction force between the belt and pulley or sheave. The power transmission capability of positive drives is limited by the strength of the belt teeth.

Flat belts transmit power through friction against flat or crowned pulleys. Belt friction is a direct result of belt tension, which must be maintained to prevent slippage. Flat belts can be run at higher speeds than V-belts due to the fact that their low profile is less affected by centrifugal force. However, flat belts have relatively low power transmission capability compared to V-belts. Flat belts are best run at high speeds in the neighborhood of 2500 ft/min (12.7 m/s) to

7500 ft/min (38.1 m/s). Lower speeds require more belt tension, which can stress bearings and support structures. Flat belts can slip and creep during normal operation, so they cannot be relied upon for timing purposes. Many manufacturers of flat belts do not provide values for power ratings, but maximum belt speed is often given. This is because, for power transmission, V-belts are often used instead of flat belts except when high speed is of primary importance. For maximum accuracy in calculations, add one belt thickness to each pulley outside diameter. This will give each pulley a pitch diameter that passes through the belt center.

V-belts are normally the belt of choice for non-synchronous power transmission. They have a trapezoidal cross section that provides self-centering in matching grooved sheaves. V-belts are designed to be run at speeds ranging from 1500 ft/min (7.6 m/s) to 6500 ft/min (33 m/s). The most common V-belt cross sections in the United States are narrow and classical, though many manufacturers also offer metric and proprietary belts as well. When power takeoff is required on both sides of the belt (as it is in serpentine drives), a double V-belt is used. The profile of narrow and classical V-belts and their designations are shown in Figures 11-30 and 11-31. The pitch diameter of a V-belt sheave is the diameter at which the tensile elements of the belt (usually fibrous cords) ride on the sheave.

Timing belts, or synchronous belts, transmit power through precisely spaced teeth on their inside surface. These teeth mesh with toothed sprockets to provide excellent synchronization and high power transmission with low belt tensions compared to flat belts or V-belts. Timing belts are not subject to slip, but are subject to tooth shearing when overloaded. These belts can be run very slowly compared to friction drive belts, because tension is not an issue.

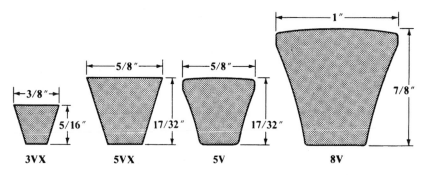

Figure 11-30: Nominal Narrow V-Belt Dimensions and Designations

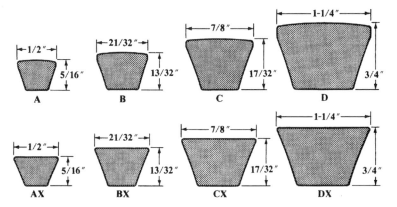

Figure 11-31: Nominal Classical V-Belt Cross Sections and Designations

Belt pitch is the distance from the center of one tooth to the center of the next. Timing belts are available with protrusions, holes, or mounting points integral to the belt for attaching carriers for moving materials. Many timing belts are proprietary, but some standard belts are also available. Standard belts governed by ANSI/RMA IP-24 are shown in Figure 11-32. The pitch diameter

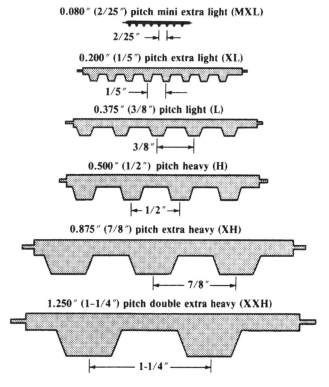

Figure 11-32: Standard Timing Belt Sections

of a timing belt sprocket is the diameter at which the belt's tensile elements (usually fibrous cords) ride on the sheave.

FLAT BELT DRIVE DESIGN & SELECTION PROCEDURE

Transmissible power of a flat belt drive is a function of friction coefficient and is calculated according to the Eytelwein's equation and the formulas in Table 11-41. Because the belt is affected by centrifugal force at the pulleys, greater speeds will require greater preloads to counteract this force and prevent slip. If objects with weight are being carried by the belt, sufficient peripheral force must be provided. Reflected torque of a conveyor belt drive on the driving shaft is detailed in Section 11.4, and can be used to calculate required drive power. When specifying a belt drive, care must be taken to account for any elevated starting and stopping torques.

Table 11-41: Flat Belt Drive Load Formulas

1: Small Pulley 2: Large Pulley	
d_p = Pitch diameter of pulley (in or mm)	F_t = Tension in tight side
μ = Coefficient of friction between belt and pulley	F_s = Tension in slack side
θ_1 = Wrap angle of belt around smaller pulley (radians)	T = Torque at drive shaft
S_f = Service factor (See Table 11-42)	

TRANSMISSIBLE POWER P = Power at shaft v = Belt linear velocity (ft/min or m/min) Check units for consistency	$P = (F_t - F_s)v$ Inch: $$hp = \frac{(F_t - F_s)v}{33000}$$ Metric: $$kW = \frac{(F_t - F_s)v}{6120}$$
DESIGN POWER SF = Safety Factor	$P_d = S_f P(SF)$
DRIVE TORQUE n = RPM or RPS of shaft	$$T_1 = \frac{P}{n_1} = (F_t - F_s)\frac{d_{p1}}{2}$$
BELT PERIPHERAL VELOCITIES **(LINEAR BELT SPEED)** v = Linear belt speed (in/time or mm/time) n = RPM or RPS of shaft Check units for consistency	$v_1 \cong v_2$ $v_1 = \pi n_1 d_{p1}$ $v_2 = \pi n_2 d_{p2}$
TENSION FORCES F_a = Allowable tension per unit belt width (catalog value) b = Belt width (in or mm)	$F_t = \dfrac{F_a b}{S_f}$ $F_s = F_t - \dfrac{P}{v}$

continued on next page

Table 11-41: Flat Belt Drive Load Formulas (Continued)

SLIP EQUATION	$$\frac{F_t - F_c}{F_s - F_c} = e^{\mu\theta_1}$$
CENTRIFUGAL HOOP TENSION v = Linear belt speed (per second) w = Belt weight per unit length g = Accel. of gravity (386 in/sec² or 9800 mm/sec²) b = Belt width (in or mm) t = Belt thickness (in or mm) γ = Belt specific weight (lbf/in³ or MPa)	$$F_c = \frac{wv^2}{g} = \frac{bt\gamma v^2}{g}$$
INITIAL TENSION	$$F_i = \frac{F_t + F_s}{2} - F_c$$
PEAK TENSION F_b = Bending Tension (function of belt construction)	$$F_{peak} = F_t + F_c + F_b$$
BENDING FREQUENCY z = Number of pulleys L = Belt length Check units for consistency	$$f_b = \frac{vz}{L}$$
FORCE EXERTED ON SHAFT F_x = Bending force on shaft along the line of centers T = Torque transmitted by belt pulley	$F_x = \sqrt{F_t^2 + F_s^2 - 2F_t F_s \cos\theta_1}$ Where $\theta_1 = \pi$ radians (180°): $$F_x = F_t + F_s \cong \frac{4T}{d_p}$$

Table 11-42: Flat Belt Service Factors

	Prime Mover		
Application Operating Conditions	**Electric Motors With Starting Torque < 1.5x Nominal Torque**	**Electric Motors With Starting Torque between 1.5x and 2.5x Nominal Torque**	**Electric Motors With Starting Torque > 2.5x Nominal Torque**
Continuous Service, Small Accelerated Masses	1.2	1.4	1.6
Interrupted Service Without Bumps, Medium-Sized Accelerated Masses	1.3	1.5	1.7
Ineterrupted Service With Bumps, Medium-Sized Accelerated Masses	1.5	1.7	1.8
Service With Severe Bumps, Large Accelerated Masses	1.6	1.8	1.9

Linear belt speed is a function of both shaft speed and pulley diameter. Because power is a function of belt speed and belt tension, an increase in belt speed (by increasing diameter, for instance) will require less belt tension for the same power output. Increasing the pulley diameter will reduce the tension requirement, which in turn reduces loads on the shafts and bearings.

Modern flat belts are typically made up of multiple layers with different characteristics. As a result, any flat belt calculations should be based on current manufacturer's data. The best way to design and size a belt drive and its components is to talk to an applications engineer at the vendor of your choice. Have the information from step 1 of the following procedure ready for discussion.

This procedure can be used when designing a flat belt drive and selecting belts:

1. <u>Determine the drive parameters.</u> These include:
   - Drive horsepower (or kW) or flat belt linear speed
   - Shaft diameters
   - Preferred speeds of both shafts
   - Preferred distance between shaft centers

   Flat belts run best at speeds ranging from 2500 ft/min (12.7 m/s) to 7500 ft/min (38.1 m/s). When selecting drive parameters from scratch, it is helpful to target a linear belt speed of 4000 ft/min (20.3 m/s) and calculate drive geometry from the belt speed using the equations in Table 11-40 and Table 11-41.

2. <u>Select a flat belt type and thickness.</u> Belt type (construction and material) is often chosen based on characteristics like coefficient of friction. Belt thickness for a particular type will typically be selected based on design power. Design power is equal to the drive horsepower (or kW) multiplied by the service factor and some safety factor. Some guideline values for flat belt service factor can be selected from Table 11-42. Consult with the manufacturer on service factors. Sometimes flat belts are rated for power or allowable tension (or stress) per unit width. In this case, calculate the tight side tension using the equations in Table 11-41 and apply the service factor as shown. The belt width b is not yet known, so leave that term in the solution for now and solve for it in subsequent steps.

3. <u>Select one pulley pitch diameter and calculate the other</u> using a manufacturer's catalog and the formulas in Table 11-40. With flat belts, add one belt thickness to each pulley diameter to calculate the pitch

diameters. If your calculated speed ratio or pulley pitch diameter is not achievable using stock pulleys, you will need to adjust your pulley speeds or diameters. Do not use a pulley that is below the minimum stock size for that belt because that will place more than the recommended bending stress on the belt.

4. <u>Select or calculate a center distance and belt length.</u> Using the manufacturer's data tables, select the manufacturer's recommended center distance closest to your desired center distance. Belt length will be given by the manufacturer for your selected center distance and pulley pitch diameters. If no data tables are available, you may calculate center distance as a function of belt length using the belt length formula in Table 11-40. This calculated or selected center distance will replace your desired center distance. Maximum center distance should be less than 20 times the pitch diameter of the small pulley. The center distance of a flat belt drive intended for high speeds should be selected such that the minimum center distance is equal to the pitch diameter of the large pulley plus half the pitch diameter of the small pulley according to the following equation:

$$d_{p2} + \frac{d_{p1}}{2} \leq C \leq 20 d_{p1}$$

For flat belts, it is often recommended that belts be specified 1% shorter than nominal to account for installation tension. Check with the manufacturer to verify this for a particular flat belt.

5. <u>Verify that the angle of wrap on the smaller pulley is greater than 120°.</u> Wrap angle can be calculated using the formulas in Table 11-40.

6. <u>Check the belt linear speed.</u> The calculated linear belt speed should not exceed the belt maximum speed listed in the manufacturer's data for that belt. Belt linear speed can be calculated using formulas from Table 11-41.

7. <u>Check the belt bending frequency.</u> The bending frequency must be within an acceptable range given by the manufacturer for the chosen belt. Bending frequency has a large impact on belt life. Bending frequency can be calculated using the formula in Table 11-41.

8. <u>Choose a belt width.</u> First determine the specific rated power per unit width for your chosen belt. This is normally done using manufacturer's data and is dependent on belt speed and belt thickness.

Based on this information, select a belt width. Select a width that yields a belt horsepower (or kW) rating that exceeds your target design power. Alternatively, belt width can be calculated based on tension requirements for power transmission using the formulas in Table 11-41. First solve for centrifugal hoop tension and leave belt width variable b in the solution. Then solve for tight side tension force and leave b in the solution. Substitute these values into the slip equation and solve for b. Choose a belt width that exceeds this value. If the calculated width is not satisfactory, recalculate based on a different belt thickness. A thicker belt will handle more stress per unit width.

9. Determine the installation and takeup allowance. The allowance, or a formula to calculate allowance, is usually provided by the manufacturer.

10. Determine the required initial belt tension. This is sometimes provided by the manufacturer and other times must be calculated using manu- facturer-supplied equations or software. A formula for initial tension can be found in Table 11-41. Static shaft load due to belt tension can then be calculated based on static belt tension and wrap angle using the formula in Table 11-41. This calculation ignores the shaft load due to pulley weight.

11. Verify that your shaft diameters are sized appropriately for your power and speed requirement. Every shaft should be designed according to professional standards to ensure safety, performance, and acceptable deflection. See Section 11.1 for information on shaft design and calcu- lation, including loads caused on shafts by belt drives. If attaching the drive pulley directly to an electric motor, it is good practice to use the NEMA standards for shaft diameter sizing. A chart of these values can be found in Figure 11-43.

12. Fit the pulleys to your shaft diameters. Some pulleys have integral hubs, but others may have additional bushings that are required to fit your shafts. Most commonly, the purchaser must machine the proper bore size and attachment features into the pulleys.

V-BELT DRIVE DESIGN AND SELECTION PROCEDURE

Transmissible power of a V-belt drive is a function of friction coefficient and is calculated according to the Eytelwein's equation and the formulas in Table 11-44. Because the belt is affected by centrifugal force at the pulleys,

Table 11-43: NEMA Shaft Sizes

Motor Frame	Shaft Diameter inches	Horsepower at Synchronous (Full Load) Speed in RPM			
		3600 (3450)	1800 (1750)	1200 (1160)	900 (870)
56	0.625	0.75 - 1	0.33 - 1	0.125 - 0.5	
143T	0.875	1.5	1	0.75	0.5
145T	0.875	2 - 3	1.5 - 2	1	0.75
182T	1.125	3 - 5	3	1.5	1
184T	1.125	5 - 7.5	5	2	1.5
213T	1.375	7.5 - 10	7.5	3	2
215T	1.375	10 - 15	10	5	3
254T	1.625	15 - 20	15	7.5	5
256T	1.625	20 - 25	20	10	7.5
284T	1.875		25	15	10
286T	1.875		30	20	15
324T	2.125		40	25	20
326T	2.125		50	30	25
364T	2.375		60	40	30
365T	2.375		75	50	40
404T	2.875		100	60	50
405T	2.875		100 - 125	75	60
444T	3.375		125 - 150	100	75
445T	3.375		150 - 200	125	100

Table 11-44: V-Belt Drive Load Formulas

1: Small Sheave 2: Large Sheave d_p = Pitch diameter of sheave (in or mm) μ = Coefficient of friction between belt and sheave θ_1 = Wrap angle of belt around smaller sheave (radians)	F_t = Tension in tight side F_s = Tension in slack side T = Torque at drive shaft
TRANSMISSIBLE POWER P = Power at shaft v = Belt linear velocity (ft/min or m/min) Check units for consistency	$P = (F_t - F_s)v$ Inch: $$hp = \frac{(F_t - F_s)v}{33000}$$ Metric: $$kW = \frac{(F_t - F_s)v}{6120}$$
DESIGN POWER SF = Safety factor	$P_d = S_f P(SF)$
CORRECTED BELT POWER C_L = Length correction factor (Table 11-51 or 11-54) C_A = Angle correction factor (Table 11-52) P_R = rated power for the belt	$P_c = C_L C_A P_R$

continued on next page

Table 11-44: V-Belt Drive Load Formulas (Continued)

NUMBER OF BELTS REQUIRED	$\#B = \dfrac{P_d}{P_c}$
DRIVE TORQUE n = RPM or RPS of shaft	$T_1 = \dfrac{P}{n_1}$
BELT PERIPHERAL VELOCITIES **(LINEAR BELT SPEED)** v = belt speed (in/time or mm/time) n = RPM or RPS of shaft Check units for consistency	$v_1 \cong v_2$ $v_1 = \pi n_1 d_{p1} \qquad v_2 = \pi n_2 d_{p2}$
TENSION FORCES F_a = Allowable tension in belt (catalog value) S_f = Service factor (See Table 11-46)	$F_t = \dfrac{F_a}{S_f} \qquad F_s = F_t - \dfrac{P}{v}$
SLIP EQUATION β = Sheave contact wedge angle (Usually 30° to 38° — see catalog)	$\dfrac{F_t - F_c}{F_s - F_c} = e^{\left[\dfrac{\mu \theta_1}{\sin(\beta/2)} \right]}$
CENTRIFUGAL HOOP TENSION v = Linear belt speed (ft/min) K_c = See Table 11-45 w = Belt weight per unit length g = Accel. of gravity (32.2 ft/sec² or 9.8 m/sec²) Check all units for consistency	$F_c = K_c \left(\dfrac{v}{1000} \right)^2$ $F_c = \dfrac{wv^2}{g}$
INITIAL TENSION	$F_i = \dfrac{F_t + F_s}{2} - F_c$
PEAK TENSION F_b = Bending Tension (function of belt construction)	$F_{peak} = F_t + F_c + F_b$
BENDING FREQUENCY z = Number of sheaves L = Belt length Check units for consistency	$f_b = \dfrac{vz}{L}$
FORCE EXERTED ON SHAFT F_x = Bending force on shaft along the line of centers T = Torque transmitted by belt sheave	$F_x = \sqrt{F_t^2 + F_s^2 - 2F_t F_s \cos \theta_1}$ Where $\theta_1 = 180°$: $F_x = F_t + F_s \cong \dfrac{3T}{d_p}$

greater speeds will require greater preloads to counteract this force and prevent slip. When specifying a belt drive, care must be taken to account for any elevated starting and stopping torques.

Linear belt speed is a function of both shaft speed and sheave diameter. Because power is a function of belt speed and belt tension, an increase in belt speed (by increasing diameter, for instance) will require less belt tension for the same power output. Increasing the sheave diameter will reduce the tension requirement, which in turn reduces loads on the shafts and bearings.

Table 11-45: V-Belt Hoop Tension Variable K_C

V-Belt Section	K_c
A	0.561
B	0.965
C	1.716
D	3.498
E	5.041
3V	0.425
5V	1.217
8V	3.288

Table 11-46: V-Belt Service Factors

		Driver					
		Reasonably Smooth			Coarse		
		Operating Period					
	Driven Machine Examples	< 8 hours	< 16 hours	> 16 hours	< 8 hours	< 16 hours	> 16 hours
Smooth	Smooth Machine Tools, Liquids Agitators, Blowers, Light Duty Conveyors, Centrifugal Pumps, Fans Up To 10hp	1.0	1.1	1.2	1.1	1.2	1.3
Light Shock	Fans Over 10hp, Uneven Machine Tools, Belt Conveyors, Line Shafts, Generators, Dough Mixers	1.1	1.2	1.3	1.2	1.3	1.4
Moderate Shock	Piston Compressors, Piston Pumps, Woodworking Machinery, Bucket Elevators, Pulverizers	1.2	1.3	1.4	1.4	1.5	1.6
Heavy Shock	Hoists, Excavators, Crushers, Shears and Heavy Presses	1.3	1.4	1.5	1.5	1.6	1.8

Many manufacturers provide software to assist in drive design and component selection. Two excellent examples of drive design and selection software can be found at www.gates.com and www.emerson-ept.com. You must register at both sites to access the online design tools. Another easy way to size a belt drive and select components is to talk to an applications engineer at the vendor of your choice. Have the information from step 1 of the following procedure ready for discussion.

This procedure can be used when designing a V-belt drive and selecting belts:

1. <u>Determine the drive parameters.</u> These include:
   - Drive horsepower (or kW)
   - Shaft diameters
   - Preferred speeds of both shafts

V-belts run best at speeds ranging from 1500 ft/min (7.6 m/s) to 6500 ft/min (33 m/s). When selecting drive parameters from scratch, it is helpful to target a linear belt speed of 4000 ft/min (20.3 m/s) and calculate drive geometry from the belt speed equations in Table 11-40 or 11-44.

2. Calculate the design power. Design power is equal to the drive horsepower (or kW) multiplied by the service factor and some safety factor. V-belt service factor can be selected from Table 11-46.

3. Select a V-belt cross section. V-belt cross section is chosen primarily according to design horsepower (or kW) rating. The horsepower ratings of narrow and classical V-belts are given in Figures 11-33 and 11-34. Horsepower ratings for these belts can also be calculated using the

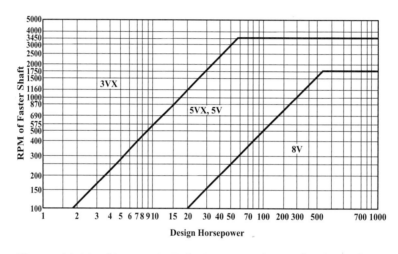

Figure 11-33: Narrow V-Belt Cross Section Selection Chart

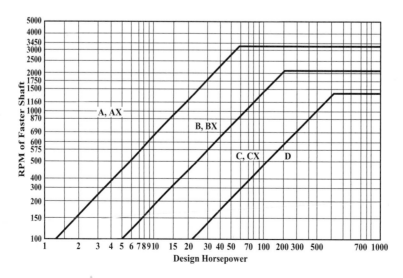

Figure 11-34: Classical V-Belt Cross Section Selection Chart

equations in Table 11-48. For other V-belt cross sections, consult the manufacturer's data. Narrow V-belts are a good choice for compact drive arrangements, and some are capable of transmitting as much or more power than classical belts. The standard power ratings for V-belts are based on a drive configuration with wrap angles of 180°, which is not usually the case. The power rating will be adjusted later based on drive geometry factors, so the selection is preliminary at this time.

Table 11-47: NEMA Minimum Sheave Sizes

	Horsepower At				Belt Sheave, Inches			
	Synchronous Speed (RPM)				A, B, C, D, E Sections		3V, 5V, 8V Sections	
Motor	3600	1800	1200	900	Minimum Pitch Diameter	Maximum Width	Minimum Outside Diameter	Maximum Width
143T	1.5	1	0.75	0.5	2.2	4.25	2.2	2.25
145T	2 - 3	1.5 - 2	1	0.75	2.4	4.25	2.4	2.25
182T	3	3	1.5	1	2.4	5.25	2.4	2.75
	5				2.6	5.25	2.4	2.75
184T			2	1.5	2.4	5.25	2.4	2.75
	5				2.6	5.25	2.4	2.75
	7.5	5			3	5.25	3	2.75
213T	7.5 - 10	7.5	3	2	3	6.5	3	3.375
215T	10		5	3	3	6.5	3	3.375
	15	10			3.8	6.5	3.8	3.375
254T	15		7.5	5	3.8	7.75	3.8	4
	20	15			4.4	7.75	4.4	4
256T	20 - 25		10	7.5	4.4	7.75	4.4	4
		20			4.6	7.75	4.4	4
284T			15	10	4.6	9	4.4	4.625
		25			5	9	4.4	4.625
286T		30	20	15	5.4	9	5.2	4.625
324T		40	25	20	6	10.25	6	5.25
326T		50	30	25	6.8	10.25	6.8	5.25
364T			40	30	6.8	11.5	6.8	5.875
		60			7.4	11.5	7.4	5.875
365T			50	40	8.2	11.5	8.2	5.875
		75			9	11.5	8.6	5.875
404T			60		9	14.25	8	7.25
				50	9	14.25	8.4	7.25
		100			10	14.25	8.6	7.25
405T			75	60	10	14.25	10	7.25
		100			10	14.25	8.6	7.25
		125			11.5	14.25	10.5	7.25
444T			100		11	16.75	10	8.5
				75	10.5	16.75	9.5	8.5
		125			11	16.75	9.5	8.5
		150					10.5	8.5
445T			125		12.5	16.75	12	8.5
				100	12.5	16.75	12	8.5
		150					10.5	8.5
		200					13.2	8.5

4. <u>Select the sheave pitch diameters.</u> Manufacturers often provide data tables that relate possible sheave pitch diameter combinations to speed ratio and horsepower (or kW) for each belt cross section. If tables are not provided, you must select a drive sheave pitch diameter and calculate the pitch diameter of the other sheave using the formulas in Table 11-40. Select a sheave closest to that calculated value, and then recalculate shaft speeds and drive geometry based on its actual pitch diameter. Do not use a sheave that is below the minimum stock size for that belt because that will place more than the recommended bending stress on the belt. When selecting between possible pitch diameters and using a direct motor drive, stay above the NEMA minimum sheave pitch diameter for that drive horsepower and speed. These minimums are set in NEMA standard MG1-14.42. NEMA minimums should be provided by the belt manufacturer, and are listed in Table 11-43. These minimums are selected to keep the maximum belt stress due to bending within an acceptable range. When selecting V-belt sheaves, you must select sheaves that are matched to the V-belt cross section.

5. <u>Select a center distance and belt length.</u> Using the manufacturer's data tables, select the manufacturer's recommended center distance closest to your desired center distance. Belt length will be normally given by the manufacturer for your selected center distance and sheave pitch diameters. If no data tables are available, you may calculate center distance as a function of belt length using the belt length formula found in Table 11-40. This calculated or selected center distance will replace your preliminary center distance. Nominal performance will result from drive geometry that follows the relationship:

$$d_{p2} < C < 3(d_{p2} + d_{p1})$$

Maximum center distance should be less than 20 times the pitch diameter of the small sheave. The center distance of a V-belt drive intended for high speeds should be selected such that the minimum center distance is equal to the pitch diameter of the large sheave plus half the pitch diameter of the small sheave according to the following equation:

$$d_{p2} + \frac{d_{p1}}{2} \leq C \leq 20 d_{p1}$$

6. <u>Verify that the angle of wrap on the smaller sheave is greater than 120°.</u> Wrap angle can be calculated using the formulas in Table 11-40.

7. Check the belt bending frequency. For narrow V-belts, a bending frequency below 60 per second is acceptable. For classical V-belts, a bending frequency below 30 per second is acceptable. Bending frequency has a large impact on belt life. Bending frequency can be calculated using the formula in Table 11-44.

8. Determine how many belts are required for the drive. First find the rated horsepower (or kW) for the belt cross section using either manufacturer's data or the equations in Table 11-48. If a rated horsepower value is found on a chart, first multiply the length and arc correction factors CL and CA by the rated horsepower (or kW) for that belt following the equation for corrected power in Table 11-44. The correction factors take into account the drive geometry determined in the previous steps. Then divide the drive design power by the corrected belt power rating to get the number of belts required for that drive. If the rated horsepower was calculated rather than retrieved from a catalog, the correction factors have already been taken into account. If calculations indicate the need for multiple belts, the selection process can be restarted using a heavier belt cross section to attempt to return to a single-belt configuration.

9. Determine the installation and takeup allowance. The allowance, or a formula to calculate allowance, is usually provided by the manufacturer.

10. Determine the required static belt tension. This is sometimes provided by the manufacturer and other times must be calculated using manufacturer-supplied equations or software. Static shaft load due to belt tension can then be calculated based on static belt tension and wrap angle using the formula in Table 11-44. This calculation ignores the shaft load due to sheave weight.

11. Verify that your shaft diameters are sized appropriately for your power and speed requirement. Every shaft should be designed according to professional standards to ensure safety, performance, and acceptable deflection. See Section 11.1 for information on shaft design and calculation, including loads caused on shafts by belt drives. If attaching the drive sheave directly to an electric motor, it is good practice to use the NEMA standards for shaft diameter sizing. A chart of these values can be found in Table 11-43.

12. Fit the sheaves to your shaft diameters. Some sheaves have integral hubs, but others may have additional bushings that are required to fit your shafts. Most commonly, the purchaser must machine the proper bore size and attachment features into the sheaves.

Table 11-48: V-Belt Horsepower Rating Formulas

NARROW V-BELT HORSEPOWER RATING FOMULA

HP = Horsepower rating of belt $\quad$ d_p= Pitch diameter of smaller sheave, inches

r = RPM of faster shaft divided by 1000 $\quad$ K_1, K_2, K_3, K_4 = Cross section parameters from Table 11-49

K_{SR} = Speed ratio correction factor from Table 11-50

C_L = Length correction factor from Table 11-51

C_A = Arc of contact correction factor from Table 11-52

$$HP = C_L C_A \left\{ d_p r \left[K_1 - \frac{K_2}{d_p} - K_3(d_p r)^2 - K_4 \log(d_p r) \right] + K_{SR} r \right\}$$

CLASSICAL V-BELT HORSEPOWER RATING FORMULAS

HP = Horsepower rating of belt $\quad$ d_p= Pitch diameter of smaller sheave, inches

r = RPM of faster shaft divided by 1000 $\quad$ K_{SR} = Speed ratio correction factor from Table 11-53

C_L = Length correction factor from Table 11-54

C_A = Arc of contact correction factor from Table 11-52

Type A:

$$HP = C_L C_A \left\{ d_p r \left[1.004 - \frac{1.652}{d_p} - 1.547 \times 10^{-4}(d_p r)^2 - 0.2126 \log(d_p r) \right] + 1.652 r \left(1 - \frac{1}{K_{SR}} \right) \right\}$$

Type AX:

$$HP = C_L C_A \left\{ d_p r \left[1.462 - \frac{2.239}{d_p} - 2.198 \times 10^{-4}(d_p r)^2 - 0.4238 \log(d_p r) \right] + 2.239 r \left(1 - \frac{1}{K_{SR}} \right) \right\}$$

Type B:

$$HP = C_L C_A \left\{ d_p r \left[1.769 - \frac{4.372}{d_p} - 3.081 \times 10^{-4}(d_p r)^2 - 0.3658 \log(d_p r) \right] + 4.372 r \left(1 - \frac{1}{K_{SR}} \right) \right\}$$

Type BX:

$$HP = C_L C_A \left\{ d_p r \left[2.051 - \frac{3.532}{d_p} - 3.097 \times 10^{-4}(d_p r)^2 - 0.5735 \log(d_p r) \right] + 3.532 r \left(1 - \frac{1}{K_{SR}} \right) \right\}$$

Type C:

$$HP = C_L C_A \left\{ d_p r \left[3.325 - \frac{12.07}{d_p} - 5.828 \times 10^{-4}(d_p r)^2 - 0.6886 \log(d_p r) \right] + 12.07 r \left(1 - \frac{1}{K_{SR}} \right) \right\}$$

Type CX:

$$HP = C_L C_A \left\{ d_p r \left[3.272 - \frac{6.655}{d_p} - 5.298 \times 10^{-4}(d_p r)^2 - 0.8637 \log(d_p r) \right] + 6.655 r \left(1 - \frac{1}{K_{SR}} \right) \right\}$$

Type D:

$$HP = C_L C_A \left\{ d_p r \left[7.16 - \frac{43.21}{d_p} - 1.384 \times 10^{-3}(d_p r)^2 - 1.454 \log(d_p r) \right] + 43.21 r \left(1 - \frac{1}{K_{SR}} \right) \right\}$$

Table 11-49: Narrow V-Belt Cross Section Parameters

Cross Section	K_1	K_2	K_3	K_4
3VX	1.1691	1.5295	0.00015229	0.15960
5VX	3.3038	7.7810	0.00036432	0.43343
5VX	3.3140	10.1230	0.00058758	0.46527
8V	8.6628	49.3230	0.00015804	1.16690

Table 11-50: Narrow V-Belt Speed Ratio Correction Factor

Speed Ratio Range	K_{SR} Cross Section	
	3VX	5VX
1.00 - 1.01	0.0000	0.0000
1.02 - 1.03	0.0157	0.0801
1.04 - 1.06	0.0315	0.1600
1.07 - 1.09	0.0471	0.2398
1.10 - 1.13	0.0629	0.3201
1.14 - 1.18	0.0786	0.4001
1.19 - 1.25	0.0944	0.4804
1.26 - 1.35	0.1101	0.5603
1.36 - 1.57	0.1259	0.6405
1.57 +	0.1416	0.7202

Speed Ratio Range	K_{SR} Cross Section	
	5V	8V
1.00 - 1.01	0.0000	0.0000
1.02 - 1.05	0.0963	0.4690
1.06 - 1.11	0.2623	1.2780
1.12 - 1.18	0.4572	2.2276
1.19 - 1.26	0.6223	3.0321
1.27 - 1.38	0.7542	3.6747
1.39 - 1.57	0.8833	4.3038
1.58 - 1.94	0.9941	4.8438
1.95 - 3.38	1.0830	5.2767
3.38 +	1.1471	5.5892

Table 11-51: Narrow V-Belt Length Correction Factor

Belt Length (Inch)	C_L			Belt Length (Inch)	C_L		
	Cross Section				Cross Section		
	3V	5V	8V		3V	5V	8V
250	0.83			1180	1.12	0.99	0.89
265	0.84			1250	1.13	1.00	0.90
280	0.85			1320	1.14	1.01	0.91
300	0.86			1400	1.15	1.02	0.92
315	0.87			1500		1.03	0.93
335	0.88			1600		1.04	0.94
355	0.89			1700		1.05	0.94
375	0.90			1800		1.06	0.95
400	0.92			1900		1.07	0.96
425	0.93			2000		1.08	0.97
450	0.94			2120		1.09	0.98
475	0.95			2240		1.09	0.98
500	0.96	0.85		2360		1.10	0.99
530	0.97	0.86		2500		1.11	1.00
560	0.98	0.87		2650		1.12	1.01
600	0.99	0.88		2800		1.13	1.02
630	1.00	0.89		3000		1.14	1.03
670	1.01	0.90		3150		1.15	1.03
710	1.02	0.91		3350		1.16	1.04
750	1.03	0.92		3550		1.17	1.05
800	1.04	0.93		3750			1.06
850	1.06	0.94		4000			1.07
900	1.07	0.95		4250			1.08
950	1.08	0.96		4500			1.09
1000	1.09	0.96	0.87	4750			1.09
1060	1.10	0.97	0.88	5000			1.10
1120	1.11	0.98	0.88	...			...

Table 11-52: V-Belt Arc of Contact Correction Factor

Arc of Contact on Small Sheave	C_A	Arc of Contact on Small Sheave	C_A
degrees		degrees	
180	1.00	133	0.87
174	0.99	127	0.85
169	0.97	120	0.82
163	0.96	113	0.80
157	0.94	106	0.77
151	0.93	99	0.73
145	0.91	91	0.70
139	0.89	83	0.65

Table 11-53: Classical V-Belt Speed Ratio Correction Factor

Speed Ratio Range	K_{SR}	Speed Ratio Range	K_{SR}
1.00 - 1.01	1.0000	1.15 - 1.20	1.0586
1.02 - 1.04	1.0112	1.21 - 1.27	1.0711
1.05 - 1.07	1.0226	1.28 - 1.39	1.0840
1.08 - 1.10	1.0344	1.40 - 1.64	1.0972
1.11 - 1.14	1.0463	1.64 +	1.1106

Table 11-54: Classical V-Belt Length Correction Factor

Belt Length (Inch)	C_L Cross Section A, AX	B, BX	C, CX	D
26	0.78			
31	0.82			
35	0.85	0.80		
38	0.87	0.82		
42	0.89	0.84		
46	0.91	0.86		
51	0.93	0.88	0.80	
55	0.95	0.89	...	
60	0.97	0.91	0.83	
68	1.00	0.94	0.85	
75	1.02	0.96	0.87	
80	1.04	...	...	
81	...	0.98	0.89	
85	1.05	0.99	0.90	
90	1.07	1.00	0.91	
96	1.08	...	0.92	
97	...	1.02	...	
105	1.10	1.03	0.94	
112	1.12	1.05	0.95	
120	1.13	1.06	0.96	0.88
128	1.15	1.08	0.98	0.89
144		1.10	1.00	0.91
158		1.12	1.02	0.93
173		1.14	1.04	0.94
180		1.15	1.05	0.95
195		1.17	1.08	0.96
210		1.18	1.07	0.98
240		1.22	1.10	1.00
270		1.24	1.13	1.02
300		1.27	1.15	1.04
330			1.17	1.06
360			1.18	1.07
390			1.20	1.09
420			1.21	1.10
480				1.13
540				1.15
600				1.17
660				1.18

SYNCHRONOUS DRIVE DESIGN AND SELECTION PROCEDURE

The relationships in Table 11-55 can be used to select and size a synchronous belt drive. Because most timing belts are proprietary, the selection will be based primarily on manufacturer's data for power and torque ratings. When specifying a belt drive, care must be taken to account for any elevated starting and stopping torques.

Many manufacturers provide software to assist in drive design and component selection. One excellent example of synchronous drive design software can be found at www.brecoflex.com under "Calculation Program." Probably the best way to size a drive and select components is to consult an applications engineer at the vendor of your choice. Have the information from step 1 in the following procedure ready for discussion.

This procedure can be used when designing a timing belt drive and sizing the components:

1. <u>Determine the drive parameters</u>. These are needed to size the drive components and are as follows:
   - Drive horsepower (or kW)
   - Shaft diameters
   - Preferred speeds of both shafts
   - Preferred distance between sprocket centers (if known)

   Reflected torque of a synchronous belt drive is detailed in Section 11.4, and can be used to calculate required drive power. Maximum center distance should be less than 20 times the pitch diameter of the small sprocket. The center distance of a timing belt drive intended for high speeds should be selected such that the minimum center distance is equal to the pitch diameter of the large sprocket plus half the pitch diameter of the small sprocket according to the following equation:

$$d_{p2} + \frac{d_{p1}}{2} \leq C \leq 20d_{p1}$$

2. <u>Calculate the design power of the drive</u>. Design power is equal to service factor multiplied by drive horsepower (or kW). Service factor can be taken from the chart in Table 11-56.
3. <u>Select the belt pitch</u>. Belt pitch is normally selected from a chart based on shaft speed and design horsepower (or kW). The faster shaft is used to make the selection. Belt pitch is related to power because the belt

Table 11-55: Synchronous Belt Equations

d_{p1} = pitch diameter of smaller sprocket	ω_1 = angular velocity of smaller sprocket
d_{p2} = pitch diameter of larger sprocket	ω_2 = angular velocity of larger sprocket
θ_1 = wrap angle of belt around driving sprocket	C = center distance between shafts
θ_2 = wrap angle of belt around driven sprocket	p = belt pitch
n_{t1} = number of teeth on smaller sprocket	
n_{t2} = number of teeth on larger sprocket	

DESIGN POWER SF = Safety factor S_f = Service factor (Table 11-56)	$P_d = S_f P(SF)$
SPEED AND DIAMETER RELATIONSHIPS To calculate speeds or pitch diameters in a compound drive (more than two shafts), see the recommended resources.	$d_{p1} = \dfrac{\omega_2 d_{p2}}{\omega_1} \quad d_{p2} = \dfrac{\omega_1 d_{p1}}{\omega_2}$ $\omega_1 = \dfrac{\omega_2 d_{p2}}{d_{p1}} \qquad \omega_2 = \dfrac{\omega_1 d_{p1}}{d_{p2}}$
DRIVE SPEED RATIO	$SR = \dfrac{d_{p2}}{d_{p1}} = \dfrac{\omega_1}{\omega_2} = \dfrac{n_{t1}}{n_{t2}}$
PITCH DIAMETER OF SPROCKET	$d_p = \dfrac{p n_t}{\pi}$
BELT LENGTH Based on drive geometry	$L \cong 2C + \dfrac{\pi(d_{p2} + d_{p1})}{2} + \dfrac{(d_{p2} - d_{p1})^2}{4C}$ $L \cong 2C + \dfrac{p}{2}(n_{t1} + n_{t2}) + \dfrac{1}{4C}\left[\dfrac{(n_{t2} - n_{t1})p}{\pi}\right]^2$
ANGLES OF WRAP On each sprocket	$\theta_1 = 180° - 2\sin^{-1}\left[\dfrac{d_{p2} - d_{p1}}{2C}\right]$ $\theta_2 = 180° + 2\sin^{-1}\left[\dfrac{d_{p2} - d_{p1}}{2C}\right]$
BENDING FREQUENCY z = Number of sprockets L = Belt length Check units for consistency	$f_b = \dfrac{vz}{L}$
TENSION FORCES F_t = Tension in tight side of belt F_s = Tension in tight side of belt F_a = Allowable tension in belt (catalog value) S_f = Service factor (See Table 11-56)	$F_t = \dfrac{F_a}{S_f} \qquad F_s = F_t - \dfrac{2T}{d_p}$

continued on next page

Table 11-55: Synchronous Belt Equations (Continued)

INITIAL TENSION	Some manufacturers: $$F_i \geq \dfrac{T}{d_p}$$
PEAK TENSION F_b = Bending Tension (function of belt construction)	$$F_{peak} = F_t + F_c + F_b$$
FORCE EXERTED ON SHAFT F_x = Bending force on shaft along the line of centers F_t = Tension in tight side of belt F_s = Tension in tight side of belt	$$F_x = \sqrt{F_t^2 + F_s^2 - 2F_tF_s \cos\theta_1}$$ Where $\theta_1 = 180°$: $$F_x = F_t + F_s$$

Table 11-56: Synchronous Belt Service Factors

	Driver					
	Smooth			Coarse		
	Operating Period					
Driven Machine	< 8 hours	< 16 hours	> 16 hours	< 8 hours	< 16 hours	> 16 hours
Instrumentation, Dispensing Equipment, Medical Equipment	1	1.2	1.4	1.2	1.4	1.6
Domestic Appliances, Office Equipment	1.2	1.4	1.6	1.4	1.6	1.8
Light Conveyors, Wood Lathes, Band Saws	1.4	1.5	1.7	1.5	1.7	1.9
Liquids Agitators, Dough Mixers, Light Metalworking Equipment	1.4	1.6	1.8	1.6	1.8	2
Agitators for Semi-Liquids, Line Shafts, Heavy Bulk Conveyors, Metalworking Equipment, Centrifugal Pumps	1.5	1.7	1.9	1.7	1.9	2.1
Bucket Conveyors, Fans, Blowers, Generators, Hoists, Saw Mill Machinery	1.6	1.8	2	1.8	2	2.2
Hammer Mills, Centrifuges, Flight Conveyors and Screw Conveyors	1.7	1.9	2.1	1.9	2.1	2.3
Positive Blowers, Pulverizers, Fans	1.8	2	2.2	2	2.2	2.4
Reciprocating Pumps and Compressors	1.9	2.1	2.3	2.1	2.3	2.6

transmits power through its teeth, and the teeth in contact with the sprocket share the loading. A larger pitch will result in fewer teeth on the sprocket. Some manufacturers will also provide a sprocket selection for the faster shaft as part of this step. Sprocket selection is normally based on the number of teeth. Do not use a sprocket that is below the minimum stock size for that belt because that will place more than the recommended bending stress on the belt, and result in less than the required number of teeth in engagement with the sprocket to carry the load.

4. Find a sprocket combination that results in the desired speed ratio calculated using the formulas in Table 11-55. If the calculated speed ratio does not exactly match what is available with selected components, revise the shaft speeds to correct the speed ratio. Always select sprockets with a tooth profile and pitch that matches the chosen timing belt.

5. Select the center distance and belt length. This can often be done using manufacturer's tables. For a given sprocket combination, select the center distance that is closest to your desired center distance. This new center distance will replace your preferred distance, and a belt length will be given. An alternative method is to calculate the center distance for a given belt length and sprocket pitch diameters. This calculation is performed using the formula for timing belt length (Table 11-55) and solving for center distance.

6. Select a belt width. This will be done on the basis of horsepower (or kW) rating for that belt pitch and sprocket size. Manufacturer's tables will be required, and usually include shaft speed, sprocket size, and horsepower (or kW) rating. If the required speed is not given in the table, interpolation will be necessary. Some manufacturers provide formulas for calculating peripheral force on the belt that are used in this step. The results should be checked against manufacturer's tables to verify that the chosen belt has a load rating that meets or exceeds the requirements. If the chosen belt does not, a wider belt or more than one belt may be required.

7. Verify the belt width based on the drive geometry. Apply any relevant power correction factors if given by the manufacturer for the chosen belt and drive geometry. Multiply the correction factors by the rated power determined in the last step. If the result exceeds your design power, more than one belt or a wider belt will be required.

8. Determine the installation and takeup distance. This is usually specified by the manufacturer.

9. Calculate the belt tensioning requirements. The tensioning requirements will vary with service factor. Tensioning formulas will be given by the manufacturer.

10. Verify that your shaft diameters are sized appropriately for your power and speed requirement. Every shaft should be designed according to professional standards to ensure safety, performance,

and acceptable deflection. See Section 11.1 for information on shaft design and calculation, including loads caused on shafts by belt drives.

11. Fit the sprockets to your shaft diameters. Sprockets can have integral hubs or may have separate bushings that are required to adapt the shaft to the stock sprocket. It is most common to purchase sprockets and then machine the proper bores into them.

CRITICAL CONSIDERATIONS: Belts

- V-belt and timing belt pulleys, sheaves, and sprockets must be mounted so that they are in the same plane. Misalignment of belt sheaves or pulleys can cause tracking problems or edge wear for the belt. Some flat belt configurations can have pulleys out of plane.
- Belts have damping capability and can protect equipment from shock loads and vibration.
- Flat belts and V-belts slip and creep. Angular position error will be present between the driving and driven shafts.
- V-belts require periodic re-tensioning.
- Keep belts away from elevated temperatures and direct sunlight.
- Belts have a limited shelf life. Check with the manufacturers and track the shelf time of all spare belts.
- Belts have finite life spans in operation. Design all belt-driven equipment to facilitate easy replacement of belts.
- Selecting the largest feasible sheaves or pulleys will reduce the required tension and minimize bearing stresses.
- Use appropriate factors of safety when sizing belt drive components.
- Belt drives are generally more economical than gear drives.
- Rotating shafts, hubs, pulleys, sheaves, sprockets, etc., and moving belts are extremely dangerous, and proper safety precautions such as guarding must be taken. See Chapter 2 for more information on machinery safety.

BEST PRACTICES: Belts

ALL BELTS
- Belt drives should be designed with an adjustable center distance to facilitate installation and allow proper tensioning. If center distance cannot be adjusted, the use of a movable idler to provide tension is recommended. If possible, place an idler on the inside of the belt near the larger pulley or sheave.
- A spring loaded belt tensioning system is recommended in cases where belt stretch is likely to be significant. Tension should be checked shortly after initial installation and regularly thereafter.
- Idlers should be placed on the slack side of the drive. Idlers should be as large as possible.
- Belt life will be better if idlers are placed such that the belt does not undergo reversing bending.
- For applications requiring two belts running in parallel, consider a matched pair. These are matched for length and should tension evenly.

FRICTION DRIVE BELTS
- It is recommended that the center distance between shafts in a friction drive be less than 15 to 20 times the pitch diameter of the smaller drive pulley or sheave.
- Friction drive belt and pulley (or sheave) combinations should be chosen that result in a belt wrap angle around the smallest drive pulley of greater than 120°. If this minimum is not met due to the geometry of the drive, an idler placed to increase wrap angle is recommended. Idlers do not have a minimum wrap angle.

V-BELTS
- Nominal performance will result from drive geometry that follows the relationship:

$$d_{p2} < C < 3(d_{p2} + d_{p1})$$

FLAT BELTS
- The speed ratio of a flat belt drive should generally be limited to 1:5.
- Flat belt velocity should be as high as possible, allowing selection of the narrowest belt possible. This reduces shaft loads.

continued on next page

BEST PRACTICES: Belts (Continued)

- Crowned pulleys are recommended to keep the belt centered on the pulley, and misalignment of the pulleys in the same plane must be avoided. For speed ratios higher than 3, the smaller pulley may be cylindrical (not crowned.)
- It is recommended that the largest pulleys rotating in the same direction be crowned, and all others be flat.

TIMING BELTS

- Timing belt and sprocket combinations should be chosen that result in a belt wrap angle around the smallest drive pulley of greater than 90°, and greater than 120° for high speed operation. If this minimum is not met due to the geometry of the drive, an idler placed to increase wrap angle is recommended. Idlers do not have a minimum wrap angle.
- The use of at least one flanged sprocket is recommended in any synchronous belt drive to prevent the belt from tracking sideways off the sprockets.
- If an idler is placed on the inside of the belt, the idler should have teeth. If placed on the outside of the belt, the idler should not have teeth.
- Timing belt tension should be kept as low as possible to prevent vibration and stress on the bearings. Timing belts require much lower initial tension than friction drive belts.
- Even "standard" timing belt profiles can vary depending on manufacturer. It is safest to purchase the belt and sprockets from the same manufacturer.
- For critical applications, select a truly endless belt rather than a spliced belt. Truly endless belts have a higher torque rating. Spliced belts do not always tension evenly and the tooth profile at the splice is generally imperfect.
- Sprockets should be selected to have an odd or prime number of teeth to ensure even wear of the belt teeth.
- For applications requiring two belts running in parallel, consider a matched pair, which are usually split from a single belt. This will ensure not only even tensioning, but also matching tooth profile and position.

11.5.2: CHAINS

Chains transmit power through precisely spaced links which engage with teeth on sprockets. Because the links positively engage the sprockets, chains require less tensioning than friction drives. One of the most common types of chains used in machinery design is the roller transmission chain. This type of chain has evenly spaced rollers which engage the sprockets. The ability of the rollers to rotate greatly reduces wear on both chain and sprockets. The main drawbacks of chains when compared to belts are their need for lubrication and tendency to wear over time. Chains and sprockets both experience wear and need periodic replacement. One advantage of chains is their high load carrying capacity.

Lubrication is essential for long chain life. Proper lubrication reduces wear, cools the chain, and can provide some damping of shock loads. Chain power ratings are based on the assumption that proper lubrication is present. Oil is generally used for chains because greases and heavy lubricants have trouble entering the chain joints. Oil must be kept clean, and periodic oil changes are recommended. There are three types of lubrication methods in general use, and the chain power ratings are dependent on the particular type being employed. The types of lubrication are as follows:

Type A: Manual or drip lubrication is a method in which oil is applied with a brush or by dripping on the chain. Lubrication is applied at least once every eight hours of operation. Lubricant should be directed at the link plate edges, and should be applied in sufficient volume such that the lubricant runs clear from the chain. This is the least effective method of lubrication.

Type B: Bath or disc lubrication is a method in which the lower strand of the chain or an application disc runs through a sump of oil. If the chain is run through the sump, the oil level should reach at least the pitch line of the chain. If an application disc is used, the disc diameter should be such that rim speeds are at least 600 fpm. This method of lubrication applies oil to the chain once per chain revolution, and is more effective than manual or drip lubrication.

Type C: Oil stream lubrication: is a method in which a steady supply of lubricant is directed at the inside of the chain loop toward the slack side. Lubricant is continually recirculated and is effective at cooling the chain. This is the most effective method of lubrication, and enables the drive to withstand high speeds and torques.

Chains are heavily standardized and governed by ANSI/ASME B29.1M. Standard metric chains are based on the inch standard. Some proprietary chains are also available. The best source of link dimensions and sprocket dimensions are manufacturers' catalogs. The standard pitches and dimensions of chains up to 3 inches are shown in Table 11-57. Chain pitch is the distance between the same point on two links.

When items are to be conveyed on a chain, the chain pins or links can be modified to provide attachment points. Some examples of these modifications are part of ANSI B29.1M and are shown in Table 11-58.

Table 11-57: ANSI Standard Roller Chain Dimensions

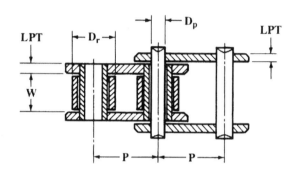

| Pitch (P) | Max. Roller Diameter (D_r) | Standard Series | | | | Heavy Series |
		Standard Chain Number	Width (W)	Pin Diameter (D_p)	Thickness of Link Plates (LPT)	Thickness of Link Plates (LPT)
0.250	(0.130)	25	0.125	0.0905	0.030	
0.375	(0.200)	35	0.188	0.141	0.050	
0.500	0.306	41	0.250	0.141	0.050	
0.500	0.312	40	0.312	0.156	0.060	
0.625	0.400	50	0.375	0.200	0.080	
0.750	0.469	60	0.500	0.234	0.094	0.125
1.000	0.625	80	0.625	0.312	0.125	0.156
1.250	0.750	100	0.750	0.375	0.156	0.187
1.500	0.875	120	1.000	0.437	0.187	0.219
1.750	1.000	140	1.000	0.500	0.219	0.250
2.000	1.125	160	1.250	0.562	0.250	0.281
2.250	1.406	180	1.406	0.687	0.281	0.312
2.500	1.562	200	1.500	0.781	0.312	0.375
3.000	1.875	240	1.875	0.937	0.375	0.500

( ) = Number represents bushing diameter. This size chain has no rollers.

Table 11-58: ANSI Standard Plate Extensions and Extended Pins

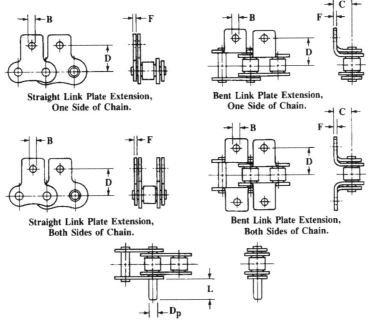

Straight Link Plate Extension, One Side of Chain.

Bent Link Plate Extension, One Side of Chain.

Straight Link Plate Extension, Both Sides of Chain.

Bent Link Plate Extension, Both Sides of Chain.

Extended Pin, One Side of Chain.

ANSI Chain Number	Straight Link Plate Extension			Bent Link Plate Extension				Extended Pin	
	B (Min.)	D	F	B (Min.)	C	D	F	D_p Nominal	L
	inch	inch	inch	inch	inch	inch	inch	inch	inch
35	0.102	0.375	0.050	0.102	0.250	0.375	0.050	0.141	0.375
40	0.131	0.500	0.060	0.131	0.312	0.500	0.060	0.156	0.375
50	0.200	0.625	0.080	0.200	0.406	0.625	0.080	0.200	0.469
60	0.200	0.719	0.094	0.200	0.469	0.750	0.094	0.234	0.562
80	0.261	0.969	0.125	0.261	0.625	1.000	0.125	0.312	0.750
100	0.323	1.250	0.156	0.323	0.781	1.250	0.156	0.375	0.938
120	0.386	1.438	0.188	0.386	0.906	1.500	0.188	0.437	1.125
140	0.448	1.750	0.219	0.448	1.125	1.750	0.219	0.500	1.312
160	0.516	2.000	0.250	0.516	1.250	2.000	0.250	0.562	1.500
200	0.641	2.500	0.312	0.641	1.688	2.500	0.312	0.781	1.875

CHAIN SELECTION AND SIZING

Chains are normally selected based on tabulated power ratings and speeds that take into account tensile forces, wear, and impacts. Static tensile load ratings for chains are also available for standard chains if needed. Some basic chain calculations are provided in Table 11-59.

Chains can be easily selected and sized with the help of a manufacturer's representative. Alternatively, the following procedure can be used to select components for a chain drive with two shafts:

1. Determine the design parameters. These are needed to size the drive components and are as follows:
   - Drive horsepower (or kW)
   - Shaft diameters
   - Speeds of both shafts
   - Distance between sprocket centers (if known)

2. Calculate the design horsepower (or kW). Design power is equal to the drive horsepower (or kW) multiplied by the service factor and some factor of safety. Service factor can be taken from Table 11-60. Most machines with non-reversing uniform loads are considered smooth. Heavy shock loads are seen by reciprocating conveyors and high impact machinery like punch presses.

3. Decide what lubrication method will be used. Lubrication method affects the power transmission capability of the chain. Power ratings of roller chain are dependent on lubrication type. The three types are as follows:

   Type A: Manual or Drip Lubrication
   Type B: Bath or Disc Lubrication
   Type C: Oil Stream Lubrication

4. Select a chain pitch and small sprocket size using power rating tables either provided by the manufacturer or available online at www.americanchainassn.com under "New Improved Power Ratings for Roller Chain." Some power rating tables are also available in *Machinery's Handbook*. Chain ratings are based on an expected life of approximately 15,000 hours and take into account tension failure of the link plates, impact of the rollers on the sprocket, and galling between the link pins and bushings. Pitch selection will be based on the speed of the drive in RPM and the design horsepower (or kW) to be transmitted. It may be necessary to reconsider the lubrication method in order to

get the desired horsepower rating in the desired sprocket size. It may also be necessary to consider a multiple strand chain if higher power is required. Multiple strand chains are subject to a multiple strand factor that further adjusts power rating.

5. <u>Verify that the small sprocket is large enough to fit the drive shaft.</u> This determination can be based on the chart in Table 11-61. The max bore given is the largest hole that can be safely made in the sprocket for the shaft.

6. <u>Select the large sprocket.</u> This is done by multiplying the speed ratio of the shafts by the number of teeth on the small sprocket. This results in the number of teeth required on the large sprocket. If this number of teeth is not a whole number, it will be necessary to round up or down. This will affect the output speed of the shaft with the large sprocket. Speed ratio is the speed in RPM of the small sprocket divided by the speed in RPM of the large sprocket. Speed ratio formulas are provided in Table 11-59.

7. <u>Calculate the chain length.</u> Chain length is calculated using the formulas given in Table 11-59. Correct the chain length to be a whole even number. This may involve rounding up or down, and will affect the shaft center distance.

8. <u>Calculate the final center distance</u> based on the whole number chain length. Center distance is calculated using the formulas provided in Table 11-59. For best results, the center distance between sprockets should not be less than 1.5 times the diameter of the larger sprocket; at least 30 times the pitch; and not more than 50 times the pitch. The following equations represent these rules of thumb:

$$C \geq 1.5 d_{p2} \quad \text{and} \quad 30p \leq C \leq 50p$$

9. <u>Verify that the angle of wrap on the smaller sprocket is greater than 120°.</u> Wrap angle can be calculated using the formulas in Table 11-59.

10. <u>Determine the installation and takeup distance.</u> Enough adjustment should be provided to easily install the chain.

11. <u>Verify that your shaft diameters are sized appropriately</u> for your power and speed requirement. Every shaft should be designed according to professional standards to ensure safety, performance, and acceptable deflection. See Section 11.1 for information on shaft design and calculation.

Table 11-59: Chain Equations

d_{p1} = pitch diameter of smaller sprocket	ω_1 = angular velocity of smaller sprocket
d_{p2} = pitch diameter of larger sprocket	ω_2 = angular velocity of larger sprocket
θ_1 = wrap angle of belt around driving sprocket	C = center distance between shafts
θ_2 = wrap angle of belt around driven sprocket	p = belt pitch
n_{t1} = number of teeth on smaller sprocket	
n_{t2} = number of teeth on larger sprocket	

DESIGN POWER SF = Safety factor S_f = Service factor (See Table 11-60)	$$P_d = S_f P(SF)$$
SPEED AND DIAMETER RELATIONSHIPS To calculate speeds or pitch diameters in a compound drive (more than two shafts), see the recommended resources.	$$d_{p1} = \frac{\omega_2 d_{p2}}{\omega_1} \qquad d_{p2} = \frac{\omega_1 d_{p1}}{\omega_2}$$ $$\omega_1 = \frac{\omega_2 d_{p2}}{d_{p1}} \qquad \omega_2 = \frac{\omega_1 d_{p1}}{d_{p2}}$$
DRIVE SPEED RATIO	$$SR = \frac{d_{p2}}{d_{p1}} = \frac{\omega_1}{\omega_2} = \frac{n_{t1}}{n_{t2}}$$
PITCH DIAMETER OF SPROCKET	$$d_p = \frac{p}{\sin\left(\dfrac{180°}{n_t}\right)}$$
CHAIN LENGTH Based on drive geometry	$$L \cong 2C + \frac{\pi(d_{p2}+d_{p1})}{2} + \frac{(d_{p2}-d_{p1})^2}{4C}$$ $$L \cong 2C + \frac{p}{2}(n_{t1}+n_{t2}) + \frac{1}{4C}\left[\frac{(n_{t2}-n_{t1})p}{\pi}\right]^2$$ $$C \cong \frac{p}{4}\left[-A + \sqrt{A^2 - 8\left(\frac{n_{t2}-n_{t1}}{2\pi}\right)}\right]$$ $$A = \frac{n_{t1}+n_{t2}}{2} - \frac{L}{p}$$
CHAIN LINEAR VELOCITY n_1 = speed of shaft in RPM or RPS	$$v = n_{t1} p \omega_1 = n_{t1} p n_1$$
FORCE EXERTED ON SHAFT F_c = Tension in tight side of chain T = Torque transmitted by chain sprocket	$$F_c = \frac{2T}{d_p}$$

Table 11-60: Chain Drive Service Factors

Type of Driven Load	Type of Input Power		
	Internal Combustion Engine with Hydraulic Drive	Electric Motor or Turbine	Internal Combustion Engine With Mechanical Drive
Smooth	1.0	1.0	1.2
Moderate Shock	1.2	1.3	1.4
Heavy Shock	1.4	1.5	1.7

Table 11-61: Recommended Roller Chain Sprocket Maximum Bore and Hub Dimensions (Sprockets Up To 25 Teeth)

Number of Teeth	Roller Chain Pitch									
	3/8		1/2		5/8		3/4		1	
	Max. Bore Diameter inch	Max. Hub Diameter inch	Max. Bore Diameter inch	Max. Hub Diameter inch	Max. Bore Diameter inch	Max. Hub Diameter inch	Max. Bore Diameter inch	Max. Hub Diameter inch	Max. Bore Diameter inch	Max. Hub Diameter inch
11	0.5938	0.8594	0.7813	1.1719	0.9688	1.4688	1.2500	1.7656	1.6250	2.3750
12	0.6250	0.9844	0.8750	1.3281	1.1563	1.6719	1.2813	2.0156	1.7813	2.7031
13	0.7500	1.1094	1.0000	1.5000	1.2813	1.8750	1.5000	2.2500	2.0000	3.0156
14	0.8438	1.2344	1.1563	1.6563	1.3125	2.0781	1.7500	2.5000	2.2813	3.3438
15	0.8750	1.3594	1.2500	1.8125	1.5313	2.2813	1.7813	2.7500	2.4063	3.6719
16	0.9688	1.4688	1.2813	1.9844	1.6875	2.4844	1.9688	2.9844	2.7188	3.9844
17	1.0938	1.5938	1.3750	2.1406	1.7813	2.6875	2.2188	3.2188	2.8125	4.3125
18	1.2188	1.7188	1.5313	2.2969	1.8750	2.8906	2.2813	3.4688	3.1250	4.6406
19	1.2500	1.8438	1.6875	2.4531	2.0625	3.0781	2.4375	3.7031	3.3125	4.9531
20	1.2813	1.9531	1.7813	2.6250	2.2500	3.2813	2.6875	3.9531	3.5000	5.2813
21	1.3125	2.0781	1.7813	2.7813	2.2813	3.4844	2.8125	4.1875	3.7500	5.5938
22	1.4375	2.8125	1.9375	2.9375	2.4375	3.6875	2.9375	4.4375	3.8750	5.9219
23	1.5625	2.3125	2.0938	3.0938	2.6250	3.8906	3.1250	4.6719	4.1875	6.2344
24	1.6875	2.4375	2.2500	3.2656	2.8125	4.0781	3.2500	4.9063	4.5625	6.5625
25	1.7500	2.5625	2.2813	3.4219	2.8438	4.2813	3.3750	5.1563	4.6875	6.8750

Number of Teeth	Roller Chain Pitch									
	1 1/4		1 1/2		1 3/4		2		2 1/2	
	Max. Bore Diameter inch	Max. Hub Diameter inch	Max. Bore Diameter inch	Max. Hub Diameter inch	Max. Bore Diameter inch	Max. Hub Diameter inch	Max. Bore Diameter inch	Max. Hub Diameter inch	Max. Bore Diameter inch	Max. Hub Diameter inch
11	1.9688	2.9688	2.3125	3.5781	2.8125	4.1719	3.2813	4.7813	3.9375	5.9844
12	2.2813	3.3750	2.7500	4.0625	3.2500	4.7500	3.6250	5.4219	4.7188	6.7969
13	2.5313	3.7813	3.0625	4.5469	3.5625	5.3125	4.0625	6.0781	5.0938	7.6094
14	2.6875	4.1875	3.3125	5.0313	3.8750	5.8750	4.6875	6.7188	5.7188	8.4219
15	3.0938	4.5938	3.7500	5.5156	4.4375	6.4531	4.8750	7.3750	6.2500	9.2188
16	3.2813	5.0000	4.0000	6.0000	4.6875	7.0156	5.5000	8.0156	7.0000	10.0313
17	3.6563	5.4063	4.4688	6.4844	5.0625	7.5781	5.6875	8.6563	7.4375	10.8438
18	3.7813	5.7969	4.6563	6.9688	5.6250	8.1406	6.2500	9.3125	8.1250	11.6406
19	4.1875	6.2031	4.9375	7.4531	5.6875	8.7031	6.8750	9.9531	9.0000	12.4375
20	4.5938	6.6094	5.4375	7.9375	6.2500	9.2656	7.0000	10.5938	9.7500	13.2500
21	4.6875	7.0000	5.6875	8.4219	6.8125	9.8281	7.7500	11.2344	10.0000	14.0469
22	4.8750	7.4063	5.8750	8.8906	7.2500	10.3906	8.3750	11.8750	10.8750	14.8438
23	5.3125	7.8125	6.3750	9.3750	7.4375	10.9375	9.0000	12.5156	11.6250	15.6563
24	5.6875	8.2031	6.8125	9.8594	8.0000	11.5000	9.6250	13.1563	13.0000	16.4531
25	5.7188	8.6094	7.2500	10.3438	8.5625	12.0625	10.2500	13.7969	13.5000	17.2500

CRITICAL CONSIDERATIONS: Chains

- Avoid using chains in applications where precise velocity or position control is required. Chordal action of the chain will prevent accurate velocity or position control, so a synchronous belt drive is recommended in those cases.
- In places where dirt and abrasive particles can get on the chain, the service life of the chain will be reduced. Dirt and other particles will interfere with the proper lubrication of the chain links.
- Chain power ratings are based on an expected life of approximately 15,000 hours.
- Chains generally cost more than belts, but less than gear drives.
- Rotating shafts, hubs, sprockets, etc., and moving chains are extremely dangerous, and proper safety precautions such as guarding must be taken. See Chapter 2 for more information on machinery safety.
- The preferred center distance for a roller chain is between 30 and 50 pitches. The maximum center distance is 80 pitches.
- The recommended maximum speed ratio for roller chain drives is 8:1. For higher reduction ratios, multiple reduction stages are recommended.
- The recommended minimum wrap angle of the chain on the smallest sprocket is 120°.

BEST PRACTICES: Chains

- When selecting a chain, it is important to remember that the smallest feasible pitch of roller chain is desired for quiet operation and high speed. However, the longest feasible chain is desired to minimize the effects of chain stretch and absorb shock.
- Use larger sprockets to increase the power capacity of the drive. For a given speed, load carrying capacity increases with the number of teeth engages with the chain on the driving sprocket.
- For best results, design a chain drive so that the sprocket axes are horizontal and the slack side of the chain is on the bottom. This configuration prevents the chain from over-wrapping the sprocket because gravity pulls it away on the slack side and tension pulls it away on the tight side. If this configuration is not possible, consider adding an idler to the slack side to prevent over-wrapping.
- It is good practice to design a chain drive such that the center distance is adjustable. This will allow proper tensioning and allow adjustment to compensate for any chain stretch over time. If center distance adjustment is not feasible, it is good practice to add an idler sprocket to take up any slack in the chain.
- When replacing a worn chain, the sprockets should also be replaced. Installation of a new chain on worn sprockets will cause premature wear and may cause the chain to jump sprocket teeth.
- Sprockets should be selected to have an odd or prime number of teeth and used with a chain with an even number of pitches to ensure even wear of the chain.

LEAD, BALL, AND ROLLER SCREWS

Screws have long been relied on to transform rotary motion into linear motion. Lead screws and ball screws are very commonly used in industry, and roller screws are relied on for heavy duty applications. Lead, ball, and roller screws are normally purchased from a manufacturer. Design of these devices is beyond the scope of this text, but this section aims to be useful in the selection and sizing of commercial screw assemblies.

RECOMMENDED RESOURCES

- R. Mott, *Machine Elements in Mechanical Design*, 5th Ed., Pearson/ Prentice Hall, Inc., Upper Saddle River, NJ, 2012
- R. L. Norton, *Machine Design: An Integrated Approach*, 4th Ed., Prentice Hall, Upper Saddle River, NJ, 2011
- Oberg, Jones, Horton, Ryffel, *Machinery's Handbook*, 28th Ed., Industrial Press, New York, NY, 2008
- Shigley, Mischke, Brown, *Standard Handbook of Machine Design*, 3rd Ed., McGraw-Hill Inc., New York, NY, 2004

SCREW CHARACTERISTICS

Screw assemblies for transforming motion are comprised of a screw and a nut that rides the screw threads. Examples of lead and ball screws are shown in Figures 11-35 and 11-36. They are normally configured such that the screw is rotated by an external driver and the nut moves linearly along the screw. The load on the nut must be guided, and the nut must be prevented from rotating with the screw. The screw shaft is normally fitted into a bearing assembly at each end and attached to a driver such as a motor, gearbox, sprocket, or other power transmission component. The bearings supporting the screw must be capable of taking both radial and thrust loads, depending on the application. Many manufacturers sell bearing assemblies that are designed to work with their screw assemblies.

Screws and screw assemblies used for linear motion are specified based on the following characteristics:

Pitch is the axial distance between like points on adjacent threads on the screw.

Figure 11-35: Lead Screws

Source: KSS Co., Ltd. (//kssballscrew.com/) Courtesy of TPA (www.tpa-us.com)

Figure 11-36: Ball Screws

Source: KSS Co., Ltd. (//kssballscrew.com/) Courtesy of TPA (www.tpa-us.com)

<u>Lead</u> is the linear distance the nut will travel per screw revolution. Lead is equal to pitch times the number of starts.

<u>The number of starts</u> on a screw is equal to the number of independent threads on the screw. Multiple starts are used to increase the lead of the screw assembly. To increase lead, increasing the number of starts is preferable to increasing screw pitch because larger pitches have smaller root diameters.

<u>Root diameter</u> is the minor diameter of the thread. This diameter is used when evaluating the strength and stiffness of the screw.

<u>Lead accuracy or error</u> is a characteristic of a screw and nut assembly. It is the difference between theoretical and actual distance traveled by the nut. Lead error or lead accuracy is usually measured in deviation per unit distance. Roll formed threads have lower precision and lower cost than machined and ground threads.

<u>Backlash</u> is a characteristic of a screw and nut assembly. It is the clearance between ball (ballscrew) or nut (leadscrew) and screw. Backlash can result in position error when forces change direction or when screw velocity changes. To reduce or eliminate backlash, one nut can be preloaded against another or preload can be built into the nut assembly.

<u>Ability to backdrive</u> is a critical characteristic of screw assemblies. Lead screws, ball screws, and roller screws all employ a helical drive screw and nut to transform rotary motion into linear motion. Rotating either part causes the other to move linearly. Normally the screw is rotated using a motor or other driver. When a linear load on the nut causes rotary motion of the screw, the condition is known as backdriving. When backdriving efficiency is negative, it means that the screw is self-locking and some torque will be required to lower a load. It is important to note that ball screws can almost always be backdriven, and marginal lead screws may backdrive under vibration. If backdriving is possible, a holding brake may be needed on the screw axis to hold a load in place. Some manufacturers will specify which screws are self-locking. ACME screws are generally considered self-locking, whereas ball screws can always be assumed to backdrive. A test for backdriving capability can be found in Table 11-62.

Table 11-62: Screw Formulas

A diagram of the variables is shown in Figures 11-37 and 11-38. Some common coefficients of friction are listed in Table 11-63. μ = Coefficient of friction between nut and screw surfaces α_n = Thread face angle normal to thread F = Operating load or applied load (lbf or N) d_m = Mean diameter of thread contact (pitch diameter)	λ = Screw lead angle α = Thread face angle in axial plane L = Screw Lead (in or mm)
LINEAR TRAVEL OF NUT S = Linear travel (in/min or mm/min) ω = Rotational speed in RPM	$S = L\omega$
POWER HP = Power in horsepower kW = Power in kilowatts T = Torque in lbf-ft or N-m ω = Rotational speed in RPM	INCH: $HP = \dfrac{T\omega}{5252}$ METRIC: $kW = \dfrac{T\omega}{9550}$
DRIVING TORQUE T_D = Drive Torque (lbf-in or N-mm) μ_c = coefficient of friction of the thrust collar d_c = thrust collar diameter	$T_D = \dfrac{FL}{2\pi E_f}$ This is torque based on loading and ignores factors like drag from support bearings or other drive components. Preload adds drag. Consult the manufacturer for preload torque or drag values. $T_D = \dfrac{Fd_m}{2}\left(\dfrac{\mu\pi d_m + L\cos\alpha_n}{\pi d_m \cos\alpha_n - \mu L}\right) + \dfrac{F\mu_c d_c}{2}$ This equation includes the drag caused by a thrust collar used with a lead screw.
DRIVING EFFICIENCY Efficiencies are often given in catalogs. Lead screw drive efficiency generally varies between 20% and 80%.	LEAD SCREW: $E_f = \tan\lambda\left(\dfrac{\cos\alpha_n - \mu\tan\lambda}{\cos\alpha_n \tan\lambda + \mu}\right)$ $\alpha_n = \arctan(\cos\lambda\tan\alpha)$ This equation neglects the friction of the thrust collar. BALL SCREW: $E_f = 0.9$ is generally assumed
TEST FOR BACKDRIVING	A screw is self-locking if either is true: $\mu \geq \dfrac{L}{\pi d_m}\cos\alpha_n \qquad \mu \geq \tan\lambda\cos\alpha_n$
BACKDRIVING TORQUE T_B = Backdriving output torque (lbf-in or N-mm)	$T_B = \dfrac{FLE_{fB}}{2\pi}$
BACKDRIVING EFFICIENCY	LEAD SCREW: $E_{fB} = \dfrac{1}{\tan\lambda}\left(\dfrac{\cos\alpha_n\tan\lambda - \mu}{\cos\alpha_n + \mu\tan\lambda}\right)$ $\alpha_n = \arctan(\cos\lambda\tan\alpha)$ BALL SCREW: $E_{fB} = 0.8$ is generally assumed

Efficiency is the actual output divided by theoretical (frictionless) output of a screw assembly. Screws have two efficiencies: direct driving and backdriving. Driving efficiency and backdriving efficiency can both be calculated using formulas in Table 11-62 if necessary, but they are usually provided by the manufacturer.

Driving torque and power to raise or lower a specified load are calculated using the formulas in Table 11-62. The relationship between torque and force depends on the lead and efficiency of the screw.

Table 11-63: Common Lead Screw Coefficients of Friction

Nut Material	Screw Material	
	Steel, Dry	Steel, Oiled
Steel	0.15 - 0.25	0.11 - 0.17
Bronze	0.15 - 0.23	0.10 - 0.16
Brass	0.15 - 0.19	0.10 - 0.15
Cast Iron	0.15 - 0.25	0.11 - 0.17
Plastic	Consult Manufacturer	Consult Manufacturer

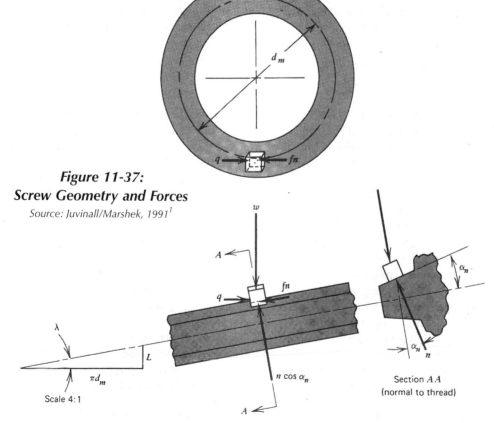

Figure 11-37:
Screw Geometry and Forces
Source: Juvinall/Marshek, 1991[1]

[1] Reprinted With Permission: Juvinall, Robert C., and Marshek, Kurt M., *Fundamentals of Machine Component Design.* John Wiley & Sons: New York, 1991.

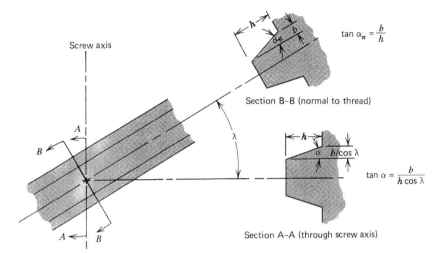

$$\tan \alpha_n = \frac{b}{h}$$

Section B-B (normal to thread)

$$\tan \alpha = \frac{b}{h \cos \lambda}$$

Section A-A (through screw axis)

Figure 11-38: Thread Angles

Source: Juvinall/Marshek, 1991[2]

SCREW STRESSES, DEFLECTION, AND BUCKLING

The load carrying capacity of a screw is determined by lead value, root diameter, and assembly stiffness. Nut preload and end mounting configuration can affect stiffness. Stiffness affects load carrying capacity due to its influence on bucking strength. Stiffness of the screw assembly is a function of the end conditions, distance between supports, and the root diameter of the screw.

Loads should be guided externally, especially for long travels, to prevent lateral or eccentric loading of the screw. Loads that are not axial to the screw should be avoided. Screws can be analyzed for stress and fatigue using the methods presented for shafts in Section 11.1 and threaded fasteners in Chapter 6. The screw root diameter will be used in those equations.

Buckling is of particular concern when screws are loaded in compression. Most screws used for linear motion fall into the slender column category and must be evaluated for buckling. Commonly used equations for axial deflection and buckling of slender columns are provided in Table 11-64. In determining the end fixity conditions for screws, a double bearing or wide single bearing can be considered as a fixed end, whereas a single narrow bearing, such as a deep groove single row bearing, can be considered as a pinned end. It is common for manufacturers to provide buckling formulas with a multiplier between 0.25 and 4 depending on end fixity. These formulas

[2] *Reprinted With Permission: Juvinall, Robert C., and Marshek, Kurt M., Fundamentals of Machine Component Design. John Wiley & Sons: New York, 1991.*

Table 11-64: Screw Deflection and Buckling Formulas

d_r = Screw Root Diameter (in or mm)	
S_y = Tensile yield strength of material (highly variable depending on material and treatment)	
E = Modulus of elasticity for screw material (Alloy steel approx.: 30 Mpsi or 206.8 GPa)	
AXIAL DEFLECTION UNDER LOAD F = Applied axial load	$\delta = \dfrac{4FL_e}{\pi d_r^2 E}$
EFFECTIVE COLUMN LENGTH L$_e$ AS A FUNCTION OF UNSUPPORTED COLUMN LENGTH L a) b) c) d)	L = Maximum Distance Between Nut and Load Carrying Bearing (in or mm) or distance between supports. a) One end fixed, one end free: L$_e$ = 2L b) Both ends pinned: L$_e$ = L c) One end fixed, one end pinned: L$_e$ = 0.7L d) Both ends fixed: L$_e$ = 0.5L
GEOMETRY FACTORS	$k = \sqrt{\dfrac{I}{A}} = \dfrac{d_r}{4} \qquad A = \dfrac{\pi d_r}{4} \qquad I = \dfrac{\pi d_r^4}{64}$
TEST FOR A SLENDER COLUMN	$\dfrac{\pi^2 E}{\left(\dfrac{L_e}{k}\right)^2} \leq \left(\dfrac{1}{2}\right) S_y$
SLENDERNESS RATIO	$SR = \dfrac{L_e}{k} = \dfrac{4L_e}{d_r}$
CRITICAL BUCKLING LOAD These equations assume that the applied force is purely axial. SF = Safety factor (3 or more is recommended) F_c = Critical Buckling Force (lbf or N) SF = Safety factor (3 or more is recommended) A = Cross sectional area of screw k = Radius of gyration of screw I = Area moment of inertia of screw	SLENDER SCREWS: $P_{cr} = \dfrac{\pi^2 EI}{(SF)L_e^2} = \dfrac{\pi^2 EA}{(SF)\left(\dfrac{L_e}{k}\right)^2}$ NON-SLENDER SCREWS: $P_{cr} = A\sigma_{cr} = A\left[S_y - \left(\dfrac{1}{E}\right)\left(\dfrac{S_y}{2\pi}\right)^2\left(\dfrac{L_e}{k}\right)^2\right]$

are derived from Euler's formula for buckling, and the end fixity factor they are providing is equal to $\frac{1}{L_e^2}$ where L$_e$ is the multiplier traditionally applied with Euler's formula to calculate equivalent length. These factors are given in Table 11-64, as is Euler's formula for slender columns and the J. B. Johnson formula for intermediate columns.

CRITICAL SPEED

Screws are essentially shafts, and are subject to critical vibration at their natural frequencies. The main mode of vibration for screws is shaft whirl due to its rotation. The critical frequency for shaft whirl is usually known as the critical speed. Critical speed determines how fast it is permissible to run the screw assembly. In general, it is recommended to keep screw speed below 80% of its critical speed. Increasing the stiffness of a screw assembly increases its critical speed. Stiffness of the screw assembly is a function of the end conditions, distance between supports, and the root diameter of the screw.

Ball and lead screw manufacturers often provide detailed instructions for calculating critical speed and permissible speed. The most commonly used formulas are presented in Table 11-65.

Table 11-65: Permissible and Critical Speed for Screws

PERMISSIBLE SPEED n_{per} = Permissible Speed (RPM)	$n_{per} = 0.8 n_c$
CRITICAL SPEED Critical speed calculations can vary slightly between manufacturers. n_c = Critical Speed (RPM) A = End Fixity Factor L = Length between bearing supports (in or mm) d_r = Screw Root Diameter (in or mm)	INCH: $n_c = (4.76 \times 10^6) A \dfrac{d_r}{L^2}$ METRIC: $n_c = (1.2 \times 10^8) A \dfrac{d_r}{L^2}$ End fixity factors A for critical speed calculations are generally provided by the manufacturer, but some common values are listed in Table 11-66.

Table 11-66: End Fixity Factors for Critical Speed

End Supports		A
	One end fixed, one end free	0.36
	Both ends supported	1.00
	One end fixed, one end supported	1.47
	Both ends fixed	2.23

LINEAR MOTION SCREW TYPES

<u>Lead screws,</u> sometimes called power screws, are simply a nut and screw assembly. The nut normally travels linearly while the screw is turned. There is sliding contact between the nut and screw threads. As a result, the efficiency and top speed are limited by friction between the screw and nut. The screw

is normally steel, but lead screw nuts can be a variety of materials, including steel, bronze, and plastic. Lead screw nuts are available with integral preloading to reduce backlash.

Ball screws are a nut and screw assembly where the nut contains ball bearings that roll along the screw threads. This configuration results in much higher speed and efficiency capabilities because there is theoretically no sliding friction between the nut and screw. The ball bearings normally recirculate through the nut to create continuous rolling motion for the bearings. Ball screw nuts are available with several types of preloading as well as several types of ball recirculation.

In a roller screw, the nut is outfitted with fixed rollers parallel to the axis of the screw. The rollers have a helical profile that engages with the screw profile to provide many load-carrying surfaces. Roller screws are an excellent option for precision linear movement under very high load conditions.

Lubrication is essential for screw life and efficiency. Lead screws are normally greased regularly on the shaft to reduce friction and wear. Ball and roller screws are normally lubricated both on the shaft and in the nut. The manufacturer will specify lubrication materials and methods.

Lead screws are generally more cost effective than ball or roller screws, but they are least efficient because sliding friction between nut and screw generates losses in the form of heat. As a result, the drive motor of a lead screw must be larger than that of a ball or roller screw for the same applied load. Ball screws replace sliding friction with rolling element friction for greater speed capacity and life expectancy. Roller screws are more costly than lead or ball screws, but have much higher stiffness and load carrying capacity. Table 11-67 can be helpful in selecting between lead, ball, and roller screws. Lead screws are most often used for manual machines, whereas ball screws are most often used in automated machinery. Applications with very high usage are better suited to a ball screw than a lead screw.

Table 11-67: Screw Selector

Screw Type	Self-Locking	Speeds	Loads
ACME Lead Screw	Yes	Low	High
Other Lead Screw	Sometimes	Moderate	High
Ball Screw	No	High	High
Roller Screw	Sometimes	Very High	Very High

LEAD SCREWS

Lead screws come in a variety of thread forms, the most common of which is the ACME thread. The ACME thread is a common choice when loads are seen in both directions. Another thread form, the buttress thread, has higher strength, but is suitable for unidirectional loading only. Manufacturer's catalogs will normally give the relevant dimensions for each thread form needed to perform sizing calculations. A cross section of an ACME lead screw assembly with preloaded nut is shown in Figure 11-39. Preloaded nuts can minimize or eliminate backlash in the assembly.

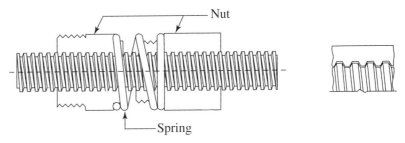

Figure 11-39: Preloaded Lead Screw Cross Section

ACME threads have the advantage of being self-locking, which makes them resistant to backdriving. ACME lead screws with plastic nuts generally have an efficiency of 40%, and ACME lead screws with bronze nuts have an efficiency of around 25%. Some proprietary lead screws are quoted by the manufacturer to have efficiencies up to 85%, depending upon lubrication, nut material, thread form, and thread form. Lead screws tend to have higher backlash than ball screws, but can be quieter running. Lead screws are also less likely to backdrive than ball screws. Life expectancy of lead screws is less predictable than that of ball screws due to complex surface wear. Lubrication needs are higher with lead screws than with ball screws. Lead screw efficiency can vary with load, and lead screw efficiencies increase with increasing lead angle. Efficiencies can be calculated using the formulas in Table 11-62.

Design load or dynamic load for a lead screw is generally defined as the maximum load that can be applied to the screw while it's in motion. This includes acceleration and friction forces. Maximum dynamic load for a lead screw is usually provided in the catalog information. Lead screw torque and power are evaluated using the formulas in Table 11-62. Assemblies containing lead screws often employ a thrust collar or washer to take the axial load on the screw, but sometimes a rolling element thrust bearing is used when friction must be minimized. The drag created by the thrust collar can be significant and should be accounted for in torque calculations. If a rolling element bearing is used instead of a sliding element, the drag can be omitted from the calculations.

Lead screws with plastic nuts should also be evaluated based on their Pressure Velocity (PV) value as well as other sizing factors. PV ratings will be given by the manufacturer and depend on nut material. PV value takes into account not only the load being carried, but the speed at which it must be moved. Calculating the projected bearing area and sliding surface velocity can be complex. Consult an applications engineer to check PV values for your application and selected screw. Many manufacturers offer software calculators to check PV. Table 11-68 provides some formulas and values used to calculate PV.

Table 11-68: PV Values for Nuts

PRESSURE VELOCITY VALUE (PV)	
F = Applied Load (lbf or N) A = Projected Bearing Area (in² or m²)	$PV = \dfrac{FV}{A}$
SLIDING SURFACE VELOCITY V = Sliding surface velocity (ft/min or m/min) ω = Angular velocity in RPM d_m = Major diameter of screw	INCH: $V = \dfrac{d_m \pi \omega}{12}$ METRIC: $V = \dfrac{d_m \pi \omega}{1000}$

BALL SCREWS

Ball screws are produced by many manufacturers and typically have an efficiency of 90% or greater. Manufacturer catalogs will normally give the lead of the screw assembly as well as the dynamic load rating needed to perform sizing calculations. A cross-section detail of a typical ball screw is shown in Figure 11-40. Stiffness of the system is determined by ball screw diameter, number and size of its load carrying balls, and the end mounting conditions of the screw.

According to the ANSI standard, dynamic load rating of a ball screw is the load under which 90% of a population of identical screws will reach or exceed 1 million inches of travel without fatigue failure. The ISO (and DIN) dynamic load rating for a ball screw is defined as the axial load under which 90% of a population of identical screws can be expected to attain or exceed 1 million revolutions without fatigue failure. The implications of the difference between ISO and ANSI rating systems are important to understand. If the lead

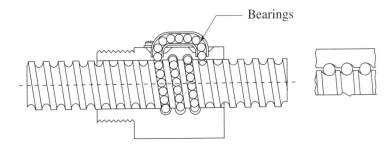

Figure 11-40: External Recirculating Ball Screw Cross Section

is smaller than 1 inch, then the load rating per ANSI definition is less than the ISO load rating for the same ball screw. The opposite is true for a ball screw with lead larger than 1 inch. In that case, the ANSI load rating will be higher than the ISO load rating for the same ball screw.

Life expectancy of a ball screw is governed by bearing principles and depends on applied forces (including acceleration and friction forces) and linear travel (number and length of loaded strokes). Formulas to calculate dynamic load rating for ball screws and required screw life are given in Table 11-69.

When the load on a screw varies over time, calculate the mean operating load applied to the screw for the application, using the formulas in Table 11-69. First calculate the load seen by the screw at each point in the cycle and estimate what percent of the cycle that load is applied. The load should be applied along the axis of the screw. Include the forces caused by acceleration of the load and the effects of friction if the load is sliding along a surface.

LEAD OR BALL SCREW SELECTION PROCEDURE

When designing a ball screw arrangement, the interrelationships in Table 11-70 are helpful to keep in mind.

The best place to find the appropriate formulas and values when selecting and sizing a screw will be in the manufacturer's catalog. Many manufacturers offer lead or ball screw sizing software, and an applications engineer will usually make short work of sizing a screw for your application. The following procedure can be used to select and size a lead or ball screw for linear motion:

1. <u>Determine the application parameters.</u> These include the following:
   - Applied axial force or load
   - Speed at which to move the load axially
   - Preliminary length between bearings
   - Preliminary end fixation type

Table 11-69: Ball Screw Formulas

DYNAMIC LOAD RATING C = Dynamic load rating (lbf or N) F_M = Mean operating load (lbf or N) L_m = Life in millions of inches (ANSI) or Millions of Revolutions (ISO)	$$C = F_M \sqrt[3]{L_m}$$ When L_m = 1 million (revs or inches), F_M = C
MEAN OPERATING LOAD APPLIED TO SCREW F_M = Mean operating load (lbf or N) q_n = Proportion of stroke or cycle (should add up to 1) F_i = Load during portion of cycle (lbf or N)	$$F_M = \sqrt[3]{\sum q_i F_i^3}$$
REQUIRED LIFE OF SCREW, INCHES (ANSI) L_m = Life in millions of inches U = Linear distance per stroke (in or mm) V = Strokes per hour W = Hours of operation per day X = Days of operation per year Y = Years of life required	$$L_m = \frac{UVWXY}{10^6}$$ Life Expectancy is often provided in graph form by ball screw manufacturers. Each model will have a line on a chart showing life expectancy vs. applied axial load. Life is affected by materials and hardnesses of both screw and nut.
REQUIRED LIFE OF SCREW, REVOLUTIONS (ISO) L_m = Life in millions of revolutions U = Linear distance per stroke (in or mm) V = Strokes per hour W = Hours of operation per day X = Days of operation per year Y = Years of life required S = Lead (in or mm per revolution)	$$L_m = \frac{UVWXY}{10^6 S}$$ Life Expectancy is often provided in graph form by ball screw manufacturers. Each model will have a line on a chart showing life expectancy vs. applied axial load. Life is affected by materials and hardnesses of both screw and nut.
LIFE EXPECTANCY AT OTHER THAN RATED LOAD L_a = Actual life expectancy L_r = Rated life expectancy F_r = Rated dynamic load F_a = Actual dynamic (or equivalent) load	$$L_a = L_r \left(\frac{F_r}{F_a} \right)^3$$

2. <u>Select a screw type</u> based on required precision or accuracy. Lead accuracy in deviation per unit distance is one criteria, and backlash is another. To avoid backlash, one nut can be preloaded against another or preload can be built into the nut assembly. Roll-formed threads have lower precision and lower cost than machined and ground threads.

3. <u>Select a lead value and screw diameter</u> based on iterative analysis of the following factors with appropriate factors of safety:

Table 11-70: Ball Screw Design Factor Interrelationships

Increase In	Affects	How
Screw Length	Critical Speed	Decreases
	Compression Load	
Screw Diameter	Critical Speed	Increases
	Inertia	
	Compression Load	
	Stiffness	
	Spring Rate	
	Load Capacity	
Lead	Drive Torque	Increases
	Load Capacity	
	Ball Diameter	
	Angular Velocity	Decreases
	Positioning Accuracy	
End Mounting Rigidity	Critical Speed	Increases
	Compression Load	
	System Stiffness	
Load	Life	Decreases
Preload	Positioning Accuracy	Increases
	System Stiffness	
	Drag Torque	
Angular Velocity	Critical Speed	Decreases
Nut Length	Load Capacity	Increases
	Stiffness	
Ball Diameter	Life	Increases
	Stiffness	
	Load Capacity	

- Static load rating of the screw assembly is the maximum static load that can be applied to the unit. Exceeding this load will cause permanent deformation of the screw assembly. This value is usually given in catalogs.
- Dynamic load rating of a <u>ball screw</u> is normally given in catalogs. This load value should not be exceeded during normal operation if full life expectancy is to be realized. Dynamic load calculations are presented in Table 11-69.

- Design load or dynamic load of a <u>lead screw</u> is the maximum load that should be applied when the screw is in motion. This value is usually given in catalogs
- Running torque required to lift the application load must be calculated and evaluated. Running torque required to lift a given load is sometimes provided by the manufacturer. To calculate torque based on efficiency and applied load, use the formulas provided in Table11-62. Efficiency is normally provided by the manufacturer. Be aware that starting torque may be many times greater than the running torque.

4. <u>Verify that the screw can run at the required speed.</u> Some manufacturers provide speed limits for their products, but in many cases you must calculate it. Critical and permissible speeds can be calculated using Table 11-65. When a plastic nut is used on a lead screw, speed will be governed by the PV relationship given in Table 11-68. Consult an applications engineer when evaluating permissible speed of a lead screw assembly with a plastic nut. Many manufacturers offer online PV calculators that can be helpful in the selection process.

5. <u>Determine if the screw can backdrive under the applied load, and with what resultant torque</u> on the screw. It is generally assumed that all ball screws can be backdriven with an efficiency of about 80%. For best results consult the manufacturer. Many lead screw catalogs will indicate which screws are self-locking (cannot be backdriven). A test for backdriving can be found in Table 11-62 along with backdriving torque formulas. Backdriving efficiency for lead screws is not normally given in catalogs; therefore, for best results, consult the manufacturer in lead screw backdriving cases. When backdriving efficiency is negative, it means that the screw is self-locking and some torque will be required to lower a load. It is important to note that marginal lead screws may backdrive under vibration. Be sure to plan for a brake to provide resisting torque if needed to prevent backdriving.

6. <u>Evaluate buckling.</u> If the screw is in compression at any position (very commonly the case) the buckling strength of the screw should be examined. Some manufacturers provide graphs of allowable compression loading for a given position and end configuration. Others provide formulas that can be used to check the design. Some buckling formulas are provided in Table 11-64.

7. Select a left hand drive or right hand drive configuration.
8. Select the screw drive end configuration and features. Manufacturers will offer a variety of end types, offering features like keyways, threads, and bearing shaft extensions. Examples of shaft end features are shown in Figure 11-41.
9. Select a nut configuration. Preloaded nuts are available. Be aware that preloaded nuts exert a drag force that can affect torque requirements.

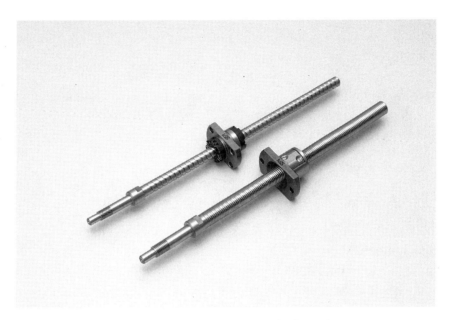

Figure 11-41: Ball Screws With Shaft End Features

Source: KSS Co., Ltd. (///kssballscrew.com/) Courtesy of TPA (www.tpa-us.com)

CRITICAL CONSIDERATIONS: Lead, Ball, and Roller Screws

- Loads that are not axial to the screw should be avoided. Loads should be guided externally, especially for long travels, to prevent lateral or eccentric loading of the screw.
- Regular and complete lubrication is critical to screw life. Dirty environments and elevated temperatures will shorten the life of a screw.
- Lead screws with plastic nuts must be evaluated for PV factor.
- Thrust collars that operate with sliding rather than rolling elements will induce drag on the assembly that will affect torque requirements.
- Evaluate every screw for buckling if the screw is in axial compression at any point during operation.
- Rotating shafts, couplings, screws, etc. are extremely dangerous, and proper safety precautions such as guarding must be taken. See Chapter 2 for more information on machinery safety.

BEST PRACTICES: Lead, Ball, and Roller Screws

- Use care to align the screw, load, and mounting elements. Misalignment shortens life.
- Radial and moment loading should be avoided and minimized. Be aware that radial and moment loads can be induced by poor alignment.
- For maximum life, load should be applied along the same axis as the screw. This evenly loads the nut and minimizes applied moments.
- Speed variations should be smooth during all parts of the screw stroke to ensure maximum life.

12

MACHINE RELIABILITY AND PERFORMANCE

Tables

12-1	Equipment Reliability Parameters	684
12-2	Quantitative Criticality Calculations	686
12-3	FMECA (Criticality) Worksheet Example	687
12-4	Severity Classifications and Probability Levels	688
12-5	FMECA (RPN) Worksheet Example	690
12-6	FMECA Worksheet RPN Score Definitions	691
12-7	Safety Category Characteristics	695
12-8	Equipment Performance Metrics	696
12-9	Overall Equipment Effectiveness (OEE) Parameters	697

Understanding machine reliability and performance is essential for good machinery design. Reliability and life expectancy greatly influence selection of machine elements and the design of components. Performance expectations drive decisions like machine type, speed, and operation requirements. Designing for reliability and performance through techniques like Failure Modes, Effects, and Criticality Analysis (FMECA) is good practice. In cases where reliability affects safety, selection of components based on safety category can ensure safe system function during failure. Both reliability engineering and performance evaluation are broad topics. This text is merely an introductory reference to topics commonly encountered in machine design; for more information, please consult the recommended resources or one of the many books on reliability engineering, quality engineering, OEE, and preventative maintenance.

RECOMMENDED RESOURCES

- R. Hansen, *Overall Equipment Effectiveness,* Industrial Press, New York, NY, 2001
- M. Lindeburg, *Mechanical Engineering Reference Manual for the PE Exam,* 11th Ed., Professional Publications, Inc., Belmont, CA, 2001
- D.Shafer, *Successful Assembly Automation,* Society of Manufacturing Engineers, Dearborn, MI, 1999
- **MIL-STD-1629A:** Procedures For Performing A Failure Mode, Effects And Criticality Analysis
- **MIL-HDBK-338B:** Military Handbook: Electronic Reliability Design Handbook (01 Oct 1998)
- **American Society for Quality, Reliability Division Web Site: www. asq.org/reliability**
- **The Reliability Information Analysis Center:** www.theriac.org
- **FMEA-FMECA** website: www.fmea-fmeca.com

MACHINE RELIABILITY

Reliability is the probability that a part or system will perform its intended function during a specified period of time under a given set of conditions. What constitutes a failure must be clearly defined for the part or system. Reliability of a part or system is usually calculated as a function of time and using an exponential distribution incorporating the rate of failure.

Reliability of equipment has many parameters, and those for a non-redundant system are found in Table 12-1. Mean time to failure (MTTF) is often an indicator of part or system life expectancy for a non-repairable item. Mean time between failures (MTBF) is the total equipment runtime divided by the number of failures. MTBF is used for repairable items. Both MTTF and MTBF can be used as design guidelines.

Calculating MTBF of a system requires adding the failure rates of the components as shown in Table 12-1. It is important to recognize that adding components or complexity to a system will lower its reliability. Consult the recommended resources and manufacturers for MTBF values of common components.

Table 12-1: Equipment Reliability Parameters

RELIABILITY (R_T)	Reliability calculated as an exponential distribution Probability that a part or system will not fail during time t To be used only when the part or system has a constant failure rate	$R_t = e^{-\lambda t}$
TIME (t)	Time period being examined	t
FAILURE RATE (λ)	Probable number of failures per unit time	$\lambda = \dfrac{failures}{t} = \dfrac{1}{MTBF}$
MEAN TIME BETWEEN FAILURES (MTBF)	Probable mean time between failures of a *repairable* part or system System MTBF is a function of its component MTBF values	$MTBF = \dfrac{1}{\lambda}$ $MTBF_{sys} = \dfrac{1}{MTBF_1} + \dfrac{1}{MTBF_2} + ...$
MEAN TIME TO FAILURE (MTTF)	Probable average, or mean, amount of time that a *non-repairable* part or system is in operational condition	$MTTF = \dfrac{1}{\lambda}$
REPAIR RATE (μ)	Probable number of repairs that can be completed per unit time	$\mu = \dfrac{1}{MTTR}$
MEAN TIME TO REPAIR (MTTR)	Probable mean time between failure of a part or system and its return to operation	$MTTR = \dfrac{1}{\mu}$
AVAILABILITY (A)	Probability that the part or system with repairable components will be operating at any given time.	$A = \dfrac{MTBF}{MTBF + MTTR}$
AVAILABLE TIME (T)	Maximum available time for operation	T
UPTIME (U)	Probable time that the part or system will be operating	$U = AT$
DOWNTIME (D)	Probable amount of time that the part or system will not be operating	$D = (1 - A)T$

One topic often associated with reliability planning is risk analysis. Risk, in this context, is often calculated in terms of potential cost per year of failure for a given part or system. Risk is evaluated by looking at the cost of failure multiplied by the likelihood of that failure occurring during the time period (usually a year). Likelihood is normally given as a percent, where one expected failure per year is represented by 100%, two failures per year as 200%, etc.

$$Risk \cong (cost\ of\ failure)(likelihood\ of\ failure\ as\ a\ percent)$$

One application of reliability data is in the planning of preventative maintenance cycles. Preventative maintenance (PM) is undertaken in an effort to prevent unscheduled downtime and thereby reduce costs. It can also prevent some catastrophic failures and reduce repair costs. Preventative maintenance is most effective when failure rates are known and have low variability. Formulating a preventative maintenance schedule is subject to many variables and is beyond the scope of this text.

FAILURE MODES, EFFECTS, AND CRITICALITY ANALYSIS

A major characteristic of a reliable process or machine is a low failure rate. It is the responsibility of the designer to deliver machines and parts that fail not only infrequently, but also repairably, inexpensively, and in a safe manner. Failure Modes, Effects, and Criticality Analysis (FMECA) is a method of assessing designs or processes with respect to the ways they can fail in an effort to improve reliability inherent in the design. FMECA requires that the evaluator think about the causes and types of failures, as well as the likelihood and effects. Part of the analysis involves coming up with a plan to address high risk areas, which can then be re-evaluated after the mitigation plan is implemented. FMECA, or some other form of failure analysis, should be performed on every machine and process to improve safety and reliability. It should be undertaken as early in the design process as possible, and repeated as the design is iterated and refined.

Failure Modes, Effects, and Criticality Analysis (FMECA) is an expansion of Failure Modes and Effects Analysis (FMEA) in which Criticality Analysis (CA) is added. Criticality analysis ranks each potential failure mode according to the combined influence of severity classification and its probability of occurrence. This analysis should be based upon the best available data. Criticality analysis can be undertaken using either a qualitative or quantitative approach. The qualitative approach is appropriate when specific failure rate data are not available.

Quantitative criticality analysis is best performed when reliability data are available for the component parts in the system or assembly being analyzed. This method allows the user to calculate a criticality value for each item on the FMECA worksheet. To calculate criticality, use the formulas in Table 12-2 and the worksheet format shown in Table 12-3.

Table 12-3 is an example worksheet for a fictional shrinkwrap machine. This machine consists of an indexing conveyor carrying products to be shrink-wrapped. The shrinkwrap itself has already been placed on the product and just needs to be heated. The machine cycle is as follows:

1. The conveyor indexes a product into position.
2. An air cylinder lowers a heater to the product and the continuous, thermostatically controlled heater warms the shrinkwrap until it conforms to the product.
3. The air cylinder then raises the heater to complete the machine cycle.

Table 12-2: Quantitative Criticality Calculations

ITEM CRITICALITY System failure rate due to item i failing in mode j	$(CR)_{ij} = \alpha_{ij}\beta_{ij}\lambda_i$
SYSTEM CRITICALITY System failure rate due to all items failing in all modes	$(CR)_s = \sum_i \sum_j (CR)_{ij}$
FAILURE MODE FREQUENCY RATIO α_{ij} = Ratio of failures of item i in mode j to all failures of item i	If all potential failure modes of a particular part or item are listed, the sum of the α values for that part or item will equal one. Individual failure mode frequencies may be derived from failure rate source data or from test and operational data. If failure mode data are not available, the α values shall represent the analyst's judgment based upon an analysis of the item's functions.
LOSS PROBABILITY β_{ij} = Probability of system failure if item i fails in failure mode j	$\beta_{ij} = 1.00$ Actual loss $0.1 < \beta_{ij} < 1.00$ Probable loss $0 < \beta_{ij} \leq 0.1$ Possible loss $\beta_{ij} = 0$ No effect
FAILURE RATE λ_i = Failure rate for item i	Stressors like environmental factors may be applied to failure rate. Failure rate data may be derived from test and operational data. If failure rate data are not available, the λ values shall represent the analyst's judgment.

Table 12-3: FMECA (Criticality) Worksheet Example

Assembly or Device Analyzed: Shrinkwrap Machine

Function	Failure Mode	Effects	Cause	Item Code	Control Method	Criticality			Total		Action Plan
						Loss Probability (β)	Frequency Ratio (α)	Failure Rate (λ) (Per Million Hours)	Criticality (CR)		
Heater lowers to part	Fails to lower	Part not shrink wrapped	Air cylinder failure	A1	Proximity sensor on cylinder	0.1	0.3	0.5	0.015		No action required
Heater raises	Fails to raise	Mechanical crash	Air cylinder failure		Proximity sensor	1	0.7	0.5	0.35		No action required
Heater heats shrinkwrap	Fails to heat to temperature	Part not shrink wrapped	Heater failure	H1	Thermostat feedback	0.1	1	1.5	0.15		No action required
Heater heats shrinkwrap	Fails to maintain temperature	Part under or over heated	Thermostat failure	T1	Thermostat feedback	0.1	1	10	1		Optional: consider failsafe thermostat
Heater maintains temperature when guard is open	Operator touches heater	Injury	Heater exposed	H1	Warning labels	1	1	10	10		**Mandatory action:** Add fixed guard to shroud heater when interlocked guard door is open

System Criticality: 11.515

For this example, consider just the heater and its positioning mechanism. In this case, it is known that the heater will not cause a problem if left in contact with the part for too long. This is an important thing to consider in many heating applications. To fill out the FMECA worksheet (Table 12-3), it is usually helpful to make a list of the steps in the sequence of operation for the machine, or part of the machine being examined. Then, for each step, list all possible failure modes. Next, list all possible causes for each failure mode. The list can become quite extensive for complex mechanisms.

When using the qualitative approach, failure modes identified in the FMECA are assessed in terms of their probability of occurrence based on engineering judgment. There are two commonly used ways of performing a qualitative criticality analysis. One way involves creating a "criticality matrix." The

Table 12-4: Severity Classifications and Probability Levels
Per MIL-STD-1629A

	Assignment	Meaning
Severity Classification	I	**Catastrophic:** A failure that could cause death or system loss.
	II	**Critical:** A failure which may cause severe injury, major property damage, or major system damage.
	III	**Marginal:** A failure which may cause minor injury, minor property damage, or minor system damage which will result in delay or loss of availability.
	IV	**Minor:** A failure not serious enough to cause injury, property damage, or system damage, but which will result in unscheduled maintenance or repair
Probability of Occurrence Level	A	**Frequent:** A high probability of occurrence during the item operating time interval. High probability may be defined as a single failure mode probability greater than 20% of the overall probability of failure during the operating time interval.
	B	**Reasonably Probable:** A moderate probability of occurrence during the operating time interval. Probable may be defined as a single failure mode probability of occurrence which is more than 10% but less than 20% of the overall probability of failure during the operating tine.
	C	**Occasional:** An occasional probability of occurrence during the operating time interval. Occasional probability may be defined as a single failure mode probability of occurrence which is more than 1% but less than 10% of the overall probability of failure during the operating time.
	D	**Remote:** An unlikely probability of occurrence during the operating time interval. Remote probability may be defined as a single failure mode probability of occurrence which is more than 0.1% but less than 1% of the overall probability of failure during the operating time.
	E	**Extremely Unlikely:** A failure whose probability of occurrence is essentially zero during the operating time interval. Extremely unlikely may be defined as a single failure mode probability of occurrence which is less than 0.1% of the overall probability of failure during the operating time.

other is to calculate the Risk Priority Number (RPN). Both are effective ways of analyzing the criticality of failure modes and their effects.

A criticality matrix is a way to qualitatively compare failure modes graphically. To create a criticality matrix, first assign scores for two criteria: severity of the effects and likelihood of occurrence. The scores and their meanings are shown in Table 12-4. The criticality matrix graphical format is shown in Figure 12-1. Plot each item on the graph, and then select the most critical items for action. This graphical format makes it easy to evaluate lots of items simultaneously.

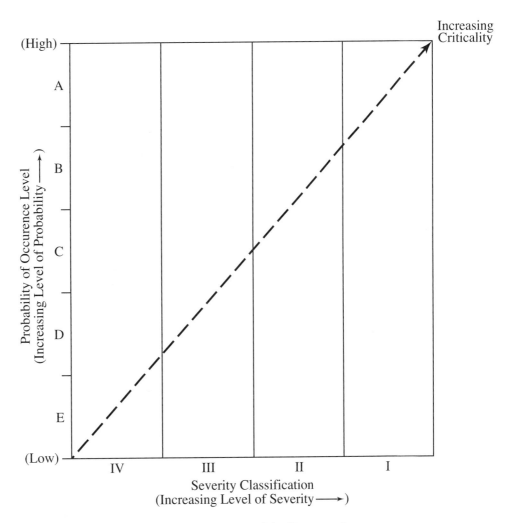

Figure 12-1: Criticality Matrix

Assembly or Device Analyzed: Shrinkwrap Machine

Table 12-5: FMECA (RPN) Worksheet Example

					RPN			Total	
Function	**Failure Mode**	**Effects**	**Cause**	**Control Method**	**SEVERITY**	**OCCURRENCE**	**CONTROL**	**RPN SCORE**	**Action Plan**
Heater lowers to part	Fails to lower	Part not shrink wrapped	Air cylinder failure	Proximity sensor on cylinder	5	3	2	30	No action required
Heater raises	Fails to raise	Mechanical crash	Air cylinder failure	Proximity sensor	8	3	2	48	No action required
Heater heats shrinkwrap	Fails to heat to temperature	Part not shrink wrapped	Heater failure	Thermostat feedback	5	3	2	30	No action required
Heater heats shrinkwrap	Fails to maintain temperature	Part under or over heated	Thermostat failure	Thermostat feedback	6	3	4	72	Optional: consider failsafe thermostat
Heater maintains temperature when guard is open	Operator touches heater	Injury	Heater exposed	Warning labels	10	3	6	180	**Mandatory action: Add fixed guard to shroud heater when interlocked guard door is open**

Risk Priority Number (RPN) is another method of qualitative criticality assessment. However, the RPN method includes an additional assessment of how well the planned control methods will reduce the probability of occurrence and/or severity of effects. Control methods are things like detectors, guards, and signs that act to detect the occurrence of failure or prevent damage to the machine or personnel. A sample RPN FMECA worksheet is shown in Table 12-5, and a list of sample individual score definitions are shown in Table 12-6. These scores are subjective, so it is best for an organization to set clear definitions for scores that are descriptive and relevant to the assemblies or devices typically being analyzed. RPN FMECA is quite common in industry.

Table 12-6: FMECA Worksheet RPN Score Definitions

	Description	Score
	No effect	1
	Very slight effect	2
	Slight effect	3
	Minor effect	4
SEVERITY	Moderate effect	5
	Significant effect	6
	Somewhat serious effect	7
	Serious effect	8
	Very Serious effect	9
	Safety violation	10
	Nearly impossible to occur	1
	Remote chance of occurring	2
	Very slight chance of occurring	3
	Slight chance of occurring	4
OCCURRENCE	Low chance of occurring	5
	Moderate chance of occurring	6
	Moderately high chance of occurring	7
	High chance of occurring	8
	Very high chance of occurring	9
	Nearly certain to occur	10
	Almost certain to prevent	1
	Very high chance of prevention	2
	High chance of prevention	3
	Somewhat high chance of prevention	4
CONTROL	Moderate chance of prevention	5
	Low chance of prevention	6
	Slight chance of prevention	7
	Very slight chance of prevention	8
	Remote chance of prevention	9
	Almost impossible to prevent	10

Three well known documents covering FMECA are MIL-STD-1629A, MIL-HDBK-338B, and SAE J1739. There is commercial software for performing and recording FMECA. Without software, FMECA can be performed using a worksheet to identify failure modes, their effects, risks, probabilities, and control (mitigation) methods. Starting at the lowest level and working upward, failures are identified and the effects traced upward. When performing a FMEA, it is important to consider not only local effects, but also higher level effects. If a failure within the subassembly being analyzed affects the machine it belongs to, that is a higher level effect that must be considered when the overall machine is analyzed.

FMECA is conducted with the following steps:

1. Gather a team. FMEA is not as effective if conducted alone.
2. Determine the subject and scope. FMEA can be conducted at the concept level, functional level, or detail level. Individual parts can be analyzed. Assemblies, whole machines, or systems can be included in the analysis.
3. Write a key characteristics list and/or sequence of operation for the subject. These can be helpful in listing all the functions being analyzed and scored. Physical functions like "lift" and "lower" or "latch" and "unlatch" are useful, as are electrical functions like "Detect" or "Sense." Many of the recommended resources suggest drawing a functional block diagram for the part, assembly, or system being examined.
4. List each possible failure mode and its effects locally and at higher levels.
5. List all possible causes for each failure mode. This is normally done using the worksheet format. This list can quickly become quite large. Each cause will have to be evaluated separately. Each cause will represent an item in a criticality analysis. Assign an item code if quantitative criticality analysis will be performed.
6. List each possible effect of each cause of each failure mode. Each effect will need to be evaluated separately.
7. Conduct criticality analyses. This is done using one of the following: Risk Priority Numbers (RPN), a criticality matrix, or quantitative criticality analysis.

   (a) RPN qualitative method:
      i. Assign scores for three criteria: severity of the effects, likelihood of occurrence, and reliability of the control method. The scores and their meanings are shown in Table 12-6.

ii. Multiply the three individual scores together to get a total RPN score for that failure mode and cause. A low score needs no action, but a high score will require an action plan to lower the total score through control methods. Low RPN scores with a high severity rating should require an action plan as well.

(b) Qualitative criticality matrix method:

i. Assign scores for two criteria: severity of the effects and likelihood of occurrence. The scores and their meanings are shown in Table 12-4.

ii. Plot the results of the scoring on the criticality matrix. The criticality matrix format is shown in Figure 12-1. Select the most critical items for action.

(c) Quantitative criticality method (appropriate if reliability data are available):

i. Determine quantitative values for three criteria: loss probability, frequency ratio, and failure rate. These are normally determined using test and operation data. More information about these values can be found in Table 12-2.

ii. Multiply the quantitative values together for each item and mode listed to arrive at a criticality number for that mode and item. Higher numbers will need action.

iii. Optional: Sum all the criticality numbers for the assembly or system being examined. This is the system criticality, or the probable rate of failure of the system due to all causes.

8. Select critical failure modes and plan design changes or mitigation methods. Re-evaluate each failure mode after corrective actions have been taken to arrive at new scores. Perform new FMECA periodically as each design evolves.

SAFETY CATEGORY

When reliability is a matter of safety, safety category is used when selecting controls components to ensure safe functioning under failure conditions. Controls components include pneumatics and hydraulics as well as electrical and safety devices. Safety categories are identified based on the amount of risk associated with failure. Components and systems for use in applications with an elevated safety category are progressively more failsafe. Redundancy is typical in devices rated for elevated safety categories. When designing a

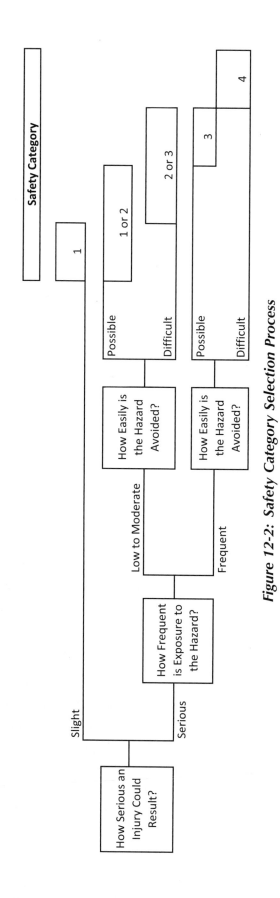

Figure 12-2: Safety Category Selection Process

Table 12-7: Safety Category Characteristics

Safety Category	Requirements	Behavior During Faults
B	Safety related components must be correctly sized and selected in accordance with relevant standards.	Loss of safety function
1	The requirements of category B apply. Known, tested components must be used.	Loss of safety function, but probability of this happening is less than category B.
2	The requirements of category 1 apply. Faults can cause loss of the safety function, but this must be detected through automatic checks conducted at suitable intervals.	Loss of safety function only between automatic checks.
3	Requirements of category 1 apply. A single fault does not result in a loss of safety function. Single faults are detected at suitable intervals.	For most faults, safety function is not lost. Not all faults are detected. Accumulation of undetected faults can cause loss of safety function.
4	Requirements of category 1 apply. Single or multiple faults do not result in loss of safety function. Faults are detected before the next demand on the safety function.	For all faults, safety function is not lost. All faults are detected. Accumulation of faults does not cause loss of safety function.

machine and selecting components, first identify the safety category and then select components rated for that safety category. Safety category can be generally determined by following the flow chart in Figure 12-2. Descriptions of the categories and their characteristics can be found in Table 12-7. This is just a quick overview of safety category terminology. A qualified safety professional and a controls professional should be consulted in any matters of component and controls system safety.

MANUFACTURING EQUIPMENT PERFORMANCE

Machinery performance is measured in many ways. When applied to systems like drive trains, efficiency is a measure of power out divided by power in. The term "machine efficiency," when applied to manufacturing equipment, is usually a reference to quality rate. Quality rate of an assembly process or task is good output divided by total output. Quality rate (Table 12-8) can be used to estimate the target output of a production process. More comprehensive measures, like overall equipment effectiveness, incorporate business

Table 12-8: Equipment Performance Metrics

POWER TRANSMISSION EFFICIENCY E = Total efficiency P_O = Power output P_I = Power input	Efficiency is power out divided by power in: $$E = \frac{P_O}{P_I}$$ To get output power P_O, you will need input power of P_I: $$P_I = \frac{P_O}{E}$$
TOTAL MACHINE EFFICIENCY E_i = Individual efficiencies	$$E = E_1 E_2 E_3 ...$$
QUALITY RATE Q = Quality rate Y = Good output X = Total output	Quality rate is good output divided by total output: $$Q = \frac{Y}{X}$$ If X good units are required at the end of a shift, then Y units will need to be 'started' during that shift. $$Y = QX$$
UNIT COST FOR A TIME PERIOD C_U = Unit cost C_P = Operating or production cost V_O = Value of good output	Unit cost is operating or production cost divided by the value of good output: $$C_U = \frac{C_P}{V_O}$$
SCRAP COST PER UNIT TIME C_S = Scrap cost in time period T = Time period Q = Quality Rate R = Production rate $\$$ = Cost per unit scrap cost for that process	$$\frac{C_S}{T} = \$(R - QR)$$

aspects of equipment performance. Many organizations have developed their own metrics for performance.

When a machine has a single process, the machine efficiency is simple to determine. However, many machines are comprised of several sub-processes arranged sequentially. In such arrangements, each sub-process acts on the good output of the previous one. Total machine efficiency is a measure of the efficiency of a machine that has several sub-processes, each with their own efficiency. When data is not yet available on total machine efficiency, it can be estimated using the individual efficiencies of the sub-processes (Table 12-8).

At many manufacturing companies, reliability is viewed as a function of cost. Two common ways to represent this are with unit cost (Table 12-8) or failure cost. These two measures can help when prioritizing improvement activities. If a particular type of failure drives up the unit cost or otherwise costs the company more, it can be identified. Higher reliability reduces the unit cost both by increasing the good output and decreasing the cost of production through scrap (lost value) reduction.

For production processes with low downtime, failure cost for a specific operation is the total cost of its bad output, usually scrap, over a given time period. This method of viewing costs is useful because a high efficiency process with high scrap cost may need attention more than a lower efficiency process with low scrap costs. Operations on higher-cost components should be more efficient than operations on lower-cost items.

One of the most common comprehensive measures of machine performance in a production environment is overall equipment effectiveness (OEE). OEE takes into account factors like scheduled downtime and operational metrics for a better measurement of equipment productivity than simply machine

Table 12-9: Overall Equipment Effectiveness (OEE) Parameters

OVERALL EQUIPMENT EFFECTIVENESS (OEE)	Comprehensive measure of equipment performance.	$OEE = MPQ$
MACHINE AVAILABILITY (M)	The actual time that the machine spent producing product divided by the time it was scheduled to produce product.	$M = \dfrac{(T-S-U)}{(T-S)}$
PERFORMANCE (P)	Ideal time the machine should take to produce a quantity of product divided by the actual time it took to produce that quantity	$P = \dfrac{O_A\left(\dfrac{1}{M_R}\right)}{(T-S-U)}$
QUALITY RATE (Q)	Quantity of good product produced divided by total parts produced	$Q = \dfrac{(O_A - R)}{O_A}$
AVAILABLE TIME (T)	Total time in period being measured	T
SCHEDULED DOWNTIME (S)	Includes breaks, scheduled maintenance, etc.	S
UNSCHEDULED DOWNTIME (U)	Includes unscheduled repairs, waits for material, etc.	U
MACHINE RATE (M$_R$)	Speed of machine based on units per time Parts per minute is a typical measurement of machine rate	$M_R = \dfrac{units}{time}$
ACTUAL OUTPUT (O$_A$)	Number of gross units produced during the measurement period	O_A
REJECTS (R)	Number of unacceptable parts produced in the measurement period	R

efficiency. OEE calculation and the parameters it is based on can be found in Table 12-9. The following OEE values and their meanings are typical in industry:

OEE < 65%	Unacceptable
OEE 65% - 75%	Passable, but needs improvement
OEE 75% – 85%	Good
OEE > 85%	Excellent for batch processes
OEE > 90%	Excellent for continuous discrete processes
OEE > 95%	Excellent for continuous stream processes

CONDITION MONITORING

When unplanned downtime must be minimized, condition monitoring may be employed. Condition monitoring uses continuous or periodic measurements to predict failure so that preventative maintenance may be scheduled. The most basic form of condition monitoring is periodic visual inspection. More sophisticated condition monitoring methods include temperature monitoring and vibration monitoring. Less common methods of condition monitoring include thermography, lubricant spectrographic analysis, and ultrasonic monitoring or inspection. These methods all have associated costs, and their benefits must be weighed against the cost of unplanned downtime or machine failure.

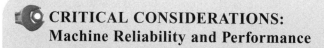

CRITICAL CONSIDERATIONS:
Machine Reliability and Performance

- Reliability is generally based on statistical probability and predicts averages. Actual reliability will vary.
- Reliability theory assumes that failure is essentially a random event. Factors that make failure more likely, like improper alignment or adverse environmental conditions, will invalidate reliability predictions.
- Fail-safe and redundant systems are recommended in critical applications, as well as in high frequency processes.
- Condition monitoring is recommended for high risk parts or systems.

BEST PRACTICES: Machine Reliability and Performance

- Remember that an assembly or process can never be more reliable than its least reliable component.
- Design for reliability. Know the time period over which a design needs to operate and under what conditions. Make design choices that ensure reliability for those conditions. Conduct FMECA early enough in the design process that changes to improve reliability are still possible. Repeat FMECA for each design iteration until the design is complete and the analysis yields acceptable results.
- Design for ease of repair and setup. Reducing repair time has a direct and positive effect on both reliability and performance.
- Design machinery to be fault tolerant, preventing downtime when routine faults occur.
- When designing assembly equipment with sequential processes, place the least efficient sub-processes as early in the line as possible, detect the failures, and stop value-added work on failed work pieces.
- Reducing variability in processes and parts will improve the reliability of the overall system.
- Eliminate adverse environmental conditions whenever possible. Environmental factors affect the reliability of equipment.
- Track the cost of preventative maintenance. In some cases, the cost of PM can exceed the cost of not doing PM.
- Conduct preventative maintenance on items that would cause catastrophic failure in a larger system. The cost of prevention in those cases is likely less than the cost of repair.
- Collect data on the failure frequency of parts and assemblies. Keep an inventory of spare parts and assemblies for high frequency of failure items.
- Lubrication, or lack thereof, is often a machinery failure driver. Make consistency and ease of lubrication a priority in design.

INDEX

A

ABEC 520–523
AC motors (see also: motors) 484
   altitude 486
   AMB 483
   asynchronous 484, 491
   bearings 488
   capacitor correction 488
   CONT 484
   cooling 484, 494
   current, locked rotor 486, 488
   design code 486
   duty or time rating 484
   efficiency 486
   ENCL (enclosure type) 484
   FLA (full-load amps) 484
   frame size 484, 493
   kVA code 486
   locked rotor current 486, 488
   MAX CORR KVAR 488
   nameplate 483, 484
   OVER TEMP 486
   performance 489, 495
   phases 484
   power factor 488
   rated frequency 484
   rated load current (FLA) 484
   rated torque 488
   rated voltage 486
   service factor 486
   shaft 488, 493
   speed control 491
   synchronous 484, 487
   synchronous speed 484, 486, 497
   TEFC 494
   TENV 494
   thermal 483, 486, 491, 495
   torque 483, 486, 494
   torque, accelerating 488
   torque, breakaway 488
   torque, breakdown 488
   torque, full-load 488
   torque, peak 488
   torque, pull-in 489
   torque, pull-out 489
   torque, pull-up 489
   torque, rated 488
   torque, starting 488
   torque, static or locked rotor 488

   types and characteristics 487
   undervoltage 483
AC power 482–484
acceleration
   gravitational 16, 18
   linear motion 31
   rotary motion 32
access, workspace 39, 44, 45, 80, 82, 87, 89
acetal 388
ACME screw 667, 672, 673
active coils (see: springs)
addendum 576, 581, 601, 608–610
adjustability, assembly 11, 133
AGMA 574, 575, 589, 590–593, 595, 596, 598
air actuator (see: pneumatics, actuator)
air cylinder (see: pneumatics, actuator)
AISI 371
alloy steel 380, 365, 371, 372, 378, 406, 418, 548
Alumina 378
aluminum (see: materials)
AMB 483
analysis
   design 1, 10
   risk 48, 49, 73, 685
angle, pressure 576, 577, 579–581, 610
angular contact ball bearings 515, 519, 520
angular misalignment 514, 516, 517, 566, 567, 569, 571
angularity 122, 173
annealing 401
anodizing 409
anthropometry (see: body dimensions)
area moments of inertia 21
areas and perimeters 20
assemblability 187
availability 684, 697
available time 684, 697
AWS 359, 361
axial stress 27, 561

B

backdrive 666–668, 673
backlash
   couplings 563, 567–570, 573

   gearboxes (see: gearboxes)
   gears 576, 580, 589, 591–595, 599, 601
   linear motion screws 666, 672, 673
ball bushings 538, 539, 541
ball screw (see also: linear motion screws) 664, 665
   backdrive 666–668
   backlash 666
   calculations 667, 670, 671, 676
   design relationships 677
   dynamic load rating 674–676
   efficiency 666–668, 672, 674
   failure 674
   life 672, 675–677
   loading 664, 666–670, 674–676
   preload 672, 677
   selection and sizing 675
   spced 671
   static load rating 677
   torque 667, 668
ball splines 540
basic dimensions 119, 120
beam coupling 567, 568, 573
beam deflection and forces 23, 24
bearing stress 27, 223
bearings 501
   clearance 506, 507, 512, 513, 521, 525
   gearbox 616, 622
   linear (see: linear bearings)
   motor 488
   plain (see: plain bearings)
   rolling element (see: rolling bearings)
   shaft 547, 549, 553, 554
Belleville springs (see also: springs) 416, 456
   allowable stress 461
   assembly considerations 459
   catalog selection steps 459
   deflection stable range 459
   deflection vs. force 457, 458
   design steps 463
   diameter ratio 460
   force vs. deflection 457, 458
   geometry 457
   height and thickness 457
   materials 418, 461
   parallel 457, 458, 461

series 457, 458, 461
sizing equations 460
snap action 459
stable range of deflection 459
stack instability 458
stacking 457, 458
stress 459, 461
yielding safety factor 461
bellows coupling 467, 571, 573
belts (see also: belts and chain) 627, 629
   drive shafts 627, 637
   forces on shafts 633, 634, 638, 650
   angle of wrap 629, 649
   best practices 653, 654
   critical considerations 652
   flat (see: flat belt)
   friction drive 629, 653
   speed recommendations 629, 630
   timing or synchronous (see: timing belt)
   v-belt (see: v-belt)
belts and chain (see also: belts) 627
   (see also: chain)
   center distance 629
   drive calculations 627–629
   length 629
   pitch diameter 628
   resources 627
   speed ratio 628
bending stress 548, 555, 600
best practices
   belts 653, 654
   chain 663
   couplings 573
   design 11, 12, 14
   fasteners 299
   fits 116
   gearboxes 626
   gears 612, 613
   heat treatment 407
   keys 227
   lead, ball, and roller screws 680
   limits 116
   limits, fits, and tolerance grades 116
   linear bearings 542
   locating techniques 212
   machine reliability and performance 699
   materials 391
   motors 499
   pins, keys, and retaining rings 227
   pipe and port threads 282
   plain bearings 510

pneumatics 481
precision locating 212
retaining rings 227
rolling bearings 534
shafts 562
springs 465, 466
surface finish 398
surface treatment 412
threaded fasteners 299
threaded inserts 308
tolerance grades 116
tolerance stack-ups 177
tolerances on drawings and GD&T 127
washers 299
welds and weldments 368
bevel gear (see also: gears) 582, 607, 585, 586
   backlash 591, 592, 594
   equations 609, 610
   miter gears 607
   lubrication 598
   mounting 594
   terminology 608–610
Bisque Alumina 378
black oxide 408, 294, 366
body
   capabilities and limitations 37, 38, 41, 42
   dimensions 39, 41, 42, 73, 79, 80, 86, 87
bolt
   preload 295, 297, 298
   stripping 295
   torque 298–301
boundary 161, 162
   calculations 163
   example problem: spring
   pin 164
   LMB 162
   MMB 161
Brinell 399
brittle materials 13
brown-out 483
buckling
   air actuator rod 473
   linear motion screws 669, 670
   springs 427, 430, 431
   column 25
bushings (see: plain bearings)
butt joint 360
button head cap screws (see: screws)

C

calculation (see: equations)
calipers 9
cap screws (see: screws, hex head cap)
capability (see: process capability)
carbon steel 372, 373, 380, 402
carburizing 402, 405, 406
Cardan joint 565
case hardening 373, 379, 401–403, 405
center distance
   belts and chain 629
   gear 578, 591–594
ceramics 376, 378
chain (see also: belts and chain) 627, 655
   angle of wrap 629
   attachments 656, 657
   best practices 663
   calculations 628, 629, 660
   critical considerations 662
   dimensions 656
   drive calculations 628, 629
   forces on shafts 660
   length 660
   lubrication 655
   pitch 656
   pitch diameter 660
   power 658, 660
   roller 655
   selection and sizing 658
   service factor 661
   shaft 661
   shaft loading 660
   speed 660
   sprocket 655, 656, 661
   standards 656
   tolerance stack-up 129
check valves 476, 479
chrome plating 408–411
circular pitch 576, 579, 600, 601, 609
circular runout 123, 124, 174
circularity 121, 168, 174
clearance 129, 146, 158, 159, 163, 181, 188, 192, 200, 217, 220
   bearing 506, 507, 512, 513, 521, 525
   gear 576, 591, 621
   linear motion screws 666
   rctaining rings 268, 270, 272, 274, 276, 277
   springs 430
   screws 302–304, 310, 312, 314, 316, 330, 332, 334, 336, 338, 340

tap drills 288
workspace 39, 43, 87
wrench 301, 304, 338, 340,
342–345
clearance fit (see also: fits) 95, 99, 183,
186, 189, 223
criterion 193, 196, 204, 205,
209, 211
keys 223
clevis pins 222
data, inch 252, 253
data, metric 254, 255
closed center circuit 467
coating (see: surface treatment)
codes and regulations, machine safety 71,
72, 85–91
coefficient of friction
linear motion screws 668
bolt tightening 299–301
coefficient of thermal expansion 18, 375,
376, 379, 384–387
coil springs (see: springs)
coil stress (see: springs)
coiled spring pins (see: spring pins)
cold work tool steel 381, 383, 385
column buckling 25, 473, 669, 670
combined tolerance stack-up 143, 144
composite tolerance 160, 173, 196,
197, 199
compound tolerance frames 126
compressed air 52, 67, 466, 467
compression ratio, pneumatic
actuator 475
compression springs (see also: springs)
Belleville washers (see: Belleville)
helical coil (see: helical coil
compression)
types 415–418
compressive strength 374
concentricity 123
conceptual design 9
condition monitoring 698
conductivity, thermal 376
cone point set screw (see: screws, hex
socket set, cone point)
conjugate action 577, 578
conversion, units 15, 17, 18
conveyor, forces on shafts 618, 550, 551
corner joint 360
corrosion resistance 375, 378, 379, 407–410
cost
scrap 696, 697
unit 696, 697

cotter pins 223
data, inch 256, 257
data, metric 258, 259
countersunk head cap screws (see:
screws, hex socket flat countersunk head
cap screws)
couplings 563
attachment 563
backlash 563, 570, 573
beam 567, 573
bellows 567, 573
best practices 573
Cardan joint 565
clamping 563, 565
critical considerations 573
disc 568, 573
flexible 567
high torque 563, 565
inertia 571
jaw 569, 573
keys 563
moment of inertia 571
Oldham 569, 573
phase misalignment 564
resources 563
rigid 564, 565
selection 570
set screws 563
shaft loading 570
shaft misalignment 564, 573
slit 567, 573
speed 564, 571
stress risers 563
torque 563, 571
types 564
u-joint 565
zero backlash 571, 573
C_{pk} (see: process capability index)
critical buckling, columns 25, 473, 670
critical considerations
belts 652
chain 662
couplings 573
design 14
ergonomics 46
fasteners 298
fits 116
gearboxes 626
gears 612
heat treatment 406
keys 226
lead, ball, and roller screws 680
limits 116

limits, fits, and tolerance grades 116
linear bearings 542
linear motion screws 680
locating techniques 212
machine reliability &
performance 698
machine safeguarding 65
machine safety - design process 51
materials 390
motors 499
pins, keys, and retaining rings 226
pipe and port threads 282
plain bearings 510
pneumatics 480, 481
precision locating 212
retaining rings 226
rolling bearings 533
safety issues – other 71
shafts 562
springs 465
surface finish 398
surface treatment 412
threaded fasteners 298
tolerance grades 116
tolerance stack-up 176
tolerances on drawings and
GD&T 126
washers 298
welds and weldments 367
critical speeds
linear motion screws 670, 671
shafts 26, 556, 559
criticality 49, 685–687, 689, 691
criticality matrix 688, 689
crossed helical gear (see: helical gear)
crossed roller ways 541, 542
cup point set screw (see: screws, hex
socket set, cup point)
C_v 472, 474, 475, 480
cylindrical roller bearings 512, 516,
519, 520
cylindrical roller thrust bearings 518, 520
cylindricity 122

D

danger signals 69, 81, 89, 90
datum 118, 119, 148–151, 162, 165,
166, 168
datum reference frame 185, 186, 147,
148, 155, 166, 169, 170, 171, 174, 198
datum shift 168–171

DC motors (see also: motors) 489
  armature rated load current 490
  base speed 490
  brush 489
  maximum speed 490
  nameplate 483, 490
  power supply 490
  rated armature voltage 490
  rated field voltage 490
  speed control 491
  terminology 490
  types and characteristics 489, 491
  winding type 490
DC power 482
decision matrix 9
dedendum 576, 601, 608, 609
deep groove ball bearings 514, 519, 520
deflection
  beams 23, 24
  Belleville springs 456–461
  helical coil compression springs 423, 426–428, 431, 432
  helical coil extension springs 423, 441–444
  helical coil torsion springs 423, 450–452
  linear motion screws 669, 670
  shafts 549, 553–555, 560
DELPHI technique 50
Delrin 376, 377, 388
design
  analysis 1, 10
  assemblability 187–189, 214
  assemblies and systems 11
  best practices 14, 15
  conceptual 9
  critical considerations 14
  detail 10
  functional specification 3
  interchangeability 133, 187, 214
  locating techniques 181, 187, 191
  machinery 3
  parts 12
  research 8
  resources 3, 8
  safety 13, 37, 38
  shafts 558
  specification 3, 9
  springs, Belleville 463
  springs, compression 437
  springs, extension 447
  springs, torsion 454
  synthesis 9

weldments 359, 368, 592
detail design 10
  assemblies and systems 11
  parts 12
deviations (see: tolerance)
diametral pitch, gear 579, 580
diamond pin (see: pins)
dimensions
  body 39–42, 73, 75, 79, 80, 86–88
  workspace 39, 40, 43–45
dimensions and tolerances 93
disc coupling 568, 573
distortion 366, 381, 391, 402, 403
distribution, normal 137–139
dog point set screw (see: screws, set, hex socket dog point)
double acting (see: pneumatics, actuator)
dowel pins 220, 219, 191, 98, 182
data, inch 228, 229, 195, 203
  data, metric 230, 231, 194, 202
  pullout (see: dowel, withdrawal)
  precision locating dimensions 194, 195, 202, 203
  stress and materials 217, 218
  withdrawal 220
  withdrawal, data, inch 232, 233
  withdrawal, data, metric 234, 235
downtime 684, 685, 697, 698
drawings
  best practices 127
  critical considerations 126
  GD&T 116
  tolerances 116
  weldments 364
ductility 375, 380, 399

E

E (modulus of elasticity) 375, 385, 387, 418, 419, 461
edge joint 360
efficiency (see also: machine efficiency)
  gearboxes 614, 615, 620, 621
  linear motion screws 666–668, 671–674
electric motors (see: motors)
electrical power 482
electroless nickel plating 366, 409–411
electroplating 408
electropolishing 408, 410, 411, 398
emergency stop devices 70, 71, 81, 90

enclosures
  controls for motors 493
  motors 484, 493, 494
  workspace 39
endurance
  bolts 297
  springs 424, 425
  strength 19
engineering units 15
epicyclic gear trains 588, 589
equations, general 18
  area moments of inertia 21
  areas and perimeters 20
  beam deflection and forces 23, 24
  buckling of columns 25
  column buckling 25
  critical speed of shafts 26
  Euler's formula for buckling 25
  intermediate column buckling 25
  lifting, NIOSH 46
  mass moments of inertia 22
  moments of inertia, area 21
  moments of inertia, mass 22
  perimeters and areas 20
  resources 18
  right triangle relationships 19
  surface areas and volumes 21
  useful values and equations 18
  volumes and surface areas 21
equipment (see: machine)
ergonomics
  access openings 39, 44, 45, 80
  books 74, 75
  capabilities and limitations 37, 38, 41–43, 86
  clearances 39, 87
  codes 72, 73
  critical considerations 46
  enclosures 39
  forces 38, 42, 79, 86, 89
  lifting 41, 43, 46, 68
  machine safety 37, 38
  regulations 72, 73
  standards 40, 46, 79, 80, 83
  web sites 84, 85
e-ring (see: retaining rings)
Euler's formula for buckling
  columns 25
  pneumatics 473
  power screws 670
exact constraint design 11, 181, 186
explosive atmospheres 68, 82
extension springs (see also: springs)

helical coil (see: helical coil extension)
types 416–418
external retaining rings (see: retaining rings)
Eytelwein's equation 632, 636

F

face of weld 360
factors of safety 13, 375, 441
    common values 14
    fatigue 29, 295, 297, 423–425
failsafe 71, 693
failure
    modes (see: FMECA)
    MTBF 684
    MTTF 684
    prediction 698
    rate 684, 685
failure modes, effects, and criticality analysis (see: FMECA)
fastener (see also: bolt)
    (see also: nuts)
    (see also: screws)
    best practices 299
    critical considerations 298
    engagement length 287, 295, 308
    grades 295
    materials 294
    resources 285
    selection criteria 294
    strength 295–297
    stripping 295
    torque 298–301
    types 294, 301
fastener threads 286–288
    tap drills 288–293
fatigue
    bolts 295, 297
    equations 28, 29
    helical coil springs 423–426
    shafts 548, 549, 555, 556, 561
    weldments 359
fault tree analysis 49
feather keys (see: keys, parallel)
feature control frame 119, 120
feeler gauge 13
fender washers (see: washers, wide)
fillet weld 360
finished hex bolts (see: screws, hex head cap)

fits 95
    ANSI 96, 100
    bearings 521–523
    best practices 116
    class 95, 96
    clearance 95
    commonly used 99
    critical considerations 116
    force 98
    grades 96, 97, 100, 101
    interference 95
    ISO 96, 97
    IT grades 101
    limits 95, 100, 101
    preferred 96
    tolerance grade 95–97, 100, 101
    transition 95
    types 95
fixed fastener 161, 183, 184
fixture, assembly 133
flame hardening 401, 402, 405, 375
flat belt (see also: belts) 629
    bending frequency 633
    calculations 628, 629, 632, 633
    pitch diameter 630
    power 632
    selection and sizing 632
    service factor 633
    shaft loading 551, 618, 633
    shafts 636, 637
    slip 629, 630, 632
    speeds 629, 634
    tension 629, 632, 633
    torque 632
    width 635
flat head cap screws (see: screws, hex socket flat countersunk head cap screws)
flat point set screw (see: screws, hex socket set, flat point)
flat washer (see: washers)
flatness 121
flexible couplings 567
floating fastener 161, 183–185
FMEA (see: FMECA)
FMECA 10, 683, 685
    procedure 692
    RPN 690, 691
    worksheet 686–688, 690
foot-pound-second system 16
force gauge 9
four-point contact ball bearings 515, 516
FPS (foot-pound-second system) 16

friction
    coefficient 299–301
    lead screws 668
    plain bearings 503–507
functional datums 118, 185
functional design specification (see: design, specification)
fusion zone 360

G

G (modulus of rigidity) 375, 385, 387, 419
gauge
    feeler or thickness 13
    force 9
    setup 11, 209
    sheet metal 389
GD&T 116, 117
    angularity 122
    basic dimensions 119, 120
    best practices 127
    circular runout 123, 124
    circularity 121
    compound tolerance frames 126, 198, 199
    concentricity 123
    critical considerations 126
    cylindricity 122
    datums 118, 148
    feature control frame 119, 120
    flatness 121
    LMC 126, 154
    MMC 126, 152
    modifier 126, 148
    parallelism 122, 123
    perpendicularity 122, 123
    position 125
    profile of a surface 124
    resources 116
    RFS 126
    runout, circular 123, 124
    runout, total 124
    standards 119, 120
    straightness 120
    symbols 119, 120
    total runout 124
gearboxes 614
    backlash 621
    bearings 616, 620, 622
    best practices 626
    calculations 617, 618, 625

configurations 614, 615, 619
critical considerations 626
direction of rotation 619
efficiency 614, 615, 620, 621
forces on 616–618
gear types 614
inertia 616–618, 625
in-line 614, 615
intermittent motion 620
lateral loads 616, 622
life 620, 622
loads on 616–618
lubrication 615, 619, 622
motors 621, 623
mounting 621
noise 620, 621, 623
overhanging moment rating 622
power 614, 621, 625
protection class 623
radial load rating 622
ratio 615, 620, 621
right angle 614, 615
rigidity 619, 621
RMC torque 620, 625
selection and sizing 619, 623
service factor 619
service life 622
shaft types 622
speed 620
stages 615
temperature 619, 623
thrust loads 616
torque 614–620, 625
torsional rigidity 619, 621
transmission error 620, 621
gears 574
attachment 589
backlash 576, 580, 589, 591–595, 599, 601
best practices 612, 613
bevel (see: bevel gear)
critical considerations 612
custom 581, 604, 605
forces on shafts 551, 552, 606, 608–610, 617
gearboxes (see: gearboxes)
helical (see: helical gear)
idler 587
law of gearing 577
lubrication 596, 598
materials 595, 597, 603
pitch line speed 591, 596, 598, 599, 602
quality ratings 589–591

rack and pinion (see: rack and pinion)
ratio 576, 578, 582, 587, 588, 592, 600
resources 574
service factor 603
shafts 589
speed, pitch line 591, 596, 598, 599, 602
speed ratio 576, 578, 582, 587, 588, 592, 600
spur (see: spur gear)
stress 603
standards 574–576
surface treatments 595
tooth thickness 592, 601, 609
torque ratio 582
types 582–586
velocity ratio 576, 578, 582, 587, 588, 592, 600
wear 595, 596, 598, 599, 601
worm (see: worm gear)
gear trains
compound 588
epicyclic (planetary) 588, 589
ratio 582, 587, 588
simple 582, 587
single stage 582, 587
two-stage 588
Geometric Dimensioning and Tolerancing (see: GD&T)
Goodman diagram 423–425
grades
fits 95–97, 100–102
threaded fasteners 295
grasping 41–43, 46
groove weld 360
grooved pins 221, 222, 217, 440
data, inch 248, 249
data, metric 250, 251
guard (see: machine safeguarding)

H

half dog point set screw (see: screws, hex socket set, half dog point)
hardenability 380, 381, 406
hardening (see also: heat treatment)
strain 401
hardness (see also: heat treatment) 375, 399
achieving 398
case hardening properties 405

common values 401
materials 384, 386, 387
on drawings 399
ranges achievable 381, 382, 404
scales 399, 400
shafts 507, 509, 549
shear 400
surface treatment 408, 409
tensile strength 375, 399
wear resistance 375, 382
hardware (see also: fasteners)
(see also: washers)
pins, keys, and retaining rings 215
pipe threads, threaded fasteners, and washers 279
hardware tables 188
hardware, locating (see: pins)
HAZ 360
hazard (see: risk)
heat
expansion (see: thermal expansion)
safety 66
heat affected zone 360, 365
heat treatment (see also: hardness) 398
best practices 407
critical considerations 406
distortion 403, 402, 406
processes 401, 402
resources 399
helical coil compression springs (see also: springs)
active coils 422, 431, 432
allowable stress 427, 433
buckling 427, 430, 431
captivation in assembly 427, 429
catalog selection steps 429
clearance around 427, 431
deflection 422, 423, 426–428, 431, 432
design steps 437
end types 427, 428
fatigue 423–426, 433
force vs. deflection 426–428, 432
geometry equations 432
length 426, 427, 431, 432
linear range 423, 426
natural frequency (surge) 423, 432
parallel 426, 427, 432
series 426, 427, 432
sizing equations 431
solid height 426, 431, 432
spring constant 422, 432
spring index 422, 431
spring surge 423

stability 430
static service 426, 433
stress 432, 433, 422–424, 427, 429
wire diameter 421, 422, 431
helical coil extension springs (see also: springs)
    active coils 422, 443
    allowable stress 423–425, 441, 446
    captivation in assembly 440
    catalog selection steps 443
    deflection 422, 423, 441–444
    design 447
    ends 440–442
    fatigue 423–426
    geometry 441, 442, 442
    hooks 440–443
    initial tension 441, 444
    length 440, 441, 443
    loops 440–443
    natural frequency 423, 444
    overload 441
    parallel 441, 442, 444
    series 441, 442, 444
    shock loading 441
    spring constant 422, 444
    spring index 422, 443
    spring sizing equations 443
    spring surge 423
    stress 423, 424, 426, 441, 442, 444, 446
    wire diameter 421, 422, 443
helical coil threaded inserts (see: threaded inserts)
helical gear (see also: gears) 584, 606
    backlash 593
    equations 607
    helix angle 606
    loads on shafts 551, 606
    lubrication 596, 598
    materials 595
helix angle 606
hex head cap screws (see: screws, hex socket head cap)
hex lock nuts (see: nuts, prevailing torque hex lock nuts)
hex socket button head cap screws (see: screws, hex socket button head)
hex socket flat countersunk head cap screws (see: screws, hex socket flat)
high collar lock washers (see: washers, high collar lock washers)
high speed tool steel 381, 383, 385
holes
        depth, tap drills 288, 293

processes, tolerances, costs 190
    tap drills 288–293
hollow shafts (see: shafts)
hook stress (see: extension springs)
hoop tension 633, 638
horsepower 33
    belts and chain 632, 637, 640, 644, 649, 658, 660
    motors 483, 497
    shafts 550
    gear 602
hot work tool steel 381, 383, 385
HP (see: horsepower)
human factors (see: ergonomics)
hypoid bevel gear 586

I

IB 161, 162, 163
idler 587
Imperial units 15
implied tolerance 117
inch-pound-second system 16
induction hardening 402, 405
inertia
    couplings (see: couplings)
    gearboxes (see: gearboxes)
    motors (see: motors)
    ratio 497
initial tension
    bolts 295, 297
    springs 441, 444
inner boundary 161, 162, 163
inserts, threaded (see: threaded inserts)
interchangeability 133, 187, 214
interference, gear 580, 581, 592
interference fit (see: fits)
intermediate column buckling 25, 670
internal retaining rings (see: retaining rings)
involute 578, 579
IPS (inch-pound-second system) 16
IT tolerance grades 97, 101

J

jam nuts (see: nuts, thin)
jaw coupling 569, 573
joints, welded 360
journal 504, 505

K

keys 223
    bearing stress 217, 218
    best practices 227
    critical considerations 226
    feather 223, 224
    length 223
    materials 217, 218
    parallel 223, 224
    parallel, data, inch 260, 261
    parallel, data, metric 262, 263
    resources 217
    shafts 219, 223, 224, 547, 548, 563, 589
    shear 217, 218
    stress 218
    surface finish 396
    tapered 224
    Woodruff 224
    Woodruff, data, inch 264, 265
    Woodruff, data, metric 266
kinematic fixturing 186
K_v 474

L

lap joint 360
laser 66
lay (see: surface finish)
lead screw (see also: linear motion screws) 665, 671–673
    ACME thread 673
    backdrive 666, 673
    buttress 673
    design or dynamic load 673
    efficiency 667, 668, 673
    forces on shafts 617, 667–669
    friction 668
    plastic nut 674
    power 667, 668
    preload 673
    PV 674
    selection and sizing 675
    speed 671, 674
    static load rating 677
    torque 667, 668
least material boundary (see: LMB)
least material condition (see: LMC)
leg of weld 360
Lewis formula with Barth revision 600, 602
lifting 38, 41, 46, 67

limits (see also: fits) 95, 96
    best practices 116
    commonly used 99
    critical considerations 116
    of size 95, 97, 104–115
linear bearings 534
    ball bushings 538, 539
    ball splines 540
    best practices 542
    critical considerations 542
    crossed roller ways 541
    loads 537, 538
    lubrication 535
    plain bearings 535, 536
    plain, boundary lubrication 537
    plain, PV 536, 537
    plain, sleeve 535
    rails and guides 540
    resources 534
    rolling element 537
    rolling element, life & loads 538
    slide configurations 536
    speed 534, 532
    torque 535, 538, 540
linear motion equations 31
linear motion screws 664
    accuracy or error 666
    backdrive 666, 667
    backlash 666
    ball (see: ball screw)
    best practices 680
    buckling 669, 670
    calculations 667, 670, 671
    characteristics 664
    critical considerations 680
    critical speeds 670, 671
    deflection 669, 670
    efficiency 666–668, 671–674
    Euler's formula for buckling 670
    geometry 668, 669
    lead 666
    lead (see: lead screw)
    number of starts 666
    pitch 664
    power 667
    resources 664
    roller 672
    root diameter 666
    selection 675
    speeds, critical 670, 671
    stress 669, 670
    torque 667, 668
linear rails and guides 540

LMB 161–163
LMC 126, 152, 154, 163, 164
loads on bolts 295–298
locating pins, stepped (see: pins, stepped locating)
locating techniques 179
    application summary matrix 213
    assemblability 184, 185,
    187–189, 214
    best practices 212
    composite tolerance 196–199
    considerations, manufacturing
    189, 190
    critical considerations 212
    datum 184, 185, 186
    design process 187
    design requirements 181
    differently sized holes 196
    dual position control frames 199
    equation summary 214
    exact constraint design 181, 186
    fixed and floating fasteners
    183–185
    functional datums 185
    GD&T 183, 185
    hardware 188, 191, 194, 195, 202,
    203, 206–208
    interchangeability 187, 214
    kinematic fixturing 186
    manufacturing considerations 190
    minimum constraint design 186
    MMC 183–185
    press-fit 182, 184, 185, 188, 214
    repeatability error 186, 196,
    210, 211
    resources 181
    RFS 185
    the two hole problem 182
    three holes / three pins / two faces
    210, 213, 214
    two holes / one round pin and
    one diamond pin / two holes 204,
    213, 214
    two holes / two pins / one hole and
    one slot 199, 213, 214
    two holes / two pins / one slot 209,
    213, 214
    two holes / two pins / two holes
    192, 213, 214
lock nuts (see: nuts, prevailing torque
hex lock nuts)
lock washers (see: washers, high collar
lock washers)

lockout/tagout 69, 81, 90
loss probability 686
low head cap screws (see: screws, low
head)
lubrication
    boundary 503, 505, 506
    bushings 506, 509
    full film 503, 504
    gears 596, 598
    hydrodynamic 504
    hydrostatic 504
    machine safety 69
    mixed film 505
    plain bearings 503, 506, 509
    plain linear bearings 535, 537
    rolling bearings 511, 530
    thin film 503

M

machinability 373, 380, 383
machine design (see: design)
machine efficiency 695, 696
    cost 696
    quality rate 695, 696
machine manuals 50, 51, 53, 63, 71, 75,
82, 90
machine performance 695
    best practices 699
    critical considerations 698
    metrics 696, 697
    resources 683
machine reliability 683
    best practices 699
    calculation 684
    critical considerations 698
    FMECA (see: FMECA)
    mean time between failures
    (MTBF) 684
    mean time to failure (MTTF) 684
    resources 683
machine safeguarding 51
    books 74
    codes 72
    critical considerations 65
    designed-into-the-machine 52
    guards, adjustable 55–57
    guards, barrier 52–54, 56–58, 61
    guards, dimensions of openings 55
    guards, fixed 55–57, 80, 88
    guards, interlocking 55 -58, 60, 77,
    80, 88

guards, self-adjusting 55–57
instructions 51, 53, 62–64, 66–68, 70, 82, 90
manuals 51, 53, 62–64, 75, 82, 90
procedural 62, 52, 53
protective devices 52, 53, 58–62, 80, 81, 87, 88
protective devices, gates 56, 58, 61
protective devices, proximity detection 58, 60
protective devices, pull-back 53, 58, 59
protective devices, restraints 58, 59
protective devices, two-hand controls 53, 58, 60, 61, 77, 81
regulations 72
resources 71
standards 76, 37, 40, 46, 47, 49, 51, 55, 56, 58, 65, 68, 70–73
warnings 51–53, 62–65, 74, 90
web sites 84, 85
machine safety (see also: machine safeguarding) 35, 47, 51, 65, 71
atmosphere 66, 68, 82
books 74
codes 72
critical considerations 51, 65, 71
design process 47
emergency stop 70, 71, 81, 90
emission of airborne substances 66, 90
explosive atmospheres 68, 82
goals 65
heat 66, 80, 86, 90
laser 66
light 66, 69
lockout/tagout 69, 81, 90
lubrication 69
moving the machine 68
noise 66, 67, 91
other issues 65
radiation 66
regulations 55, 69, 72
resources 71
signals 69, 81, 89, 90
stability 68
standards 40, 46, 47, 49, 51, 55, 56, 59, 65, 68–72, 76
vibration 66–68
warning signals 69, 81, 89, 90
web sites 84
machine screw nuts (see: nuts)
machining

at assembly 133, 134
surface finish 395
tolerances 102, 190
maintenance 39, 54, 62, 70, 133, 685
manuals 63, 75, 82, 90, 51, 53, 62
manufacturing equipment (see: machine)
manufacturing method tolerance 102, 190
mass density 374, 385, 387, 388
mass moments of inertia equations 22, 617, 618
materials 371
AISI 371, 372
alloy steel 372, 377, 380, 384, 385
aluminum 373, 377, 379, 387
bearings 506, 530
best practices 391
carbon content 371, 373, 378, 380, 406
carbon steel 373, 380, 402
ceramics 376, 378
code systems 371–373
cold rolled steel 380, 381
commonly used 218, 365, 366, 371, 376, 377
compressive strength 374, 388
corrosion resistance 375, 378, 379, 407–409
critical considerations 390
ductility 376, 380, 399
fastener 294
gears 595, 597, 603
hardness (see: hardness)
heat treatment (see: heat treatment)
hot rolled steel 380, 381
machinability 373, 380, 383
mass density 374, 385, 387, 388, 419
metals nomenclature 371–373
modulus of elasticity 375, 418, 461
modulus of rigidity 375, 386, 387, 419, 461
non-metals 376–378, 388
notch sensitivity 382
plastics 376–378, 388
properties 372, 383–388, 419, 420, 461
resources 371, 372
SAE 295, 371, 372
shafts 548, 549
shear ultimate strength 374, 383
shear yield strength 374

stainless steel 378, 386, 387, 418, 419, 461
steel numbering 371, 372
steel, alloy 372, 377, 380, 384, 385
steel, carbon 373, 380, 402
steel, cold rolled 380, 381
steel, hot rolled 380, 381
steel, properties 373–376, 380, 384, 385, 419, 420, 461
steel, stainless 378, 386, 387, 418, 419
steel, surface quality 381, 418
steel, tool 372, 381–383, 385, 403
stock sizes 389, 421
tensile yield strength 295, 374, 382, 399
thermal conductivity 376
thermal expansion 375, 376, 379
titanium 379, 386, 387
tool steel 372, 381–383, 385, 403
ultimate shear strength 374, 383
ultimate tensile strength 374, 383, 420
wear resistance 382, 373, 378, 379, 381, 400, 409, 410
weldments 365, 366
yield strength, tensile 295, 374, 382, 399
matrix, decision or Pugh 9
max material condition (see: MMC)
maximum material boundary (see: MMB)
mean 137, 138, 141–143, 159
mean time between failures (see: MTBF)
mean time to failure (see: MTTF)
metals (see: materials)
meter-in circuit 468
meter-out circuit 468
metric
prefixes 16, 17
units 16, 17
minimum constraint design 186
miter gear (see also: bevel gear) 582, 607
MMB 161, 163, 169, 170
MMC 126, 152, 153, 156, 163, 169, 170, 183–185
modified Goodman diagram 423, 556
modifier
effect on tolerance 126, 152–155, 161, 163, 169, 170
LMC 126, 152, 154, 163, 164
MMC 126, 152, 153, 156, 163
RFS 126, 185

module, gear 576, 601
modulus of elasticity 375, 418, 419
modulus of rigidity 375, 419
moment of inertia, gearboxes (see: gearboxes)
moments of inertia, area, equations 21
moments of inertia, mass, equations 22, 617, 618
Mont Carlo simulation 145, 160, 175
motors 482
 AC (see: AC motors)
 best practices 499
 C-face 493
 constant speed 484
 continuous motion 484, 486, 487, 495
 controls 490, 491, 493
 critical considerations 499
 DC (see: DC motors)
 D-flange 493
 duty cycle 484, 494
 electrical power 482
 enclosures 484, 491, 493, 494
 frames 484, 493, 494
 full-load speed (RPM) 484, 489
 gearboxes (see: gearboxes)
 horsepower 484, 486, 488
 IEC 482–484, 486, 488, 493, 494
 inertia 489, 494, 495, 497
 INSUL CLASS (insulation class) 484
 intermittent motion 484, 486, 494
 kW 483, 497
 manufacturer's type 483
 maximum ambient temperature 483
 mounting 488, 493, 494
 NEMA 482–484, 486, 487, 493, 494
 ODP 494
 overheating 483, 495
 overload protection 491
 rated horsepower 483, 488
 resources 482
 RMS torque 495, 497
 selection 487, 491
 sizing 494, 497
 speed control 489, 491
 starting 490
 stopping 491
 terminology 483, 484, 490
 torque 483, 486, 488, 489, 494, 495, 497

moving equipment 68
MTBF 684
MTTF 684

N

natural frequency
 shafts 556–559
 springs 423, 432, 444
needle roller bearings 517
needle roller thrust bearings 518
NEMA 482, 483, 484, 486, 487
Neuber's constant for notch sensitivity 28, 30, 557
NIOSH 46, 67, 85
nitride 382, 409–411, 587
noise 66, 67, 91
 of gearboxes (see: gearboxes)
non-metals 376, 378, 389
non-normal distribution 145
normal distribution 137–139
normal, standard (see: standard normal)
normalizing 401
notch sensitivity 28, 30, 382, 557
nuts (see also: fasteners) 304, 287, 294
 hex machine screw nuts 343
 hex nuts, data, inch 342, 343
 hex nuts, data, metric 344
 hex thin nuts (jam nuts) 304
 hex thin nuts, data, inch 342
 hex thin nuts, data, metric 344
 linear motion screws (see: linear motion screws)
 prevailing torque hex lock nuts 305
 prevailing torque hex lock nuts, data, inch 344
 prevailing torque hex lock nuts, data, metric 345
 stripping 295

O

OB 162, 163
OEE 697, 698
Oldham coupling 569, 573
open center circuit 467
operator capabilities and limitations 37, 41–43, 86

orientation
 factor 157, 158, 160
 tolerances 173, 122, 147, 151, 160
 variation 155, 158, 159, 171
outer boundary (see: OB)
overall equipment effectiveness (see: OEE)
overheating motors 483, 495

P

parallel keys (see: keys)
parallelism 122, 123, 173
passivation 408, 410, 411
PEEK 376–388
performance (see: machine performance)
perimeters and areas equations 20
perpendicularity 122, 123, 147, 173
physical capabilities and limitations 37, 41–43, 86
pinion 578, 580, 581, 587, 592, 594, 595
pins 219
 bearing stress 217, 218
 clevis (see: clevis pins)
 coiled spring (see: spring pins)
 cotter (see: cotter pins)
 datum: practical considerations 186
 diamond (see: stepped locating pins)
 dowel (see: dowel)
 fixed and floating fasteners 183
 grooved (see: grooved pins)
 locating (see: stepped locating pins)
 materials 218
 precision locating 182, 191
 precision locating dimensions 188, 194, 195, 202, 203, 206–208
 pullout dowel (see: dowel, withdrawal)
 resources 217
 shear 217, 218
 slotted spring (see: spring pins)
 spring (see: coiled spring pins)
 stepped locating (see: stepped locating pins)
 stress 217, 218
 types, selector 219–222
 withdrawal dowel (see: dowel, withdrawal)
pipe and port threads 281–285

pitch
  circular 576, 577, 579
  diametral 576, 577, 579, 592
  gear (coarse and fine) 579, 580
  linear motion screws 664
pitch diameter
  belts and chains 628, 630, 631
  gears 576, 577, 589
pitch circle, gears 577, 578, 581, 587, 591, 592
pitch point 577
plain bearings 503
  bearing pressure 504, 507
  best practices 510
  bushings 503, 534
  critical considerations 510
  friction 503–507
  heat & temperature 506, 507
  length 507
  linear (see: linear bearings, plain)
  loads 507, 508
  lubrication 503–506
  materials 506, 508
  oil-free 505
  plastic 506
  PV 507, 508
  resources 503
  selection and sizing 509
  shafts 505–508
  sleeve 506
  thrust 506
planetary gear trains 582, 589
plastics 376, 377, 388
plug weld 360
pneumatics 466
  actuator symbols 471
  actuator, air cylinder 468
  actuator, compression ratio 475
  actuator, Cv 474, 475
  actuator, double acting 468
  actuator, flow rate 474, 475
  actuator, force 472
  actuator, kinetic energy 472, 473, 475
  actuator, misalignment 473
  actuator, rod buckling 473
  actuator, rotary 468, 470, 472–475
  actuator, single acting 468, 478
  actuator, sizing procedure 472
  actuator, speed 468, 472, 478
  actuator, wear 473

  best practices 481
  circuits 467
  closed center circuit 467
  critical considerations 480, 481
  Cv 472, 480
  Cv calculation 474, 475
  kinetic energy, actuator 472, 473, 475
  meter-in circuit 468
  meter-out circuit 468
  open center circuit 467
  pressure 467
  regulation 467
  reservoir 467
  resources 466
  speed, actuator 468, 472, 478
  symbols 468, 469, 471, 479
  valves 476, 478, 479
  valves, ports 477, 478
  valves, positions 477, 478
  valves, sizing 480
  valves, symbols 479
  valves, vacuum 477
  valves, ways 477, 478
poka-yoke 11
port threads (see: pipe and port threads)
position tolerance 125, 151, 155, 160, 161, 169, 185
power
  AC 482, 484
  belts and chain 632, 637, 640, 644, 649, 658, 660
  DC 482
  equations 33
  linear motion screws 667, 668
  shafts 550, 553
  v-belt (see: v-belt)
power screws (see: lead screw)
power transmission devices 543
precision locating (see: locating techniques)
pre-hardening 407
preliminary hazard analysis 50
preload, bolt 295, 297, 298
press-fit criterion 214
pressure
  air 467
  angle 576, 577, 579–581, 610
prevailing torque hex lock nuts (see: nuts, prevailing torque)

preventative maintenance (see: maintenance)
procedural safeguarding (see: machine safeguarding)
process capability index 141, 142, 188
process efficiency (see: machine efficiency)
profile 124, 125, 168, 171–174
proof 295–297
protective devices (see: machine safeguarding)
PTFE 281, 376, 388, 409
Pugh matrix 9
pulley 627
pullout dowel (see: dowel, withdrawal)
PV
  lead screw 674
  plain bearings 507, 508
  plain linear bearings 536, 537

Q

quality rate 695, 696
quality ratings, gears 589–591
quenching 401

R

rack and pinion (see also: gears) 579, 585, 587
  forces on shafts 618
  interference 581
  tooth form 578–580
radiation 66
Raleigh's method 558, 559
rate, repair 684
rated horsepower, motors 483, 488
ratio
  gear 576, 577, 582, 587, 588
  gearboxes (see: gearboxes)
Rc (see: Rockwell hardness)
reaching 38, 41, 42, 46
reducer (see: gearboxes)
regardless of feature size (see: RFS)
regulator 467, 479
reliability (see: machine reliability)
repair rate 684
repeatability error 186, 196, 210
research design 8
retaining rings 225

best practices 226
clearance 268, 270, 272, 274–277
critical considerations 226
designations 225, 226
e-ring, data, inch 276
e-ring, data, metric 277
external, data, inch 272, 273
external, data, metric 274, 275
grooves 225, 268, 269–277
internal, data, inch 268, 269
internal, data, metric 270, 271
resources 217
standards 225, 226
thrust 225
types 225, 226
RFS 126, 185
right triangle equations 19
rigid coupling 564, 565, 573
risk
    and reliability 685, 689, 691
    and safety 47, 51, 62, 69, 74, 76, 82, 86, 91
    safety, analysis 48, 49, 51, 73
    safety, assessment 47–49, 51, 78, 79, 83, 85, 86, 89
    safety, evaluation 49
    safety, reduction 47–51, 53, 62, 73, 78, 83, 85, 86
risk priority number (see: RPN)
RMC torque 620, 625
RMS torque 495, 496, 497
Rockwell hardness 399, 400
roll pins (see: spring pins)
roller chain (see: chain)
roller screw 672
rolling bearings 510
    ABEC 520–523
    angular contact ball bearings 515, 519, 520
    angular misalignment 514–518, 520, 531
    applications 519
    applied load 527, 528
    arrangement design 525–527
    axial constraint 525–527
    axial displacement 512, 525
    axial load 514–520, 525, 527
    basic dynamic load rating 513, 527 - 529
    bearing arrangements 525–527
    best practices 534
    bore 512, 520
    cage 512, 518

captivation 525–527
characteristics 512
clearance, internal 513, 521, 525
combined load 515, 517, 519, 520, 525, 527, 528
critical considerations 533
cylindrical roller bearings 516
deep groove ball bearings 514
dimensions and tolerances 520–524
drag 530
dynamic load 513, 527–529
equivalent load 527, 528, 530
failure 518, 529, 530
fits 521–523
fluctuating load 528
forces 527, 528
four-point contact ball bearings 515, 516
friction 530
grease, speeds 511
heat and temperature 513
housing dimensions and tolerances 520–524
internal clearance 513, 521, 525
life 529, 530
limiting speed 513
load rating 513, 527, 529
loads 527, 528
lubrication 511, 513, 530
maintenance free 511
minimum load 528
needle roller bearings 517
non-preloaded 525
operating speed 513
precision 520, 521
preloaded 525
race 512, 521
resources 511
safety factor 530, 531
seals 511, 520
selection and sizing 519, 520, 531
self-aligning ball bearings 514, 515
separable 512, 513, 517, 518, 520
service factor 530
shaft dimensions and tolerances 520–524
shaft finish 521
shaft misalignment 514–517, 519, 520, 554
shields 511
shock 530
speed 511, 513, 514, 518, 519, 525
spherical roller bearings 516, 517

starting torque 514
static load 513, 527–529
tapered bore 512, 526
tapered roller bearings 516, 517
temperature 511, 513
thrust bearings 518–520, 528
tolerances 520–524
types 514, 519, 520
root
    gear tooth 580
    weld 360
root diameter
    gear 600, 601
    linear motion screws 666, 669, 671
root mean cubed (see: RMC)
root mean square (see: RMS)
root sum of squares method 138, 142, 146
rotary actuator (see: pneumatics, actuator)
rotary motion equations 32
rotating shafts (see: shafts)
roughness (see: surface finish)
RPN 689–691
rule 8
runout
    circular 123, 124, 174
    gears 589, 591
    total 124, 174

S

SAE 295, 371, 372
safeguarding (see: machine safeguarding)
safety (see also: machine safety)
    resources 71
    and ergonomics 38
safety category 693–695
safety factors 13, 14
scale, measuring 8
scrap 696, 697
screw selector slide chart 13
screws (see also: fasteners)
    ball (see: ball screw)
    clearance 302–304, 310, 312, 314, 316, 330, 332, 334, 336, 338, 340
    hex head cap screws 304
    hex head cap screw, data, inch 338, 339
    hex head cap screw, data, metric 340, 341

hex socket button head cap screws 304

hex socket button head cap screw, data, inch 330, 331

hex socket button head cap screw, data, metric 332, 333

hex socket flat countersunk head cap screws 304

hex socket flat countersunk head cap screw, data, inch 334, 335

hex socket flat countersunk head cap screw, data, metric 336, 337

hex socket head cap screws 301

hex socket head cap screws, data, inch 310, 311

hex socket head cap screws, data, metric 312, 313

hex socket head shoulder screws 303

hex socket head shoulder screw, data, inch 326, 327

hex socket head shoulder screw, data, metric 328, 329

hex socket set screws 302, 303

hex socket set, cone point, data, inch 318

hex socket set, cone point, data, metric 322

hex socket set, cup point, data, inch 319

hex socket set, cup point, data, metric 323

hex socket set, dog point, data metric 324

hex socket set, flat point, data, inch 321

hex socket set, flat point, data, metric 325

hex socket set, half dog point, data, inch 320

hex socket set, half dog point, data, metric 324

hex socket set, locking 303

lead (see: lead screw)

low head hex socket cap screw 302

low head hex socket cap screws, data, inch 314, 315

low head hex socket cap screws, data, metric 316, 317

roller (see: roller screw)

stripping 295

seals, rolling bearings 511, 514, 520

self-aligning ball bearings 514, 515

set screws (see: screws, hex socket set)

shaft speed, critical 26, 556, 559

shafts 547

    attachment 548, 563, 589

    axial loads 553

    axial location 525–527

    bearing, plain 504–509

    bearing, plain linear 535, 538, 540

    bearing, rotating 512, 513, 518, 522, 523, 525–527

    best practices 562

    couplings (see: couplings)

    critical considerations 562

    critical speeds 26, 556–559

    deflection 553, 554, 560

    design 558

    diameter 561

    fatigue 555, 556, 561

    forces on 549, 550, 553, 617, 618

    hardness 548, 549

    hollow 548

    keys 219, 223, 224, 547, 548, 563, 589

    materials 548, 549

    misalignment 564

    motors, NEMA 637

    natural frequency 556, 557

    notch sensitivity 557

    power 553

    resources 547

    rigidity 548, 558

    set screws 303, 548

    speed 26, 556–559

    splines 540

    stepped 553, 554

    stress 548, 555, 556, 561

    stress concentration 548, 555

    tolerancing for bearings 521–523

    vibration 557–559

    whirl 557–559

shear

    keys 217, 218, 223

    pins 217–219, 221, 222, 238, 240, 242, 244, 245, 248, 250

    shafts 554, 555

    stress 27, 217, 218

    ultimate strength 374, 383

    yield strength 374, 382, 383

sheave 627, 630, 641

shock resisting tool steel 382, 383, 385

shoulder screws (see: screws, hex socket head shoulder)

signals, warning 69, 81, 89, 90

single acting 468, 478

six-sigma 142, 146

sleeve bearings (see: plain bearings or plain linear bearings)

slender column equations 25, 670

slide chart, screw selector 13

slides (see: linear bearings)

slit coupling 567, 573

slotted spring pins (see: spring pins)

snap rings (see: retaining rings, e-rings)

socket flat head cap screws (see: screws, hex socket flat)

socket head cap screws (see: screws, hex socket head cap)

socket head shoulder screws (see: screws, hex socket head shoulder)

SolidWorks®

    fasteners and washers 309

    pins, keys, and retaining rings 278

SPC 137, 142

specification

    design 3

    limits 137, 141

speed (see also: velocity)

    bearing, plain 503, 506, 507

    bearing, rolling 511, 513, 519, 537

    belts 627–630, 632, 634, 645, 647, 649

    chain 627, 628, 655, 658, 660

    couplings 564, 565, 567, 568, 571

    gearboxes 614, 615, 620, 621, 623, 625

    gears 589, 591, 596, 598, 599, 607

    linear motion screws 670–672, 674

    shafts 26, 556–559

speed ratio, belts and chains 627, 645, 647, 649

spherical roller bearings 516, 517

spherical roller thrust bearings 518

spiral bevel gear 585, 608

split pin (see: cotter pin)

spring anchor pins (see: grooved pins)

spring pins

    coiled 221

    coiled, data, inch 238, 239

    coiled, data, metric 242, 243

    slotted 221

    slotted, data, inch 240, 241

    slotted, data, metric 244, 245, 247

springs 415

    active coils 422

    Belleville (see Belleville springs)

best practices 465, 466
clearance 430
clock, power, or motor 418
compression (see: compression springs)
constant force 417
critical considerations 465
drawbar assembly 416
endurance limit 425
extension (see: extension springs)
fatigue 423, 424, 425
garter 418
gas 418
hair or spiral 417
helical coil compression (see: helical coil compression)
helical coil extension (see: helical coil extension)
helical coil terminology 422
helical coil torsion (see: torsion)
high temperature 418, 419, 422
leaf or beam 417
materials 418–420
mean coil diameter 422
natural frequency 423, 432, 444
pitch, helical coil 426, 432
resources 415
shock loading 419, 441, 457
spring constant 422, 432, 444, 452
spring index 422, 432, 443, 451
spring rate (see: spring constant)
spring washers 416, 423, 456
stress 423–427, 429, 432, 433, 441, 442, 444, 446, 451, 452, 459–461
stress relaxation 418, 422
surface quality 418
surge 423, 441
torsion (see: torsion springs)
torsion bar 417
types 415
volute 417
wire diameters 421
sprocket
chain 627, 655, 661
timing belt 627, 630
spur gear (see also: gears) 582, 583, 599
addendum 576, 577, 581, 601
backlash 591–594
center distance 576–578, 591–594, 600, 601
circular pitch 576, 577, 579, 600, 601
clearance 576, 591

conjugate action 577, 578
contact ratio 580, 601
dedendum 576, 577, 601
diametral pitch 576, 579, 580, 592, 600, 602
equations 600–602
horsepower 602
interference 580, 581, 592
involute 578, 579
Lewis formula with Barth revision 600, 602
lubrication 592, 596, 598, 599
module 576, 601
number of teeth 576, 581, 588, 594, 600–602
pinion 580, 592, 594, 595
pitch 579, 580, 589
pitch circle 578, 581, 591, 592
pitch diameter 576, 600–602
pitch line 591, 596, 599, 602
pitch point 577
pressure angle 576, 577
root 580, 600, 601
selection and sizing 599
shafts 589, 592
stress 595, 600, 603
terminology 576, 577
tooth form 577–580, 601
tooth strength 592, 599, 600
undercutting 580–582, 601
velocity ratio 577, 578, 582, 587, 588
stack-ups (see: tolerance stack-ups)
stainless steel 378, 386, 387, 408, 418, 419, 461
standard deviation 137, 138, 141–144, 159
standard normal 138–141
statistical
tolerance stack-up 137
process control 138
steel (see: materials)
stepped locating pins 191, 220
data, metric 237
precision locating dimensions 202, 206–208
straight bevel gear (see: bevel gear)
straightness 120, 121
strain and stress 27, 375
strain hardening 401
strength (see also: materials)
endurance 29, 297, 424, 425
shear 221, 222, 303, 374, 379, 383

tensile 295, 374, 375, 379, 383, 399, 418, 422
stress
axial 27
bearing 27, 218, 223
bending 27
bolts 295–297
coil (see: springs)
hook (see: springs)
keys 217–219
pins 217, 218
shear 27, 218, 223, 295, 374, 375, 423, 432
tensile 27, 295–297
three dimensional 27
torsion 27, 423, 446
Von Mises 27
stress and strain equations 27
stress relieving 401, 403
weldments 366
stripping 295
surface areas and volumes equations 21
surface finish 391
best practices 398
common values 395–397
comparator 13
critical considerations 398
lay symbols 393
machining 397
resources 391
symbols 392–394
tolerances 394
surface roughness (see: surface finish)
surface texture (see: surface finish)
surface treatment 407
anodizing 409–411, 379
best practices 412
black oxide 408, 410, 411, 294, 366
buildup 408–410
chrome plating 408, 410, 411
corrosion resistance 375, 379, 380
critical considerations 412
electroless nickel plating 409–411
electroplating 408, 410, 411
electropolishing 408, 410, 411
friction reducing 383, 410
hardness 399, 408, 409
nitride 382, 409–411
passivation 408, 410, 411
properties 408–411
resources 407
roughness 408
selection 410, 411

titanium nitride 409–411, 382
types 408–411
wear resistance 383, 408–410
weldments 366
zinc electroplating 408, 410, 411, 295
synchronous
belt (see: timing belt)
motor 484, 487
speed 484, 486, 497
synthesis, design 9

T

tap drills 288
depths 288, 293
sizes 289–292
tapered keys 224
tapered roller bearings 516, 517
tee joint 360
tempering 401
tensile load on bolts 295–297
tensile yield strength 374
tension
chain 655, 658
flat belt drive 629, 632–634
timing belt drive 630, 649, 650
v-belt drive 638
thermal conductivity 376
thermal expansion 18, 375, 379, 384–387
thermoplastic 376
thermoset 376
thickness
gauge 13
sheet metal 389
steel stock 389
thin nuts (see: nuts)
threaded fasteners (see: fasteners)
threaded inserts 306
best practices 308
data, inch 352, 353
data, metric 354, 355
installation 307
threads, fastener (see: fastener threads)
threads, pipe and port (see: pipe and port threads)
three holes / three pins / two faces 210
three-phase power 482
throat of weld 360
through hardening 400, 402, 403

thrust bearings 506, 518–520, 527, 528, 553, 673
tightening torque 298–301
timing belt (see also: belts) 629, 630
calculations 649
forces on shafts 550, 617, 650
pitch 631
pitch diameter 649
power 649
selection and sizing 648
service factor 650
speed 628, 629, 649
sprocket 627, 630
tension 630, 649, 650
tooth shear 630
width 651
titanium 379, 386, 387
titanium nitride 382, 409–411
toe of weld 360
tolerance (see also: fits)
(see also: limits)
analysis (see: tolerance stack-ups)
and dimensions 93
assignment 133–135
best practices 116, 127, 177
critical considerations 116, 126, 176
grades 95–97, 100, 102
hardware tables 188, 194, 195, 202, 203, 206–208
implied 117
machining 102, 190
on drawings 116
resources 95, 116, 128
surface finish 394
tolerance stack-ups 127
angularity 173
application summary 175
applications 146
best practices 177
boundary example problem: spring pin 164
boundaries 161–163
calculation 128, 129, 134
chain 129, 132
combined (worst case / statistical) 143, 144
critical considerations 176
datum shift 168–171
design practice 128
floating and fixed fasteners 161
form tolerance on datum surface 148

hardware tables 188, 194, 195, 202, 203, 206–208
LMC 152, 163, 164
mean 137–139, 141–143, 159
MMC 152, 153, 156, 163
Monte Carlo 145, 146, 160, 175
multiple assembly conditions 146
normal distribution 137–139
orientation tolerances 173
parallelism 173
perpendicularity 173
plus/minus tolerancing and GD&T 147
position tolerance 151
procedure 129
profile tolerances 171
resources 128
root sum of squares 138, 142, 146
runout tolerances 174
standard deviation 137, 138, 141–144, 159
statistical 137, 138, 141–146, 159, 175
two dimensional: location and orientation variation 155
worst-case 135, 136, 143–145, 163, 175
tolerancing (see: tolerance)
tool steel 372, 381–383, 385, 403
torque
couplings (see: couplings)
gearboxes (see: gearboxes)
linear motion screws (see: linear motion screws)
tightening, bolts 298–301
torsion springs 416, 417, 450
(see also: springs)
active coils 422, 451
bending stress 452
body length 451
catalog selection steps 451
deflection 423, 451, 452
design steps 454
direction 450
end types 450
fatigue 423, 425
mandrel 451
natural frequency 423
parallel 452
sizing equations 451
spring constant 422, 452
spring index 422, 451
stress 452

torque 451, 452
    wire diameter 421, 422, 451
total machine efficiency 696
total runout 124, 174
transition fit (see: fits)
triangle equations 19
two holes / one round pin and one diamond pin / two holes 204
two holes / two pins / one hole and one slot 199
two holes / two pins / one slot 209
two holes / two pins / two holes 192
    composite tolerance 196, 198
    dual position control frames 199
    variation: differently sized holes 196
two-hole locating (see: locating techniques)

U

u-joint 565, 566
ultimate shear strength 374, 383
ultimate tensile strength 374, 383, 420
undercutting 580–582
undervoltage 483
units
    commonly used 16
    conversions 17
    engineering 15
    foot-pound-second system (FPS) 16
    Imperial systems 16
    inch-pound-second system (IPS) 16
    metric prefixes 17
    resources 15
uptime 684

V

valves (see: pneumatics, valves)
variation (see also: tolerance)
    assembly 133
v-belt (see also: belts) 627, 630
    bending frequency 638
    calculations 628, 629, 637, 638
    classical 630, 631, 640
    cross section 630, 631
    Eytelwein's equation 636, 638
    forces on shafts 550, 638
    narrow 630, 640
    pitch diameter 628, 630
    power 637, 644

selection and sizing 636
service factor 639
sheave 627, 630, 638, 641
slip 629, 638
speed 628, 630
tension 638
torque 638
velocity (see also: speed)
    linear motion 31
    ratio, gears 576, 578, 582, 587, 588, 592, 600
    rotary motion 32
vibration 66, 67, 221, 286, 305, 306, 493
    Raleigh's method 558, 559
    shafts 556–559
Vickers 399, 400
volumes and surface areas equations 21

W

warnings 51–53, 62–64, 74, 75, 90
warping (see: distortion)
washers
    best practices 299
    critical considerations 298
    fender (see: washers, wide)
    flat (see: washers, plain)
    heavy duty 305
    high collar lock washers 306
    high collar lock washers, data, inch 350
    high collar lock washers, data, metric 351
    lock washers (see: washers, high collar)
    plain regular flat washers 305
    plain regular flat washers, data, inch 346
    plain regular flat washers, data, metric 348
    resources 285
    spring (see: springs, spring washers)
    standard (see: plain regular flat)
    wide 305
    wide washers, data, inch 347
    wide washers, data, metric 349
waviness (see: surface finish)
weld (see: welds)
welded joints 360, 365, 366
weldments (see also: welds) 359
    best practices 368
    critical considerations 367

drawings 364, 365
fatigue 359, 366
materials 365, 366
resources 359
strength 359, 365
stress relieve 366
surface treatment 366
welds (see also: weldments) 359
    best practices 368
    critical considerations 367
    resources 359
    symbols 361–364
    terminology 360
    types 360
what if analysis 49
whirl 557, 558, 670, 671
wide washers (see: washers, wide)
wire diameters, spring materials 421
withdrawal dowel (see: dowel, withdrawal)
Woodruff keys 224
    data, inch 264, 265
    data, metric 266
work equations 33
workspace
    access openings 39, 43–45, 80, 87
    clearance 39, 43, 87
    dimensions 39, 43–45, 80, 87
    enclosures 39
worm gear (see also: gears) 586, 608
    efficiency 609
    equations 611
    forces on shafts 552, 610
    materials 595
worst-case tolerance stack-up 135, 136, 143 -145, 163, 175
wrench clearance 301, 304, 338, 340, 342–345

Y

yield strength 295, 374, 382, 399

Z

Zerol bevel gear (see also: bcvel gear) 586
zinc electroplating 408, 410, 411, 294
Zirconia 378